I0035078

Edible Food Packaging with Natural Hydrocolloids and Active Agents

Ahmet Yemenicioğlu

Department of Food Engineering, Izmir Institute of Technology,
Faculty of Engineering, Gulbahce Koyu
Urla, Izmir, Turkey

CRC Press
Taylor & Francis Group
Boca Raton London New York

CRC Press is an imprint of the
Taylor & Francis Group, an **informa** business

A SCIENCE PUBLISHERS BOOK

First edition published 2022
by CRC Press
6000 Broken Sound Parkway NW, Suite 300, Boca Raton, FL 33487-2742

and by CRC Press
4 Park Square, Milton Park, Abingdon, Oxon, OX14 4RN

© 2022 Taylor & Francis Group, LLC

CRC Press is an imprint of Taylor & Francis Group, LLC

Reasonable efforts have been made to publish reliable data and information, but the author and publisher cannot assume responsibility for the validity of all materials or the consequences of their use. The authors and publishers have attempted to trace the copyright holders of all material reproduced in this publication and apologize to copyright holders if permission to publish in this form has not been obtained. If any copyright material has not been acknowledged please write and let us know so we may rectify in any future reprint.

Except as permitted under U.S. Copyright Law, no part of this book may be reprinted, reproduced, transmitted, or utilized in any form by any electronic, mechanical, or other means, now known or hereafter invented, including photocopying, microfilming, and recording, or in any information storage or retrieval system, without written permission from the publishers.

For permission to photocopy or use material electronically from this work, access www.copyright.com or contact the Copyright Clearance Center, Inc. (CCC), 222 Rosewood Drive, Danvers, MA 01923, 978-750-8400. For works that are not available on CCC please contact mpkbookspermissions@tandf.co.uk

Trademark notice: Product or corporate names may be trademarks or registered trademarks and are used only for identification and explanation without intent to infringe.

Library of Congress Cataloging-in-Publication Data (applied for)

ISBN: 978-0-367-35019-2 (hbk)
ISBN: 978-1-032-37112-2 (pbk)
ISBN: 978-0-429-32989-0 (ebk)

DOI: 10.1201/9780429329890

Typeset in Times New Roman
by Innovative Processors

Preface

Due to the health concerns related to chemical food preservatives as well as environmental problems associated with fossil plastics, edible packaging with natural hydrocolloids and active agents (active edible packaging) has become increasingly popular as a food preservation method. In fact, the active edible packaging is now an interdisciplinary research field attracting interest not only from food and material scientists, but also from agronomists, microbiologists, biologists, pharmacists, and nutrition scientists. Researchers with different backgrounds have been collaborating to discover and characterize novel functions, synergetic interactions, and delivery methods of active agents, and to develop innovative and more applicable antimicrobial, antioxidant, flavor-release and bioactive edible packaging. The industrial interest in active edible packaging (especially coatings) has also been increasing continuously. Thus, to better understand the developments in the field, one should know the current content of active edible packaging (*see* Chapter 1) and factors fueling this rapidly-emerging preservation method. First of all, active edible packaging is the most sustainable packaging method since its main ingredients and functional components could be formed by natural hydrocolloids and active agents extracted mostly from agro-industrial wastes. Thus, this emerging packaging method provides an excellent opportunity for utilization of wastes into value-added products. In this book, detailed information has been provided about sources, extraction methods, the major characteristics of natural hydrocolloids (Chapter 2), and their ability to form different types of edible packaging (Chapter 3). Moreover, main sources and characteristics of natural active compounds have also been discussed in detail, and their potential as components of edible packaging has been analyzed (Chapter 4). The other major reason for increased interest in this field is that recent scientific developments, such as better understanding of synergetic interactions of active agents, and application of nanoencapsulation and controlled release technologies in their delivery, have enabled more effective use of natural phenolic compounds in active edible packaging. Developments in encapsulation technologies have also boosted the application of active packaging incorporated with probiotics, nutrients, and bioactive agents (bioactive packaging). These scientific developments and many other strategies used to enhance performance of active edible packaging have been discussed in the book in an easily comprehensible manner (Chapter 5). The book also contains basic proved methods of testing antimicrobial and antioxidant properties of edible packaging (Chapter 6), and over one hundred recent examples of active edible packaging applications (Chapter 7).

The examples have been selected carefully among the most applicable and up-to-date ones, covering a wide range of food, such as whole or minimally processed fresh fruits, vegetables, and mushrooms, nuts and seeds, raw and processed beef, pork, lamb, chicken and fish, dairy products, and bakery products, and dough food. As understood from its title and contents, this book is based on the use of natural hydrocolloids and active agents in edible packaging. Therefore, it lacks or contains minimum essential information about edible packaging of chemically-modified natural hydrocolloids, and chemical food additives. The information in this book will be of great interest, not only to researchers in academia and industry already working in the field, but also undergraduate or graduate students who are planning to start research or write a thesis in this field.

Ahmet Yemenicioğlu

I dedicate this book to

My wife Ayla and my daughter Feride Lila

for their endless support and understanding during my long writing sessions

Contents

Introduction to Active Edible Packaging

1.1 The history of edible packaging

Some people might think that the application of 'edible film' or 'edible coating' in food preservation is one of the new methods discovered in the 20th century. However, the small and large intestines of cattle and sheep have been used as natural casing material for sausages since ancient times. The earliest record about meat-stuffed casings (sausages) was discovered in almost 4,000-year-old Sumerian tablets found in Mesopotamia (Eckholm, 1985). The ripening of some cheese types in sacks made of lamb skin in Balkans and Turkey has also been applied since ancient times (Kalit *et al.*, 2010). However, the first record related to use of edible coatings in fruits is not too old as this is attributed to the Chinese, who formulated and applied edible wax coatings for preservation of oranges in the 12th century (Hardenburg, 1967). Moreover, it is also thought that the *yuba*, a proteic film formed at the surface of boiled soy milk, was discovered by Japanese in the 15th century as the first self-standing edible film applied for wrapping of food (Umaraw and Verma, 2017). In the 16th century, the coating of meat with fat (larding) in England is also an example of food preservation with edible coating (Bertuzzi and Slavutsky, 2016). After that, the use of gelatin coating in meats (Havard and Harmony, 1869) and the familiar process called 'fruit waxing' (Hoffman, 1916) were patented in the USA in the second half of 19th and at the beginning of 20th centuries, respectively. Then, the collagen casings were manufactured in Germany in the mid-1920s (Naga *et al.*, 1996). What about active edible packaging? Has this method been discovered recently by the modern food scientists? Yes, it is true that some patents exist related to incorporation of antifungal food preservatives into edible pectin and pectate films in 1950s (Owens and Schultz, 1952), but systematic studies to incorporate food preservatives into packaging had been accelerated in the 1980s and 1990s (Torres *et al.*, 1985; Guilbert *et al.*, 1996). Examples related to the use of active edible packaging in ancient times are scarce. However, it should be kept in mind that the ancient process of smoking applied to sausages causes accumulation of antimicrobial smoke components (e.g. acids, phenol, carbonyl) within the casing and on the sausage surface. Moreover, wrapping minced meat or rice and seasonings in grape leaves (causes grape leaf aromas and bioactive polyphenols to release into the filling during cooking) has been a traditional dish in Greece since ancient times (Cosme *et al.*, 2017). These historical knowledge mean that some elementary applications of edible films have been laying around us for a very long time, while active edible packaging is an emerging new method that has boosted the global interest of using edible films in food preservation.

1.2   Definition of edible packaging

In general, 'edible packaging' is defined as a continuous protective matrix made up of polysaccharides, proteins or lipids (used alone or in combination), and applied to respiring or non-respiring food as a self-standing film (used as film wrap, casing or pouch) or coating that acts as a barrier against moisture, gas, flavor, aroma or oil transfer (Kester and Fennema, 1986; Guilbert *et al.*, 1986; Guilbert *et al.*, 1996; Miller and Krochta, 1997; Park, 1999). However, rapid developments in the field showed that edible food packaging is more than film, pouch, coating and casing. For example, different emerging packaging, such as antimicrobial stickers, gel-based pads, and electrospun mats produced by using edible materials could also be considered edible packaging (Tracz *et al.*, 2018; Boyacı and Yemenicioğlu, 2020; Kavur and Yemenicioğlu, 2020; Sameen *et al.*, 2021). All film-forming components and functional additives of edible films must be food-grade non-toxic substances selected considering related national or international regulations, and foods packed with these edible films must be labeled properly for their potential allergenic constituents (*see* Chapter 2 for different allergens in natural hydrocolloids) (Rojas-Graü *et al.*, 2009). Moreover, it should also be noted that edible films refer to films having thickness less than 254 μm while thicker films are defined as sheets (Janjarasskul and Krochta, 2010).

1.3   Definition of active packaging and active edible packaging concepts

The major definitions related to active packaging, according to European Commission Regulation No. 450/2009, are seen in Table 1.1. Active edible packaging takes some specific names, such as antimicrobial, antioxidant, flavor-release or bioactive edible packaging depending on the functionality of active component(s) used in their production. An antimicrobial or antioxidant edible packaging is manufactured mostly by incorporating (rarely impregnating) antimicrobial or antioxidant substance(s) into edible packaging or by using an inherently antimicrobial or antioxidant hydrocolloid in manufacturing of edible packaging (Appendini and Hotchkiss, 2002; Shendurse *et al.*, 2018). The other active packaging concepts are flavor-release packaging and bioactive packaging that are conducted by incorporating flavor substance(s) (Marcuzzo, 2010) and health-promoting substance(s) (or food-grade probiotic microorganism) into edible packaging, respectively (Lopez-Rubio *et al.*, 2006).

1.3.1   Antimicrobial packaging

Antimicrobial packaging improves food safety by inhibiting contaminated pathogenic bacteria and/or prolonging food's shelf-life by suppressing growth of

Table 1.1: Major definitions related to active packaging according to European
Commission Regulation No. 450/2009

Term/Concept	Definition
Active materials and articles	Materials and articles that are intended to extend the shelf-life or to maintain or improve the condition of packaged food; they are designed to deliberately incorporate components that would release or absorb substances into or from the packaged food or the environment surrounding the food.
Releasing active materials and articles	Active materials and articles designed to deliberately incorporate components that would release substances into or on to the packaged food or the environment surrounding the food.
Released active substances	Substances intended to be released from releasing active materials and articles into or on to the packaged food or the environment surrounding the food and fulfilling a purpose in the food.
Active component	An individual substance or a combination of individual substances which cause the active function of a material or article, including the products of an *in situ* reaction of those substances; it does not include the passive parts, such as the material they are added to or incorporated into.

spoilage flora using minimum amount of antimicrobial agents (Appendini and Hotchkiss, 2002). Antimicrobial agents, such as antimicrobial enzymes, bacteriocins, phenolic compounds, essential oils, etc. are loaded into edible packaging by different methods, such as incorporation, impregnation, coating or immobilization (Table 1.2). However, the most frequently used method is incorporation of antimicrobials into self-standing films or coatings. The hydrophilic antimicrobials are solubilized directly in film- or coating-forming solutions while hydrophobic ones are mostly homogenized with the aid of film-forming hydrocolloid and an emulsifier to obtain emulsion-based films or coatings. The nanoencapsulation of hydrophobic antimicrobials before incorporation is also a frequently used technique to disperse hydrophobic antimicrobial substances homogenously within film- or coating-forming solutions. However, it is important to note that encapsulation is a key process for both hydrophilic and hydrophobic antimicrobials since this process also improves stability of antimicrobials and helps in sustaining their release rates. The antimicrobials can also be coated on to the surface of self-standing films by spreading, spraying or brushing, or they could be impregnated by dipping films into antimicrobial solution. The antimicrobials loaded into packaging by incorporation, impregnation or coating generally release on to packaged food's surface (also headspace of external food package, if volatile) at a certain rate, thus, showing antimicrobial activity both on the food surface and below food surface, depending on their capacity to diffuse into depths of food without losing their antimicrobial activity. In contrast, antimicrobials immobilized on to films or coatings can be used to obtain an antimicrobial effect only on the food surface (Lian *et al.*, 2012). Finally, inherently antimicrobial hydrocolloids are used in development of antimicrobial films and coatings. For example, chitosan is a unique antimicrobial hydrocolloid to be used as a self-standing film, casing, and monolayer or layer-by-layer coating

Table 1.2: Methods used to transform edible packaging into active edible packaging

Description	Name of method	Applications
Solubilization/dispersion/ emulsification of free or encapsulated active agent(s) in the edible packaging forming solution.	Incorporation	Self-standing films, coatings, casings, pouches, gel-based pads, electrospun nanofiber mats
Dipping of packaging into solution of active agent(s).	Impregnation	Self-standing films, casings
Spreading/spraying/brushing of active agent(s) solution on to edible packaging surface.	Coating	Self-standing films, coatings, casings
Creating charge – charge interaction, covalent cross-linking, hydrogen bonding, etc. between active agent(s) and hydrocolloid that form the packaging.	Immobilization	Self-standing films, coatings, casings
Use of inherently antimicrobial (e.g. chitosan) or antioxidant (e.g. milk proteins) hydrocolloids in development of packaging.	-	Self-standing films, coatings, casings

(when combined with a negatively-charged hydrocolloid). However, films and coatings of inherently antimicrobial hydrocolloids show antimicrobial effect only on the food contact surface.

1.3.2   Antioxidant packaging

This type of packaging is applied mainly by using antioxidant loaded self-standing films or coatings to inhibit lipid oxidation or enzymatic browning. Recently, there has been increased interest in using antioxidant loaded nanofiber mats for packaging (Vilchez *et al.*, 2020). In general, antioxidants, such as phenolic compounds, essential oils, carotenoids, tocopherols, ascorbic acid and derivatives, proteins and peptides, etc. are loaded into edible packaging, using similar methods described for antimicrobial packaging. However, the antioxidant compounds, capable of inhibiting lipid oxidation or enzymatic browning, are released almost always from food-contact packaging on to the food surface. Rarely some inherently antioxidant hydrocolloids (e.g. milk proteins) are employed to obtain antioxidant films or coatings effective locally on the food surface (Shendurse *et al.*, 2018). However, performances of such inherently antioxidant coatings cannot be compared with those loaded with free antioxidant agents.

1.3.3   Flavor-release packaging

This active packaging concept involves incorporation of desired flavor compounds into edible films and coatings (Marcuzzo, 2010). The application of flavor-release packaging is useful in enhancing or maintaining the desired flavor attributes of food during storage. However, to obtain maximum benefits from flavor-release packaging, it is essential to conduct encapsulation of incorporated flavor compounds, using suitable encapsulant and encapsulation methods. In some cases, the flavor compound could be encapsulated during film making by forming an emulsion between film-forming hydrocolloid and lipids (Marcuzzo, 2010).

1.3.4   Bioactive packaging

Active packaging is named bioactive packaging when its unique role is to enhance food impact on the consumer's health (Lopez-Rubio *et al.*, 2006). In bioactive edible packaging concept, probiotics and prebiotics or bioactive substances, such as phytochemicals, vitamins, marine oils, etc. are maintained on the food surface (within edible coating) or delivered on to the food surface (for solid food) or into food (for beverages). However, such a packaging must be designed considering the stability of bioactive agents during processing and storage of packaged food, and ensuring their bioavailability after consumption. In order to stabilize and improve their bioavailability, the bioactive compounds employed in packaging are treated by different technological methods (e.g. enzymatic modification, nanoencapsulation, co-encapsulation, dissolution by ultrasonication, etc.). Moreover, some compositional and structural modifications might be conducted in the food (if possible) to maximize bioavailability of delivered bioactive agents. The bioavailability of a bioactive agent is determined by limitations in its bioaccessibility (e.g. liberation from food matrix, solubilization in intestinal fluids and interactions, and insoluble complex formation), absorption (e.g. mucus layer transport, bilayer permeability, tight junction transport, active transporters, efflux transporters), and transformation (e.g. chemical degradation and metabolism) (McClements *et al.*, 2015). The definitions of basic terms used in nutrition science help to understand the challenges of bioactive packaging (Table 1.3).

1.4   Main materials used for development of active edible packaging

Proteins and polysaccharides (hydrocolloids) are the main materials used for development of active edible films and coatings. The self-standing edible films and coatings are produced mainly from (1) a single hydrocolloid (a protein or polysaccharide), (2) mixture of different hydrocolloids (blends or composites), or (3) mixture of hydrocolloids with lipids (emulsions or composites). The nanofiber mats, considered active edible packaging materials of future, are also manufactured from electrospinnable hydrocolloids (e.g. zein, gluten, gelatin, alginate) (Zhang *et al.*, 2006; Akman *et al.*, 2019; Karim *et al.*, 2020; Nie *et al.*, 2008; Fang *et al.*, 2011). However, although the lipids can be used to develop edible coatings, they cannot be utilized in manufacturing self-standing edible films due to their poor mechanical properties. Moreover, hydrophilic natural active compounds are not soluble in lipids and cannot be delivered on to the food surface with pure lipid coatings. In contrast, the majority of hydrocolloid-based packaging materials are compatible with the hydrophilic antimicrobial and antioxidant agents, and they could be utilized into

Table 1.3: Definitions of important terms necessary to understand principles
of bioactive packaging

Term	Definitions	Reference
Bioavailability	The rate and extent to which the active ingredient or active moiety is absorbed from a drug product and becomes available at the site of action.	FDA (2002)
	The fraction of ingested nutrient that is available for utilization in normal physiologic functions and for storage.	Parada and Aguilera (2007)
Bioaccessibility	The fraction of an ingested biocomponent that becomes accessible for absorption through the epithelial layer of the gastrointestinal tract (GIT).	Dima *et al.* (2020)
	Fraction that is released from food matrix and is available for intestinal absorption.	Parada and Aguilera (2007)
Bioactivity	The ability of a compound to exhibit a biological effect (e.g. antioxidant effect, antimicrobial effect, anti-inflammatory effect, etc.).	Dima *et al.* (2020)
Potency of a bioactive compound	The concentration or quantity of a biocomponent necessary to produce the corresponding biological effect.	Dima *et al.* (2020)

emulsion-based packaging to accommodate hydrophobic active agents. Therefore, this book focuses on utilization of natural hydrocolloids in development of active edible packaging materials. It is accepted that only materials having the ability to form self-standing packaging could be the future alternatives to fossil plastic films. On the other hand, the lipids have important roles in improving the moisture barrier, swelling, and sustained release properties of hydrocolloid-based films and coatings. Moreover, most of the novel delivery systems, such as nanoemulsions, solid lipid nanoparticles, nanoliposomes, nanomicelles, etc. are based on lipids. Thus, the lipids are considered essential components used to improve functional and active properties of hydrocolloid-based packaging.

1.5 Edible packaging of different natural hydrocolloids

Hydrocolloids from plant, animal, algal (seaweed), and microbial sources could be used for development of active edible packaging, such as coatings, casings, self-standing films or nanofiber mats (Table 1.4). However, extensive efforts have been spent on employing proteins and polysaccharides obtained from animal and plant sources found worldwide as agro-industrial byproducts and wastes. The major polysaccharide-based hydrocolloids obtained from plants are cellulose, starch, and pectin. The cellulose obtained from wood or non-wood sources is the

Table 1.4: Abilities of major natural hydrocolloids to form edible packaging

Hydrocolloids	Coating	Extruded casing	Solution-cast film	Compression molded film	Extruded film (D or B)[a]	Nanofiber mat
Plant-origin polysaccharides						
Starch	X[b]	-	X	X	X (D)	-
Pectin	X	X	X	X	X (D)	-
Plant-origin proteins						
Zein	X	-	X	X	X (B)	X
Gluten	X	-	X	X	X (B)	X
Soy protein	X	X	X	X	X (D)	-
Animal-origin proteins						
Gelatin	X	-	X	X	X (B)	X
Collagen	X	X	X	X	X (B)	X
Na-caseinate	X	-	X	X	X (B)	-
Whey protein	X	-	X	X	-	-
Animal-origin polysaccharides						
Chitosan	X	X	X	X	-	-
Algal-origin polysaccharides						
Alginate	X	X	X	X	-	X
Carrageenans	X	-	X	-	-	-
Microbial-origin polysaccharides						
Pullulan	X	-	X	-	-	X
Xanthan	X	-	X	-	-	-

[a] Slit-die (D) or blown (B) extruded film.
[b] The symbol "X" shows ability of hydrocolloid to form the indicated packaging type.

most abundant hydrocolloid on earth, but insolubility of natural cellulose prevents its direct use in development of edible packaging. The derivatives of cellulose obtained by chemical modification, such as cellulose ethers (e.g. carboxymethyl cellulose, methylcellulose, hydroxypropyl methylcellulose, hydroxyethyl cellulose, etc.) could be used to obtain solution-cast self-standing edible films and coatings. However, most cellulose ethers cannot be utilized alone by classical polymer processing methods because of their high viscosity between their glass transition temperature (Tg) and denaturation temperature (Tg) (Meena *et al.*, 2016). Moreover, electrospinning of cellulose ethers needs solubilization in organic solvents or use of electrospinnable carrier polymers (Frenot *et al.*, 2007). In contrast, native starches obtained from different sources, such as corn, wheat, rice, potato, or tapioca could be used effectively in development of edible packaging materials, such as coatings and solution-cast self-standing films. The self-standing edible films of starch can also be developed by compression molding and slit-die extrusion, but the development of blown extruded starch films is problematic due to their highly brittle nature (Thunwall *et al.*, 2006; Shanks and Kong, 2012). The development of electrospun mats from starch is also not practical since this needs solubilization in solvents that should be removed after nanofiber production (Kong and Ziegler, 2014). The high or low methyl ester pectins obtained mainly from citrus peels are also used in manufacturing edible coatings and solution-cast self-standing edible films. The low methoxyl pectin can be effectively cross-linked with $CaCl_2$ to obtain insoluble solution-cast self-standing films or insoluble coatings fixed at the food surface. The pectin could also be thermosplasticized and utilized in extruded sausage casings (Liu *et al.*, 2005; Liu *et al.*, 2007) or compression molded films (Gouveia *et al.*, 2019), but data about applicability of blown extruded pure pectin films are scarce. Pectin is also among the hydrocolloids that need the presence of electrospinnable carrier polymers to obtain nanofibers (Cui *et al.*, 2016).

Chitosan is the only animal-origin polysaccharide used extensively in active edible packaging. Due to its unique inherent antimicrobial properties, chitosan has become one of the most important coating materials. Moreover, the recent application of chitosan as an antimicrobial sausage casing might trigger its more extensive use by the food industry (Adzaly *et al.*, 2016). Chitosan is also used in development of self-standing films by solution-casting and compression molding, but it cannot be used in manufacture of extruded films since it is not a thermoplastic material (Van den Broek *et al.*, 2015). The high viscosity of aqueous chitosan solution interferes with its electrospinning. Therefore, electrospinning of chitosan also needs solubilization in hazardous acids (e.g. trifluoroacetic acid) with addition of toxic solvents (e.g. dichloromethane) (Ohkawa *et al.*, 2004).

The corn zein, wheat gluten, and soy proteins are major plant-origin proteins used in development of edible packaging. Zein and gluten are highly functional proteins that can be used in development of different kinds of packaging (solution-cast, extruded or compression molded self-standing films, or coatings and electrospun nanofiber mats) (Gontard and Guilbert, 1998; Padgett *et al.*, 1998; Wang and Padua, 2003; Tanada-Palmu and Grosso, 2005; Akman *et al.*, 2019; Karim *et al.*, 2020). Gluten is also one of the rare hydrocolloids to be used in development of three-dimensional materials by injection molding (Gontard and Guilbert, 1998). The soy proteins is also used in development of edible coatings or self-standing films by solution-casting and compression molding (Stuchell and Krochta, 1994; Ogale *et al.*, 2000). The thermoplasticized soy proteins were also used for development of extruded casings and self-standing films, but extruded soy protein films have limited applicability due to their poor mechanical and moisture barrier properties (Naga *et al.*, 1996; Zhang *et al.*, 2001; Koshy *et al.*, 2015). The soy proteins cannot be used to develop electrospun nanofibers without the presence of other carrier proteins (Xu *et al.*, 2012). However, the gels obtained from soy proteins and their composites and emulsions are used in delivery of bioactive compounds, such as vitamins and phenolic compounds (Hu *et al.*, 2015; Ding and Yao, 2013; Brito-Oliveira *et al.*, 2017; Marinea *et al.*, 2021).

The animal-origin proteins, such as gelatin, collagen, sodium caseinate, and whey proteins are the most extensively used proteins by the food industry as functional and nutritional ingredients. Therefore, the utilization of these proteins in development of edible packaging has also been studied extensively. The use of collagen in extruded sausage casings is probably the most important industrial edible film application in the world. The collagen suspensions prepared by acid-swelling/homogenization method could also be employed as coating, and in the manufacture of self-standing films by solution-casting (Wang *et al*., 2016) and compression molding (Andonegi *et al*., 2020). The native collagen can be converted into thermoplastic collagen (TC) by partial heat denaturation to obtain blown-extruded films, but these films have limited applicability due to their high sensitivity to moisture and poor mechanical properties in moist environments (Klüver and Meyer, 2013). Electrospinning of aqueous collagen solutions gives nanofibers, but this process transforms collagen into gelatin (Zeugolis *et al*., 2008). This interferes with the use of collagen in electrospinning since gelatin itself can be electrospun into nanofibers. However, high solubility and mechanical weakness of gelatin are great limitations to obtain its electrospun mats suitable for industrial packaging applications (Zhang *et al*., 2006). The pure gelatin is not commercially applied for casings, but is extensively used in food-coating applications. The self-standing films of gelatins can be manufactured by solution-casting, compression molding (Krishna *et al*., 2012), and blown-extrusion (Andreuccetti *et al*., 2012) methods, but widespread industrial use of these films is also limited due to their poor mechanical and water barrier properties, and high water solubility. Finally, the excellent gel-forming capacity of gelatin can be exploited to obtain active gel-based pads suitable both in absorbing drip-loss and in delivering natural antimicrobials and antioxidants on to food surface (Boyacı and Yemenicioğlu, 2020). Sodium caseinate and whey proteins are the other animal-origin proteins to be used frequently in development of edible coatings and self-standing solution-cast edible films. Both these proteins can be used for development of compression-molded edible films, but only sodium caseinate was employed successfully to develop blown extruded edible films (Belyamani *et al*., 2014a, b). Sodium caseinate and whey proteins are also among the hydrocolloids that cannot be utilized in nanofibers by electrospinning (Sullivan *et al*., 2014; Tomasula *et al*., 2016).

The major marine-origin polysaccharides extracted from seaweeds, such as sodium alginate and carrageenans (kappa-, lambda- and iota-), and microbial polysaccharides, such as pullulan and xanthan can be employed in manufacture of edible packaging. Sodium alginate is one of the most extensively used edible coating materials. Since cross-linking of sodium alginate by Ca^{++} atoms causes gelation, it is possible to develop insoluble fixed coatings or self-standing solution-cast edible films of this hydrocolloid. The combination of negatively-charged alginate with positively-charged chitosan also gives layer-by-layer coatings (Poverenov *et al*., 2014). Moreover, extruded alginate or extruded blends of alginate with gelatin, pea protein, cellulose, or starch are employed to obtain (dry or wet) sausage casings (Liu *et al*., 2007; Harper *et al*., 2015; Marcos *et al*., 2020). The alginate plasticized sufficiently with polyols is also used to obtain compression-molded self-standing films, but the development of extruded alginate films by thermal polymer processing methods is difficult due to its thermal degradation in the molten state (Gao *et al*., 2017). It is also possible to obtain electrospun alginate nanofibers when this hydrocolloid is mixed with glycerol or its entanglements are enhanced by CaCl$_2$ cross-linking (Nie *et al.*, 2008; Fang *et al.*, 2011). Therefore, the alginate is considered as one of the highly functional hydrocolloids suitable for edible packaging.

The carrageenans are also attracting industrial interest as edible coatings (Tavassoli-Kafrani *et al.*, 2016) while the application of pure carrageenans as self-standing films is very limited due to their highly hydrophilic and brittle nature. However, it was reported that the blending of carrageenans with other hydrocolloids or preparing their composites with lipids, nanoclays or nanocellulose might be employed to obtain more applicable edible films than those of pure carrageenans (Sedayu *et al.*, 2019). Moreover, the carrageenans are among the hydrocolloids that cannot be electrospun into nanofibers (Stijnman *et al.*, 2011). Finally, the potential of microbial polysaccharides, such as pullulan and xanthan as water-soluble edible coatings, should be noted (Kandemir *et al.*, 2005; Sharma and Rao, 2015). However, both pullulan and xanthan find limited edible film applications due to their high water solubility. The aqueous solutions of xanthan are not electrospinnable, but pullulan is well known for its ability to form electrospun nanofibers (Stijnman *et al.*, 2011).

References

Adzaly, N.Z., A. Jackson, I. Kang and E. Almenar (2016). Performance of a novel casing made of chitosan under traditional sausage manufacturing conditions, *Meat Sci.*, 113: 116-123.

Akman, P.K., F. Bozkurt, M. Balubaid and M.T. Yilmaz (2019). Fabrication of curcumin-loaded gliadin electrospun nanofibrous structures and bioactive properties, *Fibers Polym.*, 20: 1187-1199.

Andonegi, M., K. de la Caba and P. Guerrero (2020). Effect of citric acid on collagen sheets processed by compression, *Food Hydrocoll.*, 100: 105427.

Andreuccetti, C., R.A. Carvalho, T. Galicia-García, F. Martinez-Bustos, R. González-Nuñez and C.R. Grosso (2012). Functional properties of gelatin-based films containing *Yucca schidigera* extract produced via casting, extrusion and blown extrusion processes: A preliminary study, *J. Food Eng.*, 113: 33-40.

Appendini, P. and J.H. Hotchkiss (2002). Review of antimicrobial food packaging, *Innov. Food Sci. Emerg. Technol.*, 3: 113-126.

Belyamani, I., F. Prochazka and G. Assezat (2014a). Production and characterization of sodium caseinate edible films made by blown-film extrusion, *J. Food Eng.*, 121: 39-47.

Belyamani, I., F. Prochazka, G. Assezat and F. Debeaufort (2014b). Mechanical and barrier properties of extruded film made from sodium and calcium caseinates, *Food Packag. Shelf-Life*, 2: 65-72.

Bertuzzi, M.A. and A.M. Slavutsky (2016). Standard and new processing techniques used in the preparation of films and coatings at the lab level and scale-up, pp. 21-42. *In:* M.P.M. Garcia, M.C. Gomez-Guillen, M.E. Lopez-Caballero, G.V. Barbosa-Canovas (Eds.). *Edible Films and Coatings*, CRC Press. Boca Raton, USA.

Boyacı, D. and A. Yemenicioğlu (2020). Development of gel-based pads loaded with lysozyme and green tea extract: Characterization of pads and test of their antilisterial potential on cold-smoked salmon, *LWT-Food Sci. Technol.*, 128: 109471.

Brito-Oliveira, T.C., M. Bispo, I.C. Moraes, O.H. Campanella and S.C. Pinho (2017). Stability of curcumin encapsulated in solid lipid microparticles incorporated in cold-set emulsion filled gels of soy protein isolate and xanthan gum, *Food Res. Int.*, 102: 759-767.

Cosme, F., T. Pinto and A. Vilela (2017). Oenology in the kitchen: The sensory experience offered by culinary dishes cooked with alcoholic drinks, grapes and grape leaves, *Beverages*, 3: 42.

Cui, S., B. Yao, X. Sun, J. Hu, Y. Zhou and Y. Liu (2016). Reducing the content of carrier polymer in pectin nanofibers by electrospinning at low loading followed with selective washing, *Mater. Sci. Eng. C.*, 59: 885-893.

Dima, C., E. Assadpour, S. Dima and S.M. Jafari (2020). Bioavailability and bioaccessibility of food bioactive compounds: Overview and assessment by *in vitro* methods, *Comprehensive Rev. Food Sci. Food Saf.*, 19: 2862-2884.

Ding, X. and P. Yao (2013). Soy protein/soy polysaccharide complex nanogels: Folic acid loading, protection and controlled delivery, *Langmuir*, 29: 8636-8644.

EC (European Commission) Regulation. No. 450/2009 of 29 May 2009 on active and intelligent materials and articles intended to come into contact with food, *OJEU*, 135: 3-11.

Eckholm E. (1985). Mesopotamia: Cradle of haute cusine? *The New York Times*. https://www.nytimes.com/1985/05/15/garden/mesopotamia-cradle-of-haute-cuisine.html

Fang, D., Y. Liu, S. Jiang, J. Nie and G. Ma (2011). Effect of intermolecular interaction on electrospinning of sodium alginate. *Carbohydr. Polym.*, 85: 276-279.

FDA (US Food and Drug Administration) (2002. Guidance for Industry Bioavailability and Bioequivalence Studies for Orally Administered Drug Products-General Considerations. https://www.fda.gov/files/drugs/published/Guidance-for-Industry-Bioavailability-and-Bioequivalence-Studies-for-Orally-Administered-Drug-Products---General-Considerations.PDF

Frenot, A., M.W. Henriksson and P. Walkenström (2007). Electrospinning of cellulose-based nanofibers, *J. Appl. Polym. Sci.*, 103: 1473-1482.

Gao, C., E. Pollet and L. Avérous (2017). Properties of glycerol-plasticized alginate films obtained by thermo-mechanical mixing, *Food Hydrocoll.*, 63: 414-420.

Gontard, N. and S. Guilbert (1998). Edible and/or biodegradable wheat gluten films and coatings. pp. 324-328. *In:* Gueguen, J. and Y. Popineau (Eds.). *Plant Proteins from European Crops, Food and Non-food Applications*, Springer, Heidelberg, Germany.

Gouveia, T.I., K. Biernacki, M.C. Castro, M.P. Gonçalves and H.K. Souza (2019). A new approach to develop biodegradable films based on thermoplastic pectin, *Food Hydrocoll.*, 97: 105175.

Guilbert, S. (1986). Technology and applications of edible protective films, pp. 371-394. *In:* M. Mathlouthi (Ed.). *Food Preservation and Packaging.* Elsevier Applied Science Publications, London, UK.

Guilbert, S., N. Gontard and L.G.M. Gorris (1996). Prolongation of the shelf-life of perishable food products using biodegradable films and coatings, *LWT-Food Sci. Technol.*, 29: 10-17.

Hardenburg, R.E. (1967). Wax and related coatings for horticultural products – A bibliography. U.S. Dept. of Agricultural Research Service, *Agr. Res Bull.*, 51: 15.

Harper, B.A., S. Barbut, A. Smith and M.F. Marcone (2015). Mechanical and microstructural properties of 'wet' alginate and composite films containing various carbohydrates, *J. Food Sci.*, 80: E84-E92.

Havard, C. and M.X. Harmony (1869). *Improved Process for Preserving Meat, Fowls, Fish, etc.*, U.S. Patent, 90: 944.

Hoffman, A.F. (1916). *Preserving Fruit*, US patent, 19.160.104.

Hu, H., X. Zhu, T. Hu, I.W. Cheung, S. Pan and E.C. Li-Chan (2015). Effect of ultrasound pre-treatment on formation of transglutaminase-catalyzed soy protein hydrogel as a riboflavin vehicle for functional foods, *J. Funct. Foods*, 19: 182-193.

Janjarasskul, T. and J.M. Krochta (2010). Edible packaging materials, *Annu. Rev. Food Sci. Technol.*, 1: 415-448.

Kalit. T.M., S. Kalit and J. Havranek (2010). An overview of researches on cheeses ripening in animal skin, *Mljekarstvo*, 60: 149-155.

Kandemir, N., A. Yemenicioglu, Ç. Mecitoglu, Z.S. Elmaci, A. Arslanoglu, Y. Göksungur and T. Baysal (2005). Production of antimicrobial films by incorporation of partially purified lysozyme into biodegradable films of crude exopolysaccharides obtained from *Aureobasidium pullulans* fermentation, *Food Technol. Biotechnol.*, 43: 343-350.

Karim, M., M. Fathi and S. Soleimanian-Zad (2020). Incorporation of zein nanofibers produced by needle-less electrospinning within the casted gelatin film for improvement of its physical properties, *Food Bioprod. Process*, 122: 193-204.

Kavur, P.B. and A. Yemenicioğlu (2020). An innovative design and application of natural antimicrobial gelatin based filling to control risk of listeriosis from caramel apples, *Food Hydrocoll.*, 107: 105938.

Kester, J.J. and O.R. Fennema (1986). Edible films and coatings: A review, *Food Technol.*, 40: 47-59.

Klüver, E. and M. Meyer (2013). Preparation, processing, and rheology of thermoplastic collagen. *J. Appl. Polym. Sci.*, 128: 4201-4211.

Kong, L. and G.R. Ziegler (2014). Fabrication of pure starch fibers by electrospinning, *Food Hydrocolloids*, 36: 20-25.

Koshy, R.R., S.K. Mary, S. Thomas and L.A. Pothan (2015). Environment-friendly green composites based on soy protein isolate – A review, *Food Hydrocoll.*, 50: 174-192.

Krishna, M., C.I. Nindo and S.C. Min (2012). Development of fish gelatin edible films using extrusion and compression molding, *J. Food Eng.*, 108: 337-344.

Lian, Z.X., Z.S. Ma, J. Wei and H. Liu (2012). Preparation and characterization of immobilized lysozyme and evaluation of its application in edible coatings, *Process Biochem.*, 47: 201-208.

Liu, L., J.F. Kerry and J.P. Kerry (2005). Selection of optimum extrusion technology parameters in the manufacture of edible/biodegradable packaging films derived from food-based polymers, *J. Food Agric. Environ.*, 3: 51-58.

Liu, L., J.F. Kerry and J.P. Kerry (2007). Application and assessment of extruded edible casings manufactured from pectin and gelatin/sodium alginate blends for use with breakfast pork sausage, *Meat Sci.*, 75: 196-202.

Lopez-Rubio, A., R. Gavara and J.M. Lagaron (2006). Bioactive packaging: Turning foods into healthier foods through biomaterials, *Trends in Food Sci. Technol.*, 17: 567-575.

Marcos, B., P. Gou, J. Arnau, M.D. Guàrdia and J. Comaposada (2020). Co-extruded alginate as an alternative to collagen casings in the production of dry-fermented sausages: Impact of coating composition, *Meat Sci.*, 169: 108184.

Marcuzzo, E., A. Sensidoni, F. Debeaufort and A. Voilley (2010). Encapsulation of aroma compounds in hydrocolloidic emulsion based edible films to control flavour release, *Carbohydr. Polym.*, 80: 984-988.

Marinea, M., A. Ellis, M. Golding and S.M. Loveday (2021). Soy protein pressed gels: Gelation mechanism affects the *in vitro* proteolysis and bioaccessibility of added phenolic acids, *Foods*, 10: 154.

McClements, D.J., F. Li and H. Xiao (2015). The nutraceutical bioavailability classification scheme: Classifying nutraceuticals according to factors limiting their oral bioavailability, *Annu. Rev. Food Sci. Technol.*, 6: 299-327.

Meena, A., T. Parikh, S.S. Gupta and A.T. Serajuddin (2016). Investigation of thermal and viscoelastic properties of polymers relevant to hot melt extrusion-II: Cellulosic polymers, *J. Excip. Food Chem.*, 5: 1002.

Miller, K.S. and J.M. Krochta (1997). Oxygen and aroma barrier properties of edible films: A review, *Trends in Food Sci. Technol.*, 8: 228-237.

Naga, M., S. Kirihara, Y. Tokugawa, F. Tsuda, T. Saito and M. Hirotsuka (1996). *Process for Developing a Proteinaceous Film*, U.S. Patent No. 5, 569,482, Washington, DC: U.S.

Nie, H., A. He, J. Zheng, S. Xu, J. Li and C.C. Han (2008). Effects of chain conformation and entanglement on the electrospinning of pure alginate, *Biomacromolecules*, 9: 1362-1365.

Ogale, A.A., P. Cunningham, P.L. Dawson and J.C. Acton (2000). Viscoelastic, thermal, and microstructural characterization of soy protein isolate films, *J. Food Sci.*, 65: 672-679.

Ohkawa, K., D. Cha, H. Kim, A. Nishida and H. Yamamoto (2004). Electrospinning of chitosan, *Macromol. Rapid Commun.*, 25: 1600-1605.

Owens, H.S. and T.H. Schultz (1952). *Methods of Coating Foods with Pectinate or Pectate Films*, US Patent Office (Application July 18, 1950), Serial No. 174-564.

Padgett, T., I.Y. Han and P.L. Dawson (1998). Incorporation of food-grade antimicrobial compounds into biodegradable packaging films, *J. Food Prot.*, 61: 1330-1335.

Parada, J. and J.M. Aguilera (2007). Food microstructure affects the bioavailability of several nutrients, *J. Food Sci.*, 72: R21-R32.

Park, H.J. (1999). Development of advanced edible coatings for fruits, *Trends in Food Sci. Tech.*, 10: 254-260.

Poverenov, E., S. Danino, B. Horev, R. Granit, Y. Vinokur and V. Rodov (2014). Layer-by-layer electrostatic deposition of edible coating on fresh cut melon model: Anticipated and unexpected effects of alginate–chitosan combination, *Food Bioprocess Tech.*, 7: 1424-1432.

Rojas-Graü, M.A., R. Soliva-Fortuny and O. Martín-Belloso (2009). Edible coatings to incorporate active ingredients to fresh-cut fruits: A review, *Trends in Food Sci. Technol.*, 20: 438-447.

Sameen, D.E., S. Ahmed, R. Lu, R. Li, J. Dai, W. Qin, Q. Zhang, S. Li and Y. Liu (2021). Electrospun nanofibers food packaging: Trends and applications in food systems, *Crit. Rev. Food Sci. Nutr.* https://doi.org/10.1080/10408398.2021.1899128

Sedayu, B.B., M.J. Cran and S.W. Bigger (2019). A review of property enhancement techniques for carrageenan-based films and coatings, *Carbohydr. Polym.*, 216: 287-302.

Shanks, R. and I. Kong (2012). Thermoplastic starch, pp. 95-116. *In:* A.Z. El-Sonbati (Ed.). *Thermoplastic Elastomers*. Intech Open, Rijeka, Hr.

Sharma, S. and T.R. Rao (2015). Xanthan gum-based edible coating enriched with cinnamic acid prevents browning and extends the shelf-life of fresh-cut pears, *LWT-Food Sci. Technol.*, 62: 791-800.

Shendurse, A., G. Gopikrishna, A.C. Patel and A.J. Pandya (2018). Milk protein-based edible films and coatings–preparation, properties and food applications, *J. Nutr. Health Food Eng.*, 8: 219-226.

Stijnman, A.C., I. Bodnar and R.H. Tromp (2011). Electrospinning of food-grade polysaccharides, *Food Hydrocoll.*, 25: 1393-1398.

Stuchell, Y.M. and J.M. Krochta (1994). Enzymatic treatments and thermal effects on edible soy protein films, *J. Food Sci.*, 59: 1332-1337.

Sullivan, S.T., C. Tang, A. Kennedy, S. Talwar and S.A. Khan (2014). Electrospinning and heat treatment of whey protein nanofibers, *Food Hydrocoll.*, 35: 36-50.

Tanada-Palmu, P.S. and C.R. Grosso (2005). Effect of edible wheat gluten-based films and coatings on refrigerated strawberry (*Fragaria ananassa*) quality, *Postharvest Biol. Tech.*, 36: 199-208.

Tavassoli-Kafrani, E., H. Shekarchizadeh and M. Masoudpour-Behabadi (2016). Development of edible films and coatings from alginates and carrageenans, *Carbohydr. Polym.*, 137: 360-374.

Thunwall, M., A. Boldizar and M. Rigdahl (2006). Compression molding and tensile properties of thermoplastic potato starch materials, *Biomacromolecules*, 7: 981-986.

Tomasula, P.M., A.M. Sousa, S.C. Liou, R. Li, L.M. Bonnaillie and L. Liu (2016). Electrospinning of casein/pullulan blends for food-grade applications, *J. Dairy Sci.*, 99: 1837-1845.

Torres, J.A., Motoki, M. and M. Karel (1985). Microbial stabilization of intermediate moisture food surfaces I. Control of surface preservative concentration, *J. Food Process. Preserv.*, 9: 75-92.

Tracz, B.L., K. Bordin, K.C.P. Bocate, R.V. Hara, C. Luz, R.E.F. Macedo, G. Meca and F.B. Luciano (2018). Devices containing allyl isothiocyanate against the growth of spoilage and mycotoxigenic fungi in mozzarella cheese, *J. Food Process. Pres.*, 42: e13779.

Umaraw, P. and A.K. Verma (2017). Comprehensive review on application of edible film on meat and meat products: An eco-friendly approach, *Crit. Rev. Food Sci. Nutr.*, 57: 1270-1279.

Van den Broek, L.A., R.J. Knoop, F.H. Kappen and C.G. Boeriu (2015). Chitosan films and blends for packaging material, *Carbohydr. Polym.*, 116: 237-242.

Vilchez, A., F. Acevedo, M. Cea, M. Seeger and R. Navia (2020). Applications of electrospun nanofibers with antioxidant properties: A review, *Nanomaterials*, 10: 175.

Wang, Y. and G.W. Padua (2003). Tensile properties of extruded zein sheets and extrusion blown films, *Macromol. Mater. Eng.*, 288: 886-893.

Wang, W., Y. Liu, A. Liu, Y. Zhao and X. Chen (2016). Effect of in situ apatite on performance of collagen fiber film for food packaging applications, *J. Appl. Poly. Sci.*, 133: 44154.

Xu, X., L. Jiang, Z. Zhou, X. Wu and Y. Wang (2012). Preparation and properties of electrospun soy protein isolate/polyethylene oxide nanofiber membranes, *ACS Appl. Mater. Interfaces*, 4: 4331-4337.

Zeugolis, D.I., S.T. Khew, E.S. Yew, A.K. Ekaputra, Y.W. Tong, L.Y.L. Yung, C. Sheppard and M. Raghunath (2008). Electro-spinning of pure collagen nano-fibres – Just an expensive way to make gelatin, *Biomaterials*, 29: 2293-2305.

Zhang, J., P. Mungara and J.L. Jane (2001). Mechanical and thermal properties of extruded soy protein sheets, *Polymer*, 42: 2569-2578.

Zhang, Y.Z., J. Venugopal, Z.M. Huang, C.T. Lim and S. Ramakrishna (2006). Crosslinking of the electrospun gelatin nanofibers, *Polymer*, 47: 2911-2917.

Natural Hydrocolloids: Sources and Major Characteristics

2.1 Introduction

The molecular, physical, and chemical properties of hydrocolloids obtained from different sources might show significant differences. Therefore, hydrocolloids from different sources vary not only in their ability to form different types of packaging (e.g. compression mold, extruded or solution-cast films, extruded casings, coatings, gel-based pads, electrospun mats, etc.), but also in characteristics of their edible packaging (e.g. solubility, swelling, barrier, thermal, mechanical, and morphological properties). For example, most fish gelatin films show lower oxygen permeability than mammalian gelatin films, while mammalian gelatin films generally have superior mechanical stability than fish gelatin films (Avena-Bustillos *et al.*, 2011; Ninan *et al.*, 2010). The source of hydrocolloid also determines the sustainability and economic feasibility of packaging manufacturing. An abundant agro-industrial waste from a sustainable product is generally considered an ideal source for the hydrocolloid. However, some hydrocolloids from byproducts could also be preferred for manufacturing edible films when this supports the sustainability (e.g. the biorefinery approach). Another factor increasing the importance of hydrocolloid source is that it determines the consumer profile of food packed with obtained edible packaging. For example, due to religious concerns, some consumers do not prefer to consume foods that contain edible packaging from some mammalian hydrocolloids (e.g. porcine or bovine gelatin coatings). Some vegetarians also reject animal-origin edible coatings and prefer to consume food coated with plant-based hydrocolloids. Similarly, people suffering from fish allergy avoid food treated with fish protein coatings, while those with celiac disease cannot consume food that contain gluten coatings. Thus, as a general rule, all film-forming components of the edible films must be food-grade non-toxic substances, and foods packed with these edible films must be labelled properly for their potential allergenic constituents (Rojas-Graü *et al.*, 2009). This chapter discusses mainly molecular and physiochemical characteristics, and film-making mechanisms of different commercially-available natural hydrocolloids from animal, plant, algal, and microbial sources. The current and potential future sources, basic extraction, and preparation methods, consumer profiles, and potential negative or positive health effects of important natural hydrocolloids have also been discussed in some detail.

2.2   Natural sources of hydrocolloids

The natural hydrocolloids can be classified according to their sources that include land or marine animals, land plants, seaweed, fungi, and bacteria (Table 2.1). However, significant variations in film characteristics are expected not only for hydrocolloids from different sources, but also for hydrocolloids from different species of the same source (Ninan *et al.*, 2010; Basiak *et al.*, 2017). Although this variability brings some disadvantages in mass production of edible packaging, it can be exploited to meet the challenging demands of some packaging applications (e.g. need for coatings having different gas and moisture-barrier properties for different fruits that vary in their respiration rates). Thus, a deep knowledge about molecular and physicochemical characteristics of different hydrocolloids is essentially important to develop state-of-the-art edible films.

Table 2.1: Classification of natural hydrocolloids according to their sources[a]

Animal Source Hydrocolloids

Marine animal hydrocolloids

Proteins: Cold or hot water fish gelatin
Polysaccharides: Chitosan[1] from *Crustaceans* (e.g. crab, shrimp, lobster, krill)

Land animal hydrocolloids

Proteins: Bovine, porcine, chicken, and insect gelatins, whey proteins, caseins and caseinates, egg-white proteins

Algal Source Hydrocolloids

Seaweed hydrocolloids

Polysaccharides: Alginate[2], agar[3], carrageenan[4]

Microbial Source Hydrocolloids

Fungal hydrocolloids

Polysaccharides: Pullulan[5]

Bacterial hydrocolloids

Polysaccharides: Cellulose, xanthan[6], gellan[7]

Plant Source Hydrocolloids

Land plant hydrocolloids

Proteins: Lentil, chickpea, pea, rice, potato and soy protein isolates, wheat gluten, corn or maize zein

Polysaccharides: Wheat, potato, rice, tapioca starches[8], cellulose from wood and non-wood sources[9], citrus and apple pectins[10], gums (gum Arabic[11], karaya gum[12], guar gum[13], locust bean gum[14], gum tragacanth[15], etc.)

[a] Molecular properties of given polysaccharides: Homoglycans: 1, 5, 8, 9; Heteroglycans: 2, 3, 4, 6, 7, 10, 11, 12, 13, 14, 15; Neutral: 3, 5, 8, 9, 13, 14; Anionic: 2, 4, 6, 7, 10, 11, 12, 15; Cationic: 1; Linear: 1, 2, 3 (agarose), 4, 5, 7, 8 (amylose), 9; Branched: 3 (agaropectin), 6, 8 (amylopectin), 10, 11, 12, 13, 14, 15)

2.2.1   Animal source hydrocolloids

The major animal source hydrocolloids used in manufacturing of edible packaging include proteins, such as collagen, gelatin, egg proteins, caseins, and whey proteins, and polysaccharides, such as chitin and chitosan.

2.2.1.1   Collagen

Collagen is the main hydrocolloid of the ancient packaging materials, such as the small and large intestines of ruminant species used as natural, edible sausage casing (Eckholm, 1985), and animal skins used for manufacturing sacks employed in maturation of some traditional cheese types (Kalit *et al.*, 2010). Collagen is still a very popular packaging hydrocolloid today since it has been used extensively by the food industry to obtain artificial extruded sausage casings of different sizes (Yemenicioğlu *et al.*, 2020). Artificial casings obtained by extruding dough prepared by swelling of collagen in acid solutions are cheap, practical, uniform, and strong, but they give almost similar cooking characteristics and tenderness with natural casings (Suurs and Barbut, 2020). Moreover, innovations in the field have enabled formation of collagen casing directly on the surface of sausages during manufacturing by using co-extrusion technique (Suurs and Barbut, 2020). Therefore, collagen is one of the rare edible film materials that is produced and used extensively by the food industry.

The mammalian tissues contain many types of collagen (up to 27 types), but the major ones are Type I, II, III and IV collagens. This section refers to Type I collagen as it is the major form commercially produced and used for development of edible packaging because of its abundant presence in connective tissues, such as bones, skin, and tendons. The molecular structure of the Type 1 collagen consists of a triple helix composed of two identical alpha helixes ($2 \times \alpha_1$) and one other alpha helix (α_2). The glycine (Gly) exists in every three amino acid residues along the alpha helixes with a pattern of $(Gly-X-Y)_n$, where X and Y are formed most frequently by the amino acid proline (Pro) and hydroxyproline (Hyp), a metabolite formed by hydroxylation of proline. Thus, Gly-Pro-Hyp is the most frequently found pattern along the alpha helixes that form the collagen (Beck and Brodsky, 1998). Both α_1 and α_2 are left-handed helixes, but when the three helixes $[(\alpha_1)_2\alpha_2]$ are twisted together, they form the right-handed supercoiled structure of collagen that is stabilized by extensive interchain H-bonds formed between glycine and amide group in adjacent chains (Beck and Brodsky, 1998). The collagen molecules interact with each other to form higher level organized structures, such as microfibrils, fibrils, fibers, and bundles. The microfibrils (subfibril) are formed by binding of collagen molecules to each other while several microfibrils wrap each other to form the spiral-shaped fibril. The fiber is formed from the spiral of several fibrils, and fibrillary bundles are formed by twisting of some fiber (Schleip *et al.*, 2012). The coils in the collagen are always opposed from one stage to the next – a left-handed helix is followed by a right-handed helix and then a left-handed once again, etc. (Schleip *et al.*, 2012). Moreover, collagen molecules could form covalent crosslinks via action of enzyme lysyl oxidase. Lysyl oxidase catalyzes the conversion of specific lysine or hydroxylysine residues at short non-helical domains (telopeptides) in the NH- and

COOH-terminal of collagen into the corresponding δ-semialdehyde (allysine) that condenses spontaneously to form covalent crosslinks (Siegel *et al.*, 1970; Raghow, 2013). Therefore, the biological structures, rich in collagen, show a great tensile strength.

2.2.1.2 Gelatin

Gelatin, one of the most extensively-used edible packaging material, is a mixture of proteins and peptides obtained by partial hydrolysis of collagen. The bovine gelatin obtained by partial hydrolysis of collagen extracted from cartilage, bones and hides, and pigskin gelatin obtained by partial hydrolysis of collagen in pig skin are the major gelatin types used by the food industry. The gelatins might be Type-A, obtained by acid hydrolysis of collagen, or Type-B, obtained by alkaline hydrolysis of collagen.

Type-A and Type-B gelatins are characterized mainly by their different isoelectric points (pI) that are 8-9 and 4-5, respectively. During collagen hydrolysis, triple-helices of collagen are transformed into its coils (helix-to-coil transformation); thus, gelatin is a heterogeneous mixture of single-chain (α-chain), double-chain (named β-chain that is formed by two covalently cross-linked α-chain), and triple-chain (named γ-chain that is formed by three covalently cross-linked α-chain) polypeptides (Ramos *et al.*, 2016) (Fig. 2.1). Therefore, it is clear that not only the source of extraction, but also the hydrolysis conditions (temperature, pH, and duration of extraction) might affect the molecular, compositional, and physicochemical characteristics of gelatin and properties of its resulting films (Ramos *et al.*, 2016). Moreover, treatments, such as ultrasound and enzymatic hydrolysis, applied to

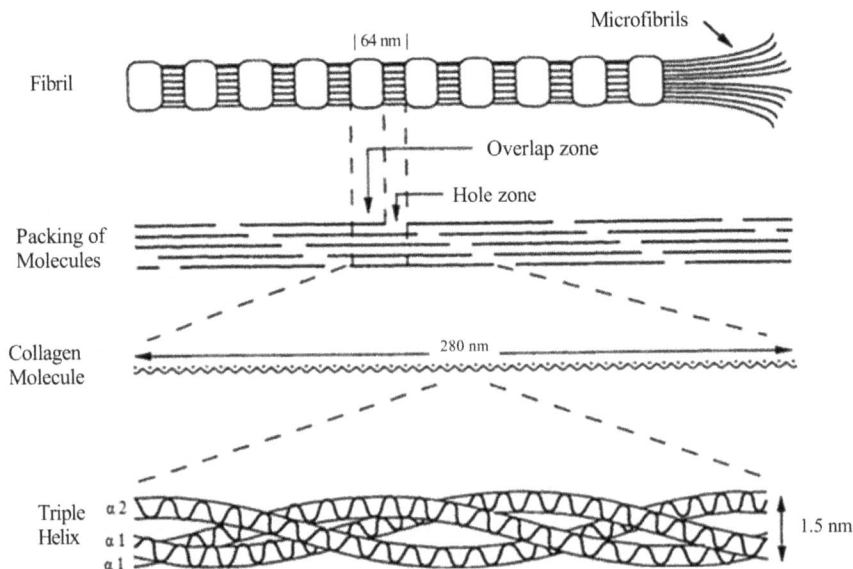

Fig. 2.1: The structural organization and characteristics of collagen molecule; reprinted with permission from Cawston (1995); copyright © 1995 Elsevier Ltd.

improve extraction yields of gelatin and/or modify its properties, also cause some conformational changes that might affect its gelation and film-making properties (Muhammad *et al.*, 2016; Cuevas-Acuna *et al.*, 2020).

The gelatin is an indispensable animal-source food hydrocolloid not only since its production transforms huge amounts of waste and byproduct from slaughtering houses into a value-added hydrocolloid, but also due to its unique film- and gel-forming capacity as well as excellent elasticity, texture, taste, and nutritive value (Yemenicioğlu *et al.*, 2020). Due to religious and health concerns associated with mammalian gelatins (e.g. risk of prions from bovine gelatins), there is also a significant interest in the use of fish gelatin obtained from processing wastes (mostly fish skins) of fisheries industry. The commercial fish gelatin is mostly obtained from cold-water fish that represents the majority of industrial fisheries (Benjakul and Kittiphattanabawon, 2019; Alfaro *et al.*, 2015). Moreover, it is also suggested that low-allergy fish (e.g. tuna) should be evaluated in gelatin production to minimize the risk of fish allergy (André *et al.*, 2003).

The gelatin from poultry (chicken, duck, and turkey) processing wastes (e.g. skin, feet, and bone) might also be a major source of gelatin since it is free form most of the religious concerns and health risks associated with mammalian and fish gelatins (Yang *et al.*, 2017; Suderman *et al.*, 2018). Finally, some edible insects, such as *Aspongopus viduatus* (melon bug) and *Agonoscelis pubescens* (sorghum bug) that are traditionally used for edible oil extraction in Sudan, could also be a potential future source of gelatin and its edible films (Mariod and Fadul, 2013). It is expected that food and food additives and ingredients (e.g. oil, protein, colorants) from insects will gain importance in the future as Earth's resources have soured due to environmental problems (Mitsuhashi, 2010; Dossey *et al.*, 2016).

2.2.1.3 Egg proteins

The food industry uses whole eggs, egg yolk, and white (albumen) in fresh, pasteurized, dried or frozen forms as a major food ingredient or as a functional additive. The development of edible films and coatings from egg proteins is attractive since these proteins show an excellent nutritional value as well as unique technological properties, such as binding (pasting), emulsification, and coagulation (Yemenicioglu *et al.*, 2020).

Due to its lipid-free nature, the egg white (or albumen) can be employed effectively to obtain protein-based films and coatings. The egg white (forms almost 63%, w/v of the eggs) consists of almost 10-11% protein and 88% water (Kovacs-Nolan *et al.*, 2005). The ovalbumin is the major egg-white protein since it forms almost half (54%) of the total egg-white proteins. The ovalbumin is a glycoprotein formed by 386 amino acids with a molecular weight (MW) of 45 kDa (Abeyrathne *et al.*, 2013). Although ovalbumin is one of the major allergen proteins in egg white (Mills and Tatham, 2003), it is an excellent source of amino acids for non-allergenic persons (Huopalahti *et al.*, 2007). Therefore, the egg-white protein films and coatings used with proper food labelling could be very suitable for active packaging purposes.

Other major protein fractions in the egg white are ovotransferrin (or conalbumin), ovomucoid, ovoglobulins, lysozyme, and ovomucin that form almost

12, 11, 8, 3.5, 3.5% of total egg-white proteins, respectively (Abeyrathne *et al.*, 2013; Kinoshita *et al.*, 2016). The bioactive properties and physiological roles of these egg-white proteins vary considerably. For example, ovotransferrin, a 76 kDa iron-binding protein, shows antimicrobial activity by limiting iron availability and by causing damage/permeability changes in membranes of different Gram-positive and Gram-negative pathogenic bacteria, such as *Pseudomonas* spp., *Salmonalla enterica* serovar Enteritidis, *Escherichia coli,* and *Staphylococcus aureus* (Kovacs-Nolan *et al.*, 2005). Ovomucoid, with a MW of 28 kDa, is a heavily glycosylated egg-white protein (25% is formed by carbohydrates) that is known as a trypsin inhibitor, and one of the major allergen proteins in the egg white (Mills and Tatham, 2003; Abeyrathne *et al.*, 2013). Ovoglobulins formed by two globulins (G2 and G3) are highly related with stability of egg-white foams, but their biological roles have not been yet clarified (Kinoshita *et al.*, 2016). Lysozyme (MW: 14.3 kDa) is an antimicrobial enzyme that shows lytic activity on peptidoglycan layer of Gram-positive bacteria, such as the very critical pathogen *Listeria monocytogenes*. Therefore, there are extensive efforts to incorporate lysozyme into active packaging as a natural antilisterial component (Min *et al.*, 2005a; Ünalan *et al.*, 2011). On the other hand, the ovomucin is another glycosylated egg-white protein with antiviral and antitumoral activities. This bioactive protein contains two protein subunits, α-ovomucin (15% is formed by carbohydrates) and β-ovomucin (50-60% is formed by carbohydrates) that MWs range between 180 and 220 kDa and 400 and 770 kDa, respectively (Strixner and Kulozik, 2011; Kovacs-Nolan *et al.*, 2005). Egg white contains also some minor protein fractions, such as avidin, ovocystatin (cystatin), ovostatin (ovomacroglobin), ovoflavoprotein (riboflavin binding protein), ovoglycoprotein, and ovoinhibitor.

On the other hand, the egg yolk (forms almost 27.5%, w/v of egg) is an emulsion that contains heavy lipids (32-35%) and some protein fractions (16-17%), such as soluble plasma (the yellow fluid) proteins and granule proteins suspended within the plasma (Kovacs-Nolan *et al.*, 2005). Depending on their lipid contents, egg-yolk proteins can be classified as low-density lipoprotein (LDL, ~20% lipid content), very low-density lipoprotein (VLDL, ~40% lipid content), and high-density lipoprotein (HDL, ~10% lipid content) (Stadelman, 2003). The majority of egg-yolk proteins show good gelation and emulsification properties (Laca *et al.*, 2014; Valverde *et al.*, 2016). However, there is a particular interest in commercial extraction and use of granule proteins in different foods since they contain less cholesterol than other egg-yolk proteins, and are less heat sensitive than egg yolk (Laca *et al.*, 2014). The precursor of major egg-yolk proteins is vitellogenin that transforms by enzymatic hydrolysis into major granule proteins, such as lipovitellins and phosvitin (Zambrowicz *et al.*, 2014). The lipovitellins (forming almost 70% of the granules and 40% of egg-yolk's protein) consist of two types of protein fractions, named α-lipovitellin and β-lipovitellin (ratio in egg yolk = 1:1.8) that are LDL group proteins characterized by their high-bound zinc content (zinc: protein ratio = 1.3 to 2) (Burley and Cook, 1961; Groche *et al.*, 2000). The α-lipovitellin is formed by four subunits with MWs of 125, 80, 40, 30 kDa while β-lipovitellin is formed by two subunits with MWs at 125 and 30 kDa (Groche *et al.*, 2000).

The phosvitin (forms almost 16% of the granules and 11-13.4% of egg-yolk proteins) is a 35kDa LDL protein that is characterized by its high-bound phosphorus content (almost 10% of protein is phosphorus) (Burley and Cook, 1961; Kovacs-Nolan *et al.*, 2005; Strixner and Kulozik, 2011; Chay Pak Ting *et al.*, 2010). Almost half of the amino acids in phosvitin are formed by serine that exists mostly in phosphorylated form (Aluko, 2015). The phosvitin in the egg yolk exists as high molecular weight complexes, such as α-phosvitin and β-phosvitin with MWs of 160 and 190 kDa, respectively (Chay Pak Ting *et al.*, 2010). Due to its high iron-binding capacity (binds almost all the iron in the egg yolk in the form of Fe^{+2} or Fe^{+3}) phosvitin shows antimicrobial and antioxidant capacity (Strixner and Kulozik, 2011; Samaraweera *et al.*, 2011). Sattar *et al.* (2000) reported that the antimicrobial activity of phosvitin increased when it was combined with a mild thermal stress at 50°C. These authors also thought that the surface activity of phosvitin and its hydrophobic domains also contributed to its antimicrobial activity by showing membrane penetration capacity. Thus phosvitin is considered as one of the potential active agents to be employed in active packaging and coating.

The plasma proteins of egg yolk contain mainly livetins that form 9.3% of the egg-yolk proteins. The livetins are lipid-free globular proteins that are formed by α-livetin, β-livetin, and γ-livetin (ratio in egg yolk = 2:5:3) (Kovacs-Nolan *et al.*, 2005; Chalamaiah *et al.*, 2017). The α-livetin, consisting of two protein fractions with MWs of 55 and 73 kDa, is an allergen protein (also known as allergen Gal d5) in the egg yolk that is identical to serum albumin protein in the blood of chicken, while β-livetin (α2-glycoprotein) consists of proteins having low molecular-weight fractions with MWs of 33 and 36 kDa (Guilmineau *et al.*, 2005; Chalamaiah *et al.*, 2017). The γ-livetin, called also immunoglobulin Y (Ig Y), is a particularly important defense protein having a quaternary structure (MW: 203 kD) formed by two heavy (~74 KDa) and two light (~24) weight subunits (Guilmineau *et al.*, 2005; Chalamaiah *et al.*, 2017). The Ig Y prevents invasion of egg yolk by microorganisms since it shows antiviral activity on viruses, such as rotavirus and coronavirus, and antibacterial activity on major pathogenic bacteria, such as *E. coli, Salmonella* spp., and *Staphylococcus aureus* (Kovacs-Nolan *et al.*, 2005). The livetins show different functional properties, such as high solubility (above 86%) at a broad pH range (pH 2.0-12.0), but their foaming and emulsifying activities are highly pH-dependent (Chalamaiah *et al.*, 2017). However, it was reported that the livetins are among egg protein fractions that transform into bioactive peptides after proteolytic hydrolysis (Zambrowicz *et al.*, 2014).

2.2.1.4   Caseins and caseinates

The caseins (forming almost 80% of total milk proteins) consist of different protein fractions, such as $α_{S1}$- (MW: 23.6 kDa), $α_{S2}$- (MW: 25.2-25.4 kDa), β- (MW: 24 kDa) γ- (MW: 11.6-20.5 kDa), and κ-caseins (MW: 19 kDa) (O'Regan *et al.*, 2009). The γ-casein is not a gene product as other caseins and it forms by posttranslational modification of β-casein by milk proteolytic enzyme, plasmin (Varnam and Shutherland, 2001). The composition of cow's feed affects the casein profile of milk, but $α_{S1}$-casein and β-casein are major fractions in bovine milk as they form 38-45% and 35-39% of total caseins, respectively (Panthi *et al.*, 2017; Stelwagen, 2016). The remaining caseins are formed mainly by κ-casein (~13%) and α_{S2}-casein (~10%) while γ-casein content (~3%) depends highly on proteolytic enzyme activity in milk (Stelwagen, 2016). The caseins exist in milk as 20 to 600 nm spherical colloidal associations (micelles) (average size: ~120 nm; average MW: ~10^8 kDa; ratio of α_{S1}-, α_{S2}-, β-, κ-caseins= 3:1:3:1) (Fig. 2.2a and 2.2b) and they are formed 92% by caseins that are stabilized by hydrogen bonds, ionic bridges, and hydrophobic interactions (Varnam and Shutherland, 2001; O'Regan *et al.*, 2009). The nano sized (~4 nm) calcium phosphate clusters are also considered among the building blocks of casein micelles since they are distributed within the network of associated casein molecules to form almost 8% of the micelle (Varnam and Shutherland, 2001; Cho and Jones, 2019). Although there are some different challenging models to explain the micelle architecture, there is an increasing consensus that the hydrophobic interactions are the main driving forces that cause the association of casein molecules to form the micelles, while the stabilization of this colloidal association is related mainly to calcium phosphate-carrying micelles that help interlocking of associated casein oligomers by calcium phosphate clusters (Qi, 2007; Nagy *et al.*, 2010).

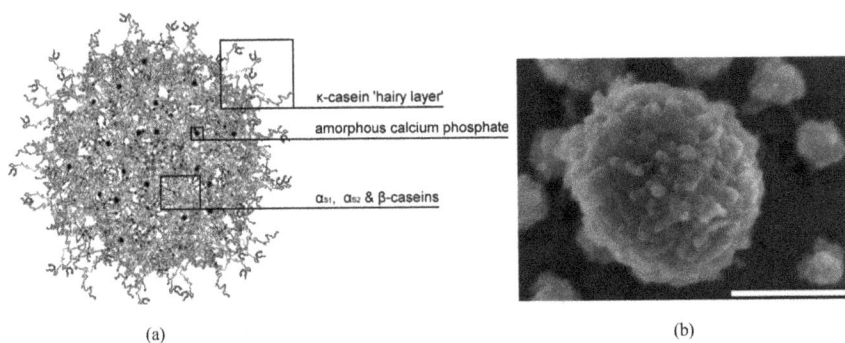

κ-casein 'hairy layer'
amorphous calcium phosphate
α_{s1}, α_{s2} & β-caseins

(a)   (b)

Fig. 2.2: Graphical representation of casein micelle model (a) (Głąb and Boratyński, 2017); © Creative commons CC. SEM micrograph of an individual casein micelle (b) (Scale bar: 200 nm) reprinted with permission from Dalgleish *et al.*, 2004; copyright © 2004 Elsevier Ltd.

The caseins, such as α_{S1}-, α_{S2}- and β-casein, are named 'calcium-sensitive' since they precipitate by binding of calcium atoms *via* their phosphoserine residues. The number of phosphoserine residues of α_{S1}-, α_{S2}- and β-caseins are 5, 7-9 and 10-13, respectively (Varnam and Shutherland, 2001). In contrast, κ-casein is named 'calcium-insensitive' since it remains soluble in the presence of calcium atoms due to its single phosphoserine residue. Thus, κ-caseins concentrate mainly on the surface of associated insolubilized α_{S1}-, α_{S2}- and β-caseins and orient in such a way that their hydrophilic and negatively-charged glycomacropeptides (also called caseinomacropeptide) form a hairy surface layer that helps to stabilize and maintain the colloidal nature of micelles by creating a steric hindrance among them (Qi, 2007). The steric hindrance provided by the hairy layer at the micelle surface is maximal (highest hairy layer height) between pH 6.0 and 7.0, but declines as pH drops below 6.0, and becomes minimal, and then collapses at pH 4.5 (lowest hairy layer height) to initiate micelle flocculation (Tuinier and Kruif, 2002). The isoelectric point (pI) of caseins is also almost at 4.6. Thus, flocculation by acidification is exploited to produce acid caseins.

In the food industry, the acid caseins are obtained from pasteurized bovine skim milk by using lactic acid-producing starters or mineral acids (e.g. hydrochloric acid, sulfuric acid). The precipitated caseins are then washed extensively to remove soluble acids, whey, lactose, and minerals, and then they are dried to obtain casein powders.

The flocculation of caseins occurs also following the action of enzyme cymosin (rennin) that causes coagulation of micelles by removal of hairy glycomacropeptides of κ-casein (called para-κ-casein after enzymatic modification). The rennet caseins obtained by enzymatic coagulation do not need acidification and are used mostly in production of all kinds of cheese. The micellar casein concentrates produced by microfiltration of skim milk, using mostly ceramic membranes, are also alternative sources of caseins (Crowley *et al.*, 2018). However, such casein concentrates still contain some residual lactose, minerals, and serum proteins (whey proteins), depending on the efficiency of microfiltration applied during processing.

Since acid caseins are not water-soluble, they should be turned into soluble caseinates by the classical alkaline neutralization process before utilization for edible films and coatings. The Na-caseinate obtained by neutralization of acid caseins (freshly prepared or dried), using mainly sodium hydroxide (sodium bicarbonate and sodium phosphate, are used less extensively) is highly water-soluble; hence, it is used extensively, both as a highly functional food ingredient (as an excellent emulsifier) and as a suitable material for edible films and coatings (Shendurse *et al.*, 2018). The Ca-caseinates obtained by neutralization of acid caseins by calcium hydroxide are also used for different food applications and production of edible films and coatings, but this type of caseinates is considerably less soluble than Na-caseinates and forms colloidal suspensions (Shendurse *et al.*, 2018). Thus, due to their high solubility, Na-caseinates show excellent functional properties (e.g. viscosity, emulsification, foaming, water binding, heat stability, etc.) while poorly soluble Ca-caseinates show limited functionality (Meena *et al.*, 2017).

2.2.1.5 *Whey proteins*

Whey protein concentrates (WPC with 35-59% or 60-80% protein) and isolates (WPI with ≥ 90% protein) obtained mainly by membrane filtration of the byproduct whey from cheese- and yogurt-making, and from production of caseins are among the most important commercial sources of protein that could be employed for manufacturing of edible films and coatings. Due to their high bioactive peptide content, whey proteins show antioxidant activity that is not only effective in controlling lipid oxidation in food, but also in reducing enzymatic browning reaction caused by polyphenol oxidase (Yi and Ding, 2015; Kumar *et al.*, 2018). Therefore, WPI films alone, or in combination with supporting acidulants, can be employed to reduce browning of minimally processed fruits (Azevedo *et al.*, 2018; Shendurse *et al.*, 2018). The commercial WPC is increasingly obtained by spray drying of retentates from whey ultrafiltration (UF) while WPI is generally produced by spray-drying of UF retentates, further purified with diafiltration (DF) (Kaur *et al.*, 2019). The WPI shows not only excellent film-making properties, but also possesses good nutritional and functional properties such as solubility, emulsification, and heat-induced gelation (Bylund, 1995). Thus, high-quality WPI has been extensively used to develop classical whey protein films that contain solely plasticizers, and more complex whey-based films, such as emulsion films, and blend or composite films that contain mixture of WPI with other hydrocolloids, lipids, emulsifiers, plasticizers, etc. (Perez-Gago and Krochta, 2000; Khwaldia *et al.*, 2004; Ramos *et al.*, 2012; Boyaci and Yemenicioğlu, 2018).

Whey forms 80 to 90% of total processed bovine milk volume, and can be classified as acid, sweet or salty whey, depending on the manufacturing methods of obtained dairy products. The whey obtained from rennet coagulation of caseins during manufacturing of soft, semi-hard or hard cheeses is called the sweet whey (pH: 5.9-6.6) while whey obtained from mineral acid coagulation of caseins is called the acid whey (pH 4.3-4.6) (Bylund, 1995). The salty whey remains from curd-salting processes applied during cheese-making, but it cannot be utilized into different products due to its high salt content. The native whey obtained specifically from unprocessed raw skim milk by microfiltration (removes caseins and residual fat) and ultrafiltration-diafiltration (removes lactose, minerals, and non-protein nitrogenous compounds) is also a whey type with unique functional and nutritional properties (Bylund, 1995). The native whey contains whey proteins in undenatured form since it receives less thermal processing than rennet and acid whey, and is free from cheese residues, starter cultures, bacteriophages, salts, acids, and glycomacropeptides produced by rennin.

Whey proteins, forming almost 20% of total bovine milk proteins, consist of different major fractions such as β-lactoglobulin (MW: 18.3 kDa), α-lactalbumin (MW: 14.0 kDa), serum albumin (MW: 63000), and immunoglobulins (MW up to 1000 kDa) (Varnam and Shutherland, 2001). The glycomacropeptide (MW: 700 kDa) is a major whey component in rennet whey, but native and acid whey lack this peptide fraction (Kumar *et al.*, 2018). β-lactoglobulin forms 50 to 60% of the total whey proteins and contains two protein fractions, β-lactoglobulin A (dominant fraction) and β-lactoglobulin B, that are responsible for the gel formation properties of WPI or WPC (Kelly *et al.*, 2009; Korhonen, 2011). The α-lactalbumin that forms almost 20% of bovine whey proteins is the second most abundant whey protein. The α-lactalbumin attracts a particular interest as an infant supplement due to its high essential amino acid content (especially tryptophan, leucine, isoleucine, valine, methionine, and lysine) (Layman *et al.*, 2018) and bioactive properties (e.g. prebiotic activity, antibacterial activity, and inhibition of bacterial adhesion on to Coco-2 cells) originating from peptides formed by its proteolytic hydrolysis in the human gastrointestinal system (Recio *et al.*, 2009). The bioactive peptides of α-lactalbumin could also be formed during fermentation or maturation of dairy products by starter cultures or nonstarter microorganism, or by use of commercial proteases (Layman *et al.*, 2018; Kamau *et al.*, 2010). Thus, enzymatic hydrolysates of α-lactalbumin having bioactive properties can be exploited to obtain bioactive packaging materials.

Moreover, the ability of α-lactalbumin to form nanotubes via self-assembly in the presence of Ca^{++} atoms might also be exploited for active packaging to encapsulate different active compounds or to obtain controlled release systems for their delivery (Assadpour and Jafari, 2019). The commercial α-lactalbumin preparations are obtained mostly from WPC using suitable combinations of selective precipitation/ aggregation (separates mainly α-lactalbumin together with some β-lactoglobulin), selective enzymatic hydrolysis (e.g. trypsin hydrolyzes β-lactoglobulin dramatically higher than α-lactalbumin), and membrane separation methods (UF alone or in combination with DF) (Kamau *et al.*, 2010).

Whey also contains some important minor bioactive proteins, such as lactoperoxidase and lactoferrin that have been purified and used as antimicrobial agents to develop active packaging materials (Min *et al.*, 2005b; Mecitoglu and Yemenicioglu, 2007). Lactoperoxidase (MW: 78 kDa), an oxidoreductase group enzyme, is responsible for production of highly reactive short-lived antimicrobial oxidation products that form the natural antimicrobial mechanism in milk effective on most bacteria, fungi, and viruses (Pakkanen and Aalto, 1997; Seifu *et al.*, 2005; Mecitoglu and Yemenicioglu, 2007). Therefore, edible films formulated, using lactoperoxidase and its natural substrates, show a broad antimicrobial spectrum (Yener *et al.*, 2009; Min *et al.*, 2005b). The commercial production of lactoperoxidase enzyme is conducted via cation exchange chromatography of sweet whey following application of microfiltration that removes fat and protein aggregates (Wang and Guo, 2019). Lactoferrin (MW ~80 kDa), another minor constituent of whey, is a member of transferrin family of non-heme iron-binding glycoproteins. This protein shows bactericidal or bacteriostatic activities on different bacteria, fungistatic activity on fungi, and antiviral activity against some human intestinal viruses which may be potential food contaminants (Marchetti *et al.*, 1999; Superti *et al.*, 1997). Due to its strong iron-binding capacity, lactoferrin may also be used as an antioxidant in foods (Huang *et al.*, 1999; Satue-Garcia *et al.*, 2000). The hydrolysis of lactoferrin with pepsin also generates a considerably more potent bactericidal peptide called lactoferricin. This peptide also shows a strong antifungal activity (De Lucca and Walsh, 1999; Liceaga-Gesualdo *et al.*, 2001). Thus, it is clear that whey is a highly critical byproduct of milk processing, not only to obtain hydrocolloids for development of edible packaging materials, but also to obtain antimicrobial and antioxidant agents that could turn edible packaging into active one.

2.2.1.6   Total milk proteins

The milk protein concentrates (MPC with almost 40 to 85% protein) and isolates (MPI with minimum 90% protein), that show similar ratio of caseins and whey proteins with bovine milk, have been increasingly produced from skim milk by membrane separation methods (e.g. microfiltration, ultrafiltration, and diafiltration) (Crowley *et al.*, 2018). The MPC and MPI contain micellar form of caseins and native form of whey proteins (O'Regan *et al.*, 2009). The ash content of MPC and MPI is relatively high since micellar caseins contain high amounts of calcium phosphate. However, reduction of ash content is possible if pH of the milk sample is reduced during UF and DF to dissociate casein molecules forming the micelles and to cause release of some calcium and phosphate from this colloidal association (Rajagopalan and Cheryan, 1991; Meena *et al.*, 2017). The limited solubility of caseins in reconstituted MPC and MPI powders is a serious problem that interferes both with their use as a functional ingredient in food and as a hydrocolloid in development of edible films and coatings. This explains why more soluble WPI and WPC as well as Na-caseinates are employed frequently during development of edible films and coatings. However, the solubility problem reduces as the protein content of MPC is reduced. Moreover, the solubility of MPC obtained can also be improved by (1) chelation of calcium in milk by addition of chelating agents, such as citric acid and phosphates; (2) reduction of calcium content of milk by UF or UF-DF followed by acidification or high pressure treatment (100-400 mPa); (3) limited proteolytic digestion; and (4) treatment of milk via high-energy methods, such as high-pressure homogenization, microfluidization, and ultrasonication (Meena *et al.*, 2017).

2.2.1.7   Chitin and chitosan

Chitin, a linear polysaccharide that consists mainly of repeating β (1-4)-linked N-acetyl-D-glucosamine residues, is the most abundant marine-origin polysaccharide, but is the second most abundant polysaccharide on Earth after cellulose (Fig. 2.3a). In nature, different chitin chains come together to form crystalline microfibrils (called also rod or crystallide), named α-, β-, or γ-chitin. The microfibril of α-chitin consists of antiparallel chitin chains while those of β-chitin and γ-chitin contain parallel and mixed chitin chains, respectively (Liu *et al.*, 2019). The chitin microfibril type in arthropods, such as crustaceans, is α-chitin that is formed approximately by 20 chitin chains with a microfibril size of 3 nm in diameter and 300 nm in length (Liu *et al.*, 2019). Such chitin microfibrils are stabilized mainly by extensive hydrogen bonding, exclusive intermolecular hydrogen bonding (single hydrogen bond), and a combination of intermolecular and intramolecular hydrogen bonding (double hydrogen bonds formed by acetamide groups) [intramolecular hydrogen bonds: C(3)—OH—O—C(5) and C(6)—OH—O=C (C=O of acetamido group); intermolecular hydrogen bonds: NH—O=C (NH and C=O of acetamido groups) and C(6)—OH—OH—C(6)] (Minke and Blackwell, 1978; Kameda *et al.*, 2005; Sikorski *et al.*, 2009).

Chitin is commercially produced from shells of crustaceans (e.g. crab, shrimp, lobster, krill, etc.) remaining as a byproduct of the fishery industry (Yadav *et al.*, 2019). The exoskeleton of crustaceans is a composite of chitin and proteins (chitin fibers embedded within proteins) with a high degree of mineralization with $CaCO_3$ deposited within this composite's matrix as calcid (major form) and amorphous forms (minor form) (Meyers and Chen, 2014). Therefore, the basic processes applied during commercial production of chitin from shells include demineralization (conducted by HCl to solubilize excessive insoluble $CaCO_3$) and deproteinization (conducted by NaOH to solubilize proteins). The crude chitin can be used to obtain chitosan or it can be further processed to obtain chitin whiskers. The chitin whiskers obtained by acid hydrolysis of crude chitin by boiling in HCl are nano-sized slender parallelepiped rods that are currently employed extensively to reinforce matrices of packaging films (Yemenicioğlu *et al.*, 2020).

The chitin is called chitosan when its degree of acetylation is reduced below 50% by the classical alkali treatment method (de-acylation process) (Van den Broek *et al.*, 2015). The de-acylation turns part of the acetamido groups (—NHCOCH$_3$) linked to C-2 of chitin into an amino group (—NH$_2$) that could be protonated (NH$_3^+$) at the acid pH. The reduction in number of acetamido groups (means less intermolecular hydrogen bonds as —NH—O═C) reduces the tendency of chitin chains to form hydrogen-bonded, rigid associations of microfibrils. Thus, this process transforms insoluble chitin into chitosan that can be solubilized in acidic solutions. There are also some different ongoing research activities to achieve transformation of chitin into chitosan by employing biotechnological methods using chitin deacetylase enzyme obtained from fungi, such as *Aspergillus nidulans*, but this enzymatic method is still insufficient to achieve a degree of deacetylation (DDA) above 50% necessary to produce chitosan (Harmsen *et al.*, 2019).

The chitosan, a soluble heteropolymer of N-acetyl-d-glucosamine and d-glucosamine residues lined through β (1-4) linkages (Fig. 2.3b), is used extensively to obtain edible films or coatings since it shows excellent film-forming ability, and unique inherent antibacterial and antifungal properties (Van den Broek

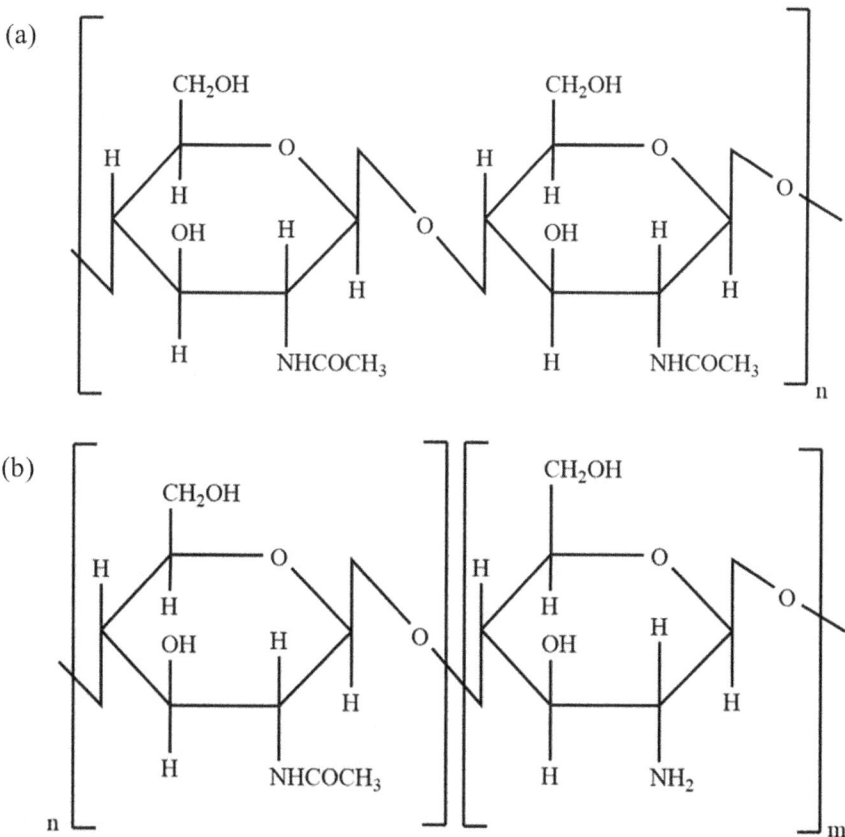

Fig. 2.3: Chemical structure of chitin (a) and chitosan (b)

et al., 2015). The antimicrobial mechanism of chitosan is not fully elucidated. However, the current knowledge suggests that the antimicrobial activity of this interesting polysaccharide originates mainly from its ability to bind negatively charged bacterial surfaces with cationic -NH_3^+ on its glucosamine residues. It is thought that the binding of chitosan on to bacterial cell walls inactivates cell since this activates secondary cellular mechanisms that lead to disruption of membrane barrier functions and membrane-bound energy generation pathways, and it causes initiation of various stress responses (Raafat *et al.*, 2008; Dutta *et al.*, 2009; Tantala *et al.*, 2019). It was proved with transmission electron micrographs that the binding of chitosan on to bacterial cell walls caused local detachments of membrane from the cell walls (Raafat *et al.*, 2008). This gave rise to formation of vacuole-like structures underneath the cell wall, triggering efflux of ions and water and causing the loss of internal bacterial pressure (Raafat *et al.*, 2008). It is also thought that the high metal-chelating capacity of chitosan is also a secondary factor that contributes its antimicrobial capacity. However, the chelating capacity of chitosan occurs when NH_3^+ groups on its glucosamine residues are deprotonated at high pH to form $-NH_2$ that acts as a chelating group because of its electron pair on nitrogen available for donation to metal ions (Ramya *et al.*, 2012). Moreover, it was reported that the low MW chitosan molecules (chitosan oligomers) also have the ability to penetrate from microbial cell walls into cytoplasm and to cause inactivation of cells by binding on their DNA, and penetrating into their nuclei to prevent their protein and mRNA synthesis (Zheng and Zhu, 2003; Tikhonov *et al.*, 2006; Van den Broek *et al.*, 2015).

Since both DDA and MW of chitosan are highly effective on its inherent antimicrobial properties, the severity of chitosan de-acetylation process (temperature and concentration of alkaline solution) during transformation of chitin to chitosan should be controlled carefully (Tolaimate *et al.*, 2000; Synowiecki and Al-Khateeb, 2003). Theoretically, for two chitosans having almost the same MW, the one with higher DDA (capable to form more NH_3^+ groups at acid pH) is expected to show higher antimicrobial activity. This was proved by Malinowska-Pańczyk *et al.* (2015) who treated the microorganisms, such as *Escherichia coli*, *Pseudomons fluorescens*, *Staphylococcus aureus*, and *Listeria innocua* with solutions of different chitosans and determined significantly higher antimicrobial activity for chitosan with MW of 759 kDa at 96% DDA than chitosan with MW of 766 kDa at 90% DDA. However, it is quite complex to estimate the antimicrobial activities of chitosans when they differ significantly both in DDA and MW. The antimicrobial performance of chitosan also shows a great variation, depending on antimicrobial test method, pH, and temperature of test medium, and type of bacteria tested. The antimicrobial mechanism of solubilized chitosan in broth, buffer or peptone water containing a microorganism includes effective surrounding of microbial surface with a layer of cationic chitosan molecules, and penetration of some chitosan from cell walls. However, in antimicrobial tests, based on contact methods (bacteria treated with insoluble chitosan films or particles), the antimicrobial activity depends mainly on ability of cationic chitosan to interact with microbial surface. The increase of pH in test medium (e.g. between pH 4.5 and 5.9) also affects the antimicrobial performance of chitosan inversely (No *et al.*, 2002). Moreover, some bacteria (e.g., *S. aureus* and

E. coli) might show high resistance against chitosan when treated at refrigeration temperatures (Malinowska-Pańczyk *et al.*, 2015). It is thought that refrigeration causes stress over bacteria and modifies chitosan-binding sites (or electronegative sites) on cell surface (Tsai and Su, 1999). Finally, the antimicrobial resistance of Gram-negative bacteria against chitosan is generally greater than that of Gram-positive bacteria (No *et al.*, 2002). Thus, extensive antimicrobial screening tests are needed to determine the most suitable chitosan form at the desired condition against target microorganisms. No *et al.* (2002) conducted antimicrobial tests on 11 bacteria with chitosan having MW between 1 and 1671 kDa and reported effective chitosan forms against these bacteria as follows; 746 kDa for *Escherichia coli;* 746 kDa for *Pseudomonas fluorescens;* 470 kDa for *Salmonella enterica* serovar Typhimurium*;* 470 kDa for *Vibrio parahaemolyticus;* 1671, 746 and 224 kDa for *Listeria monocytogenes;* 1671 and 746 kDa for *Bacillus megaterium;* 746 and 470 kDa for *B. cereus;* 1671, 746 and 470 kDa for *Staphylococcus aureus;* 1106, 746, 224 and 28 kDa for *Lactobacillus plantarum*; 1106 kDa for *L. brevis*; and 224 kDa for *L. bulgaricus*. Zheng and Zhu (2003), studied within <5 to 305 kDa MW range, found that chitosan at MWs of <5 and 48.5 kDa were most effective on *E. coli* while chitosan at MWs of 129, 166 and 305 kDa were most effective on *S. aureus*. The study of Zheng and Zhu (2003) is interesting since it suggested a direct and an inverse relationship between MW of chitosan and its antimicrobial activity against *E. coli* and *S. aureus*, respectively. Kim *et al.* (2007) studied with solutions of chitosan at different MWs in broth and reported that the ranking of antimicrobial potency for these hydrocolloids against *S. enterica* serovar Enteritidis was as follows: 282>440>1110>746 MW. Leleu *et al.* (2011) also found that *S. enterica* serovar Enteritidis was more efficiently inhibited by high MW chitosan (MW: 310-375 kDa, DDA: 75%) than low MW chitosans (MWs between 28 and 190 kDa, DDAs between 83.7 and 91.8%) tested with a plate contact method (bacteria inoculated on an agar mixed with chitosan). In contrast, Silva Jùnior *et al.* (2014) found quite similar antimicrobial activities for low and high MW chitosans (MW of 624 kDa with DDA >75%; MW of 107 kDa with DDA between 75 and 85%) tested on the critical plant pathogen *Botrytis cinerea* using a plate contact method (fungi on small agar discs were placed in Petri dishes containing different concentrations of chitosan solution at the bottom). Thus, it is clear that the use of chitosan in antimicrobial packaging needs careful selection of suitable chitosan MW form and % DDA depending on resistances of target bacteria.

2.2.2 Plant source hydrocolloids

2.2.2.1 *Cellulose and cellulose derivatives*

Cellulose a linear polysaccharide formed by β-(1-4)-linked D-Glucose units is the most abundant organic compound on Earth. Due to the β-(1-4) linkage, each D-Glucose molecule in the cellulose chain rotates 180° relative to its neighbor (Fig. 2.4). Thus, the repeating unit in the cellulose is the disaccharide cellobiose. Several parallel cellulose chains come together with hydrogen bonds to form cellulose fibrils. The long linear fibrils contain both intrachain (forms between the ring oxygen of one residue and the hydrogen of the C3 hydroxyl of a neighboring

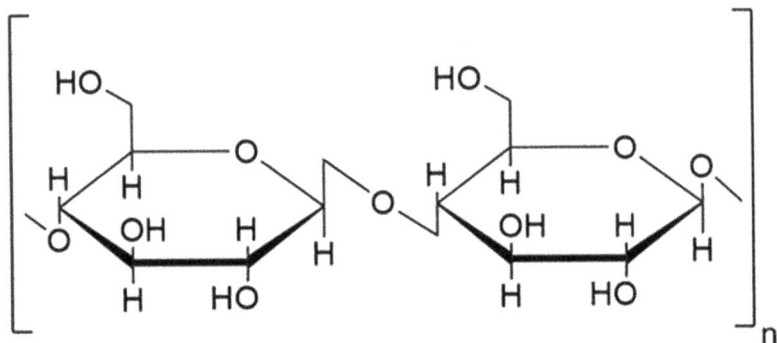

Fig. 2.4: Chemical structure of cellulose

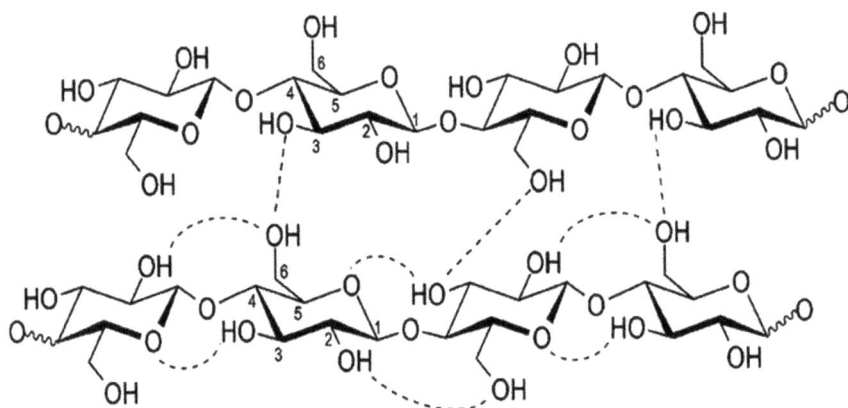

Fig. 2.5: Inter- and intra-molecular H-bonds formed by cellulose molecules; reprinted with permission from Pinkert *et al.* (2009); copyright © 2009 American Chemical Society

residue) and interchain (forms between hydrogen of hydroxyl and oxygen atoms on adjacent chains) H-bonds (Fig. 2.5) as well as van der Waal's forces that provide an additional stability to the fibril (Srivastava, 2002). The fibrils are mainly crystalline in nature since they got a precise three-dimensional order, but the crystalline regions are irregularly interrupted by disordered regions (amorphous regions) (Srivastava, 2002; Kontturi, 2018). It is assumed that the inter-sheet H-bonds among crystal sheets of fibrils are weak and disordered, but different sheets of crystal cellulose are kept together via hydrophobic interactions (Rongpipi *et al.*, 2019).

Different fibrils associate to form microfibrils while several microfibrils arrange to form macrofibrils that further organize as bundles that form cellulose fibers. The number of cellulose chains forming a microfibril is not fully clarified, but there are three different models that explain the microfibril structure with association of 18, 24 and 36 cellulose chains (Oehme *et al.*, 2015). In their recent review, Rongpipi *et al.* (2019) presented increasing focus on 18 and 24 cellulose chain models, but further studies are needed to prove which model represents the exact organization of cellulose microfibril structure.

Cellulose is generally obtained from wood using the classical chemical Kraft pulping. However, alternative non-chemical methods such as subcritical water pre-extraction of wood are needed to increase sustainability and feasibility of wood-pulping processes. Moreover, more effective use of non-wood cellulose fiber source plants, such as cotton, wheat, ramie, banana, sisal, kenaf, flax, etc. is needed to reduce the amount of wood used in cellulose production (Thakur and Thakur, 2014).

The natural cellulose cannot be used directly in manufacturing of edible films and coatings since it is water insoluble owing to its high molecular weight and crystalline nature. However, microfibrillated cellulose (MFC), nanofibrillated cellulose (NFC), microcrystalline cellulose (MCC), nanocrystalline cellulose (NCC), (*see* Fig. 2.6a to 2.6d), and different chemically modified derivatives of cellulose (cellulose ethers), such as carboxymethyl cellulose (CMC), methylcellulose (MC), hydroxypropyl methylcellulose (HPMC), and hydroxyethyl cellulose (HEC) can be

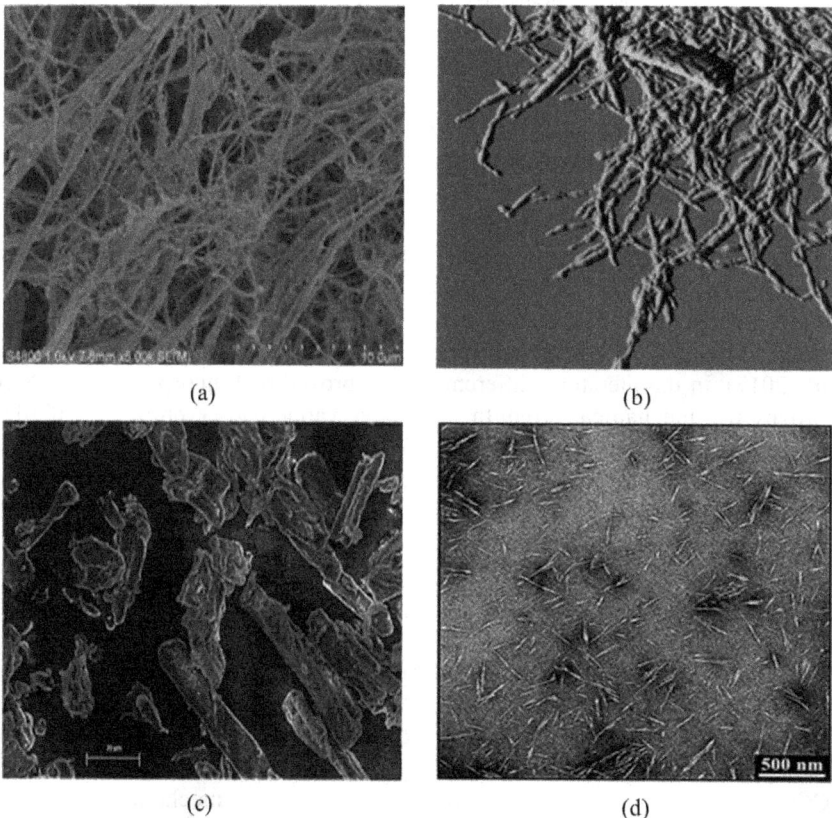

(a)   (b)

(c)   (d)

Fig. 2.6: (a) Scanning electron microscope (SEM) image of MFC from wood cellulose (scale bar: 100 μm); reprinted with permission from Kwon *et al.* (2015); copyright © 2015 Elsevier Ltd.; (b) atomic force microscope (AFM) image of NFC from wood cellulose; reprinted with permission from Henriksson *et al.* (2007); copyright © 2015 Elsevier Ltd.; (c) MCC (scale bar: 20 μm) and (d) NCC (scale bar: 500 nm) were reprinted from Cataldi *et al.* (2015); creative commons CC

used alone, in combination with each other or with other hydrocolloids to obtain edible films.

The MCC is commercially produced by treatment of natural cellulose with mineral acids as HCl to selectively hydrolyze its amorphous regions. The remaining crystalline cellulose is transformed into MCC through further mechanical treatment. There are also some patented processes that use high shear treatment (an extruder) in combination with elevated temperatures and pressures as well as chemical oxidizing agents (e.g. hydrogen peroxide) to depolymerize and transform cellulose into MCC (Kopesky and Ruszkay, 2006). However, majority of MCC production is conducted with classical acidic hydrolysis. The degree of polymerization of MCC is typically less than 400 with no more than 10% of the material has a particle size of less than 5 μm (FAO/WHO, 2017). The MCC alone does not have good film-forming properties due to its limited dispersibility, but it is an excellent filling material to obtain composite films (e.g. starch/MCC, starch/MC/MCC, and soy protein isolate/ MCC/Metal (Zn or Cu) nanocluster films) with improved mechanical, barrier and/ or thermal properties (Psomiadou et al., 1996; Li et al., 2017; Othman et al., 2019).

The NCC (3-5 nm in width and 50-500 nm in length), called also cellulose nanocrystals (CNC), cellulose whiskers or cellulose nanowhisker, is obtained by hydrolysis of amorphous regions of cellulose using sulfuric acid (Huiyu, 2017). The hydrolysis of cellulose by sulfuric acid yields negatively charged NCC due to transformation of some hydroxyl groups (-OH) at the fibril surfaces to sulfate ester groups ($-OSO_3^-$). Thus, the highly hydrophilic NCCs form stable aqueous suspensions that could be exploited to obtain homogenous nanocomposite edible films and coatings. In general, the addition of NCC into edible films causes an increase in their mechanical stability and tortuosity that is highly effective on water vapor-barrier properties of edible films (Azeredo et al., 2017). In the literature, different studies proved that incorporation of NCC improves the mechanical strength and water vapor barrier properties of films from major carbohydrate-based hydrocolloids, such as pectin, chitosan, alginate, starch, HMPC (Bilbao-Sáinz et al., 2010; Khan et al., 2012; Huq et al., 2012; Slavutsky and Bertuzzi, 2014; Chaichi et al., 2017). In contrast, the dispersion of hydrophilic NCC into lipid-based materials or hydrophobic plastic materials (e.g. thermoset plastics and thermoplastics) is very difficult and needs further chemical modifications to increase its surface hydrophobicity (Börjesson et al., 2018).

The MFC and NFC are called 'fibrillated' since the high shear treatment used during their processing causes cellulose fibers to separate into microfibrils to form a 3D network. The MFC and NFC contain both amorphous and crystalline regions since their amorphous regions are not acid hydrolyzed as MCC and NCC. The MFC is obtained by high shear mechanical treatment while NFC is obtained by chemical oxidation pretreatment and a following high pressure (2000 bar) homogenization (Kumar et al., 2014). Thus, the MFC has a considerably higher aspect ratio (10-100 nm wide and 0.5-10 mm in length) than NFC (4-20 nm in width and 500-2000 nm in length) (Huiyu, 2017). The suspensions of both MFC and NFC could be used in development of edible films and coatings when plasticized properly (e.g. with glycerol). It was reported that the

NFC gave edible films with higher tensile strength and transparency than those of MFC, but interestingly the barrier properties (water vapor and oxygen permeability) of MFC films were superior than those of NFC films (Kumar *et al.*, 2014). The NFC films were found denser than the MFC films, but it was thought that the rigid rod-like fiber geometry of NFC could be responsible for lower barrier effect of its films than those of MFC that has flexible hairy fibers (Kumar *et al.*, 2014). Thus, further studies are needed to improve the barrier properties of NFC films. Aulin *et al.* (2012) developed transparent and mechanically stable nanocomposite edible packaging materials with high gas and water vapor-barrier properties by mixing NFC with vermiculite nanoclay (5 to 20%) under high pressure homogenization. These authors reported that the use of nanoclay gave NFC films having comparable mechanical and gas barrier properties with commercial plastic packaging films. The NFC was also used as a filling material to improve different physicochemical properties of edible films. Zimmermann *et al.* (2010) employed different types of NFCs to improve the tensile strength of HPC and reported that it was not the degree of polymerization, but the homogeneity of fibrillation for NFC that improved the mechanical strength of films. Fernandes *et al.* (2010) improved mechanical and thermal properties of chitosan films by incorporation of NFC without losing film transparency while Trovatti *et al.* (2012) successfully used NFC for the same purpose in pullulan films. Hassan *et al.* (2018) also improved the mechanical stability of HPMC films by addition of NFC, but this increased the water-vapor permeability of films due to increased film hydrophilicity.

2.2.2.2 *Starch and starch derivatives*

Native starches and starch derivatives (modified starches) are extensively used by the food industry to obtain different functional properties such as coating, gelation, consistency, texture, etc. The starch in plants is stored in semi-crystalline granules having considerably different sizes and characteristic shapes, depending on the origin (Fig. 2.7a-2.7d). The term 'semi-crystalline' is used for the granule since it is formed by complex, but highly organized amorphous and crystalline regions (Liu, 2005). Angular and polygonal shaped rice starch granules, ranging between 3 and 8 μm in diameter, are among the smallest of cereal starch granules (Bao, 2019) while oval and spherical-shaped potato starches form both small and large granules with sizes ranging between 1 and 20 μm and 20 and 110 μm, respectively (Singh *et al.*, 2016). Wheat starch also forms disc-like (lenticular) small (A-type) and spherical or polygonal large (B-type) granules, ranging between 1 and 10 μm and 10 and 35 μm, respectively (Singh *et al.*, 2016).

The starch is a mixture of two polysaccharides, called amylose and amylopectin. The amylose and amylopectin form almost 20-25% and 75-80% of total starch in the semi-crystalline granules of common cereal grains (Mahmood *et al.*, 2017). However, there are also some genetically modified starches, such as waxy starches, where the granules are formed mainly from amylopectin fraction (up to 100% of total starch). The amylose is an un-branched linear chain of several hundred to ten thousand α-(1-4) linked D-glucose residues with MW ranging between 10^5 and 10^6 Da, while amylopectin is a much larger and branched version of α-(1-4)

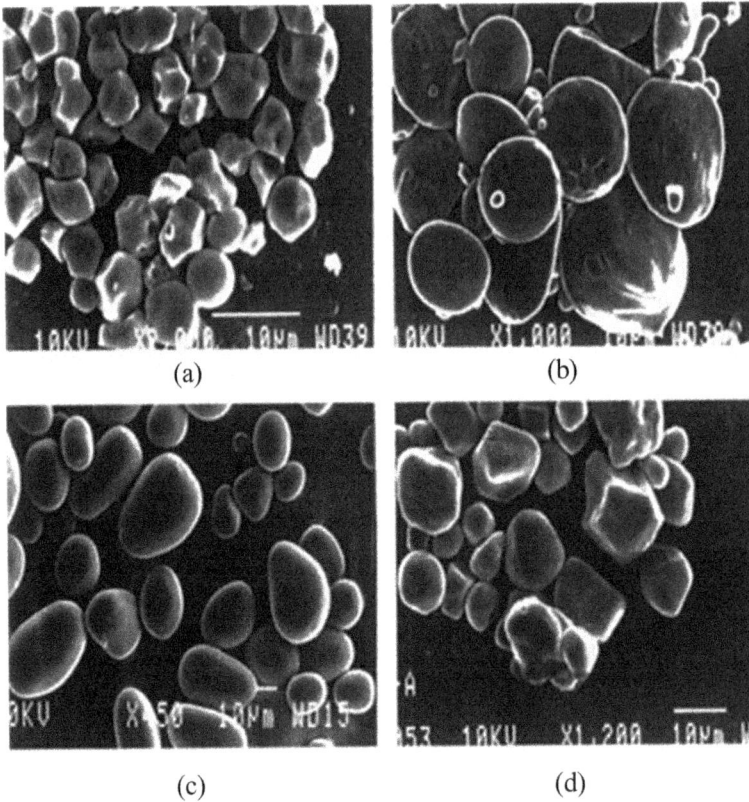

Fig. 2.7: Electron micrographs of starch granules from different sources. Rice (a), wheat (b), potato (c), and maize (d) starches; reprinted with permission from Singh *et al.* (2003); copyright © 2003 Elsevier Ltd.

linked chains of around one million D-glucose residues with MW ranging between 10^8 and 10^9 (Liu, 2005; Steel *et al.*, 2012). Both the side-chains (branches) and backbone chains of amylopectin are formed by α-(1-4) linked D-glucose residues, but the reducing ends of side-chains always attach to the backbone chain via α-(1-6) linkages that occur in every 20-30 glucose residues.

Due to its complex nature, the structure of amylopectin is not fully elucidated. However, the generally accepted cluster model is frequently used to explain the organization of different chains of amylopectin that are classified as A-, B- and C-chains (Hamaker *et al.*, 2007). Each amylopectin molecule contains multiple A- and B-chains, but there is only a single C-chain per amylopectin molecule. A-chains contain no branch and their reducing ends are linked to B- or C-chains via α-(1-6) linkage. It is generally accepted that the linear branched chains of amylopectin with degree of polymerization (DP) around 15 exist as double-helical ordered structures and form the crystalline regions in the starch granule while the branching points form the amorphous regions (Liu, 2005). The B-chains are branched with A- or other B-chains and their reducing ends are linked to C chain via α-(1-6) linkages. Thus, A- and B-chains lack any reducing end while C-chain that carry all other chains contains the only reducing end of whole amylopectin molecule (Hamaker *et al.*, 2007).

The structure of amylose is not very complex as amylopectin since it does not contain branches. However, the axial → equatorial position coupling of the (1 → 4)-linked α-d-glucopyranosyl units in amylose force these chains to form a right-handed helical (spiral) shape (BeMiller, 2018). The surface of helical amylose chain is hydrophilic since hydroxyl groups of glucose residues along the chain are positioned on the exterior of the coil, while the interior of the amylose helix, lined with hydrogen atoms, is hydrophobic (BeMiller, 2018).

The common semi-crystalline starch granules are not water soluble at room temperature, but they start to swell when their suspensions are heated. The swelling causes disruption of the highly ordered structure of granule into a disordered one. The amylose and small molecular weight fractions of amylopectin start to leach out from the granules during swelling. The viscosity of the suspension increases continuously as swelling increases the size of the starch granules (several times) and fraction of leached amylose. The continuing heating and stirring cause the disintegration (collapse) of highly swelled starch granules that are very sensitive to shear forces created by stirring. Thus some reduction in the viscosity of the mass occurs at this point. However, the final partially transparent viscous mass (called paste in food technology) contains a continuous phase of solubilized amylose and low molecular weight fractions of amylopectin, and a discontinuous phase of starch granule remnants (called also ghosts) with retained high molecular weight amylopectin fractions (BeMiller and Whistler, 1996). The process of obtaining such a partially transparent viscous paste from cloudy and opaque suspension of starch granules in hot water is called gelatinization. Bao and Bergman (2004) explained the major changes occurring in heated starch suspensions during gelatinization at four steps: (1) formation of sol state by the leached and suspended amylose molecules; (2) transformation of amylose from sol state to gel state (this gel network is reinforced by the extensive interactions among the swelled starch granules); (3) network destruction (occurs by collapse of granules due to 'melting' of the crystalline regions remaining in the swollen starch granule or disentanglement of the amylopectin molecules in the swollen granule); and (4) network strengthening (occurs due to increased networking because of interaction between amylose and leached low molecular weight amylopectin fractions) (Bao and Bergman, 2004).

The cooling of gelatinized starch leads to retrogradation that is reassociation or recrystallization of the gelatinized starch molecules (Kong and Singh, 2016). The gelation of suspended amylose fraction after heating and cooling is related to tendency of linear amylose molecules to form associated structures linked via inter-chain hydrogen bonding formed by hydroxyl groups at their monomeric units. The association of amylose molecules reduces their affinity to water, and if amylose concentration is sufficiently high, the extensive three-dimensional network formed among linear amylose molecules leads to gel formation (precipitate is formed at low amylose concentration) (Wurzburg, 1975). In the starch gels, the granules act as a filling material within the amylose gel. Amylopectin embedded within the granules contributes to increased strength of amylose gel by showing recrystallization, following heating and cooling (Miles *et al.*, 1985).

Different native starches show considerable variations in edible film characteristics since they differ in granule size, molecular weight, crystallinity, and amylose/amylopectin ratios (Table 2.2). The type of starch affects not only the properties of edible films, but also the health benefits and consumer profile of final coated food products. People suffering from gluten intolerance or celiac disease cannot consume food containing wheat starch that might contain residual amounts of gluten. Therefore, gluten-free starches, such as corn, rice, potato and tapioca starches should be employed in edible coating and film formulations of foods designed for people suffering from such diseases.

Modified starches are obtained with different physical (pre-gelatinization, thermal-treatment, freezing, etc.), chemical (esterification, etherification, cross-linking, oxidation, etc.) and biological methods (fermentation and enzymatic hydrolysis). The chemical modifications have been applied extensively: (1) to improve starch stability and functional properties, such as solubility, emulsification, viscosity, gelation, etc.; (2) to reduce the retrogradation of starch during processes, such as cooking, microwave heating, thermal processing, freezing-thawing, etc.; and (3) to obtain resistant starches. The octenyl succinylated starch (OS-starch), an amphiphilic hydrocolloid increasingly used by the food industry as an emulsifier, encapsulating agent, and fat replacer, is obtained by esterification of starch with octenyl succinic anhydride (OSA) (Altuna *et al.*, 2018). The European Union has approved 18 types of modified starches, but the changing of technological processes and huge market demand for functional product development suggest that alternative starch modification studies will continue in future. In particular, the rapidly developing functional food market encourages increased production and use of modified resistant starches that help in struggling with obesity and related diseases (Zhang *et al.*, 2015; Jain *et al.*, 2019).

Table 2.2: Amylose and amylopectin percentages of different starches

Source of starch	Amylose/Amylopectin	References
Wheat	10.7/89.3 to 25/75	Labuschagne *et al.* (2007)
Potato	17.9/82.1	Cano *et al.* (2014)
Pea	24.9/75.1	
Cassava	9/91	
Corn	23/77	Xie *et al.* (2009)
Waxy corn	0/100	
Maize	26/74	Liu *et al.* (2010)
Waxy maize	0/100	
Rice	8.7/91.3 to 19.9/80.1	Park *et al.* (2007)
Waxy rice	1/99	
Corn	19/81 to 55/45	Martens *et al.* (2018)
Potato	11/89 to 18/82	
Pea	27/73	
Wheat	23/77	
Tapioca	19.7/81.1	Herrero-Martínez *et al.* (2004)

2.2.2.3 Pectin

Pectin is a major food hydrocolloid that is highly demanded by the food industry primarily as a gelling and thickening agent. The majority of the commercial pectin (~85%) is produced from citrus peels (56% lemon peel, 30% lime peel, 13% orange peel) that contain 20-30% pectin by dry weight basis (d.w.) while rest of the pectin (~14%) is extracted from apple pomace that contains 10-15% pectin (d.w.) (Ciriminna *et al.*, 2016; Gawkowska *et al.*, 2018). However, the increased applications of pectin in food as well as recent trends to valorize different agro-industrial wastes have encouraged industry to seek alternative pectin sources such as sun-flower head residues and sugar beet pulp that contain 10-20% pectin (d.w.) (Gawkowska *et al.*, 2018). Although the pectin from sugar beet pulp shows inferior gelation properties than citrus pectin, it has unique emulsifying properties originating from its covalently-bound hydrophobic protein that helps pectin molecule to act as a surface-active hydrocolloid (Williams *et al.*, 2005; Pacheco *et al.*, 2019). On the other hand, sun-flower head pectin shows remarkable viscosity and water-binding properties (Miyamoto and Chang, 1992). The commercial pectins are generally obtained by the classical hot acidic extraction. However, the use of organic acids (e.g. citric acid) in place of mineral acids for extraction as well as application of milder pectin extraction methods, such as microwave, ultrasound or enzyme-assisted extraction have been increasingly preferred due to better environmental sustainability, economic feasibility or high yield of pectin production (Marić *et al.*, 2018).

The pectin is the most complex polysaccharide in plants since it is formed by three different fractions: homogalacturonan (HG), rhamnogalacturonan I (RG-I), and rhamnogalacturonan II (RG-II). The HG, RG-I and RG-II are attached covalently to each other (Harding *et al.*, 2017) and form almost 65, 20-35 and 10% of pectin molecule in plants, respectively (Chandrayan, 2018). The linear HG fraction forms the 'smooth regions' of pectin molecule while branched RG-I and RG-II form the 'hairy regions'. However, some other minor pectic fractions also exist, such as xylogalacturonan, apiogalacturonan, arabinan, galactan, arabinogalactan I, and arabinogalactan II (Gawkowska *et al.*, 2018).

The HG is the major pectin fraction found in primary cell walls and middle lamella of plant cells. It is a homoglycan consisting of linear chains of 1,4-linked α-D-galacturonic acid units (Watkins, 2017). The carboxyl groups at C6 of galacturonic acid residues of pectin exist in free or methyl esterified form (Fig. 2.8). The degree of esterification (DE) is an important molecular characteristic since it

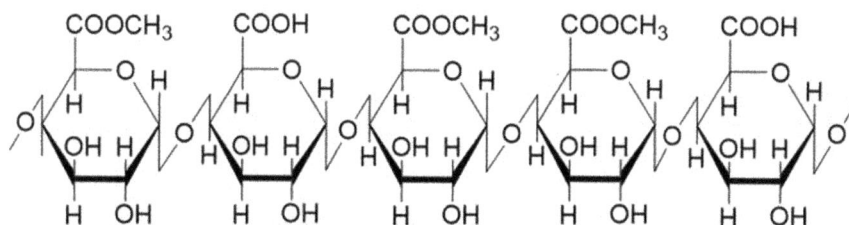

Fig. 2.8: Chemical structure of galacturonic acid polymer (backbone of HG)

highly affects the technological properties (e.g. viscosity, gelation, film-formation mechanism) of pectin. Therefore, International Pectin Producers Association (IPPA) classifies pectin depending on DE (high methoxyl pectin: DE> 50%; low methoxyl pectin: DE< 50%). The chain of galacturonic acid (GA) residues also shows some degree of acetylation (DA) (occurs with oxygen atoms at position O2 and O3 of GA) that interferes with gelation capacity of pectin (Chandrayan, 2018). However, besides DE and DA, different factors, such as molecular composition (amounts of different pectic fractions or complexed proteins) and HG chain length are also highly effective on functional properties of pectins.

RG-I is a highly complex branched heteroglycan that consists of repeating units of disaccharide formed by D-galacturonic acid and L-rhamnose [(→2)-α-L-Rha$p$-(1→4)-α-D-Gal$p$A-(1→)] (Shi *et al.*, 2017). The branching of RG-I occurs at C4 of rhamnose residues with a single sugar or different chains of arabinans, galactans or arabinogalactans (Gawkowska *et al.*, 2018). RG-II is a low molecular weight pectin fraction which consists of linear backbone chain of 1,4-linked α-D-galacturonic acid units, but it differs from HG due to its highly branched nature. It was found that the short-chained RG-II fraction contains 7-9 GA residues with four branches (A, B, C, D at C2 and C3 of GA) that chains are formed by combination of different sugar residues showing slight variation in composition, depending on the source of pectin (Yapo, 2011). For citrus pectin, the branches of RG-II are reported to contain sugars, such as rhamnose, fucose, arabinose, xylose, galactose, apiose, glucoronic acid, aceric acid, 2-*O*-methyl-fucose, 2-*O*-methyl-xylose, 2-keto-3-deoxy-D-manno-octulosonic acid, and 3-deoxy-D-lyxo-heptulosaric acid (Yapo, 2011).

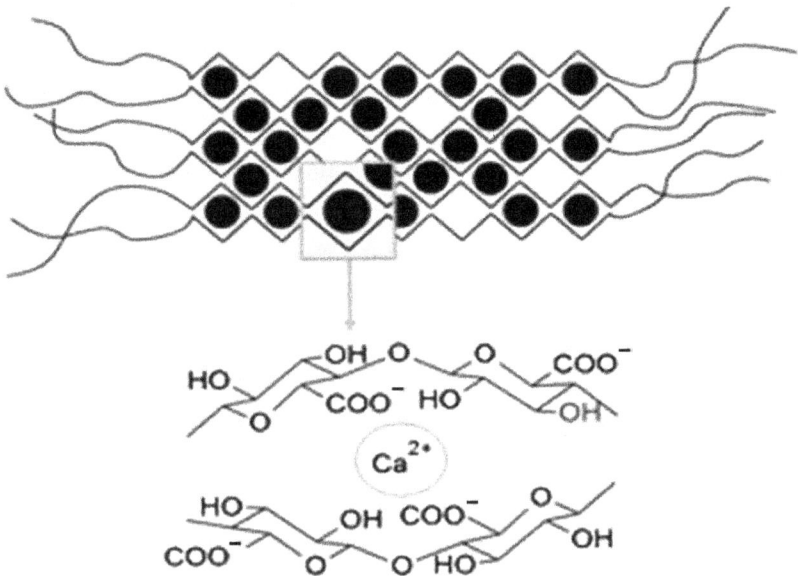

Fig. 2.9: Illustration of egg-box like structure formed by LMP; reprinted with permission from Lascol *et al.* (2016); copyright © 2016 Elsevier Ltd.

The dominant form of pectin used in food applications is high-methyl ester pectin (HMP) with a DE >50%. The HMP is also modified chemically or enzymatically to obtain low-methyl ester pectin (LMP) with DE <50% and other modified pectin forms, such as amidated low-methyl ester pectin (ALMP). The HMP is mostly used as a gelling agent at low pH (optimal at pH 2.8-3.2) and high sugar (65-72%) conditions for production of classical jams and jellies (Cemeroğlu and Acar, 1986). Moreover, it is also used extensively as a thickening, suspending, stabilizing and texturizing agent in a wide range of food products (e.g. beverages, drinks, yogurts, dressings) (Yemenicioğlu *et al.*, 2020).

The LMP is employed to obtain jams and jellies at reduced sugar concentrations (20-25%). The LMP does not require acidic conditions for gel formation, but requires the presence of sufficient amounts of divalent cations (e.g. Ca^{+2} and Mg^{+2}) (Cemeroğlu and Acar, 1986). Thus, $CaCl_2$ is mostly used to provide the Ca^{+2} ions essential for the formation of LMP-based films and coatings. The gelation induced by formation of extensive junction zones among Ca^{+2} ions and de-esterified carboxyl groups of LMP is explained by the classical 'egg-box' model (Fig. 2.9). The gel strength and gelation rate of LMP increases as pH is increased between 3.5 and 8.5 and dissociated carboxyl groups of pectin increases (Yang *et al.*, 2018). On the other hand, the ALMP is obtained by alkaline demethylation of pectin by ammonia and replacing its methyl ester groups by amino groups. The ALMP differs from common pectin with its better gelation performance and thermo-reversible gel-formation capacity (Chen *et al.*, 2015).

Pectin is one of the hydrocolloids suitable for coating of minimally processed fruits and vegetables (Rojas-Graü *et al.*, 2009). However, the continuously increasing interest in pectin is not only related to its functional properties and edible film-forming capacities. The recently discovered health benefits of pectin, such as prebiotic activity, anticarcinogenic and anti-inflammatory properties have increased the importance of this hydrocolloid as a universal coating and self-standing film-making material (Adam *et al.*, 2015; Sahasrabudhe *et al.*, 2018; Adami *et al.*, 2018). The oligosaccharides obtained by hydrolysis of pectin (POS) also show significant health benefits at the molecular level on colon cancer, diabetes, hypercholesterolemia, and inflammation (Tan *et al.*, 2018). Thus, there are also extensive studies to modify pectin via enzymatic, thermal, and chemical methods and to improve its potential health benefits as a functional food additive or ingredient (Yemenicioğlu *et al.*, 2020; Bernardino *et al.*, 2019).

2.2.2.4 *Gums*

The gums extracted from seeds or derived from exudates or sap of different plants are used extensively by the food industry. Locust bean gum (from seeds of carob three), guar gum (from guar bean), karaya gum (exudate of *Sterculia urens* tree), gum tragacanth (from sap of stems and branches of Asiatic species of *Astragalus*), and gum Arabic (exudate of different species of acacia tree) are used extensively in foods mainly for their thickening properties (Yemenicioğlu *et al.*, 2020). Due to the continuously growing plant-derived gums market, extensive studies are also conducted to increase the commercial use of emerging gums, such as durian seed gum, chia seed gum, basil seed gum, cress seed gum, sage seed gum, marshmallow flower gum, etc. (Razavi, 2019).

The majority of gums have limited and local plant resources and lack the potential to be a universal hydrocolloid, capable to meet the demands of packaging sector. However, gum Arabic (Gum acacia, Senegal gum or Sudan gum) is one of the most potential gums suitable for edible packaging since it is highly functional and a major commodity for different developing African nations in a vast geography (UNCTAD, 2018). Gum Arabic is extracted as a wound exudate from acacia trees native to western Sudan, Nigeria and Arabian peninsula, or cultivated in different areas, such as tropical areas of Africa and South Asia. Gum Arabic is used not only as a thickening agent, but also it has a remarkable emulsifying capacity originating from its covalently-bound hydrophobic protein components (Li *et al.*, 2018). The use of gum Arabic as an edible coating for preservation of fresh fruits has also attracted a considerable interest (Salehi, 2020). Gum Arabic is a heterogeneous, slightly acidic (anionic) hydrocolloid that exists as a mixture of calcium, magnesium, and potassium salt of polysaccharidic acid (Arabic acid) that consists of six carbohydrate residues, namely galactopyranose, arabinopyranose, arabinofuranose, rhamnopyranose, glucuropyranosyl uronic acid and 4-*O*-methyl glucuropyranosyl uronic acid (Islam *et al.*, 1997). The backbone structure of gum Arabic is formed mainly by 1,3-linked β-d-galactopyranosyl units with side chains of two to five 1,3-linked β-d-galactopyranosyl units linked to the main chain via 1,6-linkages (Padil *et al.*, 2019). The other sugars are distributed on to the backbone and side chains as units of α-l-arabinofuranosyl, α-l-rhamnopyranosyl, β-d-glucuronopyranosyl, and 4-*O*-methyl-β-d-glucuronopyranosyl (Padil *et al.*, 2019). The molecular weight of gum Arabic and/or its bound protein content (1 to 5%) show a great variation, depending on the used acacia species (>90% of gum is obtained from *A. senegal* and *A. seyal*), area of plantation, or even different parts of a single plant (Anderson *et al.*, 1985; Islam *et al.*, 1997). However, the molecular weight of gum Arabic from *A. senegal* and *A. seyal* were reported to be around 384 000 and 850 000 Da, respectively (Islam *et al.*, 1997). The major protein constituents of gum Arabic characterized by Mahendran *et al.* (2008) had MW of 30 and 5 kDa with approximately 250 and 45 amino acids in length, respectively. Anderson *et al.* (1985) reported that the amount of gum Arabic protein amino acid constituents, such as hydroxyproline, serine, and proline varied considerably, but amounts of alanine, cysteine, isoleucine, methionine, threonine, tyrosine, valine residues in gum Arabic proteins showed similarities.

2.2.2.5 Soy proteins

Soy protein flour (or soya protein flour) (≥50 to <65% protein, d.w.), soy protein concentrate (SPC) (≥ 65% to < 90% protein, d.w.), and soy protein isolate (SPI) (≥ 90% protein, d.w.) are extensively used in food formulations. The increasing demand of food industry to soy proteins is related to the rapidly growing market for vegan food, and unique functional, nutritional, and bioactive properties that make these proteins indispensable ingredients for food and nutritional supplements. Soy proteins contain all of the nine essential amino acids, and bioactive isoflovons that are known with their important health benefits, such as reducing risk of coronary heart disease (FDA, 2018) and anticarcinogenic effects (Sahin *et al.*, 2019). The isoflavones associated with soy proteins exist as aglycones and their β-glycosides, such as genistein and genistin, daidzein and daidzin, and glycitein and glycitin, respectively (Juturu, 2014). It was determined that SPC obtained by membrane filtration methods (e.g. diafiltration and ultrafiltration) contains 3-3.4 mg/g (d.w.) isoflavons (Batt *et al.*, 2003). Andrade *et al.* (2010) reported amounts of major aglycones in SPI as 0.30 mg/g for daidzein and 0.64 mg/g for genistein. However, aglycones form only ~19% of total isoflavons in SPI while the remaining isoflavons exist in the form of glycosides (~81%) (Andrade *et al.*, 2010).

A majority of soybean protein fractions are storage globulins filled within cotyledonous subcellular bodies called 'protein bodies' (Koshiyama, 1983). The legume-storage globulins, including those of soybeans, are formed by vicilins and legumins that are differentiated by their sedimentation coefficients mainly as 7S and 11S. (The sedimentation coefficient that is expressed in Svedberg unit (S) is a measure of the rate at which a protein molecule, suspended in a colloidal solution, sediments in an ultracentrifuge). 11S glycinin (a legumin) and 7S β-conglycinin (a vicilin) form ≥ 80% of total soybean proteins with variable ratios depending on cultivar (Nishinari *et al.*, 2014). The 11S protein, a major storage protein in most legumes, forms almost 60% of the soybean storage proteins while 7S forms the remaining 40% (Shewry *et al.*, 1995; Taski-Ajdukovic *et al.*, 2010). The monomers of 7S and 11S storage globulins in plants show a highly variable sequence, but all are known with their ability to form trimeric and hexameric quaternary structures (Shewry *et al.*, 1995). In general, the mature 11S storage proteins form six noncovalently linked subunit pairs, each pair formed by an acidic ~40 kDa subunit covalently linked to a ~20 kDa basic one with a disulfide bond (S-S bond), while 7S proteins are typically trimeric proteins lacking S-S bonds (Shewry *et al.*, 1995). β-Conglycinin is a glycoprotein (MW of trimer changes between 127 and 171 kDa) that is formed by different combinations of three subunits, α' (MW: 57–72 kDa), α (MW: 57–68 kDa) and β (MW: 45–52 kDa) (González-Pérez and Arellano, 2009; Hammond *et al.*, 2016), while glycinin is a hexamer (MW: ~320 KDa) composed of six nonrandomly paired subunits (6 × A-B) formed by six acidic (A) (MW: ~35 kDa) and six basic (B) (MW: ~20 kDa) polypeptides linked covalently with a disulfide bond (Badley *et al.*, 1975). The secondary structures of soybean 11S and 7S proteins determined by FTIR spectroscopy are formed 17 and 14.5% α-helix, 47.3 and 45.6% by β-sheet and 19.3 and 23.8% by β-turns, respectively, while the remaining structures of both proteins are disordered (Zhao *et al.*, 2008).

The SPI is generally produced by mixing/homogenizing defatted soybean flakes (called also white flakes, remaining from solvent extraction of soybean oil) or soybean flour with water or slightly alkaline solutions (pH: 7.5-9.0), and then fractionating extracts separated from nonproteinous debris by the classical isoelectric precipitation (IEP) method (Berk, 1992). During extraction, the protein yield could be improved by increasing extraction temperature up to 80°C (Berk, 1992). The IEP of obtained extract, rich in storage globulins and whey protein, is conducted simply by bringing pH to ~4.5, a pH close to isoelectric point of soy proteins. The SPI obtained by washing, neutralization, and spray-drying of storage globulins collected by the IEP is used extensively to formulate different food and nutraceutical products or to develop edible films and coatings. The protein fractions, remaining in the supernatant after IEP of storage globulins, are called whey proteins. Whey protein form 9 to 15.4% of total soy proteins, but they have no importance as food ingredients since they are composed of proteins, such as lectin and Kunitz tripsin inhibitor, and enzymes, such as lipoxygenase and α-amylase (Nishinari *et al.*, 2014).

The SPC is another soy protein product that is frequently produced by using defatted flakes or flour with 'acid-wash', 'alcohol-wash', or 'heat denaturation-water extraction' methods (Berk, 1992). The acid-wash method involves removal of soluble sugars from defatted soybean flakes or flour with an acidic solution of HCl. The extraction process of soluble sugars is conducted around pH 4.5 to keep proteins insoluble at their isoelectric point. The protein precipitates are then collected by centrifugation, and SPC is produced by spray-drying of obtained protein slurry following neutralization. On the other hand, the alcohol-wash method is based on removing soluble sugars from defatted soybean flakes or flour using 60% ethanol (or isopropanol), while keeping protein in insoluble form. The ethanol is then removed from the solid material (usually with flash-desolventizing process conducted with superheated vapors of alcohol-water mixture) and SPC is obtained by further drying and grinding processes (Berk, 1992). The third method of SPI production, the heat denaturation-water extraction process, involves heat denaturation and insolubilization of proteins in defatted flakes with humid heat, and then extraction of soluble sugars with hot water. After that, the slurry of protein rich mass is spray-dried to obtain SPC (Berk, 1992).

The methods applied to white flakes and soybean flour for production of SPC and SPI could also be employed for the same purpose in soybean meals obtained from extruding-expelling processing (EEP), a solvent-free oil extraction process. It was reported that the oil remaining in EEP meals caused reduction of final protein contents of obtained SPI and SPC, but the functional properties (dispersibility, emulsion, and foaming properties) of protein products obtained from EEP meals were similar to, or better than those of protein obtained from white flakes (Wang *et al.*, 2004). Wang *et al.* (2004) also determined that the functional properties of SPC obtained with acid-wash were superior to those of SPC obtained from alcohol-wash. These findings suggest that the acid and alcohol treatment affect the conformation of soy proteins or they cause the change of 11S/7S ratio in SPC. The 11S/7S ratio is a critical factor for soy protein preparations (SPC or SPI) since it affects the mechanical properties of soy protein films (Kunte *et al.*, 1997).

Although the production and use of soy proteins in food industry bring numerous advantages, severe criticisms (especially those about soybean allergens and GMO soybean) about these proteins have increased demand for soy-alternative plant proteins. The allergenic protein fractions in soybeans continue to cause three types of reactions for sensitive individuals: IgE-mediated reactions, non-IgE-mediated reactions, and the rarely observed, but life-threatening anaphylaxis reaction (Wilson *et al.*, 2005). There are at least 21 allergenic IgE-binding proteins identified in soybeans (Wilson *et al.*, 2005). The OECD lists 15 soy proteins designated as 'allergens' (OECD, 2012), but Gly m 1 (also known as Gly m Bd 30K or P34), that is a monomeric insoluble glycoprotein (a β-conglycinin with MW of 32 kDa), was characterized as a major allergenic protein in soybeans (Ogawa *et al.*, 1993; Wilson *et al.*, 2005). The reactivity of Gly m 1 could not be eliminated by heat treatment or thermal processing, but treatment of soy proteins by extrusion (applies both pressure and temperature), enzymatic hydrolysis, or fermentation (e.g., by *Bacillus natto*) could reduce or eliminate the reactivity of this allergenic protein (Wilson *et al.*, 2005). Some isomers of β-conglycinin (Gly m 5, with three subunits) and glycinin (Gly m 6 with five subunits) (Lu *et al.*, 2018) as well as Gly m 4 (Kosma *et al.*, 2011) are also recognized as important soy protein allergens, especially for children (Ito *et al.*, 2011). Thus, further studies are needed to identify exact allergen profiles of different soy proteins and use the least allergenic ones during development of edible films and coatings.

2.2.2.6 Pulse proteins

Protein isolates or concentrates of different pulses, such as peas, chickpeas, and lentils have attracted huge interest from the food industry (Mamone *et al.*, 2019; Yemenicioğlu *et al.*, 2020). With their current amounts of production and/or prices, these proteins cannot compete with soy proteins as source of edible film and coating material in the short-term. However, with their different taste profile, functionality, and less allergenic nature than soy proteins, these proteins could be used for development of edible films and coatings for vegans and people against GMO products.

Pea protein is the most important soy-alternative protein that has been increasingly demanded by the food industry for development of plant-based food (e.g. meat substitutes and vegan ice-cream). Peas contain 25% (d.w.) total protein, and majority of this protein (70-80%) consists of globulins such as legumin (11S), vicilin (7S), and convicilin (7-8S) (González-Pérez and Arellano, 2009). According to Barac *et al.* (2010), who studied IEP protein isolates from six different pea cultivars, amounts (and molecular weights) of legumin, vicilin, and convicilin changed between 280 and 330 g (MWs: 22.3-23.1 kDa, 40.89 kDa, and 63.6 kDa), 300 and 385 g (MWs: 31.8-37 kDa, 47.3 kDa), and 91.3 and 117.8 g (MW: 72.4-77.9 kDa) per kg of pea protein isolate, respectively. During extraction and purification of pea protein isolates by IEP, effective separation of saponins from proteins by coextraction is highly critical since this naturally occurring component in beans gives an undesired bitter and/or metallic taste to obtained protein isolate (Roland *et al.*, 2018). The pea proteins are not considered among the main food allergens, but IgE immune-detection of crude pea protein extracts suggest that a 63 kDa convicilin, and a 44 kDa vicilin and its 32 kDa proteolytic fragment could be major pea protein allergens (Sanchez-Monge *et al.*, 2004). Bogdanov *et al.* (2016) identified a ~9.4 kDa lipid transfer protein (Ps-LTP1) extracted from peas as a potential cross-reactive food allergen for LTP-sensitized patients. Therefore, suitable labelling is needed for food packed with lentil protein films and coatings.

Chickpea proteins are rated among vegans as one of the best tasting plant-based proteins. Moreover, chickpea proteins show some functional properties (e.g.

water- and oil-binding capacity, emulsion and gelation properties) comparable to or superior to those of SPI (Aydemir and Yemenicioglu, 2013a). The protein contents of chickpeas show a great variation, between 16.7 and 30.6% for desi and 12.6 and 29% for kabuli types (Wood and Grusak, 2007). Majority of chickpea proteins are formed by legumin-like 11S globulins (~60%) and 7S vicilins (~30%), while the remaining minor fractions consist of mainly 2S albumins (Oomah *et al.*, 2011; Chang *et al.*, 2012). The 2D electrophoresis profiles of chickpea protein concentrates obtained by the classical IEP procedure showed that subunits of these proteins show pI between 4.5 and 5.9 and MW between 15.0 and 76.0 kDa (dense bands concentrate mainly between 20 and 30 kDa) (Aydemir and Yemenicioglu, 2013a). Chang *et al.* (2012) used SDS-PAGE and identified chickpea legumin subunits with MWs of 39.5, 24.6, 23.5, 22.5, and 9.4 kDa, and vicilin subunits having MWs of 70.2, 50.7, 35.0, 18.9, and 15.5 kDa. Clinical studies proved that chickpea proteins contain allergenic fractions (those with MW of 26, 35, 64 and 70 kDa) that cause IgE-mediated hypersensitivity reactions (e.g. from rhinitis to anaphylaxis) (Patil *et al.*, 2001). However, the chickpeas are less allergenic than soybeans that are listed among the top eight food allergens (FDA, 2017). However, although there has been a recent trend in production and use of chickpea protein as a functional ingredient in food formulations, there are only a limited number of studies related to chickpea-based edible films (Meshkani *et al.*, 2011; Gücbilmez *et al.*, 2006).

Lentils with their 25 to 30% protein content, are major pulses extensively grown in many parts of the world, especially in South Asia, North America, Eurasia and Middle East (Çarman, 1996). Recent studies also showed that lentil is one of the richest sources of phenolic compounds and possesses the highest antioxidant activity among cool season legumes (Xu *et al.*, 2007a). Therefore, IEP lentil protein isolates are rich in polyphenols that form a complex with proteins and cannot easily be removed by repeated precipitation-washing cycles applied during IEP (Aydemir and Yemenicioğlu, 2013b). According to Bartolomé *et al.* (2000), lentil polyphenols could bind proteins via hydrogen bonds formed between phenolic hydroxyl and protein —NH and —CO groups, and hydrophobic interactions formed between nonpolar phenolic groups and protein domains. Zang *et al.* (2014) identified catechin, gallic acid, protocatechuic acid, epicatechin, and 3-hydroxycinnamic acid as major bound polyphenols by lentil hydrocolloids. Alshikh *et al.* (2015) identified different hydroxybenzoic acids (gallic acid and methyl vanillate), hydroxycinnamic acids (p-coumaric and ferulic acids), flavonoids (catechin, epicatechin and (+)-catechin-3-glucoside), and proanthocyanidins (procyanidine dimers A and B, and prodelphinidine dimer A) as major bound polyphenols of lentil hydrocolloids. However, studies about molecular and bioactive properties of polyphenols bound specifically on to lentil protein isolates are scarce.

Lentil proteins are formed by albumins, legumins, vicilins, glutelins, and prolamins that account for almost 16.8%, 44.8%, 4.2%, 11.2% and 3.5% of total lentil proteins, respectively (Gupta and Dhillon, 1993). Aydemir and Yemenicioğlu (2013a) reported that protein bands for SDS-PAGE of crude protein extracts showed distribution of most lentil protein subunits at MW between 15 and 20 kDa, 30 and 40 kDa, and 40 and 70 kDa. These authors also reported that the pI of IEP lentil protein isolates changed between 4.8 and 5.9 while major IEP protein isolate's bands determined by 2-D electrophoresis consisted of subunits between 21 and 23 kDa, and at around ~26 kDa and ~66 kDa. Ladjal-Ettoumi *et al.* (2016) determined protein bands around ~50 kDa and ~37 kDa for IEP lentil protein isolates and attributed these proteins to vicilins and legumins, respectively. Lentils are reported to contain allergenic proteins having similarities with those of soybean and peanut proteins. López-Torrejón *et al.* (2016) identified major allergenic protein in lentil as a 48 kDa vicilin that exists with its 12 to 16 kDa and 26 kDa fragments. Therefore, application of lentil protein isolates as edible film and coating needs proper labelling of the packaged food products.

2.2.2.7 Corn or maize zein

Corn contains 6-12% protein, mostly concentrated (almost 75%) in its endosperm (Shukla and Cheryan, 2001). Zein, a prolamin group hydrophobic storage protein with unique film-forming ability, forms 22-50% of total corn (or maize) proteins (Shukla and Cheryan, 2001; Zilic *et al.*, 2011). Zein contains different molecular forms, such as α-, β-, γ- and δ-zein, but the α-zein is the most abundant one that forms almost 80% of total protein fractions (Shewry and Tatham, 1990; Shewry and Halford, 2003). The α-zein is a 24.5 kDa protein that contains α-helix structure (~57%) as a dominant secondary structural form (Cabra *et al.*, 2005). It was proposed that the protein exists as asymmetric rods (length: 13 nm and axial ratio: 6:1) with its tandem helical structures connected with loops (Matsushima *et al.*, 1997). Similar hydrophobic prolamins are also found in wheat (gliadin), barley (hordein), rye (secalin), and sorghum (kafirin) (Shewry and Tatham, 1990; Shull *et al.*, 1991). However, zein is the most important prolamin since it is the major co-product of the oil and bioethanol industries (Selling *et al.*, 2008; Shukla and Cheryan, 2001; Wang *et al.*, 2007).

The corns processed by wet-milling for starch and oil production and by dry-grind process for ethanol production yield different byproducts, called 'corn gluten meal' (CGM) and 'distillers dried grains with solubles' (DDGS), respectively (Shukla and Cheryan, 2001). Both CGM and DDGS could be utilized for zein production. However, the CGM, that contains minimum 60% protein (almost half is zein), is the primary raw material used for production of commercial zein since it gives higher quality zein with higher yield and purity than DDGS that contains 27-30% protein (Shukla and Cheryan, 2001). Raw dry-milled corn (defatted or not), that contains almost 4% zein, can also be employed directly in zein production if economic issues related to solvent recovery and concentration are optimized effectively.

In a patented process Cook *et al.* (1993) obtained destarched, decolored, and deflavored zein from CGM in six steps by sequential application of enzymatic starch hydrolysis, alkaline treatment, alcohol wash, alcohol extraction, ion exchange chromatography, and precipitation (via dilution with water). According to Shukla and Cheryan (2001), the main steps of the nutrilite process, a basic extraction method that prevents zein gelation problems, are as follows: (1) CGM is mixed (1:4) with 88% isopropyl alcohol (IPA); (2) insolubilized gluten is separated while zein solubilized (~12%) is collected with centrifugation or filtration; (3) extracted zein is precipitated at -15°C; (4) zein-alcohol mixture (30% zein) is then dried to obtain zein powder with 2% oil or it is re-extracted by repeating steps 1, 2 and 3 to obtain oil-free zein powder. Anderson and Lamsal (2011) showed that change of classical initial extraction solvent (88% IPA) with 70% or 55% IPA, but increase of IPA concentration again to 88% following extraction process, could have increased α-zein yield. Xu *et al.* (2007b) developed an economically feasible method that extracts almost 44% of total zein from DDGS. In this method, DDGS is first extracted with anhydrous ethanol to remove oil and yellow color compounds. The deoiled DDGS is then extracted with 70% ethanol in the presence of reducing agent sodium sulfite (0.25%) at acidic medium at pH 2 and 78°C for 2h to obtain an extract with 10% zein content. Removal of ethanol by vacuum drying at 40°C, washing of obtained protein with water, and final drying give a light-colored zein preparation with 90% protein.

2.2.2.8   Potato proteins

Potato proteins, obtained as a byproduct of potato starch production, have attracted increased industrial interest due to their functional properties, such as solubility, edible film making, gelation, foaming and emulsification, and possible health benefits originating from bioactive properties (e.g. antioxidant, antihypertensive, anticarcinogenic, anti-inflammatory activities) of their native or hydrolyzed forms (Løkra and Strætkvern, 2009; Schäfer *et al.*, 2018; Fu *et al.*, 2020). The patatin, a glycoprotein that naturally exists as a dimer and consists of monomeric isoforms with MW ranging between 39 and 44 kDa, is the major potato protein that forms 21 to 31% of total extractable potato proteins (Barta and Bartova, 2008). The remaining potato proteins are formed by some different protease inhibitors (mainly serine, aspartate, cysteine, and Kunitz-type protease inhibitors), and some high molecular-weight protein fractions (Løkra and Strætkvern, 2009). The potato proteins have good nutritional value and contain majority of the essential amino acids at sufficient levels, with the exception of leucine (Kowalczewski *et al.*, 2019). Although the potatoes are not considered among common food allergens, patatin was identified as an IgE-binding protein with possible allergenic reactions for children (Seppälä *et al.*, 1999). Patatin also shows cross-reactivity with natural rubber latex protein Hev-b-7. Thus, it might cause allergenic reactions in latex-sensitized adults (Schmidt *et al.*, 2002).

The proteins in the potato juice recovered during starch extraction could be extracted by using different methods, such as simple solvent precipitation conducted by ethanol, salt precipitation (salting-out) using ammonium sulfate, thermal coagulation/acidic precipitation, membrane separation by UF, or chromatographic fractionation (Fu *et al.*, 2020). The thermal coagulation combined with acid precipitation is used to produce commercial potato proteins with high yield, but this method causes low protein solubility, discoloration, and off-flavor formation in the obtained preparations (Løkra and Strætkvern, 2009). The simple ethanol precipitation method is also an industrially applicable method that gives satisfactory protein yields (Fu *et al.*, 2020), while high-cost chromatographic method (expanded bad adsorption chromatography) is preferred due to its effective separation of undesired impurities, such as phenolic compounds, color pigments, and glycoalkaloids (Løkra and Strætkvern, 2009).

2.2.2.9 *Wheat gluten*

Gluten, a complex storage protein deposited in the starchy endosperm cells of wheat grain, is formed mainly by gliadin and glutenin (Shewry *et al.*, 2002). The viscoelastic properties of gluten are responsible both for the desired properties of dough that are used to obtained high quality bread and bakery products, and properties determining the characteristics of its edible films. Wieser (2007) defined gluten as a 'two-component glue' in which gliadins act as a 'plasticizer' or 'solvent' for glutenins. Both gliadin and glutenin are prolamine group proteins but show different solubility and molecular properties. The gliadins soluble in water-alcohol solutions form majority of proteins in gluten and consist of different single (monomeric) polypeptide chains (α-, β-, γ- and ω-gliadin). The molecular weights of gliadin monomers change between 39 and 55 kDa for ω-gliadins, 28 and 35 kDa for α/β gliadins, and 31 and 35 kDa for γ-gliadins (Wieser, 2007). The S-rich α-, β-, γ- gliadins form majority of gliadins (51 to 64% of gluten) and together with some low-molecular-weight (LMW) protein fractions, they have the ability to form high molecular weight oligomers (MW between 100 and 500 kDa) linked with intermolecular disulfide bonds, while S-poor ω-gliadins are minor (7 to 14% of gluten) gliadin fractions (Shewry *et al.*, 1986; Wieser, 2007). On the other hand, glutenins soluble in diluted acid or alkali solutions are giant protein aggregates (MW: 500 to 10000 kDa) linked with intermolecular disulfide bonds (Wieser, 2007). The main subunits forming these glutenin aggregates are low molecular weight (LMW) glutenins (MW: 32-39 kDa) and high molecular weight (HMW) glutenins that consist of x-type (MW: 83-88 kDa) and y-type (MW: 67-74 kDa) fractions (Wieser, 2007).

Although the wheat gluten is one of the most frequent proteins seen in human diet, and it is one of the most technologically important proteins for food industry, almost all of its fractions show some allergenic reactions. All gluten fractions show IgE-binding properties. However, the α-, β-, γ- and ω-gliadins as well as LMW glutenins are considered as major food allergens for children and adults, while HMW glutenins are minor allergens (Battais *et al.*, 2008). Besides its allergenic reactions, gluten intake can also cause some other diseases, such as celiac (or coeliac) disease (CD) and non-celiac gluten sensitivity (NCGS). Thus, the gluten-free market is one of the rapidly developing sectors owing this to people trying to avoid these health problems. The celiac disease is a serious autoimmune disease that has a prevalence of 1% in general population, with a woman predominance (Caio *et al.*, 2019). The NCGS, also called non-celiac wheat sensitivity (NCWS) due to the involvement of wheat proteins other than gluten, is a less severe disease than CD. However, although NCGS is a disease with non-specific immune response, it causes temporary and preventable intestinal and extra-intestinal symptoms associated with unbalanced diet (Roszkowska *et al.*, 2019).

The commercial production of gluten is one of the major economically important activities for the food industry. In fact, it is important to note that gluten production in volume ranks second after soy protein products (Day, 2011). The gluten was first discovered by Jacopo Bartholomeo Becarri in 1728, but the commercial gluten production process was performed first by Johannes Kesselmeyer in 1759 (This, 2018). Briefly, the basic gluten production process consists of mixing wheat flour with water to form a dough. The dough is then rest to achieve full hydration and to form gluten agglomerates. After that, the starch is separated as a co-product from gluten by washing with water and applying centrifugation or sieving processes (Day, 2011). The wet gluten extracted (~75% protein) is dried, grinded, and sieved to obtain dry power of gluten that is used extensively as an ingredient in production of different bakery products. In the continuous Martin process, the wheat dough is washed with water as it passes through a tumbling cylindrical agitator that effectively removes starch from small holes on its walls as washed dough enriched with gluten moves along the cylinder (Day, 2011; Batey and Huang, 2016). On the other hand, the discontinuous commercial Batter process involves mixing thick suspension of flour with water (batter) in a tank for several hours to separate starch from gluten, collecting gluten using a sieve, and then further washing of gluten to remove residual starch (Day, 2011; Batey and Huang, 2016). The Alfa Laval Raisio process is another alternative method that helps saving large amounts of fresh water during the process. In this method the flour is mixed with water, and then passed to splitter decanter to separate first the starch and then the gluten (Cornell and Hoveling, 1998).

2.2.2.10 Rice proteins

Rice protein isolates (RPI) and concentrates (RPC) are potential candidates for edible film production due to the rising global demand for food products from alternative plant protein sources. The rice proteins could be isolated from brown and milled rice, and rice bran that contain 7.1-8.3, 6.3-7.1, and 11-15% protein, respectively (Saleh *et al.*, 2019). Moreover, Cho *et al.* (1998) used rice-wine meal for preparation of RPC. The main feature of rice-protein concentrates and isolates is that their amino acid content and profile are comparable with those of soy protein isolate (Kalman, 2014). Rice-bran proteins are high quality proteins, rich in essential amino acids, including leucine and threonine that are generally deficient in cereals (Fabian and Ju, 2011). The glutelins' major protein fraction in brown and milled rice form 75-81% and 79-83% of total protein in these products, respectively (Amagliani *et al.*, 2017a). The remaining protein fractions in brown and milled rice are formed heavily by globulins (7-17% and 6-13%, respectively) as well as some minor albumin and prolamin fractions (each ≤ 10%) (Amagliani *et al.*, 2017a). However, the rice bran proteins contain a more balanced distribution of albumins (24-43%), globulins (13-36%), and glutelins (22-45%) while prolamins (1-5%) exist as a minor fraction (Amagliani *et al.*, 2017a). It is thought that the major rice protein fraction, glutelin, is first synthesized as a precursor, and then it is enzymatically hydrolyzed to some extent to low molecular weight fractions, named α- and β-subunits. Amagliani *et al.*, (2017b) determined with SDS-PAGE that major rice glutelin fractions consist of a 52 kDa glutelin precursor, and 28-33 and 17-21 kDa subunits corresponding to α- and β-glutelins, respectively. Amaglian i *et al.*

(2017b) also found two major subunits at 19-22 and 53-56 kDa ranges and minor subunits at 11 and 13 kDa for rice globulins; many minor bands at a wide range between 13 and 110 kDa for rice albumins; and a major band at 10 kDa and minor bands at 18 and 31-32 kDa for rice prolamins. Wang *et al.* (2016) identified rice bran glutelins with SDS-PAGE at 60, 35, 22 and 13 kDa, globulins at 63, 53, 49, 36, 21 kDa, and albumins at >35, 32, 31, 22, 17, 14 kDa.

Although rice is not among the major food allergens, allergenic properties of its proteins are well known (Watanabe, 1993). Matsuda *et al.* (1988) isolated and characterized major allergenic rice protein as a 16 kDa protein concentrated in the rice endosperm (concentration of this protein in rice endosperm is almost 5.6-fold higher than that in rice bran). Usui *et al.* (2001) also characterized a 33 kDa protein as one of the major rice protein allergens. The rice bran proteins are generally considered hypoallergenic (Fabian and Ju, 2011). However, Satoh *et al.* (2019) have recently characterized a 52 kDa protein as the major allergen in rice bran. Therefore, further studies are needed to investigate the potential allergens in rice protein isolates. A biochemical approach to solve allergen problem in rice came from Watanabe *et al.* (1990) who achieved enzymatic hydrolysis of allergenic rice proteins by impregnating actinase into rice seed during rehydration in alkaline solution. Although these workers succeeded in obtaining hypoallergenic rice, the effect of this procedure on protein purification yield and functional and edible film properties of obtained isolates should be investigated.

The classical alkaline extraction and following concentration methods, such as IEP or UF, could be employed to obtain commercial rice protein concentrates (RPC). It was reported that RPCs obtained by spray-drying of rice endosperm alkaline extracts treated by IEP and UF contained almost 86 and 71% protein, respectively, but RPC treated with UF showed much better solubility (37%) and emulsifying properties (emulsifying activity and stability) than that obtained with IEP (Paraman *et al.*, 2008). Some enzymatic methods, which target hydrolysis of starch with enzymes, such as α-amylase, glucoamylase, pullulan etc., and cell wall constituents (e.g. cellulose, hemicellulose, pectin) with enzymes, such as cellulase, hemicellulase, pectinase, xylanase, etc. are also employed to increase protein extractability from rice flour and bran, and to obtain protein isolates having protein content ≥ 90% (Amagliani *et al.* 2017a). Moreover, some physical treatments, such as ultrasound, freeze-thawing, hydrothermal cooking, colloid milling, and microfluidization, high pressure treatment, high speed blending could also be applied alone or in combination with enzymatic treatments to increase efficiency of protein extraction in brown and milled rice, and rice bran (Amagliani *et al.*, 2017a; Fabian and Ju, 2011).

2.2.3 Algal source hydrocolloids

2.2.3.1 Agar and carrageenan

Both agar and carrageenans are natural hydrocolloids found in red seaweeds (*Rhodophyceae*). The agar naturally exists as a mixture of charged (anionic) agaropectin (non-gelling fraction) and neutral agarose (gelling-fraction). However, the commercial agar preparations contain mainly agarose since an important portion of the agaropectin is removed during the commercial agar production processes (Mostafavi and Zaeim, 2020). Agarose is a linear polysaccharide made up of D-galactose and 3,6-anhydro-L-galactose linked by alternating α-1,3- and β-1,4- glycosidic bonds. Thus, there are two repeating disaccharides on the linear agarose chain named agarobiose (a disaccharide with β-1,4-glycosidic bond) and neoagarobiose (a disaccharide with α-1,3-glycosidic bond). Agaropectin is formed by natural modification of agarose. The modification of agarose occurs mainly by replacement of 3,6-anhydro-L-galactose residues with L-galactose sulphate ester (O-SO$_3^-$), and to some extent by replacement of D-galactose residues with 4-6-O-(1-carboxylidene)-D-galactose (pyruvate acetal linked to D-galactose residues), or by methylation of both monosaccharide residues of agar.

Carrageenans are anionic and linear polysaccharides that have closely related chemical structures with agars. The repeating disaccharides on linear chains of carrageenans are named carrabiose (a disaccharide with β-1,4-glycosidic bond) and neocarrabiose (a disaccharide with α-1,3-glycosidic bond). However, there are two main differences between carrageenans and agar colloids: (1) the 4-linked α-galactose residue is in D configuration instead of L, and (2) the carrageenans are more sulfated than agar colloids (Lahaye, 2001). Different types of carrageenans (15 to 20) have been isolated and characterized (Lahaye, 2001; Rosenau *et al.*, 2020), but the most industrially relevant types are kappa-, iota-, and lambda-carrageenans that contain one, two, and three sulfate groups per disaccharide repeating units, respectively (Jiao *et al.*, 2011).

Agar is extracted from agarophytes, such as *Gelidium corneum* and *Gracilaria* spp., while carrageenans are extracted mainly from carrageenophytes, such as *Kappaphycus alvarezii, K. striatum*, and *Eucheuma denticulatum* (Fleurence and Levine, 2016; Porse and Rudolph, 2017). The seaweeds used in the extraction are gathered by collecting those washed to the shore, by cutting or rooting them out from their beds, or by cultivation (Armisen and Galatas, 1987). However, due to the problems related to resource management, extensive studies are needed to develop more sustainable seaweed cultivation techniques and greener processing methods (Porse and Rudolph, 2017; Santos and Melo, 2018). The first step in production of seaweed hydrocolloids is drying of the gathered or harvested seaweeds. Reduction of the moisture level at least to 20% is essential to prevent initiation of fermentation that diminishes the hydrocolloid quality (Armisen and Galatas, 1987). Although both agar and carrageenan can be extracted from seaweeds by hot water, the hydrocolloids obtained by this procedure show weak gel strength. Therefore, alkaline treatment of seaweeds before or during hot extraction (100-120°C) is applied to enhance the gelling properties of extracted hydrocolloids. Khalil *et al.* (2018) reported that the hot alkali treatment caused reduction of extraction yields due to degradation of the hydrocolloids, but at the same time this treatment converted 6-sulfated residues and polygalactose with no or weak gelling capacity to 3,6-anhydrogalactose that increased the gel strength of obtained hydrocolloid. Therefore, a general procedure to obtain agar and carrageenan involves (1) hot water or alkaline extraction, (2) neutralization, (3) precipitation (mostly by ethanol or isopropanol), (4) filtration, (5) drying, and (6) milling (Khalil *et al.*, 2018). An alternative approach in extraction of agar and carrageenan is as follows: (1) hot water extraction, (2) filtration, (3) cooling of hydrocolloid solution to form a gel (agar shows gelation by cooling while gelation of carrageenan also needs addition of K^+ ions), (4) removal of free water in the gel by pressing or freezing-thawing, (5) drying, and (6) grounding (Stanley, 1987; Lahaye, 2001).

In the literature, reports about allergenic reactions of agar and carrageenan are scarce. Both hydrocolloids are used extensively by the food industry and are not generally considered among the common food allergens. However, there are some rare reports about IgE-mediated allergenic reactions against carrageenan. For example, Kular *et al.* (2018) reported carrageenan allergy for a pediatric patient while Tarlo *et al.* (1995) reported an allergenic case of a 26-year-old woman. There are also some reports that carrageenan might cause some adverse gastrointestinal effects in some people (Martino *et al.*, 2017). Different contradicting reports also exist related to health effects of carrageenans. For example, Mi *et al.* (2020) have recently found that intake of soluble carrageenan might induce colonic inflammation and enhanced colitis in rats fed by high-fat diet. These authors found some evidence that carrageenan might enhance growing of harmful colonic bacteria that cause inflammation. In contrast, findings of Raman and Doble (2015) suggest that carrageenan could be used as a functional food to prevent colon carcinogenesis. Rosenau *et al.* (2020) also reported that low-molecular-weight highly-sulphated carrageenan fractions showed potent antitumor activity. These authors also reported that the oligocarrageenans weakened the immune suppressing effects of the antitumor drugs. Finally, it was found that carrageenan possesses antioxidant activity and is effective in suppressing development of oxidative stress in diabetic rats (Sanjivkumar *et al.*, 2020). Different health benefits were also reported for agar. For example, Fernandez *et al.* (1989) isolated and characterized an antitumor active agar from *Gracilaria dominguensis*. Park *et al.* (2017) also determined hypolipidemic and hypoinsulinemic effects of an agar fraction in mice fed on high lipid diet. The agars and carrageenans obtained by sustainable production methods have been attracting considerable interest as packaging materials. Thus further studies are needed to clarify their potential positive and negative health effects.

2.2.3.2   Alginate

Alginate is a polysaccharide extracted from the outer cell walls of brown seaweeds (*Phaeophyceae*). It is produced mainly as Na-alginate from alginophytes, such as *Ascophyllum nodosum, Laminaria digitate, Macrocystis pyrifera*, etc. (Yemenicioğlu *et al.*, 2020). Alginate consists of linear blocs of 1-4-linked β-D-mannuronate (M) and α-L-guluronate (G) with variations in composition and sequential arrangements. The M and G monomers on the linear chain can be arranged as M, MG, or G blocks (----M-M-M-M-M----M-G-M-G-M-G---G-G-G-G-G----) with M/G ratio changing from 0.6 to 2.7, depending on the species (Gupta and Raghava, 2008). The alginate forms thermo-irreversible gels in the presence of Ca^{++} ions. These gels are much more thermostable (up to 100°C) than partially thermo-reversible gels of low-methoxyl pectin that forms weak gels at high temperatures since it undergoes chain shortening by β-elimination. During gelation, the G monomers interact with the Ca^{++} atoms to create a network, which can be explained with the classical egg-box model used also to explain gelation mechanism of low-methoxyl pectin (Cao *et al.*, 2020). Therefore, the properties of agar gels are highly affected by the M/G ratio. The alginate molecules having high amounts of G blocks form hard, brittle, and thermostable gels with high water absorption capacity while those having high amounts of M blocks form softer and more flexible gels (Cao *et al.*, 2020; Khalil *et al.*, 2018). The pH affects the gel strength and viscosity of Ca-alginate gels. Gel strength reduces while gel viscosity increases as pH is increased from 4.0 to 7.7 (Cao *et al.*, 2020).

According to Khalil *et al.* (2018), the general extraction process of alginate from brown seaweed is slightly different than those of agar and carrageenans. Briefly, after excessive washing with water, the brown seaweeds are treated with formaldehyde. This chemical treatment eliminates the color pigments and crosslinks the undesired polyphenols. The seaweeds are then treated with HCl to clarify polyphenols and to remove residual formaldehyde. After that, the alginate is extracted with hot alkali extraction (mostly by sodium hydroxide or sodium carbonate) and then neutralized (forms water soluble Na-alginate) and precipitated with alcohols or $CaCl_2$. The obtained precipitate is collected with centrifugation and the extract is dried and grinded (Khalil *et al.*, 2018).

Alginate is an indigestible viscous dietary fiber with some health benefits. It is found that the alginate might delay the digestion of lipid and protein since it shows inhibitory effect on enzymes, such as lipase and pepsin (Houghton *et al.*, 2015; Brownlee *et al.*, 2005). Thus, the use of alginate in food like bread has been suggested as a potential treatment for obesity (Houghton *et al.*, 2015). Moreover, the presence of alginate in the diet might depress the appetite and reduce blood cholesterol levels and glucose absorption (Brownlee *et al.*, 2005). The alginate is not a major food allergen, but people having fish allergy are generally suggested to avoid alginate and its derivatives. Moreover, it is also essential to confirm that alginate and its salts lack formaldehyde residues coming from the purification process and are free from residues of toxic elements, such as arsenic, cadmium, lead, and mercury (Younes *et al.*, 2017).

2.2.4   Microbial source hydrocolloids

2.2.4.1   Pullulan

Pullulan is a neutral and water-soluble exopolysaccharide obtained with fermentation from the polymorphic fungus *Aureobasidium pullulans* that is generally regarded non-pathogenic and non-toxigenic (Leathers, 2003; Kimoto *et al.*, 1997). It is an unbranched homopolysaccharide that is formed by repeating maltotriose units. The glucose molecules forming the maltotriose are linked to each other with α-1,4 glycosidic bonds while the repeating maltotriose units are linked to each other through α-1,6 glycosidic bonds [α-(1-6)-linked (1-4)-α-D-triglucosides]. The pullulan forms water-soluble films having poor mechanical properties (Xiao *et al.*, 2012; Shih *et al.*, 2011), but its films show excellent transparency and relatively low oxygen permeability (Kandemir *et al.*, 2005; Chlebowska-Śmigiel and Gniewosz, 2009). However, this polysaccharide has not found a world-wide interest in the food industry due to its high purification costs from fermentation medium (contains some other polysaccharides and colored melanins produced by

A. pullulans), variable yields and molecular properties originated from changes in *A. pullulans* morphology and/or *in situ* exopolysaccharide-degrading enzymes during fermentation (Youssef *et al.*, 1999; Seo *et al.*, 2004; Campbell *et al.*, 2003; Shingel, 2004).

2.2.4.2   Xanthan

Xanthan is an anionic and water-soluble bacterial exopolysaccharide obtained by fermentation from Gram-negative bacteria, *Xanthomonas campestris*. It has a cellulose-like linear backbone of (1-4)-linked β-D-Glucose residues substituted at O-3 of alternate glucose residues with a trisaccharide side chain (β-D-Man*p*-(1,4)-β-D-Glc*p*A-(1,2)-α-D-Man*p*-(1→) unit) (Cui *et al.*, 2013). Studies to use xanthan alone to obtain self-standing films are scarce since this hydrocolloid is a non-gelling polysaccharide (Saha and Bhattacharya, 2010). However, it has been recently demonstrated that mixtures of xanthan (2 to 10% w/w) with glycerol (20% w/w) could be complexed to form hydrogels that can be employed to obtain edible films by the classical solution-casting method (Bilanovic *et al.*, 2016). Moreover, the ability of xanthan to form strong complexes (mostly *via* noncovalent interactions, such as hydrogen bonding) with other hydrocolloids (e.g. carrageenan, gellan, gelatin, carboxymethyl cellulose, locust bean gum, starch, chitosan, protein hydrolysates) is frequently exploited to obtain blend or composite edible films with different physicochemical and mechanical properties (Harizah *et al.*, 2016; Balasubramanian *et al.*, 2018; Kurt *et al.*, 2017; Arismendi *et al.*, 2013; De Morais Lima *et al.*, 2017). It is also worth to note that xanthan can also be combined with starches of different origin to obtain composite films by using blown-extrusion method (Flores *et al.*, 2010; Melo *et al.*, 2011).

2.2.4.3   Gellan

Gellan is an anionic extracellular polysaccharide obtained by fermentation from Gram-negative bacterium *Sphingomonas elodea* (formerly known as *Pseudomonas elodea*) (Vartak *et al.*, 1995). It is a linear polysaccharide (~ 50000 sugar residues) formed by repeating units of a tetrasaccharide that consists of two D-glucose residues, one residue of D-glucuronic acid, and one residue of L-rhamnose [-4)-L-rhamnopyranosyl-(α-1-3)-D-glucopyranosyl-(β-1-4)-D-glucuronopyranosyl-(β-1-4)-D-glucopyranosyl-(β-1- with O (2) L-glyceryl and O (6) acetyl substituents on the 3-linked glucose] (Coviello *et al.*, 2007).

The gellan chains adopt strands of threefold double helix (with a pitch of 5.64 nm), each formed by two left-handed chains coiled around each other with the acetate residues on the periphery and glyceryl groups stabilizing the interchain associations (Morris *et al.*, 2012). This hydrocolloid forms gel in the presence of monovalent and divalent cations (Kang *et al.*, 2015). Divalent cations (e.g. Ca^{+2} and Mg^{+2}) form hard gels since they induce cross-liking between neighboring –COOH groups of different gellan double helices, while monovalent cations (Na^+ and K^+) form weak gels by veiling electrostatic repulsion created among negatively-charged gellan helices, and inducing their aggregation (Kang *et al.*, 2015). The degree of acylation and number of glyceryl groups also affect the strength of gellan gel network. The extensive hot alkali treatment applied during commercial production process removes bulky acyl and glyceryl groups of gellan, and this enables close association and packing of gellan chains during gelation (Mao *et al.*, 2000; Morris *et al.*, 2012). Thus, low acyl gellans cause formation of hard, non-elastic, and brittle gels while acylated gellans obtained at mild alkaline conditions form soft, elastic, and transparent gels (Kang *et al.*, 2015). The deacylated gellan could be used in development of edible films, but these films should be plasticized properly since they are quite brittle (Yang and Paulson, 2000; Paolicelli *et al.*, 2018).

References

Abeyrathne, E.D.N.S., H.Y Lee and D.U. Ahn (2013). Egg white proteins and their potential use in food processing or as nutraceutical and pharmaceutical agents – A review, *Poultry Sci.*, 92: 3292-3299.

Acquah, C., Y. Zhang, M.A Dubé and C.C. Udenigwe (2020). Formation and characterization of protein-based films from yellow pea (*Pisum sativum*) protein isolate and concentrate for edible applications, *Curr. Res. Food Sci.*, 2: 61-69.

Adam, C.L., L.M. Thomson, P.A. Williams and A.W. Ross (2015). Soluble fermentable dietary fibre (pectin) decreases caloric intake, adiposity and lipidaemia in high-fat diet-induced obese rats, *PLoS One*, 10: e0140392.

Adami, E.R., C.R. Corso, N.M. Turin-Oliveira, C.M. Galindo, L. Milani, M.C. Stipp *et al.* (2018). Antineoplastic effect of pectic polysaccharides from green sweet pepper (*Capsicum annuum*) on mammary tumor cells *in vivo* and *in vitro*, *Carbohydr. Polym.*, 201: 280-292.

Alfaro, A.T., E. Balbinot, C.I. Weber, I.B. Tonial and A. Machado-Lunkes (2015). Fish gelatin: Characteristics, functional properties, applications and future potentials, *Food Eng. Rev.*, 7: 33-44.

Alkan, B. and A. Yemenicioglu (2016). Potential application of natural phenolic antimicrobials and edible film technology against bacterial plant pathogens, *Food Hydrocolloid*, 55: 1-10.

Alshikh, N., A.C. de Camargo and F. Shahidi (2015). Phenolics of selected lentil cultivars: Antioxidant activities and inhibition of low-density lipoprotein and DNA damage, *J. Funct. Foods*, 18: 1022-1038.

Altuna, L., M.L. Herrera and M.L. Foresti (2018). Synthesis and characterization of octenyl succinic anhydride modified starches for food applications. A review of recent literature, *Food Hydrocoll.*, 80: 97-110.

Aluko, R.E. (2015). Amino acids, peptides, and proteins as antioxidants for food preservation, pp. 105-140. *In*: F. Shahidi (Ed.). *Handbook of Antioxidants for Food Preservation*. Woodhead Publishing, London, UK.

Amagliani, L., J. O'Regan, A.L. Kelly and J.A. O'Mahony (2017a). The composition, extraction, functionality and applications of rice proteins: A review, *Trends in Food Sci. Tech.*, 64: 1-12.

Amagliani, L., J. O'Regan, A.L. Kelly, and J.A. O'Mahony (2017b). Composition and protein profile analysis of rice protein ingredients, *J. Food Compos. Anal.*, 59: 18-26.

Anderson, D.M.W., J.F. Howlett and C.G.A. McNab (1985). The amino acid composition of the proteinaceous component of gum Arabic (*Acacia senegal* (L.) Willd.), *Food Addit. Contam.*, 2: 159-164.

Anderson, T.J. and B.P. Lamsal (2011). Development of new method for extraction of α-zein from corn gluten meal using different solvents, *Cereal Chem.*, 88: 356-362.

Andrade, J.E., N.C. Twaddle, W.G. Helferich and D.R. Doerge (2010). Absolute bioavailability of isoflavones from soy protein isolate-containing food in female BALB/c mice, *J. Agric. Food Chem.*, 58: 4529-4536.

André, F., S. Cavagna and C. André (2003). Gelatin prepared from tuna skin: A risk factor for fish allergy or sensitization. *Int. Arch. Allergy Imm.*, 130: 17-24.

Ardila, N., F. Daigle, M.C. Heuzey and A. Ajji (2017). Antibacterial activity of neat chitosan powder and flakes, *Molecules*, 22(100): 1-19.

Arismendi, C., S. Chillo, A. Conte, M.A. Del Nobile, S. Flores and L.N. Gerschenson (2013). Optimization of physical properties of xanthan gum/tapioca starch edible matrices containing potassium sorbate and evaluation of its antimicrobial effectiveness, *LWT-Food Sci. Technol.*, 53: 290-296.

Armisen, R. and F. Galatas (1987). Production, properties and uses of agar, pp. 1-57. *In:* D.J. McHugh. (Ed.). *Production and Utilization of Products from Commercial Seaweeds*, FAO Fish. Tech. Pap. 288 (No. 589.45 F36), Rome, Italy.

Assadpour, E. and S.M. Jafari (2019). An overview of hydrocolloid nanostructures for encapsulation of food ingredients, pp. 1-24. *In:* S.M. Jafari (Ed.). *Hydrocolloid Nanostructures for Food Encapsulation Purposes*, vol. 1. Academic Press, London, UK.

Aulin, C., G. Salazar-Alvarez and T. Lindström (2012). High strength, flexible and transparent nanofibrillated cellulose–nanoclay biohybrid films with tunable oxygen and water vapor permeability, *Nanoscale*, 4: 6622-6628.

Avena-Bustillos, R.J., B.S. Chiou, C.W. Olsen, P.J. Bechtel, D.A. Olson and T.H. McHugh (2011). Gelation, oxygen permeability, and mechanical properties of mammalian and fish gelatin films, *J. Food Sci.*, 76: E519-E524.

Aydemir, L.Y. and A. Yemenicioğlu (2013a). Potential of Turkish Kabuli type chickpea and green and red lentil cultivars as source of soy and animal origin functional protein alternatives, *LWT-Food Sci. Technol.*, 50: 686-694.

Aydemir, L.Y. and A. Yemenicioglu (2013b). Are protein-bound phenolic antioxidants in pulses unseen part of iceberg? *J. Plant Biochem. Physiol.*, 1: 1001118.

Azeredo, H.M., M.F. Rosa and L.H.C. Mattoso (2017). Nanocellulose in bio-based food packaging applications, *Ind. Crop. Prod.*, 97: 664-671.

Azevedo, V.M., M.V. Dias, H.H. de Siqueira Elias, K.L. Fukushima, E.K. Silva, J.D.D.S. Carneiro *et al.* (Borges. 2018). Effect of whey protein isolate films incorporated with montmorillonite and citric acid on the preservation of fresh-cut apples, *Food Res. Int.*, 107: 306-313.

Badley, R.A., D. Atkinson, H. Hauser, D. Oldani, J.P. Green and J.M. Stubbs (1975). The structure, physical and chemical properties of the soy bean protein glycinin, *Biochim. Biophys. Acta Protein Struct.*, 412(2): 214-228.

Balasubramanian, R., S.S. Kim and J. Lee (2018). Novel synergistic transparent k-carrageenan/xanthan gum/gellan gum hydrogel film: Mechanical, thermal and water barrier properties, *Int. J. Biol. Macromol.*, 118: 561-568.

Bao, J. and C.J. Bergman (2004). The functionality of rice starch, pp. 258-294. *In:* A.C. Eliasson (Ed.). *Starch in Food: Structure, Function and Applications.* CRC Press, Cornwal, England.

Bao, J. (2019). Rice starch, pp. 55-108. *In:* J. Bao (Ed.). *Rice, Chemistry and Technology.* AACC International Press, Cambridge, USA.

Barac, M., S. Cabrilo, M. Pesic, S. Stanojevic, S. Zilic, O. Macej and N. Ristic (2010). Profile and functional properties of seed proteins from six pea (*Pisum sativum*) genotypes, *Int. J. Mol. Sci.*, 11: 4973-4990.

Barta, J. and V. Bartova (2008). Patatin, the major protein of potato (*Solanum tuberosum* L.) tubers, and its occurrence as genotype effect: Processing versus table potatoes, *Czech J. Food Sci.*, 26: 347-359.

Bartolomé, B., I. Estrella and M.T. Hernandez (2000). Interaction of low molecular weight phenolics with proteins (BSA), *J. Food Sci.*, 65: 617-621.

Basiak, E., A. Lenart and F. Debeaufort (2017). Effect of starch type on the physico-chemical properties of edible films, *Int. J. Biol. Macromol.*, 98: 348-356.

Batey, I.L. and W. Huang (2016). Gluten and modified gluten, pp. 408-413. *In:* C.W. Wrigley, H. Corke, K. Seetharaman and J. Faubion (Eds.). *Encyclopedia of Foodgrains*. Academic Press, London, UK.

Battais, F., C. Richard, S. Jacquenet, S. Denery-Papini and D.A. Moneret-Vautrin (2008). Wheat grain allergies: An update on wheat allergens, *Eur. Ann. Allergy Clin. Immunol.*, 40: 67-76.

Batt, H.P., R.L. Thomas and A. Rao (2003). Characterization of isoflavones in membrane-processed soy protein concentrate, *J. Food Sci.*, 68: 401-404.

Beck, K. and B. Brodsky (1998). Supercoiled protein motifs: The collagen triple-helix and the α-helical coiled coil, *J. Struct. Boil.*, 122: 17-29.

BeMiller, J.N. and R.L. Whistler (1996). Carbohydrates, pp. 158-221. *In:* O.R. Fennema (Ed.). *Food Chemistry*, 3th ed. Marcel Dekker, New York, USA.

BeMiller, J.N. (2018). *Carbohydrate Chemistry for Food Scientists*. Elsevier, Woodhead Publishing, Cambridge, USA.

Benjakul, S. and P. Kittiphattanabawon (2019). Gelatin, pp. 121-127. *In:* L. Melton, F. Shahidi and P. Varelis (Eds). *Encyclopedia of Food Chemistry*, vol. 1. Elsevier, Amsterdam, NL.

Berk, Z. (1992). *Technology of Production of Edible Flours and Protein Products from Soybeans*. Agriculture Organization of the United Nations (FAO), Bulletin No. 97, Rome, Italy.

Bernardino, S., R. Prado, T.M. Shiga, H. Harazono, V.H. Hogan, A. Raz *et al.* (2019). Migration and proliferation of cancer cells in culture are differentially affected by molecular size of modified citrus pectin, *Carbohydr. Polym.*, 211: 141-151.

Bilanovic, D., J. Starosvetsky and R.H. Armon (2016). Preparation of biodegradable xanthan–glycerol hydrogel, foam, film, aerogel and xerogel at room temperature, *Carbohydr. Polym.*, 148: 243-250.

Bilbao-Sáinz, C., R.J. Avena-Bustillos, D.F. Wood, T.G. Williams and T.H. McHugh (2010). Composite edible films based on hydroxypropyl methylcellulose reinforced with microcrystalline cellulose nanoparticles, *J. Agric. and Food Chem.*, 58: 3753-3760.

Bogdanov, I.V., Z.O. Shenkarev, E.I. Finkina, D.N. Melnikova, E.I. Rumynskiy, A.S. Arseniev and T.V. Ovchinnikova (2016). A novel lipid transfer protein from the pea (*Pisum sativum*): Isolation, recombinant expression, solution structure, antifungal activity, lipid binding, and allergenic properties, *BMC Plant Biol.*, 16: 107.

Boyacı, D. and A. Yemenicioğlu (2018). Expanding horizons of active packaging: Design of consumer-controlled release systems helps risk management of susceptible individuals, *Food Hydrocoll.*, 79: 291-300.

Börjesson, M., K. Sahlin, D. Bernin and G. Westman (2018). Increased thermal stability of nanocellulose composites by functionalization of the sulfate groups on cellulose nanocrystals with azetidinium ions, *J. Appl. Polym. Sci.*, 135: 45963.

Brownlee, I.A., A. Allen, J.P. Pearson, P.W. Dettmar, M.E. Havler, M.R. Atherton and E. Onsøyen (2005). Alginate as a source of dietary fiber, *Crit. Rev. Food Sci.*, 45: 497-510.

Burley, R.W. and W.H. Cook (1961). Isolation and composition of avian egg yolk granules and their constituent α- and β-lipovitellins, *Canadian J. Biochem. Phys.*, 39: 1295-1307.

Bylund, G. (1995). *Dairy Processing Handbook*. Tetra Pak Processing Systems AB, Lund, Sweden.

Cabra, V., R. Arreguin, A. Galvez, M. Quirasco, R. Vazquez-duhalt and A. Farres (2005). Characterization of a 19 kDa α-zein of high purity, *J. Agric. Food Chem.*, 53: 725-729.

Caio, G., U. Volta, A. Sapone, D.A. Leffler, R. De Giorgio, C. Catassi and A. Fasano (2019). Celiac disease: A comprehensive current review, *BMC Med.*, 17: 1-20.

Campbell, B.S., B.M. McDougall and R.J. Seviour (2003). Why do exopolysaccharide yields from the fungus *Aureobasidium pullulans* fall during batch culture fermentation? *Enzyme Microb. Technol.*, 33: 104-112.

Cano, A., A. Jiménez, M. Cháfer, C. Gónzalez and A. Chiralt (2014). Effect of amylose: Amylopectin ratio and rice bran addition on starch films properties, *Carbohydr. Polym.*, 111: 543-555.

Cao, L., W. Lu, A. Mata, K. Nishinari and Y. Fang (2020). Egg-box model-based gelation of alginate and pectin: A review, *Carbohydr. Polym.*, 242: 116389.

Çarman, K. (1996). Some physical properties of lentil seeds, *J. Agric. Eng. Res.*, 63: 87-92.

Cataldi, A., L. Berglund, F. Deflorian and A. Pegoretti (2015). A comparison between micro- and nanocellulose-filled composite adhesives for oil paintings restoration, *Nanocomposites*, 1: 195-203.

Cawston, T.E. (1995). Proteinases and connective tissue breakdown, pp. 333-359. *In:* B. Henderson, J.C.W. Edwards, E.R. Pettipher and E.R. Pettipher (Eds.). *Mechanisms and Models in Rheumatoid Arthritis*. Academic Press, Toronto, Canada.

Cemeroglu, B. and J. Acar (1986). *Fruit and Vegetable Processing Technology*. Turkish Association of Food Technologists, 508 p.

Chaichi, M., M. Hashemi, F. Badii and A. Mohammadi (2017). Preparation and characterization of a novel bionanocomposite edible film based on pectin and crystalline nanocellulose, *Carbohydr. Polym.*, 157: 167-175.

Chalamaiah, M., Y. Esparza, F. Temelli and J. Wu (2017). Physicochemical and functional properties of livetins fraction from hen egg yolk, *Food Biosci.*, 18: 38-45.

Chandrayan, P. (2018). Biological function (S) and application (S) of pectin and pectin degrading enzymes, *Biosci. Biotechnol. Res. Asia*, 15: 87-100.

Chang, Y.W., I. Alli, A.T. Molina, Y. Konishi and J.I. Boye (2012). Isolation and characterization of chickpea (*Cicer arietinum* L.) seed protein fractions, *Food Bioprocess Tech.*, 5: 618-625.

Chay Pak Ting, B.P., Y. Pouliot, S.F. Gauthier and Y. Mine (2010). Fractionation of egg proteins and peptides for nutraceutical applications, pp. 595-618. *In:* S.S. Rizvi (Ed.). *Separation, Extraction and Concentration Processes in the Food, Beverage and Nutraceutical Industries*. Elsevier, Woodhead Publishing, Cambridge, UK.

Chen, J., W. Liu, C.M. Liu, T. Li, R.H. Liang and S.J. Luo (2015). Pectin modifications: A review, *Crit. Rev. Food Sci. Nutr.*, 55: 1684-1698.

Chlebowska-Śmigiel, A. and M. Gniewosz (2009). Effect of pullulan coating on inhibition of chosen microorganisms' growth, *ACTA Sci. Pol. Technol. Aliment*, 8: 37-46.

Cho, S.Y., J W. Park and C. Rhee (1998). Edible fims from protein concentrates of rice wine meal, *Korean J. Food Sci. Technol.*, 30: 1097-1106.

Cho, Y.H. and O.G. Jones (2019). Assembled protein nanoparticles in food or nutrition applications, pp. 47-84. *In:* L. Lim and M. Rogers (Eds.). *Advances in Food and Nutrition Research*, vol. 88. Academic Press, London, UK.

Ciriminna, R., A. Fidalgo, R. Delisi, L.M. Ilharco and M. Pagliaro (2016). Pectin production and global market, *Agro Food Ind. Hi-Tech*, 27: 17-20.

Cook, R.B., F.M. Mallee and M.L. Shulman (1993). *Purification of Zein from Corn Gluten Meal*. U.S. Patent No. 5,254,673, Washington, DC: U.S. Patent and Trademark Office.

Cornell, H. and A.W. Hoveling (1998). *Wheat: Chemistry and Utilization*. CRC Press, Boca Raton, USA.

Coviello, T., P. Matricardi, C. Marianecci and F. Alhaique (2007). Polysaccharide hydrogels for modified release formulations, *J. Control. Release*, 119: 5-24.

Crowley, S.V., E. Burlot, J.V. Silva, N.A. McCarthy, H.B. Wijayanti, M.A. Fenelon and J.A. O'Mahony (2018). Rehydration behavior of spray-dried micellar casein concentrates produced using microfiltration of skim milk at cold or warm temperatures, *Int. Dairy J.*, 81: 72-79.

Cuevas-Acuña, D.A., S. Ruiz-Cruz, J.L. Arias-Moscoso, M.A. Lopez-Mata, P.B. Zamudio-Flores, S.E. Burruel-Ibarra and H. del Carmen Santacruz-Ortega (2020). Effects of the addition of ultrasound-pulsed gelatin to chitosan on physicochemical and antioxidant properties of casting films, *Polym. Int.*, 69: 423-428.

Cui, S.W., Y. Wu and H. Ding (2013). The range of dietary fibre ingredients and a comparison of their technical functionality, pp. 96-119. *In:* J.A. Delcour and K. Poutanen (Eds.). *Fibre-rich and Wholegrain Foods: Improving Quality*. Elsevier, Woodhead Publishing, Oxford, UK.

Dalgleish, D.G., P.A. Spagnuolo and H.D. Goff (2004). A possible structure of the casein micelle based on high-resolution field-emission scanning electron microscopy, *Int. Dairy J.*, 14: 1025-1031.

Day, L. (2011). Wheat gluten: Production, properties and application, pp. 267-288. *In:* G.O. Phillips and P.A. Williams (Eds.). *Handbook of Food Proteins*. Woodhead Publishing. Cambridge, UK.

De Lucca, A.J. and T.J. Walsh (1999). Antifungal peptides: Novel therapeutic compounds against emerging pathogens, *Antimicrob. Agents Chemother.*, 43: 1-11.

De Morais Lima, M., D. Bianchini, A. Guerra Dias, E. da Rosa Zavareze, C. Prentice and A. da Silveira Moreira (2017). Biodegradable films based on chitosan, xanthan gum, and fish protein hydrolysate, *J. Appl. Polym.Sci.*, 134: 44899.

Dossey, A.T., J.A. Morales-Ramos and M.G. Rojas (2016). *Insects as Sustainable Food Ingredients: Production, Processing and Food Applications*. Elsevier, Academic Press. London, UK, 402 p.

Dutta, P.K., S. Tripathi, G.K. Mehrotra and J. Dutta (2009). Perspectives for chitosan-based antimicrobial films in food applications, *Food Chem.*, 114: 1173-1182.

Eckholm E. (1985). Mesopotamia: Cradle of haute cusine. *The New York Times*, https://www.nytimes.com/1985/05/15/garden/mesopotamia-cradle-of-haute-cuisine.html

Fabian, C. and Y.H. Ju (2011). A review on rice bran protein: Its properties and extraction methods, *Crit. Rev. Food Sci. Nutr.*, 51: 816-827.

FAO/WHO (2017). Compendium of food additive specifications, pp. 41-43. *In:* Joint FAO/WHO Expert Committee on Food Additives 84th Meeting Report, Rome, Italy.

FDA (2017). What You Need to Know about Food Allergies. https://www.fda.gov/food/buy-store-serve-safe-food/what-you-need-know-about-food-allergies

FDA (2018). Code of Federal Regulations, 21CFR101.82, Title 21, vol. 2.

Fernández, L.E., O.G. Valiente, V. Mainardi, J.L. Bello, H. Vélez and A. Rosado (1989). Isolation and characterization of an antitumor active agar-type polysaccharide of *Gracilaria dominguensis*, *Carbohydr. Res.*, 190: 77-83.

Fernandes, S.C., C.S. Freire, A.J. Silvestre, C.P. Neto, A. Gandini, L.A. Berglund and L. Salmén (2010). Transparent chitosan films reinforced with a high content of nanofibrillated cellulose, *Carbohydr. Polym.*, 81: 394-401.

Fleurence, J. and L. Levine (2016). *Seaweed in Health and Disease Prevention*. Academic Press, New York, USA.

Flores, S.K., D. Costa, F. Yamashita, L.N. Gerschenson and M.V. Grossmann (2010). Mixture design for evaluation of potassium sorbate and xanthan gum effect on properties of tapioca starch films obtained by extrusion, *Mater. Sci. Eng. C.*, 30: 196-202.

Fu, Y., W.N. Liu and O.P. Soladoye (2020). Towards potato protein utilisation: Insights into separation, functionality and bioactivity of patatin, *Int. J. Food Sci. Technol.*, 55: 2314-2322.

Gawkowska, D., J. Cybulska and A. Zdunek (2018). Structure-related gelling of pectins and linking with other natural compounds: A review, *Polymers*, 10: 762.

Głąb, T.K. and J. Boratyński (2017). Potential of casein as a carrier for biologically active agents, *Topics in Current Chemistry*, 375: 1-20.

González-Pérez, S. and J.B. Arellano (2009). Vegetable protein isolates, pp. 383-419. *In:* Phillips, G.O. and Williams, P.A. (Eds.). *Handbook of Hydrocolloids*. Elsevier, Woodhead Publishing, New York, USA.

Groche, D., L.G. Rashkovetsky, K.H. Falchuk and D.S. Auld (2000). Subunit composition of the zinc proteins α- and β-lipovitellin from chicken, *J. Protein Chem.*, 19: 379-387.

Güçbilmez, Ç.M., A. Yemenicioğlu and A. Arslanoğlu (2007). Antimicrobial and antioxidant activity of edible zein films incorporated with lysozyme, albumin proteins and disodium EDTA, *Food Res. Int.*, 40: 80-91.

Guilmineau, F., I. Krause and U. Kulozik (2005). Efficient analysis of egg yolk proteins and their thermal sensitivity using sodium dodecyl sulfate polyacrylamide gel electrophoresis under reducing and nonreducing conditions, *Journal of Agric. Food Chem.*, 53: 9329-9336.

Gupta, M.N. and S. Raghava (2008). Smart systems based on polysaccharides, pp. 129-161. *In:* R.L. Reis, N.M. Neves, J.F. Mano, M.E. Gomes, A.P. Marques and H.S Azevedo (Eds.). *Natural-based Polymers for Biomedical Applications*. Elsevier, Woodhead Publishing Ltd., Cambridge, England.

Gupta, R. and S. Dhillon (1993). Characterization of seed storage proteins of lentil (*Lens culinaris* M.), *Ann. Biol. (India)*, 9: 71-78.

Hamaker, B.R., G. Zhang and M. Venkatachalam (2007). Modified carbohydrates with lower glycemic index, pp.198-217. *In:* C.J.K. Henry (Ed.). *Novel Food Ingredients for Weight Control*. CRC Press, Woodhead Publishing Ltd. New York, USA.

Hammond, E.G., L.A. Johnson and P.A. Murphy (2016). Soybean: Grading and marketing, *Reference Module in Food Science*, Elsevier. https://doi.org/10.1016/B978-0-08-100596-5.00174-8

Harding, S.E., M.P. Tombs, G.G. Adams, B.S. Paulsen, K.T. Inngjerdingen and H. Barsett (2017). *An Introduction to Polysaccharide Biotechnology*. CRC Press, New York, USA.

Harmsen, R.A., T.R. Tuveng, S.G. Antonsen, V.G. Eijsink and M. Sørlie (2019). Can we make chitosan by enzymatic deacetylation of chitin? *Molecules*, 24: 3862.

Hassan, E.A., S.M. Fadel and M.L. Hassan (2018). Influence of TEMPO-oxidized NFC on the mechanical, barrier properties and nisin release of hydroxypropyl methylcellulose bioactive films, *Int. Biol. Macromol.*, 113: 616-622.

Hazirah, M.N., M.I.N. Isa and N.M. Sarbon (2016). Effect of xanthan gum on the physical and mechanical properties of gelatin-carboxymethyl cellulose film blends, *Food Packag. Shelf-Life*, 9: 55-63.

Henriksson, M., G. Henriksson, L.A. Berglund and T. Lindström (2007). An environmentally friendly method for enzyme-assisted preparation of microfibrillated cellulose (MFC) nanofibers, *Eur. Polym. J.*, 43: 3434-3441.

Herrero-Martínez, J.M., P.J. Schoenmakers and W.T. Kok (2004). Determination of the amylose–amylopectin ratio of starches by iodine-affinity capillary electrophoresis, *J. Chromatogr. A*, 1053: 227-234.

Houghton, D., M.D. Wilcox, P.I. Chater, I.A. Brownlee, C.J. Seal and J.P. Pearson (2015). Biological activity of alginate and its effect on pancreatic lipase inhibition as a potential treatment for obesity, *Food Hydrocoll.*, 49: 18-24.

Huang, S.W., M.T. Satué-Gracia, E.N. Frankel and J.B. German (1999). Effect of lactoferrin on oxidative stability of corn oil emulsions and liposomes, *J. Agric. Food Chem.*, 47: 1356-1361.

Huiyu, B. (2017). Preparation, structure, properties, and interactions of the PVA, pp. 275-291. *In:* V.K. Thakur, M.K. Thakur and M.R. Kessler (Eds.). *Handbook of Composites from Renewable Materials*, vol. 7. John Wiley & Sons, Massachusetts, USA.

Huopalahti, R., M. Anton, R. López-Fandiño and R. Schade (2007). *Bioactive Egg Compounds*. Springer, Berlin, Germany, 298 p.

Huq, T., S. Salmieri, A. Khan, R.A. Khan, C. Le Tien, B. Riedl and M. Lacroix (2012). Nanocrystalline cellulose (NCC) reinforced alginate based biodegradable nanocomposite film, *Carbohydr. Polym.*, 90: 1757-1763.

Islam, A.M., G.O. Phillips, A. Sljivo, M.J. Snowden and P.A. Williams (1997). A review of recent developments on the regulatory, structural and functional aspects of gum Arabic, *Food Hydrocoll.*, 11: 493-505.

Ito, K., S. Sjölander, S. Sato, R. Movérare, A. Tanaka, L. Söderström and M. Ebisawa (2011). IgE to Gly m 5 and Gly m 6 is associated with severe allergic reactions to soybean in Japanese children, *J. Allergy Clin. Immunol.*, 128: 673.

Jain, S., T. Winuprasith and M. Suphantharika (2019). Design and synthesis of modified and resistant starch-based oil-in-water emulsions, *Food Hydrocoll.*, 89: 153-162.

Jiao, G., G. Yu, J. Zhang and H.S. Ewart (2011). Chemical structures and bioactivities of sulfated polysaccharides from marine algae, *Mar. Drugs*, 9: 196-223.

Juturu, V. (2014). Polyphenols and cardiometabolic syndrome, pp. 1067-1076. *In:* R.R. Watson, V.R. Preedy and S. Zibadi (Eds.). *Polyphenols in Human Health and Disease*, vol. 2. Academic Press, Boston, USA.

Kalit, T.M., S. Kalit and J. Havranek (2010). An overview of researches on cheeses ripening in animal skin, *Mljekarstvo*, 60: 149-155.

Kalman, D.S. (2014). Amino acid composition of an organic brown rice protein concentrate and isolate compared to soy and whey concentrates and isolates, *Foods*, 3: 394-402.

Kamau, S.M., S.C. Cheison, W. Chen, X.M. Liu and R.R. Lu (2010). Alpha-lactalbumin: Its production technologies and bioactive peptides, *Compr. Rev. Food Sci. Food Saf.*, 9: 197-212.

Kameda, T., M. Miyazawa, H. Ono and M. Yoshida (2005). Hydrogen bonding structure and stability of α-chitin studied by 13C solid-state NMR, *Macromol. Biosci.*, 5: 103-106.

Kandemir, N., A. Yemenicioğlu, Ç. Mecitogwlu, Z.S. Elmaci, A. Arslanogwlu, Y. Göksungur and T. Baysal (2005). Production of antimicrobial films by incorporation of partially purified lysozyme into biodegradable films of crude exopolysaccharides obtained from Aureobasidium pullulans fermentation, *Food Technol. Biotechnol.*, 43: 343-350.

Kang, D., H.B. Zhang, Y. Nitta, Y.P. Fang and K. Nishinari (2015). Gellan. *In:* K. Ramawat and J.M. Mérillon (Eds.). *Polysaccharides*. Springer, Cham. https://doi.org/10.1007/978-3-319-03751-6_20-2.

Kaur, N., P. Sharma, S. Jaimni, B.A. Kehinde and S. Kaur (2019). Recent developments in purification techniques and industrial applications for whey valorization: A review, *Chem. Eng. Commun.*, 207: 123-138.

Kelly, P., B.W. Woonton and G.W. Smithers (2009). Improving the sensory quality, shelf-life and functionality of milk, pp. 170-231. *In:* P. Paquin (Ed.). *Functional and Speciality Beverage Technology*. Woodhead Publishing, Boca Raton, USA.

Khalil, H.P.S., T.K. Lai, Y.Y. Tye, S. Rizal, E.W.N. Chong, S.W. Yap *et al.* (2018). A review of extractions of seaweed hydrocolloids: Properties and applications, *Express Polym. Lett.*, 12: 296-317.

Khan, A., R.A. Khan, S. Salmieri, C. Le Tien, B. Riedl, J. Bouchard *et al.* (2012). Mechanical and barrier properties of nanocrystalline cellulose reinforced chitosan-based nanocomposite films, *Carbohydr. Polym.*, 90: 1601-1608.

Khwaldia, K., C. Perez, S. Banon, S. Desobry and J. Hardy (2004). Milk proteins for edible films and coatings, *Crit. Rev. Food Sci. Nutr.*, 44: 239-251.

Kim, S.H., H.K. No and W. Prinyawiwatkul (2007). Effect of molecular weight, type of chitosan, and chitosan solution pH on the shelf-life and quality of coated eggs, *J. Food Sci.*, 72: S044-S048.

Kimoto, T., T. Shibuya and S.Shiobara (1997). Safety studies of a novel starch, pullulan: Chronic toxicity in rats and bacterial mutagenicity, *Food Chem. Toxicol.*, 35: 323-329.

Kinoshita, K., T. Shimogiri, H.R. Ibrahim, M. Tsudzuki, Y. Maeda and Y. Matsuda (2016). Identification of TENP as the gene encoding chicken egg white ovoglobulin G2 and demonstration of its high genetic variability in chickens, *PloS One*, 11: e0159571.

Kong, F. and R.P. Singh (2016). Chemical deterioration and physical instability of foods and beverages, pp. 43-76. *In:* P. Subramaniam (Ed.). *The Stability and Shelf-life of Food.* Woodhead Publishing, Cambridge, USA.

Kontturi, E. (2018). Supramolecular aspects of native cellulose: Fringed-fibrillar model, leveling-off degree of polymerization and production of cellulose nanocrystals, pp. 263-276. *In:* T. Rosenau, A. Potthast and J. Hell (Eds.). *Cellulose Science and Technology: Chemistry, Analysis, and Applications.* John Wiley & Sons, Hoboken, USA.

Kopesky, R. and T. Ruszkay (2006). *Production of Microcrystalline Cellulose.* U.S. Patent. Publication No. 0020126 A1.

Korhonen, H.J. (2011). Bioactive milk proteins, peptides and lipids and other functional components derived from milk and bovine colostrum, pp. 471-511. *In:* M. Saarela (Ed.). *Functional Foods.* Woodhead Publishing, Cambridge, UK.

Koshiyama, I. (1983). Storage proteins of soybean, pp. 427-450. *In:* W. Gottschalk and H.P. Müller (Eds.). *Seed Proteins.* Springer, Dordrecht, Nederlands.

Kosma, P., S. Sjölander, E. Landgren, M.P. Borres and G. Hedlin (2011). Severe reactions after the intake of soy drink in birch pollen-allergic children sensitized to Gly m 4, *Acta Paediatr.*, 100: 305-306.

Kovacs-Nolan, J., M. Phillips and Y. Mine (2005). Advances in the value of eggs and egg components for human health, *J. Agric. Food Chem.*, 53: 8421-8431.

Kowalczewski, P.L., A. Olejnik, W. Białas, I. Rybicka, M. Zielińska-Dawidziak, A. Siger *et al.* (2019). The nutritional value and biological activity of concentrated protein fraction of potato juice, *Nutrients*, 11: 1523.

Kular, H., J. Dean and V. Cook (2018). A case of carrageenan allergy in a pediatric patient, *Ann. Allergy, Asthma Immunol.*, 121: S119.

Kumar, V., R. Bollström, A. Yang, Q. Chen, G. Chen, P. Salminen *et al.* (2014). Comparison of nano-and microfibrillated cellulose films, *Cellulose*, 21: 3443-3456.

Kumar, R., S.K. Chauhan, G. Shinde, V. Subramanian and S. Nadanasabapathi (2018). Whey proteins: A potential ingredient for food industry – A review, *Asian J. Dairy Food Res.*, 37: 283-290.

Kunte, L.A., A. Gennadios, S.L. Cuppett, M.A. Hanna and C.L. Weller (1997). Cast films from soy protein isolates and fractions, *Cereal Chem.*, 74: 115-118.

Kurt, A., O.S. Toker and F. Tornuk (2017). Effect of xanthan and locust bean gum synergistic interaction on characteristics of biodegradable edible film, *Int. J. Biol. Macromol.*, 102: 1035-1044.

Kwon, J.H., S.H. Lee, N. Ayrilmis and T.H. Han (2015). Tensile shear strength of wood bonded with urea–formaldehyde with different amounts of microfibrillated cellulose, *Int. J. Adhes. Adhes.*, 60: 88-91.

Labuschagne, M.T., N. Geleta and G. Osthoff (2007). The influence of environment on starch content and amylose to amylopectin ratio in wheat, *Starch-Stärke*, 59: 234-238.

Laca, A., B. Paredes, M. Rendueles and M. Díaz (2014). Egg yolk granules: Separation, characteristics and applications in food industry, *LWT-Food Sci. Technol.*, 59: 1-5.

Ladjal-Ettoumi, Y., H. Boudries, M. Chibane and A. Romero (2016). Pea, chickpea and lentil protein isolates: Physicochemical characterization and emulsifying properties, *Food Biophys.*, 11: 43-51.

Lahaye, M. (2001). Developments on gelling algal galactans, their structure and physico-chemistry, *J. Appl. Phycol.*, 13: 173-184.

Lascol, M., S. Bourgeois, F. Guillière, M. Hangouët, G. Raffin, P. Marote *et al.* (2016). Pectin gelation with chlorhexidine: Physico-chemical studies in dilute solutions, *Carbohydr. Polym.*, 150: 159-165.

Layman, D.K., B. Lönnerdal and J.D. Fernstrom (2018). Applications for α-lactalbumin in human nutrition, *Nutr. Rev.*, 76: 444-460.

Leathers, T.D. (2003). Biotechnological production and applications of pullulan, *Appl. Microbiol. Biotechnol.*, 62: 468-473.

Leleu, S., L. Herman, M. Heyndrickx, K. De Reu, C.W. Michiels, J. De Baerdemaeker and W. Messens (2011). Effects on *Salmonella* shell contamination and trans-shell penetration of coating hens' eggs with chitosan, *Int. J. Food Microbiol.*, 145: 43-48.

Li, K., S. Jin, H. Chen, J. He and J. Li (2017). A high-performance soy protein isolate-based nanocomposite film modified with microcrystalline cellulose and Cu and Zn nanoclusters, *Polymers*, 9: 167.

Li, X., H. Zhang, Q. Jin and Z. Cai (2018). Contribution of arabinogalactan protein to the stabilization of single-walled carbon nanotubes in aqueous solution of gum Arabic, *Food Hydrocoll.*, 78: 55-61.

Liceaga-Gesualdo, A., E.C.Y. Li-Chan and B.J. Skura (2001). Antimicrobial effect of lactoferrin digest on spores of a Penicillium sp. isolated from bottled water, *Food Res. Int.*, 34: 501-506.

Liu, P., L. Yu, X. Wang, D. Li, L. Chen and X. Li (2010). Glass transition temperature of starches with different amylose/amylopectin ratios, *J. Cereal Sci.*, 51: 388-391.

Liu, Q. (2015). Understanding starches and their role in foods, pp. 309-355. *In:* S.W. Cui (Ed.). *Food Carbohydrates: Chemistry, Physical Properties and Applications.* CRC Press, Taylor and Francis Group, New York, USA.

Liu, X., J. Zhang and K.Y. Zhu (2019). Chitin in arthropods: Biosynthesis, modification and methabolism, pp. 174-175. *In:* Q. Yang and T. Fukamizo (Eds.). *Targeting Chitin-containing Organisms.* Springer. Singapore.

Løkra, S. and K.O. Strætkvern (2009). Industrial proteins from potato juice: A review, *Food*, 3: 88-95.

López-Torrejón, G., G. Salcedo, M. Martín-Esteban, A. Díaz-Perales, C.Y. Pascual and R. Sánchez-Monge (2003). Len c 1, a major allergen and vicilin from lentil seeds: Protein isolation and cDNA cloning, *J. Allergy Clin. Immunol.*, 112: 1208-1215.

Lu, M., Y. Jin, R. Cerny, B. Ballmer-Weber and R.E. Goodman (2018). Combining 2-DE immunoblots and mass spectrometry to identify putative soybean (*Glycine max*) allergens, *Food Chem. Toxicol.*, 116: 207-215.

Mahendran, T., P.A. Williams, G.O. Phillips, S. Al-Assaf and T.C. Baldwin (2008). New insights into the structural characteristics of the Arabinogalactan – Protein (AGP) fraction of Gum Arabic, *J. Agric. Food Chem.*, 56: 9269-9276.

Mahmood, K., H. Kamilah, P.L. Shang, S. Sulaiman and F. Ariffin (2017). A review: Interaction of starch/non-starch hydrocolloid blending and the recent food applications, *Food Biosci.*, 19: 110-120.

Mamone, G., G. Picariello, A. Ramondo, M.A. Nicolai and P. Ferranti (2019). Production, digestibility and allergenicity of hemp (*Cannabis sativa* L.) protein isolates, *Food Res. Int.*, 115: 562-571.

Malinowska-Pańczyk, E., H. Staroszczyk, K. Gottfried, I. Kołodziejska and A. Wojtasz-Pająk (2015). Antimicrobial properties of chitosan solutions, chitosan films and gelatin-chitosan films, *Polimery*, 60: 11-12.

Mao, R., J. Tang and B.G. Swanson (2000). Texture properties of high and low acyl mixed gellan gels, *Carbohydr. Polym.*, 41: 331-338.

Marchetti, M., F. Superti, M.G. Ammendolia, P. Rossi, P. Valenti and L. Seganti (1999). Inhibition of poliovirus type 1 infection by iron-, manganese- and zinc-saturated lactoferrin, *Med. Microbiol. Immunol.*, 187: 199-204.

Marić, M., A.N. Grassino, Z. Zhu, F.J. Barba, M. Brnčić and S.R. Brnčić (2018). An overview of the traditional and innovative approaches for pectin extraction from plant food wastes and byproducts: Ultrasound-, microwaves- and enzyme-assisted extraction, *Trends in Food Sci. Technol.*, 76: 28-37.

Mariod, A.A. and H. Fadul (2013). Gelatin, source, extraction and industrial applications, *Acta Sci. Pol. Technol. Aliment.*, 12: 135-147.

Martens, B.M., W.J. Gerrits, E.M. Bruininx and H.A. Schols (2018). Amylopectin structure and crystallinity explains variation in digestion kinetics of starches across botanic sources in an *in vitro* pig model, *J. Anim. Sci. Biotechnol.*, 9: 91.

Martino, J.V., J. Van Limbergen and L.E. Cahill (2017). The role of carrageenan and carboxymethylcellulose in the development of intestinal inflammation, *Front. Pediatr.*, 5: 96.

Matsuda, T., M. Sugiyama, R. Nakamura and S. Torii (1988). Purification and properties of an allergenic protein in rice grain, *Agr. Biol. Chem.*, 52: 1465-1470.

Matsushima, N., G.I. Danno, H. Takezawa and Y. Izumi (1997). Three-dimensional structure of maize α-zein proteins studied by small-angle X-ray scattering, *Biochim. Biophys. Acta Protein Struct. Mol. Enzymol.*, 1339: 14-22.

Mecitoğlu, Ç. and A. Yemenicioğlu (2007). Partial purification and preparation of bovine lactoperoxidase and characterization of kinetic properties of its immobilized form incorporated into cross-linked alginate films, *Food Chem.*, 104: 726-733.

Meena, G.S., A.K. Singh, N.R. Panjagari and S. Arora (2017). Milk protein concentrates: Opportunities and challenges, *J. Food Sci. Tech. Mys.*, 54: 3010-3024.

Melo, C.D., P.S. Garcia, M.V.E. Grossmann, F. Yamashita, L.H. Dall'Antônia and S. Mali (2011). Properties of extruded xanthan-starch-clay nanocomposite films, *Braz. Arch. Biol. Technol.*, 54: 1223-1333.

Meshkani, S.M., S.A. Mortazavi, E. Milani, M. Mokhtarian and L. Sadeghian (2011). Evaluation of mechanical and optical properties of edible film from chickpea protein isolate (*Cicer arietinum* L.) containing thyme essential oil with response surface method (RSM), *Innov. Food Sci. Technol.*, 2: 25-36.

Meyers, M.A. and P.Y. Chen (2014). *Biological Materials Science: Biological Materials, Bioinspired Materials, and Biomaterials*. Cambridge University Press, Cambridge, UK.

Mi, Y., Y.X. Chin, W.X. Cao, Y.G. Chang, P.E. Lim, C.H. Xue and Q.J. Tang (2020). Native κ-carrageenan induced-colitis is related to host intestinal microecology, *Int. J. Biol. Macromol.*, 147: 284-294.

Miles, M.J., V.J. Morris, P.D. Orford and S.G. Ring (1985). The roles of amylose and amylopectin in the gelation and retrogradation of starch, *Carbohydr. Res.*, 135: 271-281.

Mills, E.N.C. and A.S. Tatham (2003). Allergens, pp. 143-150. *In:* B. Caballero, L.C. Trugo and P.M. Finglas (Eds.). *Encyclopedia of Food Sciences and Nutrition.* Academic Press, Amsterdam, The Netherlands.

Min, S., L.J. Harris, J.H. Han and J.M. Krochta (2005a). *Listeria monocytogenes* inhibition by whey protein films and coatings incorporating lysozyme, *J. Food Protect.*, 68: 2317-2325.

Min, S.L., J. Harris and J.M. Krochta (2005b). Antimicrobial effects of lactoferrin, lysozyme, and the lactoperoxidase system and edible whey protein films incorporating the lactoperoxidase system against *Salmonella enterica* and *Escherichia coli* O157: H7, *J. Food Sci.*, 70: M332-M338.

Minke, R.A.M. and J. Blackwell (1978). The structure of α-chitin, *J. Mol. Biol.*, 120: 167-181.

Mitsuhashi, J. (2010). The future use of insects as human food, pp. 115-122. *In: Forest Insects as Food: Humans Bite Back.* FAO of the United Nations Regional Office for Asia and the Pacific, Bangkok.

Miyamoto, A. and K.C. Chang (1992). Extraction and physicochemical characterization of pectin from sunflower head residues, *J. Food Sci.*, 57: 1439-1443.

Morris, E.R., K. Nishinari and M. Rinaudo (2012). Gelation of gellan – A review, *Food Hydrocoll.*, 28: 373-411.

Mostafavi, F.S. and D. Zaeim (2020). Agar-based edible films for food packaging applications – A review, *Int. J. Biol. Macromol.*, 159: 1165-1176.

Muhammad, K.M.L., F. Ariffin, H.K.B. Abd Razak and P.D.S. Sulaiman (2016). Review of fish gelatin extraction, properties and packaging applications, *Food Sci. Qual. Manag.*, 56: 47-59.

Nagy, K., A.M. Pilbat, G. Groma, B. Szalontai and F.J. Cuisinier (2010). Casein aggregates built step-by-step on charged polyelectrolyte film surfaces are calcium phosphate-cemented, *J. Biol. Chem.*, 285: 38811-38817.

Ninan, G., J. Joseph and Z. Abubacker (2010). Physical, mechanical, and barrier properties of carp and mammalian skin gelatin films, *J. Food Sci.*, 75: E620-E626.

Nishinari, K., Y. Fang, S. Guo and G.O. Phillips (2014). Soy proteins: A review on composition, aggregation and emulsification, *Food Hydrocoll.*, 39: 301-318.

No, H.K., N.Y. Park, S.H. Lee and S.P. Meyers (2002). Antibacterial activity of chitosans and chitosan oligomers with different molecular weights, *Int. J. Food Microbiol.*, 74: 65-72.

OECD (Organisation for Economic Co-operation and Development) (2012). Revised consensus document on compositional considerations for new varieties of soybean [*Glycine max* (L.) Merr.]: Key food and feed nutrients, antinutrients, toxicants and allergens, ENV/JM/MONO, 24. http://www.oecd.org/ officialdocuments /publicdisplay documentpdf/?cote=env/jm/mono(2012)24&doclanguage=en

Oehme, D.P., M.T. Downton, M.S. Doblin, J. Wagner, M.J. Gidley and A. Bacic (2015). Unique aspects of the structure and dynamics of elementary Iβ cellulose microfibrils revealed by computational simulations, *Plant Physiol.*, 168: 3-17.

Ogawa, T., H. Tsuji, N. Bando, K. Kitamura, Y.L. Zhu, H. Hirano and K. Nishikawa (1993). Identification of the soybean allergenic protein, Gly m Bd 30K, with the soybean seed 34-kDa oil-body-associated protein, *Biosci. Biotech. Bioch.*, 57: 1030-1033.

Oomah, D., A. Patras, A. Rawson, N. Singh and R. Compos-Vega (2011). Chemistry of pulses, pp. 9-55. *In:* B. Tiwari, A. Gowen and B. McKenna (Eds.). *Pulse Foods: Processing, Quality and Nutraceutical Applications.* Academic Press, London, UK.

O'Regan, J., M.P. Ennis and D.M. Mulvihill (2009). Milk proteins, pp. 298-358. *In:* G.O. Phillips and P.A. Williams (Eds.). *Handbook of Hydrocolloids*. Woodhead Publishing. Cambridge, UK.

Othman, S.H., N.A. Majid, I.S.M.A. Tawakkal, R.K. Basha, N. Nordin, I. Shapi and R. Ahmad (2019). Tapioca starch films reinforced with microcrystalline cellulose for potential food packaging application, *Food Sci. Technol.*, 39: 605-612.

Pacheco, M.T. M. Villamiel, R. Moreno and F.J. Moreno (2019). Structural and rheological properties of pectins extracted from industrial sugar beet byproducts, *Molecules*, 24: 392.

Padil, V.V.T., C. Senan and M. Černík (2019). 'Green' polymeric electrospun fibers based on tree-gum hydrocolloids: Fabrication, characterization and applications, pp. 127-172. *In:* V. Grumezescu and A.M. Grumezescu (Eds.). *Materials for Biomedical Engineering*. Elsevier, Amsterdam, The Netherlands.

Pakkanen, R. and J. Aalto (1997). Growth factors and antimicrobial factors of bovine colostrum, *Int. Dairy J.*, 7: 285-297.

Panthi, R.R., K.N. Jordan, A.L. Kelly and J.D. Sheehan (2017). Selection and treatment of milk for cheesemaking, pp. 23-50. *In:* P.L. McSweeney, P.F. Fox, P.D. Cotter and D.W. Everett (Eds.). *Cheese: Chemistry, Physics & Microbiology*. Academic Press, London, UK.

Paolicelli, P., S. Petralito, G. Varani, M. Nardoni, S. Pacelli, L. Di Muzio *et al.* (2018). Effect of glycerol on the physical and mechanical properties of thin gellan gum films for oral drug delivery, *Int. J. Pharm.*, 547: 226-234.

Paraman, I., N.S. Hettiarachchy and C. Schaefer (2008). Preparation of rice endosperm protein isolate by alkali extraction, *Cereal Chem.*, 85: 76-81.

Park, I.M., A.M. Ibáñez, F. Zhong and C.F. Shoemaker (2007). Gelatinization and pasting properties of waxy and non-waxy rice starches, *Starch-Stärke*, 59: 388-396.

Park, J.J., J.E. Kim, W.B. Yun, M.L. Lee, J.Y. Choi, B.R. Song *et al.* (2017). Hypolipidemic and hypoinsulinemic effects of dietary fiber from agar in C57BL/6N mice fed a high-fat diet, *J. Life Sci.*, 27: 937-944.

Patil, S.P., P.V. Niphadkar and M.M. Bapat (2001). Chickpea: A major food allergen in the Indian subcontinent and its clinical and immunochemical correlation, *Ann. Allergy, Asthma Immunol.*, 87: 140-145.

Perez-Gago, M.B. and J.M. Krochta (2000). Drying temperature effect on water vapor permeability and mechanical properties of whey protein-lipid emulsion films, *J. Agric. Food Chem.*, 48: 2687-2692.

Pinkert, A., K.N. Marsh, S. Pang and M.P. Staiger (2009). Ionic liquids and their interaction with cellulose, *Chemical Rev.*, 109: 6712-6728.

Porse, H. and B. Rudolph (2017). The seaweed hydrocolloid industry: 2016 updates, requirements, and outlook, *J. Appl. Phycol.*, 29: 2187-2200.

Psomiadou, E., I. Arvanitoyannis and N. Yamamoto (1996). Edible films made from natural resources; microcrystalline cellulose (MCC), methylcellulose (MC) and corn starch and polyols – Part 2, *Carbohydr. Polym.*, 31: 193-204.

Qi, P.X. (2007). Studies of casein micelle structure: The past and the present, *Le Lait*, 87: 363-383.

Raafat, D., K. von Bargen, A. Haas and H.G. Sahl (2008). Chitosan as an antibacterial compound: Insights into its mode of action, *Appl. Environ. Microbiol.*, 74: 3764-3773.

Raghow, R. (2013). Connective tissues of the subendothelium, pp. 43-69. *In:* M.A. Creager, J.A. Beckman and J. Loscalzo (Eds.). *Vascular Medicine: A Companion to Braunwald's Heart Disease*, Part-1. Elsevier, Saunders, Philadelphia, USA.

Rajagopalan, N. and M. Cheryan (1991). Total protein isolate from milk by ultrafiltration: Factors affecting product composition, *J. Dairy Sci.*, 74: 2435-2439.

Raman, M. and M. Doble (2015). κ-carrageenan from marine red algae, *Kappaphycus alvarezii* – A functional food to prevent colon carcinogenesis, *J. Funct. Foods*, 15: 354-364.

Ramos, M., A. Valdes, A. Beltran and M.C. Garrigós (2016). Gelatin-based films and coatings for food packaging applications, *Coatings*, 6: 41.

Ramos, Ó.L., J.C. Fernandes, S.I. Silva, M.E. Pintado and F.X. Malcata (2012). Edible films and coatings from whey proteins: A review on formulation, and on mechanical and bioactive properties, *Crit. Rev. Food Sci. Nutr.*, 52: 533-552.

Ramya, R., J. Venkatesan, S.K. Kim and P.N. Sudha (2012). Biomedical applications of chitosan: An overview, *J. Biomater. Tiss. Eng.*, 2: 100-111.

Razavi, S.M. (2019). *Emerging Natural Hydrocolloids: Rheology and Functions*. John Wiley & Sons, Hoboken, New Jersey, USA.

Recio, I., F.J. Moreno and R. López-Fandiño (2009). Glycosylated dairy components: Their roles in nature and ways to make use of their biofunctionality in dairy products, pp. 170-211. *In:* M. Corredig (Ed.). *Dairy-Derived Ingredients*. Woodhead Publishing, Cambridge, UK.

Rojas-Graü, M.A., R. Soliva-Fortuny and O. Martín-Belloso (2009). Edible coatings to incorporate active ingredients to fresh-cut fruits: A review, *Trends in Food Sci. Tech.*, 20: 438-447.

Roland, W.S., L. Pouvreau, J. Curran, F. van de Velde and P.M. de Kok (2017). Flavor aspects of pulse ingredients, *Cereal Chem.*, 94: 58-65.

Rongpipi, S., D. Ye, E.D. Gomez and E.W. Gomez (2019). Progress and opportunities in the characterization of cellulose – An important regulator of cell wall growth and mechanics, *Front. in Plant Sci.*, 9: 1894.

Rosenau, T., M. Khotimchenko, V. Tiasto, A. Kalitnik, M. Begun, R. Khotimchenko *et al.* (2020). Antitumor potential of carrageenans from marine red algae, *Carbohydr. Polym.*, 246: 116568.

Roszkowska, A., M. Pawlicka, A. Mroczek, K. Bałabuszek and B. Nieradko-Iwanicka (2019). Non-celiac gluten sensitivity: A review, *Medicina*, 55: 222.

Saha, D. and S. Bhattacharya (2010). Hydrocolloids as thickening and gelling agents in food: A critical review, *J. Food Sci. Tech. MYS*, 47: 587-597.

Sahasrabudhe, N.M., M. Beukema, L. Tian, B. Troost, J. Scholte, E. Bruininx *et al.* (2018). Dietary fiber pectin directly blocks toll-like receptor 2-1 and prevents doxorubicin-induced Ileitis, *Front. Immunol.*, 9: 383.

Sahin, I., B. Bilir, S. Ali, K. Sahin and O. Kucuk (2019). Soy isoflavones in integrative oncology: Increased efficacy and decreased toxicity of cancer therapy, *Integr. Cancer Ther.*, 18: 1-11.

Saleh, A.S., P. Wang, N. Wang, L. Yang and Z. Xiao (2019). Brown rice versus white rice: Nutritional quality, potential health benefits, development of food products, and preservation technologies, *Compr. Rev. Food Sci. Food Saf.*, 18: 1070-1096.

Salehi, F. (2020). Edible coating of fruits and vegetables using natural gums: A review, *Int. J. Fruit Sci.*, 20: S570-S589.

Samaraweera, H., W.G. Zhang, E.J. Lee and D.U. Ahn (2011). Egg yolk phosvitin and functional
phosphopeptides, *J. Food Sci.*, 76: R143-R150.

Sanchez-Monge, R., G. Lopez-Torrejón, C.Y. Pascual, J. Varela, M. Martin-Esteban and G. Salcedo (2004). Vicilin and convicilin are potential major allergens from pea, *Clin. Exp. Allergy*, 34: 1747-1753.

Sanjivkumar, M., M.N. Chandran, A.M. Suganya and G. Immanuel (2020). Investigation on bio-properties and *in-vivo* antioxidant potential of carrageenans against alloxan induced oxidative stress in Wistar albino rats, *Int. J. Biol. Macromol.*, 151: 650-662.

Santos, R. and R.A. Melo (2018). Global shortage of technical agars: Back to basics (resource management), *J. Appl. Phycol.*, 30: 2463-2473.

Satoh, R., I. Tsuge, R. Tokuda and R. Teshima (2019). Analysis of the distribution of rice allergens in brown rice grains and of the allergenicity of products containing rice bran, *Food Chem.*, 276: 761-767.

Sattar, K.M.A., S. Nakamura, M. Ogawa, E. Akita, H. Azakami and A. Kato (2000). Bactericidal action of egg yolk phosvitin against *Escherichia coli* under thermal stress, *J. Agric. Food Chem.*, 48: 1503-1506.

Satué-Gracia, M.T., E.N. Frankel, N. Rangavajhyala and J.B. German (2000). Lactoferrin in infant formulas: Effect on oxidation, *J. Agric. Food Chem.*, 48: 4984-4990.

Schäfer, D., M. Reinelt, A. Stäbler and M. Schmid (2018). Mechanical and barrier properties of potato protein isolate-based films, *Coatings*, 8: 58.

Schleip, R., T.W. Findley, L. Chaitow and P.A. Huijing (2012). Book Review of Fascia: The Tensional Network of the Human Body, *Elsevier Health Sciences*, London, UK.

Schmidt, M.H., M. Raulf-Heimsoth and A. Posch (2002). Evaluation of patatin as a major cross-reactive allergen in latex-induced potato allergy, *Ann. Allergy Asthma Im.*, 89: 613-618.

Seifu, E., E.M. Buys and E.F. Donkin (2005). Significance of the lactoperoxidase system in the dairy industry and its potential applications: A review, *Trends in Food Sci. Technol.*, 16: 137-154.

Selling, G.W., K.K. Woods, D. Sessa and A. Biswas (2008). Electrospun zein fibers using glutaraldehyde as the crosslinking reagent: Effect of time and temperature, *Macromol. Chem. Phys.*, 209: 1003-1011.

Seo, H.P., C.W. Son, C.H. Chung, D.I. Jung, S.K. Kim, R.A. Gross *et al.* (2004). Production of high molecular weight pullulan by Aureobasidium pullulans HP-2001 with soybean pomace as a nitrogen source, *Bioresour. Technol.*, 95: 293-299.

Seppälä, U., H. Alenius, K. Turjanmaa, T. Reunala, T. Palosuo and N. Kalkkinen (1999). Identification of patatin as a novel allergen for children with positive skin prick test responses to raw potato, *J. Allergy Clin. Immunol.*, 103: 165-171.

Shendurse, A.M., G. Gopikrishna and A.C. Patel (2018). Milk protein based edible films and coatings – Preparation, properties and food applications, *J. Nutr. Heal Food Eng.*, 8: 219-226.

Shewry, P.R., A.S. Tatham, J. Forde, M. Kreis and B.J. Miflin (1986). The classification and nomenclature of wheat gluten proteins: A reassessment, *J. Cereal Sci.*, 4: 97-106.

Shewry, P.R. and A.S. Tatham (1990). The prolamin storage proteins of cereal seeds: Structure and evolution, *Biochem. J.*, 267: 1-12.

Shewry, P.R., J.A. Napier and A.S. Tatham (1995). Seed storage proteins: Structures and biosynthesis, *Plant Cell*, 7: 945.

Shewry, P.R., N.G. Halford, P.S. Belton and A.S. Tatham (2002). The structure and properties of gluten: An elastic protein from wheat grain, *Philos. T. R. Soc. B*, 357: 133-142.

Shewry, P.R. and N.G. Halford (2003). The prolamin storage proteins of sorghum and millets, *AFRIPRO*, Pretoria, South Africa. http://www.afripro.org.uk/papers/Paper03Shewry.pdf.

Shi, H., L. Yu, Y. Shi, J. Lu, H. Teng, Y. Zhou and L. Sun (2017). Structural characterization of a rhamnogalacturonan I domain from ginseng and its inhibitory effect on galectin-3, *Molecules*, 22: 1016.

Shih, F.F., K.W. Daigle and E.T. Champagne (2011). Effect of rice wax on water vapour permeability and sorption properties of edible pullulan films, *Food Chem.*, 127: 118-121.

Shingel, K.I. (2004). Current knowledge on biosynthesis, biological activity, and chemical modification of the exopolysaccharide, pullulan, *Carbohydr. Res.*, 339: 447-460.

Shukla, R. and M. Cheryan (2001). Zein: The industrial protein from corn, *Ind. Crops Prod.*, 13: 171-192.

Shull, J.M., J.J. Watterson and A.W. Kirleis (1991). Proposed nomenclature for the alcohol-soluble proteins (*kafirins*) of Sorghum bicolor (*L. Moench*) based on molecular weight, solubility, and structure, *J. Agric. Food Chem.*, 39: 83-87.

Siegel, R.C., S.R. Pinnell and G.R. Martin (1970). Cross-linking of collagen and elastin: Properties of lysyl oxidase, *Biochemistry*, 9: 4486-4492.

Sikorski, P., R. Hori and M. Wada (2009). Revisit of α-chitin crystal structure using high resolution X-ray diffraction data, *Biomacromolecules*, 10: 1100-1105.

Silva Jùnior, S., N.P. Stamford, M.A.B. Lima, T.M.S. Arnaud, M. Pintado and B.F. Sarmento (2014). Characterization and inhibitory activity of chitosan on hyphae growth and morphology of *Botrytis cinerea* plant pathogen, *Int. J. Appl. Res. Nat. Prod.*, 7: 31-38.

Singh, N., J. Singh, L. Kaur, N.S. Sodhi and B.S. Gill (2003). Morphological, thermal and rheological properties of starches from different botanical sources, *Food Chem.*, 81: 219-231.

Singh, J., R. Colussi, O. J. McCarthy and L. Kaur (2016). Potato starch and its modification, pp. 195-247. *In:* J. Singh and L. Kaur (Eds.). *Advances in Potato Chemistry and Technology*. Academic Press, San Francisco, USA.

Slavutsky, A.M. and M.A. Bertuzzi (2014). Water barrier properties of starch films reinforced with cellulose nanocrystals obtained from sugarcane bagasse, *Carbohydr. Polym.*, 110: 53-61.

Srivastava, L.M. (2002). *Plant Growth and Development: Hormones and Environment*. Academic Press, New York, USA.

Stadelman, W.J. (2003). Structure and composition, pp. 2005-2009. *In:* B. Caballero, L.C. Trugo and P.M. Finglas (Eds.). *Encyclopedia of Food Sciences and Nutrition*. Academic Press, Amsterdam, The Netherlands.

Stanley, N. (1987). Production, properties and uses of carrageenan, pp. 116-146. *In:* D.J. McHugh (Ed.). *Production and Utilization of Products from Commercial Seaweeds*. FAO Fisheries Technical Paper, 288 (No. 589.45 F36), Rome, Italy.

Steel, C.J., M.G.V. Leoro, M. Schmiele, R.E. Ferreira and Y.K. Chang (2012). Thermoplastic extrusion in food processing, pp. 272- 290. *In:* A. El-Sonbati (Ed.). *Thermoplastic Elastomers*. InTech, Rijeka, Croatia.

Stelwagen, K. (2016). Mammary gland, milk biosynthesis and secretion: Milk protein, pp. 359-367. *In: Encyclopedia of Food Sciences*, 2nd ed., Elsevier.

Strixner, T. and U. Kulozik (2011). Egg proteins, pp. 150-209. *In:* G.O. Phillips and P.A. Williams (Eds.). *Handbook of Food Proteins*, Woodhead Publication Series, Oxford, UK.

Suderman, N., M.I.N. Isa and N.M. Sarbon (2018). Characterization on the mechanical and physical properties of chicken skin gelatin films in comparison to mammalian gelatin films. *In: IOP Conference Series: Materials Science and Engineering*, IOP Publishing, 440: 012033.

Superti, F., M.G. Ammendolia, P. Valenti and L. Seganti (1997). Antirotaviral activity of milk proteins: Lactoferrin prevents rotavirus infection in the enterocyte-like cell line HT-29, *Med. Microbiol. Immunol.*, 186: 83-91.

Suurs, P. and S. Barbut (2020). Collagen use for co-extruded sausage casings – A review, *Trends in Food Sci. Technol.*, 102: 91-101.

Synowiecki, J. and N.A. Al-Khateeb (2003). Production, properties, and some new applications of chitin and its derivatives, *Crit. Rev. Food Sci. Nutr.*, 43: 145-171.

Tan, H., W. Chen, Q. Liu, G. Yang and K. Li (2018). Pectin oligosaccharides ameliorate colon cancer by regulating oxidative stress- and inflammation-activated signaling pathways, *Front. Immunol.*, 9: 1504.

Tantala, J., K. Thumanu and C. Rachtanapun (2019). An assessment of antibacterial mode of action of chitosan on *Listeria innocua* cells using real-time HATR-FTIR spectroscopy, *Int. J. Biol. Macromol.*, 135: 386-393.

Tarlo, S.M., J. Dolovich and C. Listgarten (1995). Anaphylaxis to carrageenan: A pseudo-latex allergy, *J. Allergy Clin. Immunol.*, 95: 933-936.

Taski-Ajdukovic, K., V. Djordjevic, M. Vidic and M. Vujakovic (2010). Subunit composition of seed storage proteins in high-protein soybean genotypes, *Pesqui. Agropecu. Bras.*, 45: 721-729.

Thakur, V.K. and M.K. Thakur (2014). Processing and characterization of natural cellulose fibers/thermoset polymer composites, *Carbohydr. Polym.*, 109: 102-117.

This, H. (2018). Who discovered the gluten and who discovered its production by lixiviation? *Notes Académiques de l'Académie d'Agriculture de France (N3AF)*, 3: 1-11. Tikhonov, V.E., E.A. Stepnova, V.G. Babak, I.A. Yamskov, J. Palma-Guerrero, H.B. Jansson *et al.* (2006). Bactericidal and antifungal activities of a low molecular weight chitosan and its N-/2(3)-(dodec-2-enyl) succinoyl/-derivatives, *Carbohydr. Polym.*, 64: 66-72.

Tolaimate, A., J. Desbrieres, M. Rhazi, A. Alagui, M. Vincendon and P. Vottero (2000). On the influence of deacetylation process on the physicochemical characteristics of chitosan from squid chitin, *Polymer*, 41: 2463-2469.

Trovatti, E., S.C. Fernandes, L. Rubatat, D. da Silva Perez, C.S. Freire, A.J., Silvestre and C.P. Neto (2012). Pullulan-nanofibrillated cellulose composite films with improved thermal and mechanical properties, *Compos. Sci. Technol.*, 72: 1556-1561.

Tsai, G.J. and W.H. Su (1999). Antibacterial activity of shrimp chitosan against *Escherichia coli*, *J. Food Prot.*, 62: 239-243.

Tuinier, R. and C.G. De Kruif (2002). Stability of casein micelles in milk, *J. Chem. Phys.*, 117: 1290-1295.

Ünalan, İ.U., F. Korel and A. Yemenicioğlu (2011). Active packaging of ground beef patties by edible zein films incorporated with partially purified lysozyme and Na$_2$EDTA, *Int. J. Food Sci. Technol.*, 46: 1289-1295.

UNCTAD (United Nations Conference on Trade and Development) (2018). *Gum Arabic: Growing Demand Means New Opportunities for African Producers.* http://unctad.org/en/Pages/SUC/Commodities/SUC-Commodities-at-a-Glance.aspx (accessed 12 March, 2018).

Usui, Y., M. Nakase, H. Hotta, A. Urisu, N. Aoki, K. Kitajima and T. Matsuda (2001). A 33-kDa allergen from rice (*Oryza sativa* L. *Japonica*) cDNA cloning, expression, and identification as a novel glyoxalase I, *J. Biol. Chem.*, 276: 11376-11381.

Valverde, D., A. Laca, L.N. Estrada, B. Paredes, M. Rendueles and M. Díaz (2016). Egg yolk and egg yolk fractions as key ingredient for the development of a new type of gels, *Int. J. Gastron. Food Sci.*, 3: 30-37.

Van den Broek, L.A., R.J. Knoop, F.H. Kappen and C.G. Boeriu (2015). Chitosan films and blends for packaging material, *Carbohydr. Polym.*, 116: 237-242.

Varnam, A. and J.P. Sutherland (2001). *Milk and Milk Products: Technology, Chemistry and Microbiology*, vol. 1. Aspen Publishers, Inc. Gaithersburg, Maryland, USA.

Vartak, N.B., C.C. Lin, J.M. Cleary, M.J. Fagan and M.H. Saier Jr. (1995). Glucose metabolism in *Sphingomonas elodea*: Pathway engineering via construction of a glucose-6-phosphate dehydrogenase insertion mutant, *Microbiology*, 141: 2339-2350.

Wang, C., F. Xu, D. Li and M. Zhang (2016). Physico-chemical and structural properties of four rice bran protein fractions based on the multiple solvent extraction method, *Czech J. Food Sci.*, 33: 283-291.

Wang, G. and M. Guo. Manufacturing technologies of whey protein products, pp. 13-37. *In:* M. Guo (Ed.). *Whey Protein Production, Chemistry, Functionality, and Applications.* John Wiley and Sons, INC., Hoboken, USA.

Wang, H., L.A. Johnson and T. Wang (2004). Preparation of soy protein concentrate and isolate from extruded-expelled soybean meals, *J. Am. Oil Chem. Soc.,* 81: 713-717.

Wang, H.J., S.J. Gong, Z.X. Lin, J.X. Fu, S.T. Xue, J.C. Huang and J.Y. Wang (2007). *In vivo* biocompatibility and mechanical properties of porous zein scaffolds, *Biomaterials,* 28: 3952-3964.

Watanabe, M., J. Miyakawa, Z. Ikezawa, Y. Suzuki, T. Hirao, T. Yoshizawa and S. Arai (1990). Production of hypoallergenic rice by enzymatic decomposition of constituent proteins, *J. Food Sci.,* 55: 781-783.

Watanabe, M. (1993). Hypoallergenic rice as a physiologically functional food, *Trends in Food Sci. Tech.,* 4: 125-128.

Watkins, C.B. (2017). Postharvest physiology of edible plant tissues, pp. 1017-1085. *In:* S. Damodaran and K.L. Parkin (Eds.). *Fennema's Food Chemistry,* CRC Press, Boca Raton, USA.

Wieser, H. (2007). Chemistry of gluten proteins, *Food Microbiol.,* 24: 115-119.

Williams, P.A., C. Sayers, C. Viebke, C. Senan, J. Mazoyer and P. Boulenguer (2005). Elucidation of the emulsification properties of sugar beet pectin, *J. Agric. Food Chem.,* 53: 3592-3597.

Wilson, S., K. Blaschek and E.G. De Mejia (2005). Allergenic proteins in soybean: Processing and reduction of P34 allergenicity, *Nutr. Rev.,* 63: 47-58.

Wood, J.A. and M.A. Grusak (2007). Nutritional value of chickpea, pp. 101-142. *In:* S.S. Yadav and W. Chen (Eds.). *Chickpea Breeding and Management,* CABI, Cromwell Press, Trowbridge, UK.

Wurzburg, O.B. (1975). Starch in the food industry, pp. 361-395. *In:* T.E. Furia (Ed.). *CRC Handbook of Food Additives,* CRC Press, Boca Raton, USA.

Xiao, Q., L.T. Lim and Q. Tong (2012). Properties of pullulan-based blend films as affected by alginate content and relative humidity, *Carbohydr. Polym.,* 87: 227-234.

Xie, F., L. Yu, B. Su, P. Liu, J. Wang, H. Liu and L. Chen (2009). Rheological properties of starches with different amylose/amylopectin ratios, *J. Cereal Sci.,* 49: 371-377.

Xu, B.J., S.H. Yuan and S.K.C. Chang (2007a). Comparative analyses of phenolic composition, antioxidant capacity, and color of cool season legumes and other selected food legumes, *J. Food Sci.,* 72: 167-177.

Xu, W., N. Reddy and Y. Yang (2007b). An acidic method of zein extraction from DDGS, *J. Agric. Food Chem.,* 55: 6279-6284.

Yadav, M., P. Goswami, K. Paritosh, M. Kumar, N. Pareek and V. Vivekanand (2019). Seafood waste: A source for preparation of commercially employable chitin/chitosan materials, *Bioresour. Bioprocess.,* 6: 1-20.

Yang, L. and A.T. Paulson (2000). Effects of lipids on mechanical and moisture barrier properties of edible gellan film, *Food Res. Int.,* 33: 571-578.

Yang, S.Y., K.Y. Lee, S.E. Beak, H. Kim and K.B. Song (2017). Antimicrobial activity of gelatin films based on duck feet containing cinnamon leaf oil and their applications in packaging of cherry tomatoes, *Food Sci. Biotechnol.,* 26: 1429-1435.

Yang, X., T. Nisar, D. Liang, Y. Hou, L. Sun and Y. Guo (2018). Low methoxyl pectin gelation under alkaline conditions and its rheological properties: Using NaOH as a pH regulator, *Food Hydrocoll.,* 79: 560-571.

Yapo, B.M. (2011). Pectin rhamnogalacturonan II: On the 'Small Stem with Four Branches' in the primary cell walls of plants, *Int. J. Carbohydr. Chem.,* 2011: 964521

Yemenicioğlu, A., S. Farris, M. Turkyilmaz and S. Gulec (2020). A review of current and future food applications of natural hydrocolloids, *Int. J. Food Sci. Tech.*, 55: 1389-1406.

Yener, F.Y., F. Korel and A. Yemenicioğlu (2009). Antimicrobial activity of lactoperoxidase system incorporated into cross-linked alginate films, *J. Food Sci.*, 74: M73-M79.

Yi, J. and Y. Ding (2015). Dual effects of whey protein isolates on the inhibition of enzymatic browning and clarification of apple juice, *Czech J. Food Sci.*, 32: 601-609.

Younes, M., P. Aggett, F. Aguilar, R. Crebelli, M. Filipic, M.J. Frutos *et al.* (2017). Scientific opinion on the re-evaluation of alginic acid and its sodium, potassium, ammonium and calcium salts (E 400–E 404) as food additives, *EFSA J.*, 15: 5049.

Youssef, F., T. Roukas and C.G. Biliaderis (1999). Pullulan production by a non-pigmented strain of Aureobasidium pullulans using batch and fed-batch culture, *Process Biochem.*, 34: 355-366.

Zambrowicz, A., A. Dąbrowska, L. Bobak and M. Szołtysik (2014). Egg yolk proteins and peptides with biological activity, *Post. Hig. Med. Dosw*, 68: 1524-1529.

Zhang, B., Z. Deng, Y. Tang, P.X. Chen, R. Liu, D.D. Ramdath *et al.* (2014). Effect of domestic cooking on carotenoids, tocopherols, fatty acids, phenolics, and antioxidant activities of lentils (*Lens culinaris*), *J. Agric. Food Chem.*, 62: 12585-12594.

Zhang, L., H.T. Li, S.H.E.N. Li, Q.C. Fang, L.L. Qian and W.P. Jia (2015). Effect of dietary resistant starch on prevention and treatment of obesity-related diseases and its possible mechanisms, *Biomed. Environ. Sci.*, 28: 291-297.

Zhao, X., F. Chen, W. Xue and L. Lee (2008). FTIR spectra studies on the secondary structures of 7S and 11S globulins from soybean proteins using AOT reverse micellar extraction, *Food Hydrocoll.*, 22: 568-575.

Zheng, L.Y. and J.F. Zhu (2003). Study on antimicrobial activity of chitosan with different molecular weights, *Carbohydr. Polym.*, 54: 527-530.

Zilic, S., M. Milasinovic, D. Terzic, M. Barac and D. Ignjatovic-Micic (2011). Grain characteristics and composition of maize specialty hybrids, *Span. J. Agric. Res.*, 9: 230-241.

Zimmermann, T., N. Bordeanu and E. Strub (2010). Properties of nanofibrillated cellulose from different raw materials and its reinforcement potential, *Carbohydr. Polym.*, 79: 1086-1093.

CHAPTER

3

Ability of Major Natural Hydrocolloids to Form Edible Packaging

3.1 Introduction

Different hydrocolloids can be used to develop edible packaging, such as films, coatings, pads, stickers, electrospun mats, casings, sachets, etc. In most of the research studies, proteins or polysaccharides, or their mixtures with each other, are used in development of self-standing films by employing the classical solution-casting method. However, the solution-casting is not a highly efficient method as thermoplastic processing for large-scale manufacturing of edible films (Hernandez-Izquierdo and Krochta, 2008). Therefore, edible films obtained by thermoplastic polymer processing methods (e.g. slit-die extrusion, blown extrusion or compression molded films, extruded casings) still maintain their importance as a major research topic. The colloidal solutions of polysaccharides and/or proteins or their emulsions with lipids are also extensively employed by the industry to develop edible coatings for different food products. Moreover, combinational use of two oppositely-charged hydrocolloids to obtain layer-by-layer food coatings is also a developing method adapted for preservation of foods (Sowmyashree et al., 2021). Furthermore, recent developments in the field have also shown that extruded active edible casings from polysaccharides will compete with extruded collagen casings in the market soon. Besides films, coatings, and casings, alternative emerging active edible packaging, such as pads obtained from hydrogels or aerogels, mats obtained from electrospun nanofibers, and stickers obtained by different methods have also been increasingly tested on different foods (Lopes et al., 2018; Boyaci and Yemenicioğlu, 2020; Hemmati et al., 2021). This chapter discusses mainly the ability of major natural proteins and polysaccharides to form different types of edible packaging. Particular emphasis has also been laid on critical factors necessary for film formation and promising research studies conducted in the field to improve critical properties (e.g. thermal, mechanical, barrier properties) of packaging. Moreover, major advantages and disadvantages of packaging obtained from important hydrocolloids have been discussed, and different approaches that help to increase their industrial applicability have been introduced.

3.2 Protein-based edible packaging

3.2.1 Packaging from zein

Zein has been extensively used as a coating and to obtain self-standing films using the classical solution-casting method. Due to its unique solubility in solvents, like ethanol, its excellent film-forming ability, controlled-release properties, and compatibility with most of the natural antimicrobial and antioxidant compounds (e.g. antimicrobial enzymes, bacteriocins, polyphenols, and essential oils), zein is one of the most promising active packaging materials (Yemenicioglu, 2016). Therefore, the edible film characteristics of zein have been studied extensively. The characteristic film structure of zein consists of a meshwork composed of doughnut structures formed by asymmetric rods joined to each other (Guo *et al.*, 2005). The mechanism of film formation is attributed to extensive hydrophobic interactions among asymmetric zein rods following heat denaturation (Yemenicioğlu, 2016). However, it is also thought that strong intermolecular disulfide bonds contribute in keeping the zein rods together (Argos *et al.*, 1982; Guo *et al.*, 2005). The gas, water vapor, and aroma barrier properties of zein films could vary, depending on the film-making method. The classical zein films obtained by the solution-casting method generally show a porous surface (Fig. 3.1a) (Padgett *et al.*, 1998; Güçbilmez *et al.*, 2007; Wang *et al.*, 2005a) while compression-molded zein films have a non-porous surface with no visible pores (Padgett *et al.*, 1998). However, the current applications in the food industry concentrate mainly on coating of candies, fresh and dried fruits, and nuts (Bai *et al.*, 2003; Lai and Padua, 1997; Shukla and Cheryan, 2001). Moreover, since the O_2 permeability values determined for zein films were lower than those of different plastic films, such as low and high density polyethylene, polypropylene, polystyrene, and polyvinyl chloride (Gennadios *et al.*, 1993) application of zein coatings on to the surface of plastic films has been suggested to improve their gas barrier properties (Tihminlioglu *et al.*, 2010). The encapsulation

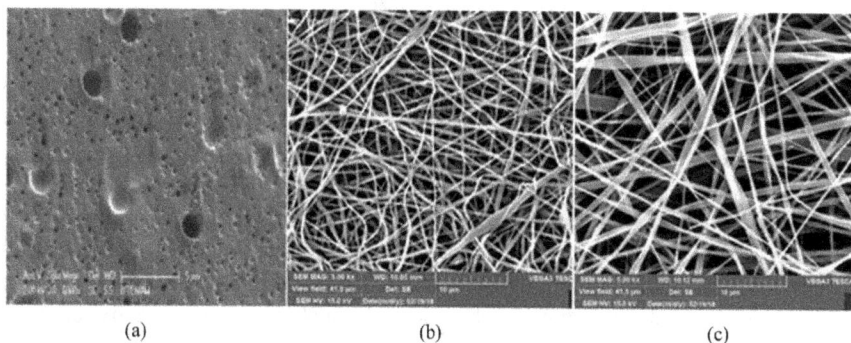

(a)   (b)   (c)

Fig. 3.1: Scanning electron micrographs of (a) solution-cast zein film surface (scale bar 5 μm); reprinted with permission from Arcan and Yemenicioğlu (2011); copyright © 2011 Elsevier Ltd., (b) electrospun zein nanofibers (scale bar 10 μm), and (c) electrospun zein ribbon-like nanofibers (scale bar 10 μm); reprinted with permission from Leena et al. (2020); copyright © 2020 Elsevier Ltd.

studies with zein also suggested that materials from zein could provide good barrier properties against desired or undesired aroma and flavor compounds (Torres-Giner *et al*., 2010; Fabra *et al*., 2014). However, similar to most protein-based edible films, zein films suffer from their inferior water vapor permeability than those of plastic films (Mauri *et al*., 2016). Moreover, another disadvantage of zein films is their classical brittleness problem originating from highly hydrophobic nature of this protein (Arcan and Yemenicioğlu, 2011). Therefore, plasticization of zein films has been studied extensively by different workers. Different plasticizers used for zein include glucose, galactose, and fructose (Ghanbarzadeh *et al*., 2006), dibutyl tartrate, triethylene glycol, polyethylene glycol (300), levulinic acid, glycerol, and oleic acid (Lawton, 2004), synergistic mixtures of oleic acid and glycerol (Xu *et al*., 2012a), phenolic acids, and flavonoids, such as gallic acid, p-hydroxybenzoic acid, ferulic acid, and catechin (Arcan and Yemenicioğlu, 2011), and essential oils, such as eugenol, carvacrol, and thymol (Alkan and Yemenicioğlu, 2016; Boyaci *et al*., 2019). The use of phenolic substances for zein plasticization is more advantageous than other plasticizers since poylphenols also provide antimicrobial and antioxidant activity (Unalan *et al*., 2013). However, food application of these films has some limitations, originating from strong aroma and/or flavor of polyphenols. Recently, Sözbilen *et al*. (2022) also showed that organic acids, such as lactic and malic acids are effective plasticizers of zein films. The use of organic acids as plasticizer also gives antimicrobial films, but highly acidic taste of these films is again a disadvantage that limits some food applications of the films.

Thermoplastic processing methods can also be employed to obtain zein films. Zein and corn gluten meal, a zein-rich protein byproduct separated during corn milling, could be thermoplasticized at 80°C. Thus they could be processed into edible films by compression molding by using suitable plasticizers (Di Gioia and Guilbert, 1999; Di Maio *et al*., 2010). Wang and Padua (2003) employed single- and double-screw extruded zein resin to obtain blown films. It was reported that the twin-screw extrusion reduced the number and size of voids in blown films since it squeezed out the moisture entrapped in the resin effectively. However, loss of moisture during twin-extrusion resulted with reduced elongation at break values of blown films. The heat treatment of dry samples was also found effective in eliminating voids of the blown films. It was reported that the optimal moisture content of extrudate before blowing into cylindrical films was 14–15% while temperatures at different barrel zones were between 20 and 35°C, and the temperature at the blowing head was 45°C (Wang and Padua, 2003). Oliviero *et al*. (2010) also produced blown films by using zein thermoplasticized in a twin counter-rotating mixing chamber that applied heat at 70°C under shear stress. It was reported that the thermoplasticization process was beneficial in obtaining blown films since it modified zein conformation by reducing its β-sheet content, while almost unchanging its α-helix content. Luecha *et al*. (2010) obtained single-screw extruded zein resin incorporated with montmorillonite clay to obtain composite blown films. It was reported that the blown zein films incorporated with 1% montmorillonite clay showed better thermal stability and superior mechanical (~2.8-fold higher elongation at break) and barrier properties (1.7-fold lower water vapor permeability) than control blown films (Luecha *et al*., 2010).

Zein is also one of the popular hydrocolloids used to obtain electrospun mats. Yao *et al.* (2007) electrospun zein at 20% (w/v) in ethanolic solutions to obtain nanofibers with a diameter of ~500 nm, but when they increased the zein concentration to 30, 40, and 50% (w/v), they obtained ribbon-like nanofibers that showed increase in diameter (between 1-6 μm) in a concentration-dependent manner. According to Yao *et al.* (2007), the obtained nanofiber mats were quite fragile. Thus, to increase their mechanical strength, these workers applied chemical cross-linking of obtained nanofibers with hexamethylene diisocyanate. The transformation of zein nanofiber into ribbon-like nanostructures was also observed when phenolic compounds were added into electrospun zein solutions (Fig 3.1b and 3.1c). Leena *et al.* (2020) reported formation of ribbon-like structures in electrospun zein nanofibers incorporated with resveratrol while Li *et al.* (2014) observed similar structures in quercetin loaded zein nanofibers. Park *et al.* (2013) obtained zein-montmorillonite clay composite nanofibers by electrospinning zein at 30% (w/w) in ethanolic solutions containing suspended montmorillonite clay (MMT) at 1 to 5% (w/w of zein). It was reported that the addition of montmorillonite increased the thermal stability and hydrophilicity of composite nanofibers. The electrospun zein nanofibers provide an excellent tool for encapsulation and controlled release of natural volatile active compounds. For example, essential oils of daphne, rosemary (Göksen *et al.*, 2020), and thyme (Ansarifar *et al.*, 2021), and cumin essential oil component – cumin aldehyde (Hajjari *et al.*, 2021) have been incorporated into zein nanofibers obtained by needle-based electrospinning devices. A recent study by Karim *et al.* (2021) is also very promising since these authors produced cinnamaldehyde loaded zein nanofibers for controlled release of this natural agent using needle-less electrospinning that is more suitable for large-scale production than needle-based electrospinning. The large-scale production of zein nanofibers also enhances the development of novel packaging materials not only to achieve unique controlled release properties for natural active compounds, but also to improve mechanical and barrier properties of edible films. For example, Karim *et al.* (2020) incorporated zein nanofibers obtained by needle-less electrospinning into gelatin films to improve their poor mechanical and moisture barrier properties.

3.2.2 Packaging from gluten

Gluten is one of the potential candidates to develop packaging materials alternative to fossil plastics. The solutions of gluten is used to obtain edible food coating materials or can be employed to develop self-standing films by the classical solvent-casting method (Tanada-Palmu and Grosso, 2005). The film-forming solutions of gluten are generally obtained by suspending this protein complex in slightly acidic (pH ~4) or alkaline (pH ~10) ethanolic solutions together with a plasticizer like glycerol. However, it was reported that the most transparent, but least water-soluble gluten films are obtained when ethanol concentration in film-forming solution is between 32.5 and 45% (v/v) and its pH is between 2 and 4 (Gontard *et al.*, 1992). The glutenins purified from gluten or gluten separated into gliadin- or glutenin-rich fractions can also be used to obtained edible films. Hernández-Muñoz *et al.* (2004) reported that the glutenin films prepared by the classical solution-casting method (using 7.5% (w/w) glutenin in 50% (v/v) ethanol at pH 5.0) were homogenous, flexible, and translucent when films were plasticized with low molecular weight plasticizers, such as glycerol, sorbitol, and triethanol amine (33 g plasticizer/100 g dry protein). However, these authors also reported that plasticization with sorbitol gave films with the highest tensile strength and water vapor barrier properties that remained stable (unlike films with glycerol) during 16-day storage at 23°C and 50% RH. Mangavel *et al.* (2002), who developed edible films from gluten and its gliadin- and glutenin-rich fractions, reported that differences in glutenin/gliadin ratios did not cause considerable changes in the characteristic highly-flexible nature of gluten-based films except the slightly higher tensile strengths of films obtained with glutenin-rich fraction. Guo *et al.* (2012) used composites of wheat gluten with corn zein to obtain films showing different mechanical and barrier properties. These authors found that increased zein weight in composites gave films showing higher tensile strength (minimum zein:glutenin ratio = 40:60) and water vapor barrier properties (minimum zein:glutenin ratio = 20:80).

Gluten is one of the amorphous hydrocolloids that undergoes glass transition at low temperatures. Therefore, it can be used to obtain two-dimensional packaging materials through extrusion or three-dimensional materials by injection molding (Gontard and Guilbert, 1998). It was reported that extrusion of gluten at alkaline conditions obtained by using ammonium hydroxide helped to reduce protein solubility and to improve mechanical and gas barrier properties of its films (Ullsten *et al.*, 2010). Türe *et al.* (2012) obtained gluten/montmorillonite clay nanocomposite films by a solvent-free method using a single-screw extruder equipped with a flat sheet die. It was demonstrated that addition of 5% montmorillonite clay improved the water vapor (1.3-fold) and gas barrier (1.9-fold) properties of extruded gluten films, but the composite films were less extensible than the gluten films. Gluten films could also be obtained by compression molding. Heating of gluten during processing (up to 75°C) results with unfolding of mainly its glutenin fraction (Schofield *et al.*, 1983). The unfolding exposes the –SH groups of glutenin and oxidation of these groups causes polymerization of glutenin fractions. The sulphydryl/disulphide (–SH/SS) interchange reactions between glutenin and gliadin accelerates mainly at temperatures exceeding 90°C, thus the newly produced free SH-groups can further react with either gliadin or glutenin (Abedi and Pourmohammadi, 2021). Sun *et al.* (2008) reported that as the compression molding temperature was increased between 65 and 105°C, the tensile strength and elongation at break of obtained gluten films showed a parallel increase. Further increase of compression molding temperature to 125°C maximized film tensile strength due to extensive –SH/SS interactions, but too much increase in film mechanical strength at this high temperature caused a sharp decline in elongation at break values of films (Sun *et al.*, 2008). Zubeldía *et al.* (2015) suggested that the processing temperatures of gluten during compression molding should not be exceeded 100°C. According to these authors, compression-molding temperature of 100°C formed sufficient cross-linking within gluten film matrix. It was also reported that processing temperatures above 100°C caused extensive darkening of obtained films (Zubeldía *et al.*, 2015). Ansorena *et al.* (2016) produced thyme essential oil loaded active gluten films by compression molding conducted at 100°C and 100 kg/m² for 10 min. in a hydraulic heated press.

It was reported that almost 73.2–80.4% of thymol and 72.9–75.4% of carvacrol in the thyme essential oil (EO) were retained in the films after processing. The films showed antimicrobial activity on different test bacteria, such as *Escherichia coli* and *Pseudomonas aeruginosa* at 3.5% thyme EO (w/w of film), and *Listeria innocua* and *Staphylococcus aureus* at 10% thyme EO. The thyme EO at 10% EO caused significant changes in film mechanical properties (reduced tensile strength, but increased elongation at break), but it did not considerably affect film water vapor permeability (Ansorena *et al.*, 2016).

The gluten and its component gliadin can also be processed into nanofiber mats by electrospinning. Zhang *et al.* (2020) produced antimicrobial gluten nanofiber mats by electrospinning glycerol monolaurate loaded gluten solutions. Sharif *et al.* (2018a) developed electrospun gliadin nanofibers loaded with ferulic acid and hydroxypropyl-beta-cyclodextrin. These authors reported that ferulic acid and hydroxypropyl-beta-cyclodextrin formed an inclusion complex, which improved the photo stability and solubility of highly sensitive ferulic acid. Akman *et al.* (2019) also employed electrospun nanofibers of gliadin for encapsulation of curcumin. It was determined that the gliadin concentration in electrospun solution should exceed 20% (w/v) to obtain smooth gliadin nanofibers free from microparticles. The gliadin nanofibers were found useful in controlled release of curcumin and they showed a protective effect on antioxidant and antimicrobial activity of curcumin.

3.2.3 Packaging from collagen

Collagen, the main component of small and large intestines of cattle, pig, sheep, and goat, is an ancient edible packaging material that has been extensively used to provide natural sausage casings. The food industry also uses collagen obtained from bovine hides, bones, and connective tissues to obtain artificial extruded sausage casings of different sizes (Yemenicioglu *et al.*, 2020). Artificial collagen casings obtained by extruding collagen dough prepared by swelling of collagen in acid solutions are not only cheap, practical, uniform, and strong, but they also give almost similar cooking characteristics and tenderness with natural casings (Suurs and Barbut, 2020). Moreover, recent innovations in the field enabled formation of collagen casing directly on the surface of sausages during manufacturing by using co-extrusion technique (Suurs and Barbut, 2020). The developed casings can also be used in antimicrobial packaging by incorporating natural antimicrobials, such as nisin into casing by vacuum impregnation (Batpho *et al.*, 2017). However, different studies are continuing to improve the mechanical and barrier properties of classical artificial collagen casings. The mechanical strength of artificial casing should be strong enough to hold the meat batter during cooking process and mechanical handling while barrier properties are important to prevent weight loss by drying and control oxidative changes in sausages (Simelane and Ustunol, 2005; Suurs and Barbut, 2020). A basic physical method applied to improve desired properties of collagen films includes heat treatment that affects protein (triple helices) in many different ways (e.g. conformation change, orientation, degradation, denaturation, depolymerization, and cross-linking). Xu *et al.* (2020) investigated the effect of heat treatment between 25 and 90°C for 30 min. on the mechanical and barrier properties of solution-cast collagen films. These workers determined that both the tensile strength and elongation at break as well as water vapor and O_2 permeability of collagen films reduced gradually through increase of heat treatment temperature. The aging of collagen films at 60-70°C and 60-70% RH for one to three days is also a green physical cross-linking method that improves films' Young's modulus in dry, wet, and boiled form, and tensile strength in wet and boiled form (Shi *et al.*, 2019). The aging process also improves the cooking performance (decreases water absorption and shrinkage) of collagen films. Wang *et al.* (2015) also employed some physical methods, such as ultraviolet irradiation (UV) (at 365 nm for 30 min. at 9.43 mW/cm^2), dehydrothermal treatment (DHT) (at 105°C for 24 h under vacuum at 0.05 bar), and combination of these two methods to improve mechanical properties of extruded collagen casings. All three methods were effective in improving the tensile strength of collagen casings, but DHT alone gave casings with the highest tensile strength. An innovative physical method to improve mechanical and barrier properties of collagen films is high-pressure homogenization. The high-pressure homogenization applied between 20 and 35 psi turned collagen macrofibers into micro- or nanofibers with diameters ranging between 1 μm and 83 nm, depending on the severity (pressure and time) of the process (Ma *et al.*, 2020). It was determined that the use of mechanically modified collagen nanofibers having diameter between 20.5-36.5 nm range gave highly transparent edible films with 380% higher density, 80% lower water vapor permeability, and 690% higher tensile strength than pristine collagen fiber films (Ma *et al.*, 2020). Some composite film-making methods could also be used to improve desired properties of collagen casings. For example, Wang *et al.* (2018) achieved improvements in tensile strength, and water vapor and O_2 barrier properties of solution-cast collagen films by preparing composites of collagen with carboxylated cellulose nanofibers (≤50 g nanofiber was added per kg of collagen). It was reported that the obtained composite films were homogenous and had a compatible spatial network due to the electrostatic interactions between collagen and carboxylated cellulose nanofibers (Wang *et al.*, 2018). Long *et al.* (2018) also improved the tensile strength, and water vapor and O_2 barrier properties of extruded collagen casings by incorporating cellulose nanocrystals at 4% (w/w of film-forming solution). These workers attributed the improvements in film properties to dense film network formed as a result of increased collagen cross-linking by cellulose nanocrystals via hydrogen bonding and electrostatic interactions. Wu *et al.* (2017) applied transglutaminase cross-linking of collagen with proteins, such as casein, keratin or soy protein isolate at 1:1 ratio. These workers obtained more thermostable and mechanically strong solution-cast composite films than that of collagen. However, the collagen-casein composite films showed most remarkable increases in tensile strength (~41.3 kPa of collagen increased to 55 kPa) and thermal stability (peak temperatures determined by DSC for collagen and collagen-casein composite films were 82.8 and 130.8°C, respectively) (Wu *et al.*, 2017).

Although the edible film studies about collagen have focused mainly on development of extruded casings, self-standing films of collagen obtained by solution-casting or compression molding can also be employed for edible food packaging (Wang *et al.*, 2016; Andonegi *et al.* 2020). The native collagen can also be converted into thermoplastic collagen (TC) by partial heat denaturation, and properly plasticized TC could be further processed into edible films using blown-extrusion method (Klüver and Meyer, 2013). However, the films of TC have limited applicability since they are highly sensitive to moisture and show poor mechanical properties in moist environments (Klüver and Meyer, 2013). Collagen is also one of the hydrocolloids that can be electrospun into nanofibers, but this process is not attractive since electrospinning applied for aqueous solutions of collagen causes denaturation and then transformation of this hydrocolloid into gelatin (Zeugolis *et al.*, 2008).

3.2.4 Packaging from gelatin

Gelatin is one of the hydrocolloids used widespread in development of edible films and coatings. The films of gelatin show different characteristics depending on the source of gelatin. Films of gelatins from warm- and cold-water fish show different mechanical and barrier properties from each other and from films of mammalian gelatins due to differences in their gel-setting temperature that affects the molecular conformation of proteins forming their film matrix (Avena-Bustillos *et al.*, 2011). It was determined that the warm-fish gelatin (e.g. catfish) and mammalian gelatins of bovine and porcine origin form helical structures within their film matrices, while cold-fish gelatins (e.g. pink salmon, cod, pollock or haddock) form amorphous forms within their film matrix due to their significantly lower gel-setting temperature than other gelatins (Avena-Bustillos *et al.*, 2011). Most fish gelatin films show lower oxygen permeability than mammalian gelatin films, while mammalian gelatin films generally show the highest mechanical stability (tensile strength and puncture deformation) followed by warm-fish gelatin and cold-fish gelatin (Avena-Bustillos *et al.*, 2011; Ninan *et al.*, 2010). The chicken gelatin, an emerging source of gelatin, has also been tasted as a film-forming hydrocolloid. It was found that edible chicken gelatin films show significantly lower water vapor permeability, but higher tensile strength than bovine and porcine gelatins (Suderman *et al.*, 2018). A recent study by Yang *et al.* (2017) also showed that duck feet gelatin also gives edible films showing comparable mechanical strength with chicken gelatin films.

The edible packaging of gelatin can be incorporated by different natural phenolic compounds, such as pure polyphenols, phenolic extracts, and essential oil, or protein and peptide-based antimicrobials (Etxabide *et al.*, 2017; Luciano *et al.*, 2021). Gelatin is a highly hydrophilic material that contains many ionizable and H-bonding groups. Therefore, it should be considered that a portion of natural antimicrobials are immobilized within the gelatin matrix due to extensive charge-charge attraction and/or H-binding. Boyaci and Yemenicioğlu (2020) reported that ≥90% of green tea polyphenols and lysozyme added into bovine gelatin gel-based active pads were immobilized by the gelatin matrix at pH 6.0. It was also noted that the addition of green tea extract caused cross-linking of gelatin gel matrix. Thus, this further increased the amount of immobilized lysozyme. However, lysozyme and green tea extract loaded active gel pads placed on both sides of cold-smoked salmon slices inhibited the growth of *Listeria innocua* inoculated on these samples (Boyaci and Yemenicioğlu, 2020). Ma *et al.* (2013) also reported that the free lysozyme in gelatin films reduced as concentration of natural cross-linking agent, genipin, increased. However, these workers also showed that free lysozyme in gelatin films increased as pH was reduced from 7.0 to 3.8 – a sufficiently low pH value to eliminate negative charges of gelatin (pI: 5-7.5) matrix that bound positively charged lysozyme (pI: 11.4). In contrast, Benbettaïeb *et al.* (2016) reported that only 23% of total quercetin incorporated into fish gelatin-chitosan composite films was immobilized (gelatin:chitosan ratio = 3:1).

Although the use of gelatin-based edible films and coatings brings many advantages, such as perfect film forming ability, low oxygen permeability, glossy appearance, etc., some drawbacks of gelatin also exist, such as low mechanical strength, poor moisture barrier properties, high solubility, and rapid swelling. Thus, during development of edible films and coatings, gelatin was increasingly combined with other hydrocolloids or nano-sized fillers (e.g. clay nanoparticles or nanocrystalline cellulose) to obtain composite or blend materials with minimum drawbacks. However, this is a highly challenging task since composite and blend film making interferes with the desired properties (e.g. transparency and glossy appearance) of gelatin films. For example, Farahnaky *et al.* (2014) incorporated montmorillonite clay nanoparticles into solution-cast bovine gelatin film at 18% (w/w of gelatin). These workers reported that the incorporation of clay nanoparticles reduced water vapor permeability of gelatin film from 0.86 to 0.42 g mm/kPa m^2 h. In contrast, the addition of montmorillonite clay did not cause a considerable increase in the tensile strength of gelatin films. Moreover, clay nanoparticles caused darkening and reduced transparency of gelatin films (Farahnaky *et al.*, 2014). Wang *et al.* (2017) also tried to improve mechanical and barrier properties of solution-cast bovine gelatin films by incorporating different maize starches that vary in their amylose/amylopectin ratios. The normal maize starch, high amylose starch, and waxy maize starch used by these authors contained 27, 72 and 0% amylose, respectively. It was reported that the addition of normal starch and high amylose starch at 50% (w/w of gelatin) improved the tensile strength of gelatin films most effectively at 75% relative humidity (12.6 kPa of control increased to 21 and 29 kPa, respectively). In contrast, no considerable benefits in tensile strength of films were determined by addition of waxy starch. Moreover, addition of 20 to 50% of normal starch, 30% of waxy starch, or 40% of high amylose starch caused most significant reductions in water vapor permeability of films (almost 2, 1.6, and 1.45-fold lower than that of control gelatin film, respectively) at 90% RH (Wang *et al.*, 2017). These findings suggested that the most significant benefits in both mechanical and water vapor barrier properties of gelatin-starch composite films were provided by native maize starch that has an amylose/amylopectin ratio of 2.7. The cross-linking of gelatin with natural functional agents is another strategy that has been employed to improve mechanical strength and water vapor barrier properties of solution-cast gelatin films. Choi *et al.* (2018) employed oxidized phenolic compounds, such as tannic acid, caffeic acid, and green tea polyphenols for cross-linking of turmeric loaded active bovine gelatin films. These workers determined that low concentrations of caffeic acid were most effective as an oxidized phenolic compound to improve tensile strength and water vapor barrier properties of solution-cast gelatin films. The addition of caffeic acid at 10 mg/g of gelatin-turmeric mixture increased the tensile strength of turmeric loaded gelatin films from 7.3 to almost 12 MPa, while water vapor permeability of films reduced from

0.64 to 0.47 g.mm/kPa.m^2.h. The increase of oxidized caffeic acid concentrations (20 or 30 mg/g of gelatin-turmeric mixture) reduced the beneficial effects of this polyphenol, but the cross-linked films still showed significantly better tensile strength and water vapor permeability than control films. Leite *et al.* (2021) applied an effective combinational strategy (phenolic cross-linking and nanoparticles) by incorporating both tannic acid and nanocrystalline cellulose (NCC) at 4% (w/w of gelatin) into solution-cast bovine gelatin films. These workers found that the degree of cross-linking in oxidized and non-oxidized tannic acid loaded gelatin films was 53 and 36%, respectively. However, they preferred cross-linking of gelatin films with non-oxidized tannic acid to maintain its original antioxidant and antimicrobial activity. It was reported that the incorporation of tannic acid and NCC increased the tensile strength of gelatin films from almost 17 kPa to 30 kPa. The water vapor permeability of gelatin films also reduced dramatically from almost 16 to 0.06 g.mm/kPa.m^2.h by addition of tannic acid and NCC (Leite *et al.*, 2021). Finally, the properties of gelatin films can also be modified by using enzyme transglutaminase that catalyzes the formation of a covalent bond between a free amine group of protein-bound lysine and the γ-carboxamide group of protein-bound glutamine (Zhu *et al.*, 2019). De Carvalho and Grosso (2004) reduced the solubility and water vapor permeability (from 0.198 to 0.120 g.mm/kPa.m^2.h) of bovine gelatin films by transglutaminase treatment, but the enzyme treatment did not affect the mechanical properties of films considerably. Liu *et al.* (2017) also modified bovine gelatin by transglutaminase and obtained edible solution-cast films with reduced solubility and improved mechanical properties, such as increased elongation at break and toughness, but the enzyme treatment did not improve the tensile strength of films. Kołodziejska and Piotrowska (2007) also determined that the treatment of fish gelatin with transglutaminase was effective in reducing solution-cast fish gelatin-chitosan (4:1) composite films' solubility and increasing their flexibility dramatically (elongation at break up to 420% in the presence of glycerol). However, these authors also reported that enzyme modification caused significant reductions in tensile strength of fish gelatin-based composite films, and made no positive contributions to improving water vapor permeability of these films.

Gelatin has also been used extensively to obtain different types of extruded films. Hanani *et al.* (2014) obtained heat sealable bovine gelatin-corn oil composite films using twin-screw co-rotating extruder equipped with a die working at processing temperatures between 90 and 130°C. These authors determined that the tensile strength (changed between 1.43 and 5.37 MPa) and puncture strength of extruded films increased as processing temperature increased, but the elongation at break values of the obtained films (changed between 1.68 and 2.1%) were not considerably affected by the processing temperature. It was also found that the processing temperatures of 120 and 130°C gave films with lowest water vapor and oxygen permeability values, respectively (Hanani *et al.*, 2014). Andreuccetti *et al.* (2012) obtained blown films of porcine gelatin from pellets prepared with a single screw extruder. These workers incorporated extracts of *Yucca schidigera* – a source of natural surfactant saponin and antioxidant polyphenols into blown films (Fig. 3.2). It was reported that proper plasticization of blown gelatin films with glycerol was very critical since non-plasticized films obtained by this method turned highly brittle during storage (Andreuccetti *et al.*, 2012). Krishna *et al.* (2012) developed fish gelatin films plasticized with 20 or 25% glycerol (w/w of gelatin) using twin-screw extrusion at 110 or 120°C followed by compression molding at 80°C. According to these authors, the compression molded gelatin films (thickness: 0.34-0.58 mm) showed significantly lower tensile strength (4 to 12 fold), but higher elongation at break (5-10 fold) than classical solution-cast films (thickness: 0.1 mm). Krishna *et al.* (2012) also determined that the water vapor permeability values of different compression molded gelatin films (ranging between 15 and 29 g.mm/ kPa.m^2.h) were comparable with those of solution-cast gelatin films (ranging between 19 and 25 g mm/kPa.m^2.h). Prodpran *et al.* (2017) investigated

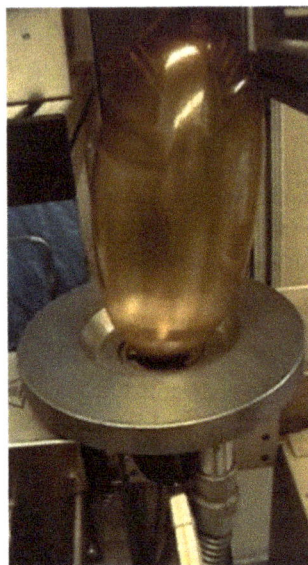

Fig. 3.2: Blown extruded bovine gelatin film incorporated with *Yucca schidigera* extract; reprinted with permission from Andreuccetti *et al.* (2012); copyright © 2012 Elsevier Ltd.

properties of compression molded fish and bovine gelatin films, but they reported limited differences between tensile strengths and water vapor permeability values of these films (ranging between 11.42 and 20.82 mPa, and 2.25 and 3.29 x 10^{-10} g.m/m^2.s.Pa, respectively).

The precast gels of gelatin can also be used as a pad in packaging of food. A precast sheet of hydrogel can be employed not only as an absorbent pad to bind drip-loss fluids from food, but also as a high capacity reservoir to deliver antimicrobials, antioxidants, or flavor compounds on to the food surface (Batista *et al.*, 2018). Boyaci and Yemenicioglu (2020) improved the mechanical properties of gelatin gels by preparing composites of gelatin with rice starch or carnauba wax (each was added at 7.5%, w/w of gelatin solution). The developed lysozyme and green tea extract loaded precast composite gelatin gel sheets (thickness: 0.5 cm) placed on both sides of cold-smoked salmon slices inhibited the growth of *Listeria innocua* during 15 days cold-storage at 4°C. However, further studies are needed in this field to show the benefits of such pads in meat and poultry.

Gelatin is also one of the hydrocolloids that can be electrospun to obtain nanofibers, but high solubility and mechanical weakness of gelatin is a severe limitation in obtaining electrospun mats suitable for industrial applications (Zhang *et al.*, 2006). Thus, different studies have been conducted to improve water resistance and mechanical stability of gelatin nanofibers using natural cross-linking agents. For example, Tavassoli-Kafrani *et al.* (2017) employed oxidized phenolic compounds, such as tannic, caffeic, gallic and ferulic acids as cross-linking agents to reduce water solubility of bovine gelatin nanofibers. It was reported that tannic acid was the most effective phenolic cross-linker for gelatin (degree of crosslinking: 13.3%) and its application caused significant reductions in nanofiber solubility (30-50% of control nanofibers reduced to 8-12% for cross-linked ones) between pH 3.8 and 9.0. Kwak *et al.* (2021) achieved sugar-cross-linking of fish gelatin nanofibers by exploiting the Maillard reaction. It was reported that incubation of electrospun gelatin nanofibers with fructose at 100°C for 4h gave highly insoluble cross-linked nanofibers (degree of cross-linking: 63.4%) that resisted against hydrolytic degradation in phosphate-buffered saline (PBS) for 10 days. In contrast, gelatin nanofibers treated with glucose and sucrose showed 42 and 7.6% cross-linking and almost 50% and > 95% solubility in PBS within 10 days, respectively.

3.2.5 Packaging from soy proteins

Soy protein isolates have been used extensively to develop solution-cast edible films. The film formation process with soy proteins involves mostly solubilization of proteins in an alkaline solution and a following heat treatment that causes unfolding of proteins to enhance formation of intermolecular hydrophobic interactions and bonds (e.g. disulfide bonds and H-bonds) essential to obtain mechanically stable films (Rangavajhyala *et al.*, 1997; Subirade *et al.*, 1998; Rhim *et al.*, 2000). The soy protein films show high oxygen barrier properties at low moisture conditions, but they do not show good moisture barrier properties because of their hydrophilic nature (Foulk and Bunn, 2001). Therefore, different studies have been conducted to improve moisture barrier properties of solution-cast soy protein films. Stuchell and Krochta (1994) reported that heat treatment was an effective method to reduce water vapor permeability of soy protein films, but they found that the enzymatic cross-linking with peroxidase did not affect the water vapor permeability of soy protein films. Monedero *et al.* (2009) developed composites of soy proteins with oleic acid-beeswax mixture (soy protein:oleic acid-beeswax mixture ratio: 1:0.25 or 1:0.5) and reduced water vapor permeability of soy protein-based films by 30-50% (from ~6.7 to 3.5-4.8 g.mm/kPa.m^2.h). Similarly, Rhim *et al.* (2005) and Li *et al.* (2016) also achieved almost 50% reduction in water vapor permeability by incorporating bentonite clay and carbon nanoparticles into soy protein films, respectively. Insaward *et al.* (2015) used oxidized phenolic compounds, such as caffeic, gallic and ferulic acids at 0.5-1.5% (w/w of protein), to cross-link soy proteins, but this strategy caused a limited reduction (maximum 20%) in water vapor permeability of the obtained films. In contrast, oxidized phenolic cross-linking improved the tensile strength of films by almost 1.5-fold. Friesen *et al.* (2015) employed non-oxidized rutin and epicatechin as cross-linking agents to improve the properties of soy protein films. Rutin caused a moderate reduction (from 1.7 to 1.2 g.mm/kPa.m^2.h), while epicatechin increased the water vapor permeability of soy protein films. However, both rutin and epicatechin caused considerable improvements in tensile strength of soy protein films (9.3 kPa of control was increased to 35 kPa with rutin and 22 kPa with epicatechin). The incorporation of montmorillonite clay (5%, w/w of dry film) into solution-cast soy protein films also improved the tensile strength and elongation at break of these films by 2.8- and 5.5-fold, respectively, while reducing their water vapor permeability values from 3.8 to 2.96 g.mm/kPa. m^2.h (Kumar *et al.*, 2010).

The mechanical properties of pristine soy protein films are affected from their protein composition (11S/7S ratio). The 11S soy proteins give much more mechanically stable (high tensile strength) edible films than 7S proteins (Kunte *et al.*, 1997). It was found that the 11S/7S protein ratio larger than 1 (11S higher than 50%) increased the tensile strength of SPI films, while 11S/7S ratios lower than 1 reduced film tensile strength (Zhao *et al.*, 2011). However, it was reported that no significant differences in water vapor permeability values occurred between films of fractionated 11S and 7S proteins (Kunte *et al.*, 1997) or between high and low 11S/7S ratio protein films (Zhao *et al.*, 2011).

The soy proteins have also been used for the development of compression molded films. A detailed thermal and structural analysis of Ogale *et al.* (2000) revealed that the maximum compression temperature for soy protein films is 150°C. According to Guerrero *et al.* (2010), the use of plasticizer glycerol was essential to obtain transparent compression molded (at 150°C) soy protein films with suitable mechanical properties. The control films without glycerol were too brittle, white colored, and non-transparent (Fig. 3.3a). The films with soy protein/ glycerol ratios at 90:10 and 80:20 were also too brittle, but the films turned more transparent starting from 80:20 soy protein/glycerol ratio (Fig. 3.3b and 3.3c). The films showed optimal mechanical properties and good transparency at 70:30 and 60:40 soy protein/glycerol ratios (Fig. 3.3d and 3.3e), but further increase of glycerol turned films quite sticky. It was also reported that compression molding gave soy protein films 1.9 to 2.4-fold higher tensile strength than classical solution-cast films (Guerrero *et al.*, 2010). Ciannamea *et al.* (2016) obtained transparent

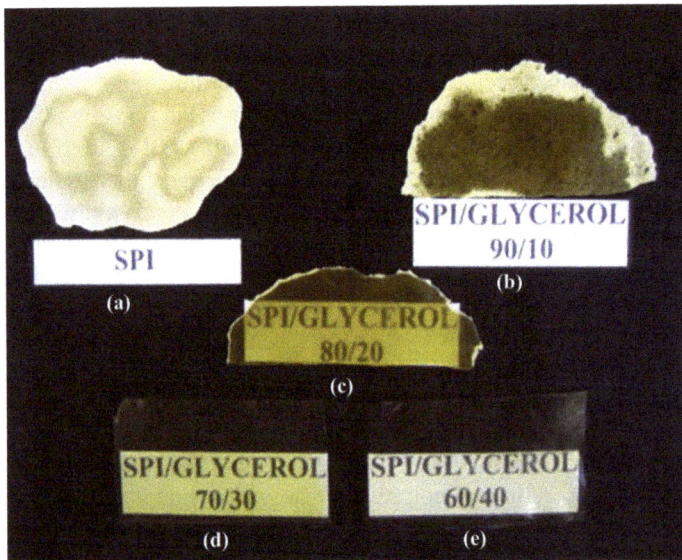

Fig. 3.3. Photos of compression molded soy protein isolate (SPI) films with different SPI/ glycerol ratios. Control film without glycerol (a), and films with 90:10 (b), 80:20 (c), 70:30 (d) and 60:40 (e) SPI/glycerol ratios; reprinted with permission from Guerrero *et al.* (2010); copyright © 2010 Elsevier Ltd.

compression molded active soy protein films by using glycerol at 30% (w/w of soy proteins) and incorporating red grape extract at 2–10% (w/w of soy proteins). It was found that the addition of red grape extract at 10% did not affect the mechanical properties of films considerably while a slight increase was determined in water vapor permeability of soy protein films. In contrast, the red grape extract caused a 37% reduction in oxygen permeability of films (Ciannamea *et al.*, 2016).

The thermoplasticized soy proteins were also used for development of extruded casings (Naga *et al.*, 1996) and self-standing films (Zhang *et al.*, 2001), but extruded soy protein films have limited applicability due to their poor mechanical and moisture barrier properties (Zhang *et al.*, 2001; Koshy *et al.*, 2015).

Due to their non-spinnable nature, soy proteins are mostly combined with other hydrocolloids that enhance their spinnability and fiber quality. In general, the poly (ethylene glycol) (synonym: poly (ethylene oxide)) is employed as a co-spinning polymer to obtain soy protein nanofibers (Xu *et al.*, 2012b). Salas *et al.* (2014) developed nanofibers by electrospinning aqueous alkaline solutions of soy protein isolate and soy glycinin with lignin in the presence of poly(ethylene glycol). Kolbasov *et al.* (2016) applied industrially scalable electrospinning of soy protein isolate-poly(ethylene glycol) blends as aqueous solutions, using a spinneret with eight rows with 41 concentric annular nozzles. It was reported that the nanofiber mats collected on a drum (0.1–1 m^2 mat was formed in about 10 s) were uniform and had nanofiber diameter of about 500–600 nm. Wang *et al.* (2013) also obtained anthocyanin-rich red raspberry (*Rubus strigosus*) extract loaded electrospun soy protein isolate (SPI) nanocomposite fiber mats by using poly(ethylene glycol). Moreover, Raeisi *et al.* (2021) obtained uniform, smooth, and bread-less nanofiber morphology by electrospinning soy protein-gelatin (70:30) mixtures containing *Zataria multiflora* and *Cinnamon zeylanicum* essential oils. It was reported that the soy-based nanofibers loaded with indicated essential oils showed antimicrobial effect on different bacterial pathogens, such as *E. coli*, *Salmonella enterica* serovar Typhimurium, *L. monocytogenes*, *B. cereus*, and *S. aureus*.

The gels obtained from soy proteins and their composites and emulsions have also been studied for delivery of bioactive substances, such as riboflavin (Hu *et al.*, 2015), folic acid (Ding and Yao, 2013), curcumin (Brito-Oliveira *et al.*, 2017), protocatechuic and coumaric acids (Marinea *et al.*, 2021), etc. Thus, such gel-based systems can also be exploited to develop bioactive coatings and pads.

3.2.6 Packaging from Na-caseinate

Na-caseinate is one of the most extensively used animal-origin proteins utilized in edible films and coatings. Similar to most other edible protein films, Na-caseinate films show appropriate mechanical and oxygen barrier properties, but suffer from poor water vapor permeability. The addition of plasticizers also brings additional challenges in controlling water vapor permeability of Na-caseinate films, but application of these agents is important to obtain and maintain the desired mechanical properties of self-standing films during food applications. It was reported that incorporation of glycerol into solution-cast Na-caseinate films at 10, 20 and 30% (by weight of Na-caseinate) gave films having 3.3, 14 and

27-fold higher water vapor permeability (at 90% RH and 25°C) than pristine Na-caseinate films (Khwaldia *et al.*, 2004a). Thus, methods of reducing water vapor permeability have been extensively applied to Na-caseinate films. Avena-Bustillos and Krochta (1993) applied different strategies, such as adjustment of pH to 4.6 (a pH close to pI of caseinates to increase protein-protein interactions), Ca^{+2} ion cross-linking, combination of adjustment of pH to 4.6 with Ca^{+2} ion cross-linking, and emulsion film formation with beeswax (Na-caseinate:beeswax ratio at 75:25) and achieved 36, 42, 43, and 90% reduction in water vapor permeability of solution-cast Na-caseinate films ($WVP_{control}$ = 1.53 g.mm/kPa.m^2.h at 81% RH and 25°C), respectively. Fabra *et al.* (2008) also exploited the emulsifying capacity of Na-caseinate and developed lipid-based solution-cast composite films with a 1:0.3:0.5 protein:glycerol:lipid ratio (lipid is a mixture of oleic acid and beeswax at a ratio of 70:30). The composite films showed considerably lower tensile strength, but higher elongation at break than control films. Moreover, they showed almost 66% lower water vapor permeability than control Na-caseinate films ($WVP_{control}$ = 4.1 g.mm/kPa.m^2.h at 58% RH and 5°C) (Fabra *et al.*, 2008). Pereda *et al.* (2015) applied the emulsion film-formation strategy by developing Na-caseinate-linseed oil resin (linseed oil:protein weight ratio: 0.05-0.12) films having up to 60% lower water vapor permeability than control Na-caseinate films. In contrast, emulsion formation of Na-caseinate with Tung oil (Tung oil : protein weight ratio: 0.05-0.15) (Pereda *et al.*, 2010) or with milk fat (milk fat : protein weight ratio: 0.1-0.3) (Khwaldia *et al.*, 2004b) failed to cause any reduction in water vapor permeability of obtained solution-cast films. Another strategy to improve water vapor barrier properties of Na-caseinate films is to obtain nanocomposites with other more hydrophobic hydrocolloids. Wang *et al.* (2019) applied this strategy by developing nanocomposite of Na-caseinate with zein. These workers formed solution-cast nanocomposite films by alkali treatment of different ratios of Na-caseinate/zein mixtures to induce dissociation of Na-caseinate and dissolution of zein. The alkali solution was then neutralized to form colloidal Na-caseinate-zein associations, having particle size between 131 and 184 nm. The solution-cast edible films of associated Na-caseinate-zein nanocomposite structures prepared at Na-caseinate:zein ratios of 75:25 and 50:50 showed almost 50 and 65% lower water vapor permeability than control Na-caseinate films, respectively. Moreover, the nanocomposite films could be turned into antioxidant films by loading curcumin (1 mg/ml of film forming solution) into films without impairing their water vapor barrier properties (Wang *et al.*, 2019). The nanocomposite film technology can also be employed effectively to improve mechanical strength of solution-cast Na-caseinate films. For example, Pereda *et al.* (2011) increased tensile strength of solution-cast Na-caseinate films almost three-fold by incorporating nanocellulose fibers at 3% (w/w of dry weight). The addition of nanocellulose increased the surface hydrophilicity of Na-caseinate films, but the rigid Na-caseinate nanocomposite films showed almost similar water vapor permeability with control films.

The Na-caseinate is also one of the hydrocolloids to be used for development of extruded films. Belyamani *et al.* (2014a) successfully used twin-screw extruder (Fig. 3.4a) (barrel temperature along the screw ranged between 40 and 80°C) to obtain Na-caseinate pellets (Fig. 3.4b) and blown films (Fig. 3.4c). The obtained blown films were water soluble and showed a water vapor permeability of 1.15, 1.91 and 4.50 g.mm/kPa.m^2.h (100-60% RH gradient at 30°C) at 20, 25 and 33% (w/w) glycerol content, respectively. The blown Na-caseinate films showed slightly lower tensile strength and water vapor barrier performance than blown Ca-caseinate films, but they were considerably more flexible than Ca-caseinate films (Belyamani *et al.*, 2014b). Colak *et al.* (2015) showed that it is possible to develop blown films of Na-caseinate by incorporating lysozyme into these films. It was reported that almost 41 and 26% of lysozyme activity survived from pellet and blown film making, respectively. Hartwig (2010) intended manufacturing of edible pouches and developed Na-caseinate-methylcellulose (4:1) composite films by compression molding at 88-93°C range. These authors investigated the sealing properties of composite films at 74-90.5°C range and determined that complete fusion of sealing for the films occurred at 90.5°C. Chevalier *et al.* (2018) compression molded Na-caseinate sheets extruded at 75°C to obtain edible films, but these films were highly sensitive to moisture and showed complete water solubility. Thus, natural modification strategies, such as enzymatic cross-linking of Na-caseinate as applied by Juvonen *et al.* (2011) using tyrosinase and transglutaminase to obtain insoluble solution-cast films, should also be tested for extruded films.

The pure Na-caseinate is not among hydrocolloids, such as gelatin and zein that are capable of forming nanofibers in pure form by electrospinning. Thus, it needs a carrier polymer that enables its electrospinning through solution viscosity

(a)

(b)   (c)

Fig. 3.4: Blown film from Na-caseinate. Co-rotating twin-screw extruder (a), sodium caseinate pellets (b) and blown Na-caseinate film (c); reprinted with permission from Belyamani *et al.* (2014a); copyright © 2014 Elsevier Ltd.

enhancement, lowering of surface tension, or electrical conductivity (Tomasula *et al.*, 2016). Pullulan and gelatin are some of the hydrocolloids that are capable of acting as a carrier for Na-caseinate to obtain some electrospun nanofibers (Nieuwland *et al.*, 2014; Tomasula *et al.*, 2016), but the soluble nature of these hydrocolloids interferes with their use in development of nanofiber mats suitable for widespread food packaging applications.

3.2.7   Packaging from whey proteins

Whey, obtained as a by-product of cheese and casein production, is among the most abundant sources of film-forming proteins. Similar to most other proteins, the ability of whey proteins to form an edible film depends on proper heat treatment that unfolds protein, exposes different hydrophobic amino acid side-chains and reactive amino acids (e.g. half-cystine and lysine-residues), and enhances formation of intermolecular (also intramolecular) interactions and bonds (e.g. hydrophobic bonds, disulfide bonds, hydrogen bonds, ionic interactions) necessary for film formation (deWit and Klarenbeek, 1984; Gounga *et al.*, 2010). The main advantage of whey proteins over Na-caseinate films is that they can be used as self-standing edible films for both low- and high-moisture food since they give almost water insoluble films at a broad pH range between 4.0 and 8.0 following heat denaturation (Pérez-Gago *et al.*, 1999). According to McHugh *et al.* (1994), the best conditions for solution-cast film formation are achieved at neutral pH with 10% (w/w) whey protein solution heated for 30 min. at 90°C. Chae and Heo (1997) also studied edible film properties of whey protein concentrate solutions heated for 10 min. between 75 and 95°C at 6.75-9.0 pH range. The change of heating temperature and pH gave solution-cast edible films with tensile strength and elongation at break between 1.3 and 3.8 MPa, and 13 and 39%, and water vapor permeability between 1.07 and 1.54 kPa.m^2.h, respectively. In general, the increase of heating temperature increased the tensile strength of films, but reduced their elongation at break values. Moreover, the increase of pH reduced the tensile strength, but increased the elongation at break of films (Chae and Heo, 1997). Heating at 90°C and pH 6.75 gave the highest tensile strength, while heating at 80°C and pH 9.0 gave the most flexible films. In contrast, the increased heating temperature at different pH did not cause a considerable reduction in water vapor permeability of films with the exception of pH 8.0 or 9.0 at 95°C that gave films with slightly lower water vapor permeability (Chae and Heo, 1997). Pérez-Gago *et al.* (1999) did not determine any considerable differences among the water vapor permeability of heat denatured (for 30 min. at 90°C) solution-cast whey protein films obtained at pH 6.0, 7.0, and 8.0, but films produced at pH 4.0 and 5.0 showed 40 to 70% higher water vapor permeability than films obtained at higher pHs, respectively. The plasticization is an important step during preparation of whey protein films since they turn brittle during storage as in most heat-denatured protein films. Glycerol at whey:glycerol ratios of 1:0.3 and 1:0.4 are frequently applied for plasticization of whey protein films to obtain acceptable film mechanical properties (Chae and Heo, 1997; Pérez-Gago *et al.*, 1999), but it was reported that the sorbitol gave films having higher tensile strength than those plasticized with glycerol (Chae and Heo, 1997).

Although the classical heat denaturation method gives insoluble whey protein films with acceptable mechanical and barrier properties, enzymatic crosslinking with transglutaminase is also highly effective in polymerizing whey proteins and improving their mechanical and barrier properties. It was reported that crosslinking of whey proteins with transglutaminase caused five-fold reduction in oxygen permeability and two-fold reduction in water vapor transmission rate of cross-linked solution-cast whey protein films (Schmid *et al.*, 2014). Jiang *et al.* (2019) applied transglutaminase cross-linking for solution-cast composite films of heat denatured (80°C for 10 min.) whey protein with nanocrystalline cellulose (NCC). It was reported that the addition of NCC at 10 and 15% (w/w of whey protein) without enzymatic crosslinking improved the tensile strength and water vapor permeability of control whey films by 50 and 60%, and 25 and 39%, respectively. The transglutaminase cross-linking did not make a considerable contribution in the tensile strength of composite films, but improved their flexibility and water vapor permeability (Jiang *et al.*, 2019). Di Pierro *et al.* (2013) applied transglutaminase cross-linking strategy to composite films of heat denatured (80°C for 30 min.) whey protein with pectin used at the level of 25% (w/w of whey protein). It was determined that the combination of whey protein with pectin at pH 5.1 in the presence of transglutaminase formed a whey-pectin complex that gave films possessing better mechanical (50 and 100% higher tensile strength and elongation at break, respectively) and barrier properties (36% lower oxygen and water vapor permeability) than uncross-linked composite films. Marquez *et al.* (2014) proved that the transglutaminase cross-linked whey protein-pectin composite coating having similar formulation with that of Di Pierro *et al.* (2013) acted as a water vapor barrier when it was applied to biscuits stored at 25°C and 50% RH. Moreover, it was also reported that the whey-pectin composite films were effective in preventing water loss and oil uptake of French-fries and doughnuts during frying.

The whey proteins can also be utilized into edible films by compression molding. Sothornvit *et al.* (2007) reported that the minimum temperature for sheet formation of whey proteins in compression molding was 104°C, while maximum temperature of compression molding without thermal degradation was 140°C. These workers also determined that the compression molding temperature (104, 127 or 140°C) and pressure (0.81 or 2.25 MPa) had a limited effect on mechanical properties of films. The whey protein films compression molded at 140°C were transparent and showed higher tensile strength than classical solution-cast whey protein films (Sothornvit *et al.*, 2007). Azevedo *et al.* (2017) obtained whey protein-corn starch (1:1) composite pellets in a twin-screw extruder, and processed these pellets into sheets in a single-screw extruder. The sheets were then compression molded at 115°C, using a hydraulic press to obtain whey protein-starch composite films. The whey-starch composite films showed 47% lower water vapor permeability (0/75% RH gradient at 25°C) than compression molded corn starch films, but composite film making caused significant reductions in film flexibility and puncture strength (Azevedo *et al.*, 2017).

Similar to most of the proteins, the globular structure of whey proteins interferes with their electrospinning due to lack of entanglements and sufficient intermolecular interactions necessary during electrospinning. Therefore, whey proteins should be mixed with carrier hydrocolloids to obtain electrospun nanofibers. For example, the blending of whey protein isolate with pullulan at proper ratios (e.g. 50:50, 30:70 and 20:80) increased viscosity and lowered conductivity of the whey protein solutions. Thus it gave uniform bread-free edible nanofibers with diameters of around 231 nm (Drosou *et al.*, 2018). Sullivan *et al.* (2014) also obtained uniform electrospun nanofibers within 312-690 nm range by mixing whey protein isolate with polyethylene glycol at different ratios. Moreover, Aslaner *et al.* (2021) also developed electrospun nanofibers with antioxidant activity by mixing whey protein concentrate at 4% (w/v) with rye flour at 4 or 6% (w/v), polyethylene glycol at 2% (w/v), and grape seed extract at 20% (w/w). The use of polyethylene glycol in electrospun whey protein nanofibers interferes with their edible food packaging applications, but these nanofibers can be evaluated as food contact packaging if their polyethylene glycol levels are optimized according to regulations (Younes *et al.*, 2018).

3.2.8   Packaging from pulse proteins

Pea proteins are the most extensively studied pulse proteins for their edible films. The majority of the pea protein-film studies are related to solution-cast films. It was determined that the pea proteins treated by the classical heat denaturation process (exceeding 5 min. at 90°C) gave strong, but flexible edible films when they were plasticized with glycerol at 70:30 or 60:40 protein/plasticizer ratios (Choi and Han, 2001; Choi and Han, 2002). The mechanism of film formation for heat-denatured pea protein isolates was attributed mainly to intermolecular disulfide bond formation (Choi and Han, 2001). Acquah *et al.* (2020) reported similar flexibilities for pea protein and whey protein isolate films plasticized with glycerol (protein/plasticizer ratios: 50/50), but the pea protein isolate film showed almost 2.6-fold lower tensile strength than whey protein isolate film. Viroben *et al.* (2000) reported that the commercial pea proteins obtained with isoelectric precipitation (containing mainly globulins) and ultrafiltration (containing both albumins and globulins) gave solution-cast edible films with similar mechanical properties and surface hydrophobicity. The pea proteins could also be used to obtain extruded films. For example, Liu *et al.* (2010) applied compression molding at 140°C to obtain pea protein films that showed promising flexibility and mechanical strength (elongation at break: 62.6%, tensile strength: 7.1 MPa). These authors also blended pea proteins with zein to improve flexibility of compression molded zein films, but the desired flexible films were obtained only when pea protein weight in blends dominated over zein weigh (zein:pea protein ratios at 40:60, 30:70, 20:80 and 10:90). In contrast, Klüver and Meyer (2015) reported that compression molding at 160°C failed to give coherent films from extruded pea protein sheets. Huntrakul *et al.* (2020) incorporated pea proteins up to 20% into blown acetylated starch films. These authors reported that the incorporated pea protein reduced the stickiness of extruded starch, and improved the barrier properties and structural integrity of blown starch films. Although classical solution-casting and compression molding are applicable to obtain pure pea protein films, the globular nature of pea proteins interferes with their electrospinning used to obtain nanofiber mats. However, recent studies have shown that pea protein-pullulan blends are electrospinnable and give smooth nanofiber mats (Aguilar-Vázquez *et al.*, 2018; Jia *et al.*, 2020). It was also determined that the pea protein content of a blend with pullulan must be kept below 70% to obtain an electrospinnable material having the desired nanofiber morphology (Jia *et al.*, 2020). Jia *et al.* (2020) also effectively used thermal cross-linking (3 h at 120°C) to reduce the solubility of nanofibers obtained from pea protein-pullulan blends.

Chickpea proteins are also among the potential pulse sources for edible films. However, although there has been a recent trend in production and use of chickpea protein as a functional ingredient in food formulations, there are only limited reports about development of edible films from chickpea proteins (Meshkani *et al.*, 2011; Meshkani *et al.*, 2012; Gücbilmez *et al.*, 2007). However, chickpea flour, that is mainly a mixture of protein and starch, has recently attracted an increased interest as an edible film-making material (Diaz *et al.*, 2019; Kocakulak *et al.*, 2019). Further studies are needed to understand the characteristics of pure chickpea protein film and its films with other hydrocolloids and lipids.

Different studies suggest that lentil protein isolates or concentrates can also be potential sources for edible films and coatings. Bamdad *et al.* (2006) obtained edible films from lentil protein concentrate (~69% protein) using glycerol as plasticizer (protein:glycerol ratio: 2:1) and reported that these films showed comparable flexibility and tensile strength with soy and pea protein films (protein:glycerol ratio: 7:3). De Apodaca *et al.* (2020) showed that use of sorbitol instead of glycerol for plasticization of lentil protein concentrate (63% protein) films gave more mechanically stable films with almost 9.5 and 60-fold lower water vapor and oxygen permeability rates, respectively. The food applications of edible lentil protein coatings are scarce. However, Zurlini *et al.* (2018) successfully employed lentil protein-based edible composite coatings (6% lentil protein + 2.5% linoleum oil or 6% lentil protein + 2.5% linoleum oil + 1% chitosan + 0.5% lactic acid) to prevent weight loss and spoilage of fresh truffles without interfering with their desired flavor characteristics. Further studies are needed to increase the use of pulse proteins in development of edible films and coatings.

3.2.9 Packaging from potato proteins

The high functionality and increased production of commercial potato protein products have initiated studies related to the use of these proteins as a source of edible films. However, recent findings indicate that potato protein films produced by classical plasticization and solution-casting methods suffer from brittleness and brown discoloration. Schäfer *et al.* (2018) reported that the potato protein isolates (PPIs) obtained by ultrafiltration and diafiltration of potato juice gave solution-cast films with satisfactory mechanical properties only when they were plasticized by using high amounts of glycerol (potato protein isolate:glycerol ratio $\geq$ 1:1, w/w). However, these authors also reported that plasticizers, like ethylene glycol, propylene glycol, sorbitol or propylene glycol 400 were not compatible with PPI and gave highly brittle or creamy films, depending on their concentrations. Moreover, PPI gave brown colored solution-cast films having significantly inferior mechanical and barrier properties than solution-cast whey protein films (Schäfer *et al.*, 2018). Du *et al.* (2005) obtained compression molded (at 155°C and 7.5-10 mPa) films from different PPI (showed variations in MW profiles) preparations plasticized with 30 to 50% (w/w) glycerol. These workers also modified the PPI films with alkali treatment (at pH 12 for 2 h at room temperature) or with reducing agent (Na_2SO_3). The films from PPI, rich in high molecular weight proteins (mainly > 37 MW), gave significantly higher tensile strength and Young's modulus than films from PPI rich in low molecular weight proteins (mainly < 25 kDa). The reducing agents did not cause dramatic changes in mechanical properties of PPI films, while alkali treatment caused opposite effects, depending on MW profile (high MW profile PPI gave more elastic films while low MW profile PPI gave stiffer films) (Du *et al.*, 2005). Newson *et al.* (2015) also used compression molding (at 100 to 190°C and 100 kN) to obtain edible potato protein concentrate (PPC) films, using 15-25% glycerol as plasticizer. These workers reported that heat induces modifications in PPC at 170°C in the presence of 25% glycerol yielded films showing some flexibility (with ~40% elongation at break). Muneer *et al.* (2019) showed that pressing temperature that gave extensible PPC films can be reduced by almost 20°C (from 150 to 130°C) in the presence of 30% glycerol when PPC was pre-treated at 75°C for 30 min. at alkaline medium (at pH 10). The FTIR of these films suggested that the alkaline treatment caused secondary structural changes (mainly reduced β-sheets) in potato proteins (Muneer *et al.*, 2019). Further studies are needed to improve the characteristics and color problems of compression-molded potato protein films and to show their potential food applications.

3.2.10 Packaging from rice proteins

The development of edible films from rice proteins has some difficulties since these films suffer from their poor mechanical and barrier properties. Shih (1996) combined rice protein concentrate with pullulan, but mechanical properties and water vapor permeability of films improved only slightly to moderately when pullulan weight was increased in the rice protein concentrate (rice protein:pullulan ratio: 3:4) and canola oil coating was applied on the film surface. Cho *et al.* (1998) found that rice protein concentrate films with highest tensile strength and lowest water vapor permeability were obtained at pH 11.0 when glycerol (0.2 g/g of rice protein concentrate) was used as a plasticizer. Recently, rice protein concentrate films with or without essential oils were successfully employed as coating for hen eggs to improve their internal quality and shell strength, and to reduce their weight loss (da Silve Pires *et al.*, 2019; da Silva Pires *et al.*, 2020a). Da Silva Pires *et al.* (2020b) reported that the highest egg quality was obtained when rice protein concentrate coating was plasticized with sorbitol, instead of glycerol and propylene glycol.

The rice bran protein concentrate (RBPC) was also used for development of edible films. For example, Adebiyi *et al.* (2008) developed edible RBPC films, using glycerol as plasticizer. These authors determined minimum solubility of films at pH 3.0. Moreover, they found that mechanical stability (puncture strength) of films increased as pH increased up to 8.0. Schmidt *et al.* (2015) also worked with RBPC edible films plasticized with glycerol (0.2 or 0.4 g glycerol/g RBPC) and found that water solubility of these films changed between 25 and 37%. These authors tried to improve the mechanical properties and water vapor permeability of RBPC films by addition of montmorillonite clay (01 g clay/g RBC), but no improvements were obtained in the properties of composite films. Shin *et al.* (2011) improved the tensile strength of RBPC films significantly by employing fructose (2% of film forming solution) as a plasticizer and by using composite film making with gelatin (RBPC:gelatin ratio: 1:1), but this strategy was not successful in improving the water vapor permeability of films. There are also some efforts to employ RBPC as food- coating material. Shin (2012) employed grapefruit seed extract loaded RBPC coatings to improve post-harvest quality of strawberries, while Kim *et al.* (2004) coated brown rice with RBPC to increase their storage stability. Further studies are needed to develop mechanical and barrier properties of rice protein-based films and to evaluate these underutilized proteins as food packaging and coating material.

3.3 Polysaccharide-based edible packaging

3.3.1 Packaging from starch

Starch is the most important polysaccharide source for edible films and coatings. However, the understanding of factors effective on characteristics of starch edible films is difficult since starches obtained from different sources might show great variations in their properties (e.g. granule shape, molecular weight, crystallinity, and amylose/amylopectin ratio, etc.). In some starch types, increased amylose content affects the mechanical (mostly increases tensile strengths while reducing elongation at break values) or barrier properties of their edible films. Lourdin *et al.* (1995) reported that solution-cast films from wheat (26% amylose) and pea starches (33% amylose) had similar tensile strengths, but these films showed significantly higher tensile strengths than potato starch (19% amylose) films. These authors also showed positive contribution of amylose content on tensile strength for different maize starch films prepared by blending waxy maize starch (0% amylose) and high-amylose maize starch (70% amylose). Interestingly, increased amylose content of maize-starch blends caused also an unexpected increase in their elongation at break values. Alves *et al.* (2007) also showed that the enrichment of cassava starch with pure cassava amylose gave solution-cast films having higher tensile strengths. Domene-López *et al.* (2019) also found a positive correlation between amylose contents of different starch types, such as corn (24.8% amylose), wheat (24.5% amylose), potato (20.5%), and rice (16.9%) starch and tensile strengths of their solution-cast edible films. Basiak *et al.* (2017) determined that edible solution-cast films of potato starch with 20% amylose showed significantly lower water vapor permeability (at 0-75% RH gradient) than solution-cast films of wheat and corn starch having 25 and 27% amylose, respectively. In contrast, Talja *et al.* (2008) did not find any significant relationship between amylose contents (varied between 11.9 and 20.1%) of starches from 11 different potato cultivars and mechanical properties or water vapor permeability of their solution-cast edible films. Cano *et al.* (2014) did not also find considerable differences among the water vapor and oxygen permeability of solution-cast films from pea, potato, and cassava starches that had 24.9, 17.9 and 9% amylose content, respectively. Similarly, no significant differences were found among mechanical and water vapor permeability values of solution-cast films of gelatinized corn, cassava, and yam starches having 25, 19 and 29% amylose contents, respectively (Mali *et al.*, 2006). Thus, in some starches, factors other than amylose content, such as molecular weight or crystallinity, or granule size should also be effective in mechanical and barrier properties of films. Analysis of studies with 100% amylose and amylopectin films might help to better understand the effects of different factors on mechanical and barrier properties of starch films. Rindlav-Westling *et al.* (1998) showed that tensile strength as well as gas and water vapor barrier properties of solution-cast potato amylose films are significantly better than those of potato amylopectin films. They thought that differences between barrier properties of films were related to their ability to form stable crystals within their matrices and variations in their other structural features. They found that amylose films were denser than amylopectin films, and contained stable amylose crystals (~34%) independent from presence of plasticizer glycerol. In contrast, these workers also determined that amylopectin existed in amorphous form in its unplasticized dry films and formed crystals only in the presence of glycerol at high RH film-drying conditions (Rindlav-Westling *et al.*, 1998). Forssell *et al.* (2002) studied with solution-cast potato amylose and waxy maize amylopectin films showed that these films had excellent oxygen barrier properties close to that of polyethylene laminated ethylene vinyl alcohol films at dry conditions (films with less than 15% moisture content at RH less than 60%). However, increased moisture content of films or RH above 80% caused loss of film gas barrier properties. It was also reported that the amylopectin films showed considerably higher sensitivity against RH than amylose films (Forssell *et al.*, 2002). These reports suggest better solution-cast film-making properties of amylose than amylopectin, but the majority of starches contain amylopectin as a major fraction, and this weakens the desired properties of the obtained films.

    Starch is also one of the promising hydrocolloids used to obtain extruded films. Pure starch in dry form cannot be melt due to its semi-crystalline structure. However, in the presence of plasticizers, like water at moderate levels between 10 and 30% (w/w of starch), extensive hydrogen-bonding of diffused water with starch causes loss of crystalline structure. Thus, melting of starch under elevated temperature and pressure is possible (thermoplastic starch is formed) (Shanks and Kong, 2012). The amylose/amylopectin ratio is highly effective on properties of extruded starch films. It was found that high amylose starches are more suitable for extrusion processing since linear and low molecular weight amylose molecules flow better than branched high molecular weight amylopectin (Shanks and Kong, 2012). Van Soest and Essers (1997) mixed different starches, waxy corn starch (<1% amylose) and high amylose starches (70% amylose), and glycerol at 30% (w/w of starch) to obtain extruded starch films having nine different amylose/amylopectin ratios. These workers aged their films at 20°C for two weeks between 40 and 65% RH before mechanical tests. It was reported that similar to solution-cast films, the gradual increase of amylose content (from 0 to 70%) improved the tensile strength (from 0.5 to 8.5 MPa), but reduced the elongation at break (from 453 to 53%) of extruded starch films having 12% moisture content. Both amylose and amylopectin showed crystallization during aging of the extruded materials, but it was proposed that the crystallization of amylose into single and double helical structures was highly related to increased mechanical strength of aged extruded starch materials (Van Soest and Essers, 1997). Li *et al.* (2011) also obtained extruded films using starches having different amylose contents (waxy, regular, Gelose 50, and Gelose 80 starches with 4.3, 29.0, 61.5 and 77.4% amylose contents, respectively), and employing water as a plasticizer, but they aged their films at room temperature for one week at 75% RH (Fig. 3.5). The results of these authors also proved that there is a positive correlation between amylose content of starch and tensile strength of obtained films (varying between 12.68 and 23.82 KPa). In contrast, they determined that only a slight increase in elongation at break (changed between 2.28 and 4.95%) of films occurred by increased amylose content. Li *et al.* (2011) attributed the improved mechanical strength of films by increased amylose content to easy entanglement of long linear amylose chains within the film matrix and self-reinforcement effect of retained granular structures in high-amylose films. The starch films can also be obtained by the blown film extrusion method, but due to the fragile nature of blown starch films, more mechanically stable composite blown films of starch were developed with different hydrocolloids, such as chitosan (Dang *et al.*, 2015), gelatin-beeswax (Cheng *et al.*, 2021), and gelatin-cellulose (Rodríguez-Castellanos *et al.*, 2015).

Different workers also developed compression molded starch films. The compression molded films obtained from plasticized high amylose starches are also generally more mechanically stable than those films from normal thermoplastic starches (Thunwall *et al.*, 2006). The composite film-making method can be

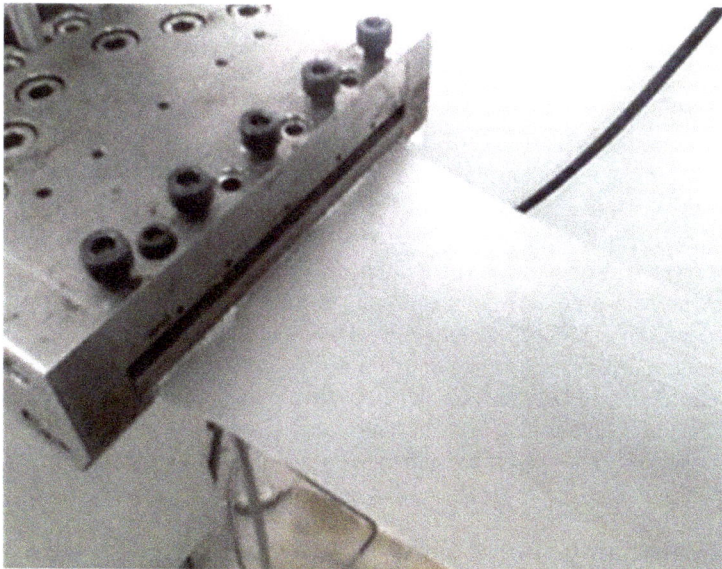

Fig. 3.5: A typical slit-die extruded corn starch film; reprinted with permission from Li *et al.* (2011); copyright © 2011 Elsevier Ltd.

employed to improve the mechanical properties of compression molded starch films. An example study of using this method is that of Hietala *et al.* (2013), who increased the tensile strength of extruded compression molded starch films by 34 to 100% by incorporation of 5 to 20% (w/w of composite film) cellulose nanofibers. Recent trends show that compression molded starch-based composites might also serve valorization of agro-industrial waste as edible packaging materials. For example, Luchese *et al.* (2018) employed compression molding at 150°C to obtain colorful cassava starch-based composite films with reduced water solubility and improved thermal stability by using blueberry waste obtained from fruit juice production as a film component. A similar approach was also applied by Collazo-Bigliardi *et al.* (2018), who used compression molding at 160°C to develop more mechanically stable starch-based composites with cellulosic material obtained from cocoa or rice husk. Compression molding can also be used to develop active edible packaging materials if the active components are protected from thermal degradation and complexation reactions by encapsulation. Talón *et al.* (2019) developed active starch films by compression molding of a blend composed of corn starch, microencapsulated eugenol, and glycerol (weight ratio: 1:0.35:0.3) at 150°C. The incorporation of eugenol prepared with a whey protein-based encapsulant into starch films gave stiffer active films with reduced flexibility, but encapsules improved the O_2 gas and water vapor barrier properties of films by 61 and 28%, respectively.

The electrospinning of pure starch into nanofibers is a challenging process since it needs the solubilization of starch in proper solvents (e.g. dimethyl sulfoxide). Thus, an additional process, called electro-wet-spinning (or wet electrospinning), is applied to remove undesired solvents from electrospun nonofibers (Kong and Ziegler, 2012). This process involves electrospinning of starch solubilized in a solvent (e.g. dimethyl sulfoxide) and depositing obtained electrospun nanofibers in a bath of non-solvent (e.g. ethanol) combined with the electrospinning device. The nanofibers formed within the non-solvent bath are then collected,and are further washed with the non-solvent to remove solvent residues before drying under vacuum (Kong and Ziegler, 2012). Moreover, the use of high amylose starches is also more beneficial than using normal starch since linear amylose molecules more easily involve in entanglements during the electrospinning process (Kong and Ziegler, 2012; Kong and Ziegler, 2014). Fonseca *et al.* (2019) obtained water-soluble carvacrol-loaded starch nanofibers with diameters ranging between 73 and 95 nm by preparing potato starch solutions in formic acid solutions. The carvacrol increased not only the electrospinnability of starch, but it also helped in developing active nanofiber mats showing antioxidant and antimicrobial activity. These authors also used the same method to develop thyme-incorporated starch nanofibers (Fonseca *et al.*, 2020). However, further studies are needed to improve solvent removal methods from starch nanofibers obtained by using dimethyl sulfoxide and formic acid.

3.3.2 Packaging from chitosan

Due to its unique inherent antimicrobial activity, chitosan is used extensively to obtain edible films and coatings. The chitosan is also a perfect material to form blend or composite films and coatings with other compatible hydrocolloids (e.g. gelatin, caseinates, whey proteins, alginate, carrageenan, and pectin) (Elsabee and Abdou, 2013). Moreover, research studies related to composites of chitosan with nanoclays have attracted interest since cation exchange and H-bonding of clays with cationic chitosan create strong interactions, and give stiffer and more mechanically stable composite films than pure chitosan films. Wang *et al.* (2005b) combined chitosan and montmorillonite clay (at 2.5-10%, w/w of film forming solution) to obtain solution-cast nanocomposite films having 1.1-1.4-fold higher elastic modulus and firmness than control films. Lavorgna *et al.* (2010) also determined improvements in the tensile strength and water vapor barrier properties of solution-cast chitosan nanocomposite films with montmorillonite clay (at 3 or 10%, w/w of chitosan) prepared with or without using plasticizer glycerol. Moreover, Neji *et al.* (2020) combined chitosan with kaolinite clay (at 5 to 10%, w/v of film forming solution) to obtain solution-cast nanocomposite films showing 2.5 to 6.7-fold higher tensile strength than the control chitosan films. The composite film making with kaolinite reduced films' flexibility, and caused formation of translucent or opaque films, depending on clay concentration. The antibacterial activity of chitosan-kaolinite composite films tested against five different bacteria did not change considerably, but a reduction was observed in antifungal activity of films tested against three different fungi (Neji *et al.*, 2020).

An important development related to chitosan film studies is that this hydrocolloid has been recently tested as a potential sausage casing as an alternative to collagen. Adzaly *et al.* (2015) reported that chitosan films containing 50% (w/w of chitosan) glycerol, 2.2% (w/v of film forming solution) cinnamaldehyde, and 0.2% (w/w of chitosan) Tween 80 showed similar mechanical properties, but lower water solubility, and superior transparency and UV-light barrier properties than collagen films. These authors also presented a continuous chitosan casing production line that includes solution mixing, casting, film drying, and peeling sections, and a final casing-shaping section that seal one side of the rolled film with acetic acid and pressure (Adzaly *et al.*, 2015). Further studies by Adzaly *et al.* (2016) also showed considerably lower water vapor, oxygen and smoke permeability of chitosan casing than collagen casing. Moreover, these workers also conducted sausage production from the developed chitosan casings (Fig. 3.6).

Pure chitosan cannot be used in production of extruded films, and its films cannot be heat-sealed since it is not a thermoplastic material (it degrades before it melts) (van den Broek *et al.*, 2015). Thus, chitosan is mostly mixed with thermoplastic materials, like starch, to obtain extruded films (Llanos *et al.*, 2021). However, self-standing chitosan films can be obtained by heat pressing. Epure *et al.* (2011) developed heat-pressed (two times pressing at 110°C) chitosan films by using glycerol at 30% (w/w of chitosan) and acetic acid at 2% (w/v of film forming mixture). Guerrero *et al.* (2019) used citric acid at 10 and 20% (w/w of chitosan) as a cross-linking agent and developed heat-pressed (at 125°C) chitosan films showing 11 and 32-fold higher elongation at break and 1.1 and 1.8-fold higher tensile strength than control chitosan films obtained without citric acid, respectively.

The inherent antimicrobial activity of chitosan on different bacteria, fungi, and viruses makes this unique hydrocolloid the most promising coating material for

Fig. 3.6: Sausages produced using chitosan casing. Before (a) and after cooking (b); reprinted with permission from Adzaly *et al.* (2016); copyright © 2016 Elsevier Ltd.

almost all food categories (e.g. fresh fruit and vegetables, meat, pork, poultry, fish and their products, dairy products, and dough food). In fresh fruits and vegetables, chitosan also shows multiple other benefits, such as positive contributions in abiotic stresses (e.g. drought and heat resistance) at the pre-harvest stage, and in physiological properties at pre- and post-harvest stages (Sharif *et al.*, 2018b; Adiletta *et al.*, 2021). The chitosan used in combination with anionic hydrocolloids (e.g. alginate, pectin, carrageenan, xanthan, gellan, etc.) is also an indispensable part of layer-by-layer coatings used in food coating applications. Moreover, bioactive substances and nutrients prepared by using chitosan as a nanoencapsulant can be incorporated into active packaging to obtain different benefits, such as sustained release, increased bioavailability, and stability (Akbari-Alavijeh *et al.*, 2020, Yemenicioğlu *et al.*, 2020).

The electrospinning of pure chitosan is very difficult due to its high viscosity. To overcome this problem, the chitosan should be dissolved in hazardous acids (e.g. trifluoroacetic acid) with addition of toxic solvents (e.g. dichloromethane) (Ohkawa *et al.*, 2004). The viscosity problem can also be overcome by alkaline hydrolysis that reduces chitosan molecular weight and increases its electrospinnability (Homayoni *et al.*, 2009). However, to obtain acceptable nanofiber quality (average diameter: 140-284 nm), the alkaline hydrolysis should last for 48 h and the treated chitosan should be solubilized in 70-90% acetic acid solution. The difficulties in electrospinning of chitosan and alternative methods of obtaining chitosan nanofibers have been discussed by de Farias *et al.* (2019) with details. However, further studies are needed to develop more applicable methods of obtaining edible nanofiber chitosan mats suitable for food packaging.

3.3.3 Packaging from pectin

Pectin is one of the most suitable hydrocolloids for coating of minimally processed fruits and vegetables (Rojas-Graü *et al.*, 2009). The pectin coatings, incorporated with antimicrobials, enzymatic browning inhibitors (e.g. ascorbic acid and derivatives, malic and citric acids, etc.), and firming agents (e.g. $CaCl_2$) are effectively used to improve the shelf-life of minimally processed fruits and vegetables (Yemenicioğlu, 2017). Solutions of high methoxyl pectin can be applied by dipping, brushing or spraying to form the desired coating on the food surface. After draining of excessive film solutions on the product surface, such films might be fixed on to the product surface by applying mild drying. Alternatively, the films might be left wet to act as a reservoir for water and to prevent water loss from fresh product by evaporation. In this case, the use of low methoxyl pectin and fixation of film solution on to the food surface simply by chemical crosslinking using $CaCl_2$ (applied by dipping or spraying) could be more suitable. Due to their anionic properties, low- or high-methoxyl pectins are also frequently employed to obtain layer-by-layer food coatings with oppositely-charged hydrocolloids, like cationic chitosan (Brasil *et al.*, 2012; Wei *et al.*, 2018). Moreover, pectin can also be used to obtain emulsion-based coatings applied mainly for preservation of whole or minimally processed fruits. The use of essential oils in emulsion film formation is very advantageous since this provides antimicrobial activity while use of waxes in emulsion films helps mainly to improve water vapor barrier properties of films. Some examples of pectin-based emulsion coatings and applied products are: pectin-beeswax emulsion coating for whole avocadoes (Maftoonazad and Ramaswamy, 2008), pectin-orange essential oil emulsion coating for fresh-cut oranges (Radi *et al.*, 2018), pectin-eugenol emulsion coating for whole melons (Çavdaroğlu *et al.*, 2020), and pectin-cinnamon leaf oil coating for grapes (Melgarejo-Flores *et al.*, 2013).

Although studies about edible pectin films have concentrated mainly on coatings, this hydrocolloid can also be used to obtain solution-cast or extruded self-standing films or casings. Pure pectin can be thermoplasticized by blending with glycerol and water at different ratios. For example, de Oliveira *et al.* (2021) obtained compression molded (at 110°C) films from thermoplasticized pectin prepared by blending pectin:glycerol:water (pH 7.0) or pectin:glycerol:citric acid solution (pH 2.0) in the ratio of 49:30:21. These authors reported that the pH was not an effective factor on water vapor barrier properties of films. The pH did not also affect the tensile strength of films, but films obtained at pH 7.0 showed almost two-fold greater puncture strength than those obtained at pH 2.0 (de Oliveira *et al.*, 2021). Gouveia *et al.* (2019) also developed compression-molded (at 120°C) films from thermoplastic pectin obtained by mixing pectin with choline chloride-glycerol plasticizer mixture (1:2) used at 30% (w/w of pectin). The thermoplasticized pectin can also be utilized in extruded sausage casings (Liu *et al.*, 2005). Liu *et al.* (2007) obtained pectin-based extruded casings from emulsions of pectin with corn oil or olive oil at 2.5 or 5% (w/w). It was reported that the sausages filled into pectin-corn oil casings showed better overall sensory properties than those filled into pectin-olive oil casings at the end of six days of cold storage. Moreover, sausages filled into pectin-based casing with corn oil showed considerably lower lipid oxidation than other casings. The study of Gamboni *et al.* (2019) showed that blending pectin with starch (1:1) provides different advantages, such as possibility of using higher extrusion temperatures and obtaining better flexibility for blend films than pure pectin films.

Pectin is among the hydrocolloids that need the presence of carrier polymers (e.g. polyethylene oxide) to obtain nanofibers by the electrospinning process (Cui *et al.*, 2016). However, blends of pectin with electrospinnable hydrocolloids, such as pullulan, can be employed to obtain nanofibers and their mats suitable for delivery of active compounds or probiotic bacteria (Liu *et al.*, 2016).

3.3.4   Packaging from sodium alginate

Edible coatings of sodium alginate have long been used as an ideal coating for meat, poultry, and fish since these coatings prevent surface dehydration of these products by acting as a water reservoir (Lindstrom *et al.*, 1992). The foods might be coated with sodium alginate by dipping into film-forming solution or by spraying of film-forming solution on to the food surface. The film formation occurs by applying $CaCl_2$ and causing gelation of the coating at the food surface. Sodium alginate is also one of the edible materials suitable for coating minimally processed fruits and vegetables. For inhibition of enzymatic browning, sodium alginate films are generally incorporated with polyphenol oxidase (PPO) enzyme inhibitors (e.g. ascorbic acid and derivatives) and calcium salts that improve tissue integrity and firmness of plant tissues (Yemenicioğlu, 2016; Rojas-Graü *et al.*, 2009). Sodium alginate has also been combined with chitosan by different workers to obtain a layer-by-layer antimicrobial coating for different products, such as fresh-cut fruits (e.g. melon, mango) (Poverenov *et al.*, 2014; Souza *et al.*, 2015) or seafood (e.g. shrimp) (Kim *et al.*, 2018). There are also studies to obtain composite or blend self-standing films by combining sodium alginate with other hydrocolloids, especially with gelatin, that suffers from brittleness or poor mechanical properties. For example, Dou *et al.* (2018) developed edible gelatin-based films by adding a small amount of sodium alginate (3%, w/w of gelatin) into gelatin (4% w/v) film-forming solutions. It was reported that the incorporation of tea polyphenols into gelatin-alginate blend films between 0.4 and 2.0% (w/w of gelatin) improved the tensile strength and water vapor barrier properties as well as the antioxidant activity of these films in a concentration-dependent manner. Syarifuddin *et al.* (2017) also obtained some gelatin-sodium alginate blend films by mixing larger amounts of sodium alginate with gelatin (sodium alginate:gelatin ratios: 0.5, 1 and 2.5). These authors reported that the tensile strength of blend films improved as sodium alginate content in the films increased. It is also noteworthy to report the work by Liu *et al.* (2007) who developed extruded casings from gelatin-sodium alginate blends (gelatin:sodium alginate ratio: 0.25) and applied these casings to obtain pork sausages. However, shrinkage of such blend casing during storage of sausages was a problem that interfered with their industrial application. Moreover, extruded alginate or its extruded blends with potato starch, pea proteins, and cellulose have also been used commercially as (wet or dry) sausage casing (Harper *et al.*, 2015; Marcos *et al.*, 2020), but this application needs careful minimization of cross-linking agent $CaCl_2$ to control formation of efflorescence on the product surface during storage (Hilbig *et al.*, 2020).

One of the most important applications of alginate involves its use as an encapsulant for probiotic cultures. The microcapsules based on alginates, blends of alginates with milk or whey proteins, and chitosan-coated alginates obtained by using layer-by-layer strategy have been employed to improve viability and colonization of probiotics under gastric conditions (Nishinari *et al.*, 2018; Shori,

2017). Thus, there is also a growing interest in the use of alginate-based films and coatings in bioactive packaging loaded with probiotics and prebiotics. Some of the examples for many different studies in this field include those of Shahrampour *et al.* (2020), who developed and characterized alginate-pectin probiotic edible films containing *Lactobacillus plantarum* KMC 45 and Bambace *et al.* (2019), who used *Lactobacillus rhamnosus* CECT 8361 with alginate and prebiotics, like inulin and oligo fructose, to develop probiotic, prebiotic, and antimicrobial coatings suitable for fresh-blueberries. Moreover, Ye *et al.* (2018) also loaded *Lactococcus lactis* into sodium alginate-sodium carboxymethyl cellulose films to obtain protective films against pathogenic growth on food surfaces.

The use of thermal processing methods for development of extruded alginate films is difficult since thermal degradation of alginate occurs before reaching its 'molten state' (Gao *et al.*, 2017a). Therefore, it is necessary to obtain thermoplastic sodium alginate by use of polyol plasticizers, such as glycerol and sorbitol. Gao *et al.* (2017b) obtained flexible edible films by employing thermomechanical mixing (at 80°C) of sodium alginate-glycerol and sodium alginate-sorbitol mixtures (alginate:polyol ratios between 1 and 4) prior to compression molding at 120°C. It was reported that the elongation at break of compression molded alginate films plasticized with glycerol or sorbitol increased up to 73-74% when alginate:polyol ratio was 1 (Gao *et al.*, 2017b). Thus, incorporation of compression molded alginate films with different natural antimicrobials and/or antioxidants might be a promising method to obtain active packaging materials.

The electrospinning of sodium alginate solutions to obtain nanofibers is a challenging process since this hydrocolloid's rigid extended conformation in aqueous solutions lacks the necessary entanglements (Nie *et al.*, 2008). Nie *et al.* (2008) reported that the use of glycerol solutions (glycerol:water ratio: 2) as a solvent for sodium alginate reduced the rigidity (increased flexibility) of alginate chains and created essential entanglements for its electrospinning. Fang *et al.* (2011) also improved the electrospinnability of sodium alginate by adding some $CaCl_2$ into its solutions and creating some entanglements among alginate chains. Therefore, there is some potential in development of alginate nanofiber mats loaded with active agents.

3.3.5 Packaging from agar and carrageenan

The polysaccharides obtained from seaweeds grown with sustainable methods have been attracting increased interest as packaging materials. The characteristics of films from major seaweed polysaccharides, agar and carrageenan, are affected from their gelation conditions. Heating and cooling of kappa-carrageenan and iota-carrageenan solutions cause thermo-reversible gel formation in the presence of specific cations, K^+ and Ca^{++}, respectively. In contrast, lamda-carrageenan cannot form gels, but has the ability to form viscous solutions (Stanley, 1987). The carrageenans are soluble in boiling water and their melting and gelation temperatures are between 50 and 70°C, and 30 and 50°C, respectively. The exact gel-formation mechanism of carrageenans has not been fully clarified, but it is thought that the gel formation occurs by transformation of carrageenan structure from coil to helix form by heating, and subsequent formation of a three-dimensional network during cooling by the interacted helical forms in the presence of cations (Geonzon *et al.*, 2020). However, the presence of cations is not essential to obtain edible carrageenan films.

Agar is also soluble in boiling water and has a melting point between 85°C and 95°C, and a gelation temperature between 32°C and 43°C. Its gels are thermo-reversible, but they do not melt unless they are heated above 85°C. Moreover, agar gels show two to 10 times greater gel strength than those of carrageenan (Khalil *et al.*, 2018). The gelation of agar occurs due to aggregation of double helical structures of agarose molecules that form a three-dimensional network by hydrogen bonding. During the production of solution-cast edible films of agar, the temperature of the film-forming solution and casting surface temperature should be above the gelling temperature of agar to prevent uncontrolled gelation during casting (Mostafavi and Zaeim, 2020). Agar forms transparent edible films that can be employed for food packaging purposes, but these films have some major drawbacks, such as brittleness, high water vapour permeability and poor thermal properties (Mostafavi and Zaeim, 2020). The kappa-carrageenan films are also transparent and show similar drawbacks with agar films, but possess good oxygen barrier properties comparable to some of the plastic films (e.g. PLA and LDPE films) (Sedayu *et al.*, 2019). Rhim (2012) compared mechanical and water vapor permeability of agar and kappa-carrageenan films plasticized with glycerol (hydrocolloid: glycerol ratio: 1:1). This worker did not find dramatic differences between tensile strengths and water vapor permeability of carrageenan and agar films while the latter gave considerably more flexible films (agar films showed a two-fold higher elongation at break than carrageenan films). However, it is important to note that due to their good film-forming abilities, both the hydrocolloids attract a significant interest as active packaging materials. Active agar and carrageenan films with polyphenols and essential oils are quite promising since these natural agents give not only antimicrobial and antioxidant properties to these films, but may also modify the film matrix to improve their water vapor and/or gas barrier properties (Shojaee-Aliabadi et al., 2013; Shojaee-Aliabadi *et al.*, 2014; Liu *et al.*, 2020; Campa-Siqueiros *et al.*, 2020; Moreno *et al.*, 2020). Finally, aqueous solutions of agar and carrageenan cannot be transformed into nanofibers by the electrospinning method (Stijnman *et al.*, 2011). Pure agar can be transformed into microfibers by the wet-electrospinning (Bao *et al.*, 2010; Liu *et al.*, 2018), but the use of this process for development of edible fiber-based materials faces some difficulties since frequently used solvents (e.g. dimethyl sulfoxide) used in this method cause health concerns.

References

Abedi, E. and K. Pourmohammadi (2021). Physical modifications of wheat gluten protein: An extensive review, *J. Food Process Eng.*, 44: e13619.

Acquah, C., Y. Zhang, M.A Dubé and C.C. Udenigwe (2020). Formation and characterization of protein-based films from yellow pea (*Pisum sativum*) protein isolate and concentrate for edible applications, *Curr. Res. Food Sci.*, 2: 61-69.

Adebiyi, A.O., D.H. Jin, T. Ogawa and K. Muramoto (2008). Rice bran protein-based edible films, *Int. J. Food Sci. Technol.*, 43: 476-483.

Adiletta, G., M. Di Matteo and M. Petriccione (2021). Multifunctional role of chitosan edible coatings on antioxidant systems in fruit crops: A review, *Int. J. Mol. Sci.*, 22: 2633.

Adzaly, N.Z., A. Jackson, R. Villalobos-Carvajal, I. Kang and E. Almenar (2015). Development of a novel sausage casing, *J. Food Eng.*, 152: 24-31.

Adzaly, N.Z., A. Jackson, I. Kang and E. Almenar (2016). Performance of a novel casing made of chitosan under traditional sausage manufacturing conditions, *Meat Sci.*, 113: 116-123.

Aguilar-Vázquez, G., G. Loarca-Piña, J.D. Figueroa-Cárdenas and S. Mendoza (2018). Electrospun fibers from blends of pea (*Pisum sativum*) protein and pullulan, *Food Hydrocoll.*, 83: 173-181.

Akbari-Alavijeh, S., R. Shaddel and S.M. Jafari (2020). Encapsulation of food bioactives and nutraceuticals by various chitosan-based nanocarriers, *Food Hydrocoll.*, 105: 105774.

Akman, P.K., F. Bozkurt, M. Balubaid and M.T. Yilmaz (2019). Fabrication of curcumin-loaded gliadin electrospun nanofibrous structures and bioactive properties, *Fibers Polym.*, 20: 1187-1199.

Alkan, B. and A. Yemenicioglu (2016). Potential application of natural phenolic antimicrobials and edible film technology against bacterial plant pathogens, *Food Hydrocoll.*, 55: 1-10.

Alves, V.D., S. Mali, A. Beléia and M.V.E. Grossmann (2007). Effect of glycerol and amylose enrichment on cassava starch film properties, *J. Food Eng.*, 78: 941-946.

Andonegi, M., K. de la Caba and P. Guerrero (2020). Effect of citric acid on collagen sheets processed by compression, *Food Hydrocoll.*, 100: 105427.

Andreuccetti, C., R.A. Carvalho, T. Galicia-García, F. Martinez-Bustos, R. González-Nuñez and C.R. Grosso (2012). Functional properties of gelatin-based films containing *Yucca schidigera* extract produced via casting, extrusion and blown extrusion processes: A preliminary study, *J. Food Eng.*, 113: 33-40.

Ansarifar, E. and F. Moradinezhad (2021). Preservation of strawberry fruit quality via the use of active packaging with encapsulated thyme essential oil in zein nanofiber film, *Int. J. Food Sci. Tech.*, 56: 4239-4247.

Ansorena, M.R., F. Zubeldía and N.E. Marcovich (2016). Active wheat gluten films obtained by thermoplastic processing, *LWT-Food Sci. Technol.*, 69: 47-54.

Arcan, I. and A. Yemenicioğlu (2011). Incorporating phenolic compounds opens a new perspective to use zein films as flexible bioactive packaging materials, *Food Res. Int.*, 44: 550-556.

Argos, P., K. Pedersen, M.D. Marks and B.A. Larkins (1982). A structural model for maize zein proteins, *J. Biol. Chem.*, 257: 9984-9990.

Aslaner, G., G. Sumnu and S. Sahin (2021). Encapsulation of grape seed extract in rye flour and whey protein-based electrospun nanofibers, *Food Bioproc. Tech.*, 14: 1118-1131.

Avena-Bustillos, R.J. and J.M. Krochta (1993). Water vapor permeability of caseinate-based edible films as affected by pH, calcium crosslinking and lipid content, *J. Food Sci.*, 58: 904-907.

Avena-Bustillos, R.J., B.S. Chiou, C.W. Olsen, P.J. Bechtel, D.A. Olson and T.H. McHugh (2011). Gelation, oxygen permeability, and mechanical properties of mammalian and fish gelatin films, *J. Food Sci.*, 76: E519-E524.

Azevedo, V.M., S.V. Borges, J.M. Marconcini, M.I. Yoshida, A.R.S. Neto, T.C. Pereira and C.F.G. Pereira (2017). Effect of replacement of corn starch by whey protein isolate in biodegradable film blends obtained by extrusion, *Carbohydr. Polym.*, 157: 971-980.

Bai, J., V. Alleyne, R.D. Hagenmaier, J.P. Mattheis and E.A. Baldwin (2003). Formulation of zein coatings for apples (*Malus domestica* Borkh), *Postharvest Biol. Tec.*, 28: 259-268.

Bambace, M.F., M.V. Alvarez and M. del Rosario Moreira (2019). Novel functional blueberries: Fructo-oligosaccharides and probiotic lactobacilli incorporated into alginate edible coatings, *Food Res. Int.*, 122: 653-660.

Bamdad, F., A.H. Goli and M. Kadivar (2006). Preparation and characterization of proteinous film from lentil (*Lens culinaris*): Edible film from lentil (*Lens culinaris*), *Food Res. Int.*, 39: 106-111.

Bao, X., K. Hayashi, Y. Li, A. Teramoto and K. Abe (2010). Novel agarose and agar fibers: Fabrication and characterization, *Mater. Lett.*, 64: 2435-2437.

Basiak, E., A. Lenart and F. Debeaufort (2017). Effect of starch type on the physico-chemical properties of edible films, *Int. J. Biol. Macromol.*, 98: 348-356.

Batista, R.A., P.J.P. Espitia, J.D. S.S. Quintans, M.M. Freitas, M.Â. Cerqueira, J.A. Teixeira and J.C. Cardoso (2018). Hydrogel as an alternative structure for food packaging systems, *Carbohydr. Polym.*, 205: 106-116.

Batpho, K., W. Boonsupthip and C. Rachtanapun (2017). Antimicrobial activity of collagen casing impregnated with nisin against foodborne microorganisms associated with ready-to-eat sausage, *Food Control*, 73: 1342-1352.

Belyamani, I., F. Prochazka and G. Assezat (2014a). Production and characterization of sodium caseinate edible films made by blown-film extrusion, *J. Food Eng.*, 121: 39-47.

Belyamani, I., F. Prochazka, G. Assezat and F. Debeaufort (2014b). Mechanical and barrier properties of extruded film made from sodium and calcium caseinates, *Food Packag. Shelf-Life*, 2: 65-72.

Benbettaïeb, N., O. Chambin, T. Karbowiak and F. Debeaufort (2016). Release behavior of quercetin from chitosan-fish gelatin edible films influenced by electron beam irradiation, *Food Control*, 66: 315-319.

Boyacı, D., G. Iorio, G.S. Sozbilen, D. Alkan, S. Trabattoni, F. Pucillo *et al.* (2019). Development of flexible antimicrobial zein coatings with essential oils for the inhibition of critical pathogens on the surface of whole fruits: Test of coatings on inoculated melons, *Food Packag. Shelf-Life*, 20: 100316.

Boyacı, D. and A. Yemenicioğlu (2020). Development of gel-based pads loaded with lysozyme and green tea extract: Characterization of pads and test of their antilisterial potential on cold-smoked salmon, *LWT-Food Sci. Technol.*, 128: 109471.

Brasil, I.M., C. Gomes, A. Puerta-Gomez, M.E. Castell-Perez and R.G. Moreira (2012). Polysaccharide-based multilayered antimicrobial edible coating enhances quality of fresh-cut papaya, *LWT-Food Sci. Technol.*, 47: 39-45.

Brito-Oliveira, T.C., M. Bispo, I.C. Moraes, O.H. Campanella and S.C. Pinho (2017). Stability of curcumin encapsulated in solid lipid microparticles incorporated in cold-set emulsion filled gels of soy protein isolate and xanthan gum, *Food Res. Int.*, 102: 759-767.

Campa-Siqueiros, P.I., I. Vargas-Arispuro, P. Quintana-Owen, Y. Freile-Pelegrín, J.A. Azamar-Barrios and T.J. Madera-Santana (2020). Physicochemical and transport properties of biodegradable agar films impregnated with natural semiochemical based-on hydroalcoholic garlic extract, *International J. Biol. Macromol.*, 151: 27-35.

Cano, A., A. Jiménez, M. Cháfer, C. Gónzalez and A. Chiralt (2014). Effect of amylose: amylopectin ratio and rice bran addition on starch films properties, *Carbohydr. Polym.*, 111: 543-555.

Çavdaroğlu, E., S. Farris and A. Yemenicioğlu (2020). Development of pectin–eugenol emulsion coatings for inhibition of Listeria on webbed-rind melons: A comparative study with fig and citrus pectins, *Int. J. Food Sci. Tech.*, 55: 1448-1457.

Chae, S.I. and T.R. Heo (1997). Production and properties of edible film using whey protein, *Biotechnol. Bioprocess Eng.*, 2: 122-125.

Cheng, Y., W. Wang, R. Zhang, X. Zhai and H. Hou (2021). Effect of gelatin bloom values on the physicochemical properties of starch/gelatin–beeswax composite films fabricated by extrusion blowing, *Food Hydrocoll.*, 113: 106466.

Chevalier, E., G. Assezat, F. Prochazka and N. Oulahal (2018). Development and characterization of a novel edible extruded sheet based on different casein sources and influence of the glycerol concentration, *Food Hydrocoll.*, 75: 182-191.

Cho, S.Y., J.W. Park and C. Rhee (1998). Edible films from protein concentrates of rice wine meal, *Korean J. Food Sci. Technol.*, 30: 1097-1106.

Choi, W.S. and J.H. Han (2001). Physical and mechanical properties of pea-protein-based edible films, *J. Food Sci.*, 66: 319-322.

Choi, W.S. and J.H. Han (2002). Film-forming mechanism and heat denaturation effects on the physical and chemical properties of pea-protein-isolate edible films, *J. Food Sci.*, 67: 1399-1406.

Choi, I., S.E. Lee, Y. Chang, M. Lacroix and J. Han (2018). Effect of oxidized phenolic compounds on cross-linking and properties of biodegradable active packaging film composed of turmeric and gelatin, *LWT-Food Sci. Technol.*, 93: 427-433.

Ciannamea, E.M., P.M. Stefani and R.A. Ruseckaite (2016). Properties and antioxidant activity of soy protein concentrate films incorporated with red grape extract processed by casting and compression molding, *LWT-Food Sci. Technol.*, 74: 353-362.

Colak, B.Y., P. Peynichou, S. Galland, N. Oulahal, G. Assezat, F. Prochazka and P. Degraeve (2015). Active biodegradable sodium caseinate films manufactured by blown-film extrusion: Effect of thermo-mechanical processing parameters and formulation on lysozyme stability, *Ind. Crop. Prod.*, 72: 142-151.

Collazo-Bigliardi, S., R. Ortega-Toro and A. Chiralt Boix (2018). Reinforcement of thermoplastic starch films with cellulose fibers obtained from rice and coffee husks, *J. Renew. Mater.*, 6: 599-610.

Cui, S., B. Yao, X. Sun, J. Hu, Y. Zhou and Y. Liu (2016). Reducing the content of carrier polymer in pectin nanofibers by electrospinning at low loading followed with selective washing, *Mater. Sci. Eng. C*, 59: 885-893.

Dang, K.M. and R. Yoksan (2015). Development of thermoplastic starch blown film by incorporating plasticized chitosan, *Carbohydr. Polym.*, 115: 575-581.

Da Silva Pires, P.G.S., G.S. Machado, C.H. Franceschi, L. Kindlein and I. Andretta (2019). Rice protein coating in extending the shelf-life of conventional eggs, *Poult. Sci.*, 98: 1918-1924.

Da Silva Pires, P.G.S., A.F.R. Leuven, C.H. Franceschi, G.S. Machado, P.D.S. Pires, P.O. Moraes *et al.* (2020a). Effects of rice protein coating enriched with essential oils on internal quality and shelf-life of eggs during room temperature storage, *Poult. Sci.*, 99: 604-611.

Da Silva Pires, P.G.S., C. Bavaresco, A.F.R. Leuven, B.C.K. Gomes, A.K. de Souza, B.S. Prato, *et al.* (2020b). Plasticizer types affect quality and shelf-life of eggs coated with rice protein, *J. Food Sci. Technol.*, 57: 971-979.

Day, L. (2011). Wheat gluten: Production, properties and application, pp. 267-288. *In:* G.O. Phillips and P.A. Williams (Eds.). *Handbook of Food Proteins.* Woodhead Publishing, Cambridge, UK.

De Apodaca, E.D., A. Montanari, L. Fernandez-de Castro, E. Umilta, L. Arroyo, C. Zurlini, and M.C. Villaran (2020). Lentil byproducts as a source of protein for food packaging applications, *Am. J. Food Technol.*, 15: 1-10.

De Carvalho, R.A. and C.R.F. Grosso (2004). Characterization of gelatin based films modified with transglutaminase, glyoxal and formaldehyde, *Food Hydrocol.*, 18: 717-726.

De Farias, B.S., T.R.S.A.C. Junior and L.A. de Almeida Pinto (2019). Chitosan-functionalized nanofibers: A comprehensive review on challenges and prospects for food applications, *Int. J. Biol. Macromol.*, 23: 210-220.

De Oliveira, A.C.S., L.F. Ferreira, D. de Oliveira Begali, J.C. Ugucioni, A.R.de Sena Neto, M.I. Yoshida and S.V. Borges (2021). Thermoplasticized pectin by extrusion/thermo-compression for film industrial application, *J. Polym. Environ.*, 29: 2546-2556.

De Wit, J.N. and G. Klarenbeek (1984). Effects of various heat treatments on structure and solubility of whey proteins, *J. Dairy Sci.*, 67: 2701-2710.

Díaz, O., T. Ferreiro, J.L. Rodríguez-Otero and Á. Cobos (2019). Characterization of chickpea (*Cicer arietinum* L.) flour films: Effects of pH and plasticizer concentration, *Int. J. Mol. Sci.*, 20: 1246.

Di Gioia, L. and S. Guilbert (1999). Corn protein-based thermoplastic resins: Effect of some polar and amphiphilic plasticizers, *J. Agric. Food Chem.*, 47: 1254-1261.

Di Maio, E., R. Mali and S. Iannace (2010). Investigation of thermoplasticity of zein and kafirin proteins: Mixing process and mechanical properties, *J. Polym. Environ.*, 18: 626-633.

Ding, X. and P. Yao (2013). Soy protein/soy polysaccharide complex nanogels: Folic acid loading, protection, and controlled delivery, *Langmuir*, 29: 8636-8644.

Di Pierro, P., G. Rossi Marquez, L. Mariniello, A. Sorrentino, R. Villalonga and R. Porta (2013). Effect of transglutaminase on the mechanical and barrier properties of whey protein/pectin films prepared at complexation pH, *J. Agric. Food Chem.*, 61: 4593-4598.

Domene-López, D., J.C. García-Quesada, I. Martin-Gullon and M.G. Montalbán (2019). Influence of starch composition and molecular weight on physicochemical properties of biodegradable films, *Polymers*, 11: 1084.

Dou, L., B. Li, K. Zhang, X. Chu and H. Hou (2018). Physical properties and antioxidant activity of gelatin-sodium alginate edible films with tea polyphenols, *Int. J. Biol. Macromol.*, 118: 1377-1383.

Drosou, C., M. Krokida and C.G. Biliaderis (2018). Composite pullulan-whey protein nanofibers made by electrospinning: Impact of process parameters on fiber morphology and physical properties, *Food Hydrocoll.*, 77: 726-735.

Du, Y., F. Chen, Y. Zhang, C. Rempel, M.R. Thompson and Q. Liu (2015). Potato protein isolate-based hydrocolloids, *J. Appl. Polym. Sci.*, 132: 42723.

Elsabee, M.Z. and E.S. Abdou (2013). Chitosan based edible films and coatings: A review, *Mater. Sci. Eng. C*, 33: 1819-1841.

Epure, V., M. Griffon, E. Pollet and L. Avérous (2011). Structure and properties of glycerol-plasticized chitosan obtained by mechanical kneading, *Carbohydr. Polym.*, 83: 947-952.

Etxabide, A., J. Uranga, P. Guerrero and K. De la Caba (2017). Development of active gelatin films by means of valorisation of food processing waste: A review, *Food Hydrocoll.*, 68: 192-198.

Fabra, M., Talens, J.P. and A. Chiralt (2008). Tensile properties and water vapor permeability of sodium caseinate films containing oleic acid–beeswax mixtures, *J. Food Eng.*, 85: 393-400.

Fabra, M.J., A. Lopez-Rubio and J.M. Lagaron (2014). Nanostructured interlayers of zein to improve the barrier properties of high barrier polyhydroxyalkanoates and other polyesters, *J. Food Eng.*, 127: 1-9.

Fang, D., Y. Liu, S. Jiang, J. Nie and G. Ma (2011). Effect of intermolecular interaction on electrospinning of sodium alginate, *Carbohydr. Polym.*, 85: 276-279.

Farahnaky, A., S.M.M. Dadfar and M. Shahbazi (2014). Physical and mechanical properties of gelatin-clay nanocomposite, *J. Food Eng.*, 122: 78-83.

Fonseca, L.M., C.E. dos Santos Cruxen, G.P. Bruni, Â.M. Fiorentini, E. da Rosa Zavareze, L.T. Lim and A.R.G. Dias (2019). Development of antimicrobial and antioxidant electrospun soluble potato starch nanofibers loaded with carvacrol, *Int. J. Biol. Macromol.*, 139: 1182-1190.

Fonseca, L.M., M. Radünz, H.C. dos Santos Hackbart, F.T. da Silva, T.M. Camargo, G.P. Bruni *et al.* (2020). Electrospun potato starch nanofibers for thyme essential oil encapsulation: Antioxidant activity and thermal resistance, *J. Sci. Food Agric.*, 100: 4263-4271.

Forssell, P., R. Lahtinen, M. Lahelin and P. Myllärinen (2002). Oxygen permeability of amylose and amylopectin films, *Carbohydr. Polym.*, 47: 125-129.

Foulk, J.A. and J.M. Bunn (2001). Physical and barrier properties of developed bilayer protein films, *Appl. Eng. Agric.*, 17: 635.

Friesen, K., C. Chang and M. Nickerson (2015). Incorporation of phenolic compounds, rutin and epicatechin, into soy protein isolate films: Mechanical, barrier and cross-linking properties, *Food Chem.*, 172: 18-23.

Gamboni, J.E., A.M. Slavutsky and M.F. Bertuzzi (2019). Starch-pectin films obtained by extrusion and compression molding, *J. Multidiscip. Eng. Sci. Technol.*, 6: 10175-10183.

Gao, C., E. Pollet and L. Avérous (2017a). Properties of glycerol-plasticized alginate films obtained by thermo-mechanical mixing, *Food Hydrocoll.*, 63: 414-420.

Gao, C., E. Pollet and L. Avérous (2017b). Innovative plasticized alginate obtained by thermo-mechanical mixing: Effect of different biobased polyols systems, *Carbohydr. Polym.*, 157: 669-676.

Gennadios, A., C. Weller and R.F. Testin (1993). Temperature effect on oxygen permeability of edible protein-based films, *J. Food Sci.*, 58: 212-214.

Geonzon, L.C., F.B.A. Descallar, L. Du, R.G. Bacabac and S. Matsukawa (2020). Gelation mechanism and network structure in gels of carrageenans and their mixtures viewed at different length scales – A review, *Food Hydrocoll.*, 108: 106039.

Ghanbarzadeh, B., A.R. Oromiehie, M. Musavi, Z.E. D-Jomeh, E.R. Rad and J. Milani (2006). Effect of plasticizing sugars on rheological and thermal properties of zein resins and mechanical properties of zein films, *Food Res. Int.*, 39: 882-890.

Göksen, G., M.J. Fabra, H.I. Ekiz and A. López-Rubio (2020). Phytochemical-loaded electrospun nanofibers as novel active edible films: Characterization and antibacterial efficiency in cheese slices, *Food Control*, 112: 107133.

Gontard, N., S. Guilbert and J.L. Cuq (1992). Edible wheat gluten films: Influence of the main process variables on film properties using response surface methodology, *J. Food Sci.*, 57: 190-195.

Gontard, N. and S. Guilbert (1998). Edible and/or biodegradable wheat gluten films and coatings, pp. 324-328. *In:* J. Gueguen and Y. Popineau (Eds.). *Plant Proteins from European Crops, Food and Non-food Applications*, Springer, Heidelberg, Germany.

Gounga, M.E., S.Y. Xu and Z. Wang (2010). Film forming mechanism and mechanical and thermal properties of whey protein isolate-based edible films as affected by protein concentration, glycerol ratio and pullulan content, *J. Food Biochemistry*, 34: 501-519.

Gouveia, T.I., K. Biernacki, M.C. Castro, M.P. Gonçalves and H.K. Souza (2019). A new approach to develop biodegradable films based on thermoplastic pectin, *Food Hydrocoll.*, 97: 105175.

Güçbilmez, Ç.M., A. Yemenicioğlu and A. Arslanoğlu (2007). Antimicrobial and antioxidant activity of edible zein films incorporated with lysozyme, albumin proteins and disodium EDTA, *Food Res. Int.*, 40: 80-91.

Guerrero, P., A. Retegi, N. Gabilondo and K. De la Caba (2010). Mechanical and thermal properties of soy protein films processed by casting and compression, *J. Food Eng.*, 100: 145-151.

Guerrero, P., A. Muxika, I. Zarandona and K. De La Caba (2019). Crosslinking of chitosan films processed by compression molding, *Carbohydr. Polym.*, 206: 820-826.

Guo, Y.C., Z.D. Liu, H.J. An, M.Q. Li and J. Hu (2005). Nano-structure and properties of maize zein studied by atomic force microscopy, *J. Cereal Sci.*, 41: 277-281.

Guo, X., Y. Lu, H. Cui, X. Jia, H. Bai and Y. Ma (2012). Factors affecting the physical properties of edible composite film prepared from zein and wheat gluten, *Molecules*, 17: 3794-3804.

Hajjari, M.M., M.T. Golmakani, N. Sharif and M. Niakousari (2021). *In-vitro* and In-silico characterization of zein fiber incorporating cuminaldehyde, *Food Bioprod. Process.*, 128: 166-176.

Hanani, Z.N., J.A. O'Mahony, Y.H. Roos, P.M. Oliveira and J.P. Kerry (2014). Extrusion of gelatin-based composite films: Effects of processing temperature and pH of film forming solution on mechanical and barrier properties of manufactured films, *Food Packag. Shelf-Life*, 2: 91-101.

Harper, B.A., S. Barbut, A. Smith and M.F. Marcone (2015). Mechanical and microstructural properties of 'wet' alginate and composite films containing various carbohydrates, *J. Food Sci.*, 80: E84-E92.

Hartwig, K. (2010). *Characterization of Compression Molded Sodium Caseinate-based Films, Clemson University Libraries, Master Thesis, All Thesis 894.* https://tigerprints. clemson.edu/all_theses/894

Hemmati, F., A. Bahrami, A.F. Esfanjani, H. Hosseini, D.J. McClements and L. Williams (2021). Electrospun antimicrobial materials: Advanced packaging materials for food applications, *Trends in Food Sci. Technol.*, 111: 520-533.

Hernández-Muñoz, P., A. López-Rubio, V. del-Valle, E. Almenar and R. Gavara (2004). Mechanical and water barrier properties of glutenin films influenced by storage time, *J. Agric. Food Chem.*, 52: 79-83.

Hernandez-Izquierdo, V.M. and J.M. Krochta (2008). Thermoplastic processing of proteins for film formation – A review, *J. Food Sci.*, 73: R30-R39.

Hietala, M., A.P. Mathew and K. Oksman (2013). Bionanocomposites of thermoplastic starch and cellulose nanofibers manufactured using twin-screw extrusion, *Eur. Polym. J.*, 49: 950-956.

Hilbig, J., K. Hartlieb, K. Herrmann, J. Weiss and M. Gibis (2020). Influence of calcium on white efflorescence formation on dry fermented sausages with co-extruded alginate casings, *Food Res. Int.*, 131: 109012.

Homayoni, H., S.A.H. Ravandi and M. Valizadeh (2009). Electrospinning of chitosan nanofibers: Processing optimization, *Carbohydr. Polym.*, 77: 656-661.

Hu, H., X. Zhu, T. Hu, I.W. Cheung, S. Pan and E.C. Li-Chan (2015). Effect of ultrasound pre-treatment on formation of transglutaminase-catalysed soy protein hydrogel as a riboflavin vehicle for functional foods, *J. Funct. Foods*, 19: 182-193.

Huntrakul, K., R. Yoksan, A. Sane and N. Harnkarnsujarit (2020). Effects of pea protein on properties of cassava starch edible films produced by blown-film extrusion for oil packaging, *Food Packag. Shelf-Life*, 24: 100480.

Insaward, A., K. Duangmal and T. Mahawanich (2015). Mechanical, optical, and barrier properties of soy protein film as affected by phenolic acid addition, *J. Agric. Food Chem.*, 63: 9421-9426.

Jia, X.W., Z.Y. Qin, J.X. Xu, B.H. Kong, Q. Liu and H. Wang (2020). Preparation and characterization of pea protein isolate-pullulan blend electrospun nanofiber films, *Int. J. Biol. Macromol.*, 157: 641-647.

Jiang, S.J., T. Zhang, Y. Song, F. Qian, Y. Tuo and G. Mu (2019). Mechanical properties of whey protein concentrate based film improved by the coexistence of nanocrystalline cellulose and transglutaminase, *Int. J. Biol. Macromol.*, 126: 1266-1272.

Juvonen, H., M. Smolander, H. Boer, J. Pere, J. Buchert and J. Peltonen (2011). Film formation and surface properties of enzymatically crosslinked casein films, *J. Appl. Polym. Sci.*, 119: 2205-2213.

Karim, M., M. Fathi and S. Soleimanian-Zad (2020). Incorporation of zein nanofibers produced by needleless electrospinning within the casted gelatin film for improvement of its physical properties, *Food Bioprod. Process.*, 122: 193-204.

Karim, M., M. Fathi and S. Soleimanian-Zad (2021). Nanoencapsulation of cinnamic aldehyde using zein nanofibers by novel needle-less electrospinning: Production, characterization and their application to reduce nitrite in sausages, *J. Food Eng.*, 288: 110140.

Khalil, H.P.S., T.K. Lai, Y.Y. Tye, S. Rizal, E.W.N. Chong, S.W. Yap *et al.* (2018). A review of extractions of seaweed hydrocolloids: Properties and applications, *Express Polym. Lett.*, 12: 296-317.

Khwaldia, K., S. Banon, C. Perez and S. Desobry (2004a). Properties of sodium caseinate film-forming dispersions and films, *J. Dairy Sci.*, 87: 2011-2016.

Khwaldia, K., S. Banon, S. Desobry and J. Hardy (2004b). Mechanical and barrier properties of sodium caseinate–anhydrous milk fat edible films, *Int. J. Food Sci. Technol.*, 39: 403-411.

Kim, K.M., I.S. Jang, S.D. Ha and D.H. Bae (2004). Improved storage stability of brown rice by coating with rice bran protein, *Korean J. Food Sci. Technol.*, 36: 490-500.

Kim, J.H., W.S. Hong and S.W. Oh (2018). Effect of layer-by-layer antimicrobial edible coating of alginate and chitosan with grapefruit seed extract for shelf-life extension of shrimp (*Litopenaeus vannamei*) stored at 4°C, *Int. J. Biol. Macromol.*, 120: 1468-1473.

Klüver, E. and M. Meyer (2013). Preparation, processing, and rheology of thermoplastic collagen, *J. Appl. Polym. Sci.*, 128: 4201-4211.

Klüver, E. and M. Meyer (2015). Thermoplastic processing, rheology, and extrudate properties of wheat, soy, and pea proteins, *Polym. Eng. Sci.*, 55: 1912-1919.

Kocakulak, S., G. Sumnu and S. Sahin (2019). Chickpea flour-based biofilms containing gallic acid to be used as active edible films, *J. Appl. Polym. Sci.*, 136: 47704.

Kolbasov, A., S. Sinha-Ray, A. Joijode, M.A. Hassan, D. Brown, B. Maze *et al.* (2016). Industrial-scale solution blowing of soy protein nanofibers, *Ind. Eng. Chem. Res.*, 55: 323-333.

Kołodziejska, I. and B. Piotrowska (2007). The water vapor permeability, mechanical properties and solubility of fish gelatin–chitosan films modified with transglutaminase or 1-ethyl-3-(3-dimethylaminopropyl) carbodiimide (EDC) and plasticized with glycerol, *Food Chem.*, 103: 295-300.

Kong, L. and G.R. Ziegler (2012). Role of molecular entanglements in starch fiber formation by electrospinning, *Biomacromolecules*, 13: 2247-2253.

Kong, L. and G.R. Ziegler (2014). Fabrication of pure starch fibers by electrospinning, *Food Hydrocolloids*, 36: 20-25.

Koshy, R.R., S.K. Mary, S. Thomas and L.A. Pothan (2015). Environment friendly green composites based on soy protein isolate – A review, *Food Hydrocoll.*, 50: 174-192.

Krishna, M., C.I. Nindo and S.C. Min (2012). Development of fish gelatin edible films using extrusion and compression molding, *J. Food Eng.*, 108: 337-344.

Kumar, P., K.P. Sandeep, S. Alavi, V.D. Truong and R.E. Gorga (2010). Preparation and characterization of bio-nanocomposite films based on soy protein isolate and montmorillonite using melt extrusion, *J. Food Eng.*, 100: 480-489.

Kunte, L.A., A. Gennadios, S.L. Cuppett, M.A. Hanna and C.L. Weller (1997). Cast films from soy protein isolates and fractions, *Cereal Chem.*, 74: 115-118.

Kwak, H.W., J. Park, H. Yun, K. Jeon and D.W. Kang (2021). Effect of crosslinkable sugar molecules on the physico-chemical and antioxidant properties of fish gelatin nanofibers, *Food Hydrocoll.*, 111: 106259.

Lai, H.M. and G.W. Padua (1997). Properties and microstructure of plasticized zein films, *Cereal Chem.*, 74: 771-775.

Lavorgna, M., F. Piscitelli, P. Mangiacapra and G.G. Buonocore (2010). Study of the combined effect of both clay and glycerol plasticizer on the properties of chitosan films, *Carbohydr. Polym.*, 82: 291-298.

Lawton, J.W. (2004). Plasticizers for zein: Their effect on tensile properties and water absorption of zein films, *Cereal Chem.*, 81: 1-5.

Leena, M.M., K.S. Yoha, J.A. Moses and C. Anandharamakrishnan (2020). Edible coating with resveratrol loaded electrospun zein nanofibers with enhanced bioaccessibility, *Food Biosci.*, 36: 100669.

Leite, L.S.F., C. Pham, S. Bilatto, H.M. Azeredo, E.D. Cranston, F.K. Moreira *et al.* (2021). Effect of tannic acid and cellulose nanocrystals on antioxidant and antimicrobial properties of gelatin films, *ACS Sustain. Chem. Eng.*, 9: 8539-8549.

Li, M., P. Liu, W. Zou, L. Yu, F. Xie, H. Pu *et al.* (2011). Extrusion processing and characterization of edible starch films with different amylose contents, *J. Food Eng.*, 106: 95-101.

Li, X.Y., C.J. Shi, D.G. Yu, Y.Z. Liao and X. Wang (2014). Electrospun quercetin-loaded zein nanoribbons, *Biomed. Mater. Eng.*, 24: 2015-2023.

Li, Y., H. Chen, Y. Dong, K. Li, L. Li and J. Li (2016). Carbon nanoparticles/soy protein isolate bio-films with excellent mechanical and water barrier properties, *Ind. Crops Prod.*, 82: 133-140.

Lindstrom, T.R., K. Morimoto and C.J. Cante (1992). Edible films and coatings, pp. 659-663. *In:* Y.H. Hui (Ed.). *Encyclopedia of Food Science and Technology*, vol. 2. John Wiley and Sons Inc., New York, USA.

Liu, L., J.F. Kerry and J.P. Kerry (2005). Selection of optimum extrusion technology parameters in the manufacture of edible/biodegradable packaging films derived from food-based polymers, *J. Food Agric. Environ.*, 3: 51-58.

Liu, L., J.F. Kerry and J.P. Kerry (2007). Application and assessment of extruded edible casings manufactured from pectin and gelatin/sodium alginate blends for use with breakfast pork sausage, *Meat Sci.*, 75: 196-202.

Liu, C., Y. Chen, X. Wang, J. Huang, P.R. Chang and D.P. Anderson (2010). Improvement in physical properties and cytocompatibility of zein by incorporation of pea protein isolate, *J. Mater. Sci.*, 45: 6775-6785.

Liu, S.C., R. Li, P.M. Tomasula, A.M. Sousa and L. Liu (2016). Electrospun food-grade ultrafine fibers from pectin and pullulan blends, *Food Nutr. Sci.*, 7: 636-646.

Liu, F., B.S. Chiou, R.J. Avena-Bustillos, Y. Zhang, Y. Li, T.H. McHugh and F. Zhong (2017). Study of combined effects of glycerol and transglutaminase on properties of gelatin films, *Food Hydrocoll.*, 65: 1-9.

Liu, J., Z. Xue, W. Zhang, M. Yan and Y. Xia (2018). Preparation and properties of wet-spun agar fibers, *Carbohydr. Polym.*, 181: 760-767.

Liu, Y., X. Zhang, C. Li, Y. Qin, L. Xiao and J. Liu (2020). Comparison of the structural, physical and functional properties of κ-carrageenan films incorporated with pomegranate flesh and peel extracts, *Int. J. Biol. Macromol.*, 147: 1076-1088.

Llanos, J.H.R., C.C. Tadini and E. Gastaldi (2021). New strategies to fabricate starch/chitosan-based composites by extrusion, *J. Food Eng.*, 290: 110224.

Long, K., R. Cha, Y. Zhang, J. Li, F. Ren and X. Jiang (2018). Cellulose nanocrystals as reinforcements for collagen-based casings with low gas transmission, *Cellulose*, 25: 463-471.

Lopes, L.F., G. Meca, K.C. Bocate, T.M. Nazareth, K. Bordin and F.B. Luciano (2018). Development of food packaging system containing allyl isothiocyanate against

Penicillium nordicum in chilled pizza: Preliminary study, *J. Food Process. Pres.*, 42: e13436.

Lourdin, D., G. Della Valle and P. Colonna (1995). Influence of amylose content on starch films and foams, *Carbohydr. Polym.*, 27: 261-270.

Luchese, C.L., J. Uranga, J.C. Spada, I.C. Tessaro and K. de la Caba (2018). Valorisation of blueberry waste and use of compression to manufacture sustainable starch films with enhanced properties, *Int. J. Biol. Macromol.*, 115: 955-960.

Luciano, C.G., L. Tessaro, R.V. Lourenço, A.M.Q.B. Bittante, A.M. Fernandes, I.C.F. Moraes and P.J. do Amaral Sobral (2021). Effects of nisin concentration on properties of gelatin film-forming solutions and their films, *Int. J. Food Sci. Technol.*, 56: 587-599.

Luecha, J., N. Sozer and J.L. Kokini (2010). Synthesis and properties of corn zein/montmorillonite nanocomposite films, *J. Mater. Sci.*, 45: 3529-3537.

Ma, W., C.H. Tang, S.W. Yin, X.Q. Yang and J.R. Qi (2013). Genipin-crosslinked gelatin films as controlled releasing carriers of lysozyme, *Food Res. Int.*, 51: 321-324.

Ma, Y., A. Teng, K. Zhao, K. Zhang, H. Zhao, S. Duan *et al.* (2020). A top-down approach to improve collagen film's performance: The comparisons of macro, micro and nano sized fibers, *Food Chem.*, 309: 125624.

Maftoonazad, N. and H.S. Ramaswamy (2008). Effect of pectin-based coating on the kinetics of quality change associated with stored avocados, *J. Food Process. Preserv.*, 32: 621-643.

Mali, S., M.V.E. Grossmann, M.A. García, M.N. Martino and N.E. Zaritzky (2006). Effects of controlled storage on thermal, mechanical and barrier properties of plasticized films from different starch sources, *J. Food Eng.*, 75: 453-460.

Mangavel, C., J. Barbot, E. Bervas, L. Linossier, M. Feys, J. Gueguen and Y. Popineau (2002). Influence of prolamin composition on mechanical properties of cast wheat gluten films, *J. Cereal Sci.*, 36: 157-166.

Marcos, B., P. Gou, J. Arnau, M.D. Guàrdia and J. Comaposada (2020). Co-extruded alginate as an alternative to collagen casings in the production of dry-fermented sausages: Impact of coating composition, *Meat Sci.*, 169: 108184.

Marinea, M., A. Ellis, M. Golding and S.M. Loveday (2021). Soy protein pressed gels: Gelation mechanism affects the *in vitro* proteolysis and bioaccessibility of added phenolic acids, *Foods*, 10: 154.

Marquez, G.R., P. Di Pierro, M. Esposito, L. Mariniello and R. Porta (2014). Application of transglutaminase-crosslinked whey protein/pectin films as water barrier coatings in fried and baked foods, *Food and Bioproc. Tech.*, 7: 447-455.

Mauri A.N., P.R. Salgado, M.C. Condés and M.C. Añon (2016). Films and coatings from vegetable protein, pp. 67-87. *In:* M.P.M. García, M.C. Gómez-Guillén, M.E. López-Caballero and G.V. Barbosa-Cánovas (Eds.). *Edible Films and Coatings: Fundamentals and Applications*, CRC Press, Boca Raton, USA.

McHugh, T.H., J.F. Aujard and J.M. Krochta (1994). Plasticized whey protein edible films: Water vapor permeability properties, *J. Food Sci.*, 59: 416-419.

Melgarejo-Flores, B.G., L.A. Ortega-Ramírez, B.A. Silva-Espinoza, G.A. González-Aguilar, M.R.A. Miranda and J.F. Ayala-Zavala (2013). Antifungal protection and antioxidant enhancement of table grapes treated with emulsions, vapors, and coatings of cinnamon leaf oil, *Postharvest Biol. Technol.*, 86: 321-328.

Meshkani, S.M., S.A. Mortazavi, E. Milani, M. Mokhtarian and L. Sadeghian (2011). Evaluation of Mechanical and optical properties of edible film from chickpea protein isolate (*Cicer arietinum* L.) containing thyme essential oil with response surface method (RSM), *Innov. Food Sci Technol.*, 2: 25-36.

Meshkani, S.M., S.A. Mortazavi, E. Milani and F. Bakhshi Moghadam (2012). Effect of physical properties and optimized formulation of edible film with using form chickpea protein isolate (*Cicer Arietinum* L.), *Iranian J. Food Sci. Technol.*, 9: 109-117.

Monedero, F.M., M.J. Fabra, P. Talens and A. Chiralt (2009). Effect of oleic acid-beeswax mixtures on mechanical, optical and water barrier properties of soy protein isolate based films, *J. Food Eng.*, 91: 509-515.

Moreno, M.A., H. Bojorges, I. Falcó, G. Sánchez, G. López-Carballo, A. López-Rubio *et al.* (2020). Active properties of edible marine polysaccharide-based coatings containing *Larrea nitida* polyphenols enriched extract, *Food Hydrocoll.*, 102: 105595.

Mostafavi, F.S. and D. Zaeim (2020). Agar-based edible films for food packaging applications – A review, *Int. J. Biol. Macromol.*, 159: 1165-1176.

Muneer, F., E. Johansson, M.S. Hedenqvist, T.S. Plivelic and R. Kuktaite (2019). Impact of pH modification on protein polymerization and structure – Function relationships in potato protein and wheat gluten composites, *Int. J. Mol. Sci.*, 20: 58.

Naga, M., S. Kirihara, Y. Tokugawa, F. Tsuda, T. Saito and M. Hirotsuka (1996). *Process for Developing a Proteinaceous Film*, U.S. Patent No. 5,569,482, Washington, DC: U.S.

Neji, A.B., M. Jridi, M. Nasri and R.D. Sahnoun (2020). Preparation, characterization, mechanical and barrier properties investigation of chitosan-kaolinite nanocomposite, *Polym. Test*, 84: 106380.

Newson, W.R., F. Rasheed, R. Kuktaite, M.S. Hedenqvist, M. Gällstedt, T.S. Plivelic and E. Johansson (2015). Commercial potato protein concentrate as a novel source for thermoformed bio-based plastic films with unusual polymerisation and tensile properties, *RSC Advances*, 5: 32217-32226.

Nie, H., A. He, J. Zheng, S. Xu, J. Li and C.C. Han (2008). Effects of chain conformation and entanglement on the electrospinning of pure alginate, *Biomacromolecules*, 9: 1362-1365.

Nieuwland, M., P. Geerdink, P. Brier, P. Van Den Eijnden, J.T. Henket, M.L. Langelaan *et al.* (2014). Reprint of 'Food-grade electrospinning of proteins', *Innov. Food Sci. Emerg. Technol.*, 24: 138-144.

Ninan, G., J. Joseph and Z. Abubacker (2010). Physical, mechanical, and barrier properties of carp and mammalian skin gelatin films, *J. Food Sci.*, 75: E620-E626.

Nishinari, K., Y. Fang, N. Yang, X. Yao, M. Zhao, K. Zhang and Z. Gao (2018). Gels, emulsions and application of hydrocolloids at Phillips Hydrocolloids Research Centre, *Food Hydrocoll.*, 78: 36-46.

Ogale, A.A., P. Cunningham, P.L. Dawson and J.C. Acton (2000). Viscoelastic, thermal, and microstructural characterization of soy protein isolate films, *J. Food Sci.*, 65: 672-679.

Ohkawa, K., D. Cha, H. Kim, A. Nishida and H. Yamamoto (2004). Electrospinning of chitosan, *Macromol. Rapid Commun.*, 25: 1600-1605.

Oliviero, M., E. Di Maio and S. Iannace (2010). Effect of molecular structure on film blowing ability of thermoplastic zein, *J. Appl. Polym. Sci.*, 115: 277-287.

Padgett, T., I.Y. Han and P.L. Dawson (1998). Incorporation of food-grade antimicrobial compounds into biodegradable packaging films, *J. Food Prot.*, 61: 1330-1335.

Park, J.H., S.M. Park, Y.H. Kim, W. Oh, G.W. Lee, M.R. Karim *et al.* (2013). Effect of montmorillonite on wettability and microstructure properties of zein/montmorillonite nanocomposite nanofiber mats, *J. Compos. Mater.*, 47: 251-257.

Pereda, M., M.I. Aranguren and N.E. Marcovich (2010). Caseinate films modified with tung oil, *Food Hydrocoll.*, 24: 800-808.

Pereda, M., G. Amica, I. Rácz and N.E. Marcovich (2011). Structure and properties of nanocomposite films based on sodium caseinate and nanocellulose fibers, *J. Food Eng.*, 103: 76-83.

Pereda, M., N.E. Marcovich and M.A. Mosiewicki (2015). Sodium caseinate films containing linseed oil resin as oily modifier, *Food Hydrocoll.*, 44: 407-415.

Pérez-Gago, M.B., P. Nadaud and J.M. Krochta (1999). Water vapor permeability, solubility, and tensile properties of heat-denatured versus native whey protein films, *J. Food Sci.*, 64: 1034-1037.

Poverenov, E., S. Danino, B. Horev, R. Granit, Y. Vinokur and V. Rodov (2014). Layer-by-layer electrostatic deposition of edible coating on fresh cut melon model: Anticipated and unexpected effects of alginate–chitosan combination, *Food Bioprocess Tech.*, 7: 1424-1432.

Prodpran, T., K. Chuaynukul, M. Nagarajan, S. Benjakul and S. Prasarpran (2017). Impacts of plasticizer and pre-heating conditions on properties of bovine and fish gelatin films fabricated by thermo-compression molding technique, *Ital. J. Food Sci.*, 29: 487-504.

Radi, M., S. Akhavan-Darabi, H.R. Akhavan and S. Amiri (2018). The use of orange peel essential oil microemulsion and nanoemulsion in pectin-based coating to extend the shelf-life of fresh-cut orange, *J. Food Process. Preserv.*, 42: e13441.

Raeisi, M., M.A. Mohammadi, O.E. Coban, S. Ramezani, M. Ghorbani, M. Tabibiazar, and S.M.A. Noori (2021). Physicochemical and antibacterial effect of soy protein isolate/gelatin electrospun nanofibers incorporated with *Zataria multiflora* and *Cinnamon zeylanicum* essential oils, *J. Food Meas. Charact.*, 15: 1116-1126.

Rangavajhyala, N., V. Ghorpade and M. Hanna (1997). Solubility and molecular properties of heat-cured soy protein films, *J. Agric. Food Chem.*, 45: 4204-4208.

Rhim, J.W., A. Gennadios, A. Handa, C.L. Weller and M.A. Hanna (2000). Solubility, tensile, and color properties of modified soy protein isolate films, *J. Agric. Food Chem.*, 48: 4937-4941.

Rhim, J.W., J.H. Lee and H.S. Kwak (2005). Mechanical and water barrier properties of soy protein and clay mineral composite films, *Food Sci. Biotechnol.*, 14: 112-116.

Rhim, J.W. (2012). Physical-mechanical properties of agar/κ-carrageenan blend film and derived clay nanocomposite film, *J. Food Sci.*, 77: N66-N73.

Rindlaw-Westling, A., M. Stading, A.M. Hermansson and P. Gatenholm (1998). Structure, mechanical and barrier properties of amylose and amylopectin films, *Carbohydr. Polym.*, 36: 217-224.

Rodríguez-Castellanos, W., F. Martínez-Bustos, D. Rodrigue and M. Trujillo-Barragán (2015). Extrusion blow molding of a starch–gelatin polymer matrix reinforced with cellulose, *Eur. Polym. J.*, 73: 335-343.

Rojas-Graü, M.A., R. Soliva-Fortuny and O. Martín-Belloso (2009). Edible coatings to incorporate active ingredients to fresh-cut fruits: A review, *Trends in Food Sci. Technol.*, 20: 438-447.

Salas, C., M. Ago, L.A. Lucia and O.J. Rojas (2014). Synthesis of soy protein–lignin nanofibers by solution electrospinning, *React. Funct. Polym.*, 85: 221-227.

Schäfer, D., M. Reinelt, A. Stäbler and M. Schmid (2018). Mechanical and barrier properties of potato protein isolate-based films, *Coatings*, 8: 58.

Schmid, M., S. Sängerlaub, L. Wege and A. Stäbler (2014). Properties of transglutaminase crosslinked whey protein isolate coatings and cast films, *Packag. Technol. Sci.*, 27: 799-817.

Schmidt, C.G., M.A. Cerqueira, A.A. Vicente, J.A. Teixeira and E.B. Furlong (2015). Rice bran protein-based films enriched by phenolic extract of fermented rice bran and montmorillonite clay, CyTA-J, *Food*, 13: 204-212.

Schofield, J.D., R.C. Bottomley, M.F. Timms and M.R. Booth (1983). The effect of heat on wheat gluten and the involvement of sulphydryl-disulphide interchange reactions, *J. Cereal Sci.*, 1: 241-253.

Sedayu, B.B., M.J. Cran and S.W. Bigger (2019). A review of property enhancement techniques for carrageenan-based films and coatings, *Carbohydr. Polym.*, 216: 287-302.

Shahrampour, D., M. Khomeiri, S.M.A Razavi and M. Kashiri (2020). Development and characterization of alginate/pectin edible films containing *Lactobacillus plantarum* KMC 45, *LWT-Food Sci. Technol.*, 118: 108758.

Shanks, R. and I. Kong (2012). Thermoplastic starch, pp. 95-116. *In:* A.Z. El-Sonbati (Ed.). *Thermoplastic Elastomers*, Intech Open, Rijeka, Hr.

Sharif, N., M.T. Golmakani, M. Niakousari, S.M.H. Hosseini, B. Ghorani and A. Lopez-Rubio (2018a). Active food packaging coatings based on hybrid electrospun gliadin nanofibers containing ferulic acid/hydroxypropyl-beta-cyclodextrin inclusion complexes, *Nanomaterials*, 8: 919.

Sharif, R., M. Mujtaba, M. Ur Rahman, A. Shalmani, H. Ahmad, T. Anwar *et al.* (2018b). The multifunctional role of chitosan in horticultural crops: A review, *Molecules*, 23: 872.

Shi, D., F. Liu, Z. Yu, B. Chang, H.D. Goff and F. Zhong (2019). Effect of aging treatment on the physicochemical properties of collagen films, *Food Hydrocoll.*, 87: 436-447.

Shih, F.F. (1996). Edible films from rice protein concentrate and pullulan, *Cereal Chem.*, 73: 406-409.

Shin, Y.J., S.A. Jang and K.B. Song (2011). Preparation and mechanical properties of rice bran protein composite films containing gelatin or red algae, *Food Sci. Biotechnol.*, 20: 703-707.

Shin, Y.J., H.Y. Song and K.B. Song (2012). Effect of a combined treatment of rice bran protein film packaging with aqueous chlorine dioxide washing and ultraviolet-C irradiation on the postharvest quality of 'Goha' strawberries, *J. Food Eng.*, 113: 374-379.

Shojaee-Aliabadi, S., H. Hosseini, M.A. Mohammadifar, A. Mohammadi, M. Ghasemlou, S.M. Ojagh *et al.* (2013). Characterization of antioxidant-antimicrobial κ-carrageenan films containing Satureja hortensis essential oil, *Int. J. Biol. Macromol.*, 52: 116-124.

Shojaee-Aliabadi, S., H. Hosseini, M.A. Mohammadifar, A. Mohammadi, M. Ghasemlou, S.M. Hosseini, and R. Khaksar (2014). Characterization of κ-carrageenan films incorporated plant essential oils with improved antimicrobial activity, *Carbohydr. Polym.*, 101: 582-591.

Shori, A.B. (2017). Microencapsulation improved probiotics survival during gastric transit, *HAYATI J. Biosci.*, 24: 1-5.

Shukla, R. and M. Cheryan (2001). Zein: The industrial protein from corn, *Ind. Crops Prod.*, 13: 171-192.

Simelane, S. and Z. Ustunol (2005). Mechanical properties of heat-cured whey protein-based edible films compared with collagen casings under sausage manufacturing conditions, *J. Food Sci.*, 70: E131-E134.

Sothornvit, R., C.W. Olsen, T.H. McHugh and J.M. Krochta (2007). Tensile properties of compression-molded whey protein sheets: Determination of molding condition and glycerol-content effects and comparison with solution-cast films, *J. Food Eng.*, 78: 855-860.

Souza, M.P., A.F. Vaz, M.A. Cerqueira, J.A. Texeira, A.A. Vicente and M.G. Carneiro-da-Cunha (2015). Effect of an edible nanomultilayer coating by electrostatic self-assembly on the shelf-life of fresh-cut mangoes, *Food Bioprocess Tech.*, 8: 647-654.

Sowmyashree, A., R.R. Sharma, S.G. Rudra and M. Grover (2021). Layer-by-layer coating of hydrocolloids and mixed plant extract reduces fruit decay and improves postharvest life of nectarine fruits during cold storage, *Acta Physiologiae Plantarum*, 43: 1-10.

Sözbilen, G.S., E. Çavdaroğlu and A. Yemenicioğlu (2022). Incorporation of organic acids turns classically brittle zein films into flexible antimicrobial packaging materials, *Packag. Technol. Sci.*, 35: 81-95.

Stanley, N. (1987). Production, properties and uses of carrageenan, pp.116-146. *In:* D.J. McHugh (Ed.). *Production and Utilization of Products from Commercial Seaweeds*, FAO Fisheries Technical Paper, 288, (No. 589.45 F36), Rome, Italy.

Stijnman, A.C., I. Bodnar and R.H. Tromp (2011). Electrospinning of food-grade polysaccharides, *Food Hydrocoll.*, 25: 1393-1398.

Stuchell, Y.M. and J.M. Krochta (1994). Enzymatic treatments and thermal effects on edible soy protein films, *J. Food Sci.*, 59: 1332-1337.

Subirade, M., I. Kelly, J. Guéguen and M. Pézolet (1998). Molecular basis of film formation from a soybean protein: Comparison between the conformation of glycinin in aqueous solution and in films, *Int. J. Biol. Macromol.*, 23: 241-249.

Suderman, N., M.I.N. Isa and N.M. Sarbon (2018). Characterization on the mechanical and physical properties of chicken skin gelatin films in comparison to mammalian gelatin films. *In: IOP Conference Series: Materials Science and Engineering*, IOP Publishing, 440: 012033.

Sullivan, S.T., C. Tang, A. Kennedy, S. Talwar and S.A. Khan (2014). Electrospinning and heat treatment of whey protein nanofibers, *Food Hydrocoll.*, 35: 36-50.

Sun, S., Y. Song and Q. Zheng (2008). Thermo-molded wheat gluten plastics plasticized with glycerol: Effect of molding temperature, *Food Hydrocoll.*, 22: 1006-1013.

Suurs, P. and S. Barbut (2020). Collagen use for co-extruded sausage casings – A review, *Trends in Food Sci. Technol.*, 102: 91-101.

Syarifuddin, A., A. Dirpan and M. Mahendradatta (2017). Physical, mechanical, and barrier properties of sodium alginate/gelatin emulsion based-films incorporated with canola oil. *In: IOP Conference Series: Earth and Environmental Science*, IOP Publishing, 101: 012019.

Talja, R.A., M. Peura, R. Serimaa and K. Jouppila (2008). Effect of amylose content on physical and mechanical properties of potato-starch-based edible films, *Biomacromolecules*, 9: 658-663.

Talón, E., M. Vargas, A. Chiralt and C. González-Martínez (2019). Eugenol incorporation into thermoprocessed starch films using different encapsulating materials, *Food Packag. Shelf-Life*, 21: 100326.

Tanada-Palmu, P.S. and C.R. Grosso (2005). Effect of edible wheat gluten-based films and coatings on refrigerated strawberry (*Fragaria ananassa*) quality, *Postharvest Biol. Tech.*, 36: 199-208.

Tavassoli-Kafrani, E., S.A.H. Goli and M. Fathi (2017). Fabrication and characterization of electrospun gelatin nanofibers crosslinked with oxidized phenolic compounds, *Int. J. Biol. Macromol.*, 103: 1062-1068.

Thunwall, M., A. Boldizar and M. Rigdahl (2006). Compression molding and tensile properties of thermoplastic potato starch materials, *Biomacromolecules*, 7: 981-986.

Tihminlioglu, F., İ.D. Atik and B. Özen (2010). Water vapor and oxygen-barrier performance of corn-zein coated polypropylene films, *J. Food Eng.*, 96: 342-347.

Tomasula, P.M., A.M. Sousa, S.C. Liou, R. Li, L.M. Bonnaillie and L. Liu (2016). Electrospinning of casein/pullulan blends for food-grade applications, *J. Dairy Sci.*, 99: 1837-1845.

Torres-Giner, S., A. Martinez-Abad, M.J. Ocio and J.M. Lagaron (2010). Stabilization of a nutraceutical omega-3 fatty acid by encapsulation in ultrathin electrosprayed zein prolamine. *J. Food Sci.*, 75: N69-N79.

Türe, H., T.O. Blomfeldt, M. Gällstedt and M.S. Hedenqvist (2012). Properties of wheat-gluten/montmorillonite nanocomposite films obtained by a solvent-free extrusion process, *J. Polym. Environ.*, 20: 1038-1045.

Ullsten, N.H., M. Gällstedt, G.M. Spencer, E. Johansson, S. Marttila, R. Ignell and M.S. Hedenqvist (2010). Extruded high quality materials from wheat gluten, *Polym. from Renew. Resour.*, 1: 173-186.

Ünalan, İ.U., I. Arcan, F. Korel and A. Yemenicioğlu (2013). Application of active zein-based films with controlled release properties to control *Listeria monocytogenes* growth and lipid oxidation in fresh Kashar cheese, *Innov. Food Sci. Emerg.*, 20: 208-214.

Van den Broek, L.A., R.J. Knoop, F.H. Kappen and C.G. Boeriu (2015). Chitosan films and blends for packaging material, *Carbohydr. Polym.*, 116: 237-242.

Van Soest, J.J.G. and P. Essers (1997). Influence of amylose-amylopectin ratio on properties of extruded starch plastic sheets, *J. Macromol. Sci. Pure Appl. Chem.*, 34: 1665-1689.

Viroben, G., J. Barbot, Z. Mouloungui and J. Guéguen (2000). Preparation and characterization of films from pea protein, *J. Agric. Food Chem.*, 48: 1064-1069.

Wang, Y. and G.W. Padua (2003). Tensile properties of extruded zein sheets and extrusion blown films, *Macromol. Mater. Eng.*, 288: 886-893.

Wang, Y., F.L. Filho, P. Geil and G.W. Padua (2005a). Effects of processing on the structure of zein/oleic acid films investigated by X-ray diffraction, *Macromol. Biosci.*, 5: 1200-1208.

Wang, S.F., L. Shen, Y.J. Tong, L. Chen, I.Y. Phang, P.Q. Lim and T.X. Liu (2005b). Hydrocolloid chitosan/montmorillonite nanocomposites: Preparation and characterization, *Polym. Degrad. Stab.*, 90: 123-131.

Wang, S., M.F. Marcone, S. Barbut and L.T. Lim (2013). Electrospun soy protein isolate-based fiber fortified with anthocyanin-rich red raspberry (*Rubus strigosus*) extracts, *Food Res. Int.*, 52: 467-472.

Wang, W., Y. Zhang, R. Ye and Y. Ni (2015). Physical crosslinkings of edible collagen casing, *Int. J. Biol. Macromol.*, 81: 920-925.

Wang, W., Y. Liu, A. Liu, Y. Zhao and X. Chen (2016). Effect of in situ apatite on performance of collagen fiber film for food packaging applications, *J. Appl. Poly. Sci.*, 133: 44154.

Wang, K., W. Wang, R. Ye, J. Xiao, Y. Liu, J. Ding *et al.* (2017). Mechanical and barrier properties of maize starch–gelatin composite films: Effects of amylose content, *J. Sci. Food Agr.*, 97: 3613-3622.

Wang, W., X. Zhang, C. Li, G. Du, H. Zhang and Y. Ni (2018). Using carboxylated cellulose nanofibers to enhance mechanical and barrier properties of collagen fiber film by electrostatic interaction, *J. Sci. Food Agric.*, 98: 3089-3097.

Wang, L., J. Xue and Y. Zhang (2019). Preparation and characterization of curcumin loaded caseinate/zein nanocomposite film using pH-driven method, *Ind. Crop. Prod.*, 130: 71-80.

Wei, F., F. Ye, S. Li, L. Wang, J. Li and G. Zhao (2018). Layer-by-layer coating of chitosan/ pectin effectively improves the hydration capacity, water suspendability and tofu gel compatibility of okara powder, *Food Hydrocoll.*, 77: 465-473.

Wu, X., Y. Liu, A. Liu and W. Wang (2017). Improved thermal-stability and mechanical properties of type I collagen by crosslinking with casein, keratin and soy protein isolate using transglutaminase, *Int. J. Biol. Macromol.*, 98: 292-301.

Xu, H., Y. Chai and G. Zhang (2012a). Synergistic effect of oleic acid and glycerol on zein film plasticization, *J. Agric. Food Chem.*, 60: 10075-10081.

Xu, X., L. Jiang, Z. Zhou, X. Wu and Y. Wang (2012b). Preparation and properties of electrospun soy protein isolate/polyethylene oxide nanofiber membranes, *ACS Appl. Mater. Interfaces*, 4: 4331-4337.

Xu, J., F. Liu, H.D. Goff and F. Zhong (2020). Effect of pre-treatment temperatures on the film-forming properties of collagen fiber dispersions, *Food Hydrocoll.*, 107: 105326.

Yang, S.Y., K.Y. Lee, S.E. Beak, H. Kim and K.B. Song (2017). Antimicrobial activity of gelatin films based on duck feet containing cinnamon leaf oil and their applications in packaging of cherry tomatoes, *Food Sci. Biotechnol.*, 26: 1429-1435.

Yao, C., X. Li and T. Song (2007). Electrospinning and crosslinking of zein nanofiber mats, *J. Appl. Polym. Sci.*, 103: 380-385.

Ye, J., D. Ma, W. Qin and Y. Liu (2018). Physical and antibacterial properties of sodium alginate – Sodium carboxymethylcellulose films containing *Lactococcus lactis*, *Molecules*, 23: 2645.

Yemenicioğlu, A. (2016). Zein and its composites and blends with natural active compounds: Development of antimicrobial films for food packaging, pp. 503-513. *In:* J. Barros-Velazquez (Ed.). *Antimicrobial Food Packaging*, Academic Press, London, UK.

Yemenicioğlu, A. (2017b). Basic strategies and testing methods to develop effective edible antimicrobial and antioxidant coating, pp. 63-88. *In:* A. Tiwari (Ed.). *Handbook of Antimicrobial Coatings*, 1st ed., Elsevier, Amsterdam, The Netherlands.

Yemenicioğlu, A., S. Farris, M. Turkyilmaz and S. Gulec (2020). A review of current and future food applications of natural hydrocolloids, *Int. J. Food Sci. Tech.*, 55: 1389-1406.

Younes, M., P. Aggett, F. Aguilar, R. Crebelli, B.M. Dusemund, M.J. Filipič *et al.* (2018). Refined exposure assessment of polyethylene glycol (E 1521) from its use as a food additive, *EFSA Journal*, 16: e05293.

Zeugolis, D.I., S.T. Khew, E.S. Yew, A.K. Ekaputra, Y.W. Tong, L.Y.L. Yung, *et al.* (2008). Electro-spinning of pure collagen nano-fibres – Just an expensive way to make gelatin? *Biomaterials*, 29: 2293-2305.

Zhang, Y.Z., J. Venugopal, Z.M. Huang, C.T. Lim and S. Ramakrishna (2006). Crosslinking of the electrospun gelatin nanofibers, *Polymer*, 47: 2911-2917.

Zhang, J., P. Mungara and J.L. Jane (2001). Mechanical and thermal properties of extruded soy protein sheets, *Polymer*, 42: 2569-2578.

Zhang, Y., L. Deng, H. Zhong, J. Pan, Y. Li and H. Zhang (2020). Superior water stability and antimicrobial activity of electrospun gluten nanofibrous films incorporated with glycerol monolaurate, *Food Hydrocoll.*, 106116.

Zhao, X.Y., K.C. Guo, Zhang, Y. Ma and X.H. Li (2011). Effect of 11S/7S ratios on mechanical and barrier properties of edible films based on soybean protein isolate, *Adv. Mat. Res.*, 335: 312-319.

Zhu, D., Q. Wu and L. Hua (2019). Industrial enzymes, pp. 1-13. *In:* M. Moo-Young (Ed.). *Comprehensive Biotechnology*, 3rd ed., vol. 3, Pergamon, Amsterdam, The NL.

Zubeldía, F., M.R. Ansorena and N.E. Marcovich (2015). Wheat gluten films obtained by compression molding, *Polym. Test.*, 43: 68-77.

Zurlini, C., E. Umiltà, A. Montanari and A. Brutti (2018). A combined application of edible coatings and passive refrigeration to extend the shelf-life of fresh truffles, *Int. J. Adv. Res.*, 6: 354-365.

Natural Active Agents: Sources, Major Characteristics and Potential as Edible Packaging Components

4.1 Introduction

The health concerns of the consumers associated with chemical food additives have increased the application of natural active agents to improve safety, quality, and functional properties of food products. Antimicrobial enzymes (e.g. lysozyme, lactoperoxidase, glucose oxidase, and chitinase), proteins and peptides (e.g. phosvitin, lactoferrin, lactoferricin) extracted from different animal sources, and peptides obtained by microbial fermentation (e.g. nisin, pediocin, polylysine) have been extensively employed as active agents in antimicrobial packaging (Santos *et al.*, 2018; Yemenicioğlu, 2016a). Moreover, antimicrobial and/or antioxidant secondary plant metabolites, such as phenolic compounds (e.g. purified green tea catechins, quercetin, curcumin or crude phenolic extracts of green tea, grape seed, sage, rosmarine, grapefruit seed, pomegranate peel), and essential oils (e.g. oregano, citrus, sage, thyme, anise, and cinnamon essential oils) and their active phenolic constituents (e.g. eugenol, carvacrol, geraniole, thymol, citral, limonene) are frequently used in active packaging (Yemenicioğlu, 2016a; Chibane *et al.*, 2019). Furthermore, prebiotics and probiotics, important for human health, protective microorganisms capable of producing natural antimicrobials in food have been combined with different hydrocolloids to develop active edible food packaging and coating materials (Gialamas *et al.*, 2010; Pavli *et al.*, 2018; Neri-Numa *et al.*, 2020). This chapter discusses biochemical properties, current and novel sources, mechanisms of antimicrobial and antioxidant action, antimicrobial spectrum, example uses in different edible packaging as well as available regulatory status of natural active agents. The information provided in this chapter shows both advantages and disadvantages of employing natural active agents in edible packaging, and describes basic methods for overcoming difficulties in their effective utilization in active packaging.

4.2 Antimicrobial enzymes

4.2.1 Lysozyme

Lysozyme (LYS, EC 3.2.1.17), called also muramidase, is one of the most potential candidates for use in antimicrobial edible packaging since it shows high stability in aqueous or ethanolic film solutions, in edible films at the dry state, or in different foods at refrigerated conditions (Boyaci and Yemenicioğlu, 2018; Sozbilen and Yemenicioğlu, 2020; Mecitoglu *et al.*, 2006). The enzyme shows its antimicrobial activity by splitting the bonds between the N-acetylmuramic acid and N-acetylglucosamine of the peptidoglycan (PG) layer in Gram-positive bacterial cell walls. Although some Gram-positive bacteria, such as *Staphylococcus aureus* and *Clostridium sporogenes,* show a high resistance against LYS (Sudağıdan and Yemenicioğlu, 2012; Ávila *et al.*, 2014), the enzyme is highly effective on *Bacillus* spp., lactic acid bacteria, *C. tyrobutyricum*, and critical pathogenic bacteria *Listeria monocytogenes* (Duan *et al.*, 2007; Min *et al.*, 2005a; Sozbilen and Yemenicioğlu, 2020; Wu *et al.*, 2019a). Although majority of the current commercial food applications of LYS target control or inactivation of spoilage bacteria, the use of enzyme in antimicrobial packaging targets generally the inactivation of *L. monocytogenes* (Duan *et al.*, 2007; Ünalan *et al.*, 2013; Min *et al.*, 2005a; Boyaci and Yemenicioğlu, 2018). The Gram-negative bacteria are not affected from lytic action of LYS since their PG layer is surrounded by a protective lipopolysaccharide (LPS) layer. However, the combination of LYS with chelating agents (e.g. disodium EDTA) causes the removal of Ca^{+2} atoms stabilizing the protective LPS layer of Gram-negative bacteria. Thus, the enzyme becomes effective against critical pathogens, such as *S. enterica* serovar Typhimurium and *Escherichia coli* O157:H7 (Ünalan *et al.*, 2011a). Moreover, the antimicrobial spectrum of LYS can also be improved by modification of its molecular properties. Most of these methods aim at increasing surface active properties of LYS to ease overcoming LPS barrier of Gram-negative bacteria. Nakamura *et al.* (1991, 1992) showed that conjugates of LYS with dextran or galactomannan obtained by natural Maillard reaction exhibited good emulsifying activities and antimicrobial activity against major Gram-negative pathogenic bacteria, such as *E. coli, Vibrio parahaemolyticus, Klebsiella pneumoniae, Aeromonas hdrophila* and *Proteus mirabilis* when applied in combination with mild heating at 50°C. The ability of LYS to penetrate LPS of Gram-negative bacteria could also be improved by its chemical modification through lipophilization. Ibrahim *et al.* (1993) found that LYS lipophilized with fatty acids, such as palmitic, myristic and stearic acids showed antimicrobial activity against *E. coli* at room temperature, but lipophilization of LYS with these fatty acids caused significant losses in lytic activity of this enzyme. The low recovery problem of lipophilization process might be overcome by stabilization of enzyme by Maillard-type glycosylation before lipophilization process (Liu *et al.*, 2000a). Moreover, it was also determined that the lipophilization with short-chained fatty acids, such as caproic, capric, and myristic acids prevented significant activity losses and gave modified enzyme effective on *E. coli* at 20°C (Liu *et al.*, 2000b).

The modification of LYS conformation by partial heat denaturation is another method to improve its antibacterial spectrum. This strategy targets unfolding of LYS to expose its hydrophobic groups that show affinity on LPS layer of Gram-negative bacteria. It was reported that the heat-denatured LYS (heated at 80°C at pH 7.0 or at pH 6.0 above 90°C) was free from lytic activity, but it showed antibacterial activity on both *E. coli* and *S. aureus* by membrane penetration mechanism (Ibrahim *et al.*, 1996a). The most potent form of heat-treated LYS on *E. coli* and *S. aureus* is the partially denatured form (with 50% of lytic activity) that is obtained by heating in pH 6.0 buffer at 80°C for 20 min., but bacterium like *S. enterica* serovar Enteritidis shows an apparent resistance to partially denatured LYS (Ibrahim *et al.*, 1996b). Recently, effectiveness of heat-denatured LYS on food-borne viruses, such as hepatitis A and norovirus, has also been demonstrated by Takahashi *et al.* (2018). These authors employed heat-denatured LYS for antiviral treatment of blueberries and a mixture of strawberries and raspberries obtained more than 3 log reduction in hepatitis A and norovirus in these samples.

The hen egg white LYS (MW: 14.3 kDa, pI: 11.4) used in food applications is obtained commercially by using the classical salt crystallization method. The crystallization is applied repeatedly to reach the desired purity level; thus, this method works slowly and requires almost a week. Therefore, some faster methods, such as affinity membrane chromatographic methods and ultrafiltration are also developed for purification of hen egg white lysozyme (Arıca *et al.*, 2004; Grasselli *et al.*, 1999; Ghosh and Cui, 2000; Hou and Lin, 1997; Jiang *et al.*, 2001). Moreover, application of some partial purification methods based on selective precipitation or partitioning of lysozyme from hen egg white or selective precipitation of undesired egg white proteins (e.g. ovalbumin and conalbumin) while maintaining lysozyme solubility in the egg white have also been developed. Some of these methods include (1) partitioning of lysozyme by polyethylene glycol/ salt aqueous two-phase system (Su and Chiang, 2006), (2) selective precipitation of LYS with anionic surfactant di-(2-ethylhexyl) sodium sulfosuccinate (AOT) and its following recovery with acetone (Shin *et al.*, 2003), (3) selective precipitation of non-lysozyme proteins in the egg white by heat-induced denaturation and gelation (Chang *et al.*, 2000), (4) incubation in the presence of 30% ethanol to precipitate non-lysozyme egg proteins with or without applying following dialysis (Jiang *et al.*, 2001; Mecitoglu *et al.*, 2006). Although, the hen egg-white is currently the only source of commercial LYS for food applications, there are also some other LYS types which could be used in food or other antimicrobial applications (e.g. biomedical or pharmaceutical purposes). For example, Datta *et al.* (2008) purified oyster LYS and used it in alginate coatings to suppress *L. monocytogenes* and *S. anatum* in cold-smoked salmon. These authors determined similar effectiveness of oyster and hen egg-white LYSs against target bacteria. Other LYSs include human LYS and bovine stomach LYS that show antimicrobial activity towards both Gram-positive and Gram-negative bacteria as well as fungi. Wilcox *et al.* (1997) showed that bovine stomach LYS genetically introduced into tobacco plant may be recovered and purified to 93% homogeneity from the transgenic plants by an easily scalable process. Takaichi and Oeda (2000) also reported the accumulation of genetically-introduced human lysozyme in transgenic carrots. These studies showed that the commercial production of alternative LYS from transgenic plants is possible. However, future industrial uses of such LYSs are questionable due to the existing concerns about GMOs.

The use of hen egg white LYS directly in food generally requires declaration of the enzyme on the product label since it causes sensitization in individuals suffering from egg allergy (Liburdi *et al.*, 2014). In Japan, hen egg white LYS is used for preservation of different food, such as vegetables, seafood, pasta, and salads (Davidson *et al.*, 2012). The U.S. Food and Drug Administration (FDA) designated hen egg white LYS as a generally recognized as safe (GRAS) agent for use in ripened cheeses to prevent late blowing caused by *C. tyrobutyricum* (FDA, 1998). The enzyme is also used extensively in Europe for the same purpose in ripened cheeses (*quantum satis* according to EPCD 1995), such as parmesan, edam, and gouda (Davidson *et al.*, 2012; De Roos *et al.*, 1998). The European Union (EU) also allows use of enzyme to control malolactic fermentation in wines (max. 500 mg/L according to EC 2001) caused by lactic acid bacteria, such as *Oenococcus oeni, Lactobacillus* spp., *Pediococcus damnosus, P. parvulus* (Liburdi *et al.*, 2014). The enzyme could also be employed to prolong the shelf-life of unpasteurized beer since it is able to delay the growth of lactic acid-spoilage bacteria, such as *P. damnosus* and *Lactobacillus brevis* (Makki and Durance, 1996). There is also a significant potential to employ LYS in the preservation of beverages since it is a natural antimicrobial with a sweet taste (Masuda *et al.*, 2001). A recent study by Sozbilen *et al.* (2018) showed that LYS in combination with nisin (NIS) could be effectively used to control growth of natural lactic acid flora in the fermented traditional beverage, boza. The combination of LYS with NIS directly in food or packaging materials is a rising trend in food preservation since these two natural antimicrobial agents show synergetic action not only against lactic acid spoilage bacteria, but also against pathogenic bacteria, such as *L. monocytogenes* (Chung and Hancock, 2000; Gill and Holley, 2000; Mangalassary *et al.*, 2008; Bhatia and Bharti, 2015: Sozbilen and Yemenicioğlu, 2020).

The LYS is the most frequently employed antimicrobial enzyme for antimicrobial edible films. The enzyme has been incorporated into protein-based edible films, such as whey (Min *et al.*, 2005a), zein (Mecitoğlu *et al.*, 2006), and soy protein films (Padgett *et al.*, 1998), and polysaccharide-based edible films, such as alginate and carrageenan (Cha *et al.*, 2002), chitosan (Park *et al.*, 2004), pullulan (Kandemir *et al.*, 2005), low methoxyl pectin (Bayarri *et al.*, 2014), and starch films (Fabra *et al.*, 2014). Moreover, the enzyme was also used to develop antimicrobial composite films, such as zein-wax or zein-fatty acid films (Arcan and Yemenicioğlu, 2013), Ca-alginate-chitosan films (Wei *et al.*, 2011), rectorite clay-chitosan films (Li *et al.*, 2017), whey-fatty acid and whey-wax films (Boyacı *et al.*, 2016), zein-soy protein and zein-lentil protein films (Boyaci and Yemenicioğlu, 2018). Finally, the enzyme was used to obtain antimicrobial cellulose acetate nanofiber mats (CANFM) intended for food packaging. These mats were obtained simply by layer-by-layer coating of electrospun CANFM with positively and negatively charged molecules and hydrocolloids, such as LYS-chitosan-rectorite clay mixture and Na-alginate (Huang *et al.*, 2012), LYS-chitosan chloride mixture and Na-alginate (Huang *et al.*, 2013), LYS and gold nanoparticles (Zhou *et al.*, 2014), and LYS and pectin (Zhang *et al.*, 2015), respectively.

4.2.2 Lactoperoxidase

Lactoperoxidase (LPS, EC 1.11.1.7) is an oxidoreductase group enzyme that forms part of the natural antimicrobial mechanism in milk. The mechanism of the antimicrobial action of LPS depends on its ability to convert thiocynate (SCN⁻) to antimicrobial products, such as hipothiocyanite (OSCN⁻) ion, hypothiocyanous acid (HOSCN), and some other highly reactive and short-lived oxidation products in the presence of H_2O_2 (Pruitt *et al.*, 1982). Therefore, the activation of enzyme needs the addition of thiocyanate and H_2O_2, or thiocyanate and glucose oxidase that is capable of generating the desired H_2O_2 essential for the reaction (Popper and Knorr, 1997). Due to the broad antimicrobial spectrum of its reaction products, LPS shows bactericidal effect on Gram-negative bacteria, such as coliforms, *Salmonella*, *Shigella*, and *Pseudomonas*, and bacteriostatic effect on Gram-positive bacteria, such as streptococci and lactobacilli (Seifu *et al.*, 2005; Jooyandeh *et al.*, 2011). It also shows antifungal activity on different yeasts, such as *Rhodutorula rubra* and *Saccharomyces cerevisiae*, dimorphic (mold-yeast) fungus *Mucor rouxii*, and molds, such as *Aspergillus niger* and *Byssochlamys fulva* (Popper and Knorr, 1997). The antiviral activity of LPS against viruses, such as influenza, hepatitis C, human immunodeficiency virus, poliovirus, and herpes simplex virus has also been reported (Tripathi and Vashishtha, 2006; Redwan *et al.*, 2015; Gingerich *et al.*, 2017; El-Fakharany *et al.*, 2017).

The use of LPS in non-dairy products needs supplementation of food with enzyme purified from bovine whey, using cation exchange or affinity chromatography. The large-scale extraction of bovine LPS (MW: 75-79 kDa, pI: 9.6) is generally conducted by cation-exchange chromatography from sweet or acid whey obtained from skimmed milk (Yoshida, 1991; Mecitoğlu and Yemenicioğlu, 2007). A concentration step or selective precipitation step for globulins by salting-out before column chromatography is beneficial in improving the performance of the separation process (Yoshida, 1991).

The activation of naturally occurring LPS in milk by addition of thiocyanate and H_2O_2 has long been employed to increase the safety and to prolong the shelf-life of bovine milk in underdeveloped areas where refrigeration is a problem. According to FAO (1999) that conducted a global LPS program, the activation of LPS in bovine milk by addition of 10 ppm thiocyanate and 8-9 ppm H_2O_2 (added in the form of granulated sodium carbonate peroxyhydrate) could increase the shelf-life of bovine milk by several hours that is very beneficial for the transportation of milk from small farm to larger ones or collection points. The LPS system in cheese milk could also be activated by addition of thiocyanate and H_2O_2 to improve microbial quality of resulting cheese (Seifu *et al.*, 2004; Seifu *et al.*, 2005). The LPS has also been used to improve the microbial quality of meat, fish, and fruit and vegetable products (Kennedy *et al.*, 2000a; Elliot *et al.*, 2004; Touch *et al.*, 2004; Jooyandeh *et al.*, 2011). It is important to note that the food standards in Australia and New Zealand allow the use of LPS as a processing aid to improve the microbial quality on the meat surfaces (Australia and New Zealand Act, 1991).

Although the LPS could be exploited easily to obtain a potent antimicrobial mechanism, its utilization in edible films as an active agent has gained less interest than LYS since it requires addition of thiocyanate, H_2O_2 or H_2O_2 generating enzymes, like glucose oxidase into films and/or food products. However, results of different studies related to whey protein films incorporated with LPS are quite promising since such films showed broad antimicrobial spectrum and high applicability on different food products. For example, different antimicrobial tests with LPS loaded whey protein films showed the antimicrobial potential of enzyme against very critical pathogens, such as *S. enterica* and *E. coli* O157:H7 (Min *et al.*, 2005a), and *L. monocytogenes* (Min *et al.*, 2005b) as well as spoilage fungi, such as *Penicillium commune* (Min and Krochta, 2005). Further studies to develop food applications also clearly proved the antimicrobial benefits of LPS loaded whey protein coatings to control growth of *L. monocytogenes* in smoked salmon (Min *et al.*, 2005c), fish specific spoilage microorganism such as *Shewanella putrefaciens* and *Pseudomonas fluorescence* in cold-stored rainbow trout fillets and pike-perch fillets (Shokri *et al.*, 2015; Shokri and Ehsani, 2017; Rostami *et al.*, 2017), *S. enterica* and *E. coli* O157:H7 in coated roasted turkey (Min *et al.*, 2006), and *Pseudomonas aeruginosa* in chicken thigh meat (Molayi *et al.*, 2018). The addition of LPS into chitosan films also has a good potential since this enhanced inherent antimicrobial activity of chitosan without affecting its water vapor barrier and mechanical properties considerably (Mohamed *et al.*, 2013). The chitosan films or coatings loaded with LPS have also been employed to inhibit bacterial spoilage flora of carb burger (Ehsani *et al.*, 2019), fish burger (Ehsani *et al.*, 2020), and rainbow trout fillets (Jasour *et al.*, 2015), and to suppress fungal spoilage flora in mango (Cissé *et al.*, 2015). Some studies also exist on incorporating LPS into alginate (Yener *et al.*, 2009), gelatin (Ehsani *et al.*, 2020) and defatted soybean meal films (Lee and Min, 2013). However, further antimicrobial tests are needed to evaluate long-term effectiveness of LPS system in edible films and to optimize levels of different antimicrobial components of this antimicrobial enzyme.

4.2.3 Glucose oxidase

Glucose oxidase (GOD, EC 1.1.3.4) is an oxidoreductase group enzyme that owes it antimicrobial activity to its ability to oxidase β-D-glucose to D-glucono-delta-lactone and to generate antimicrobial product H_2O_2 in the presence of molecular oxygen used in reaction as an electron acceptor (Ferri *et al.*, 2011). The commercial GOD is obtained mostly by fermentation from fungi, such as *P. amagasakiense* (MW of its GOD: 150 MW) and *A. niger* (MW of its GOD: 152 kDa) (Bankar *et al.*, 2009). The GOD has also been used effectively in active packaging as an oxygen scavenger to minimize oxidative changes in food (Dubey *et al.*, 2017). For oxygen scavenging purposes, the GOD is generally combined with catalase to eliminate residual H_2O_2 generated by the enzyme. In contrast, when the enzyme is applied in antimicrobial packaging, its activity is enhanced to form H_2O_2. The H_2O_2 is not only an effective antimicrobial compound, but is also one of the essential components of LPS system. Thus, application of GOD in active packaging is based on use of this enzyme as (1) an O_2 scavenger, (2) an antimicrobial enzyme-producing H_2O_2, or (3) a processing aid supporting LPS antimicrobial system.

The application of GOD in combination with LPS has been demonstrated for chitosan, whey protein, alginate, and gelatin films (Min *et al.*, 2005b; Jasour *et al.*, 2015). Hanušová *et al.* (2013) immobilized GOD in different polymeric films via the aid of chemical crosslinking agents and exploited its ability to generate H_2O_2 for inhibition of *E. coli, L. innocua* and *P. fluorescens*. The enzyme alone is also adapted for antimicrobial packaging with edible films. For example, Murillo-Martínez *et al.* (2013) developed GOD (0.024 or 0.5 mg/mL film-forming protein solution) incorporated whey protein films that showed antimicrobial activity against *L. innocua, Brochothrix thermosphacta, E. coli* and *Enterococcus faecalis* at pH 5.5. Lopes *et al.* (2011) also developed GOD incorporated chitosan films to enhance inherent antimicrobial activity of chitosan films. However, studies related to food application of GOD-loaded antimicrobial edible packaging are limited due to the potential toxic effects and oxidative changes of generated H_2O_2.

4.2.4   Chitinase

Chitinases are enzymes that show antifungal activity via hydrolyzing (1-4) β-linkages of N-acetylglucosamine (GlcNAc) of chitin that is the structural polysaccharide of fungi (Felse and Panda, 2000). The chitinases are classified as (1) endochitinases (EC 3.2.1.14) and (2) exochitinases that consist of chitobiosidases (EC 3.2.1.29) and 1-4-β-glucosaminidases (EC 3.2.1.30) (Hamid *et al.*, 2013). The endochitinases hydrolyze chitin at internal sites to form the dimer di-acetylchitobiose, chitotriose, and chitotetraose. The chitobiosidases catalyze sequential hydrolysis of di-acetylchitobiose starting from the non-reducing end of the chitin, while 1-4-β-glucosaminidases hydrolyze oligomeric products of endochitinases and exochitinases to form monomers of GlcNAc (Hamid *et al.*, 2013). There is a growing interest in using chitinase and chitinase-producing microorganism in crop protection against fungi and insects as environmentally friendly biocontrol agents. Thus, extensive studies have been conducted for production of chitinase from fungal and bacterial sources (e.g. *Trichoderma* spp., *Aspergillus* spp., *Bacillus* spp., etc.) by fermentation (Sandhya *et al.*, 2004; Gunalan *et al.*, 2012; Martínez-Zavala *et al.*, 2020). Some studies also exist on producing recombinant chitinases from GMO microorganism (e.g. *Pichia pastoris*) (Banani *et al.*, 2015) or from insects or transgenic plants that express insect chitinases (Kramer and Muthukrishnan, 1997).

Recent trends suggest that the antimicrobial edible coatings have a great potential to control bacterial plant pathogens on trees or in crops at the pre-harvest stages (Alkan and Yemenicioğlu, 2016). Thus, there is a good potential to employ chitinase loaded edible coatings against fungal plant pathogens in crops and insects causing their infestation. Although studies related to use of chitinase in antimicrobial edible packaging are scarce, there are several studies that employ chitinases in biodegradable films. For example, Silva *et al.* (2011, 2012) employed immobilized or free forms of chitinases extracted form *Trichoderma asperellum* in composite films of cashewnut gum/polyvinyl alcohol and cassava starch/poly (butylene adipate-co-terephtalate) and achieved inactivation of fungal pathogens, such as *Sclerotinia sclerotiorum, A. niger* and *Penicillium* sp. These authors reported that immobilized enzymes in biodegradable films caused morphological modifications (e.g. hyphae disruption and change of spore shape) on vegetative and germ structures of fungi. However, widespread application of chitinases in biocontrol purposes and antimicrobial food packaging needs development of more economically feasible methods for chitinase production (Das *et al.*, 2012, Shivalee *et al.*, 2018).

4.2.5 Polyphenoloxidase

Polyphenoloxidases (PPOs), tyrosinase, catecholoxidase, laccase, and phloroglucinoloxidase are oxidoreductase group metalloenzymes found in plants, fungi, animals, and bacteria (Yemenicioğlu, 2016b; Mayer, 1986). These enzymes use polyphenols as substrates and catalyze oxidation and/or hydroxylation of these compounds in different ways. For example, tyrosinase catalyzes hydroxylation of monophenols to *o*-diphenols (monophenolase activity) and oxidation of *o*-diphenols to *o*-quinones (diphenolase activity). The catecholoxidase catalyzes oxidation of *o*-diphenols to *o*-quinones while laccase catalyzes oxidation of both *o*-diphenols and p-diphenols to corresponding quinones (Yemenicioğlu, 2016b; Vamos-Vigyazo, 1981), and phloroglucinoloxidase catalyzes the oxidation of tree hydroxyl phenols without showing activity on diphenols (Rahman *et al.*, 2012). The quinons formed by enzymatic oxidation of phenolic compounds undergo nonenzymatic polymerization reactions, resulting in the formation of dark colored melanins that cause browning during processing of vegetables, fruits, mushrooms, and some crustacean species (Şimşek and Yemenicioğlu, 2007).

The PPOs are responsible for the defense reactions against pathogens in plants since melanins formed by these enzymes show antimicrobial activity (Boeckx *et al.*, 2015; Li and Steffens, 2002). Liu *et al.* (2019a) exploited mulberry PPO's oxidation products from chlorogenic acid and obtained antimicrobial effects against various bacteria, such as *E. coli*, *P. aeruginosa*, *S. aureus*, *Botrytis cinerea*, and *Sclerotina sclerotiorum*. It was determined that the PPO oxidation products show antimicrobial effects by causing damages at cell membranes and cell walls of bacteria (Liu *et al.*, 2019a). The studies related to incorporation of PPO into edible films intended for antimicrobial packaging are scarce. However, Elegir *et al.* (2008) developed cellulosic antimicrobial paper effective on *Staphylococcus aureus* and *Escherichia coli* by grafting laccase oxidized caffeic acid and isoeugenol on to unbleached Kraft liner fibers. There are also extensive studies related to immobilization of PPO on different supports, such as glass, zeolite, activated carbon, carbon nanotubes, synthetic polymers, and hydrocolloids (e.g. chitosan, gelatin, alginate, agarose, gum Arabic, cellulose, carboxymethyl cellulose, etc.), but these research studies aim at use of enzyme in different technological purposes (e.g. production of enzymatically modified antioxidant polyphenols, development of biosensors for detection of polyphenols, removal/modification of undesired polyphenols in food or toxic polyphenols in waste waters) (Meschini *et al.*, 2018; Zdarta *et al.*, 2018; Durán *et al.*, 2002). Further studies are needed to adopt PPO as a potential antimicrobial compound in active packaging.

4.3 Antimicrobial proteins and peptides

4.3.1 Lactoferrin

Lactoferrin (LF), an 80 kDa non-heme iron-binding glycoprotein found in milk, is a member of transferrin family proteins. LF is important for active packaging since it shows a broad antimicrobial spectrum. It shows antimicrobial activity on major food pathogenic bacteria, such as *L. monocytogenes, S. aureus, S. enterica* serovar Typhimurium, *E. coli O157: H7, Camphylobacter jejuni,* and *Bacillus cereus* (Jenssen and Hancock, 2009; Jahani *et al.* 2015; Niaz *et al.*, 2019). Moreover, it shows antifungal activity (Andersson *et al.*, 2000), and exhibits antiviral activity against some human intestinal viruses (e.g. rotaviruses, adenoviruses, polioviruses, and enteroviruses) which are potential food contaminants (Superti *et al.*, 1997; Marchetti *et al.*, 1999; Jenssen and Hancock, 2009). The LF owes its antimicrobial spectrum to its multiple antimicrobial mechanisms originating from its unique molecular structure. The bacteriostatic and fungistatic activities of LF have been attributed mainly to its ability to bind and reduce the availability of iron for the microorganisms (Andersson *et al.*, 2000). It was reported that the LF's affinity constant for iron is 300 times higher than that of transferrin (Sanchez *et al.*, 1992). The LF shows its iron-binding capacity when it is in 'apo' form (iron-deficient), thus its antibacterial activity weakens when it is in the 'halo' form (iron-saturated) (Ellison *et al.*, 1988; Farnaud *et al.*, 2003). The bactericidal activity of LF is attributed to its direct interaction with bacterial membranes. It is thought that the LF binds particularly to outer cell-wall proteins of Gram-negative bacteria (Naidu, 2002). It was also demonstrated that the LF has the ability to bind divalent cations (Ca^{+2} and Mg^{+2}) stabilizing the protective LPS layer surrounding the PG layer on the cell walls of Gram-negative bacteria (Ellison *et al.*, 1988; Sanchez *et al.*, 1992). Moreover, LF also blocks the microbial attachment factors (e.g. fimbriae, adhesins) (Naidu, 2002). The mechanism of the antiviral activity of LF has not been fully understood, but it has been determined that the antiviral activity of LF is maintained even when it is in halo form (Jenssen and Hancock, 2009). Thus, it seems that the antiviral activity of LF might be related to its complex formation ability with virus particles as specified by Farnaud *et al.* (2003).

Besides its high potential as an antimicrobial agent, LF is also an important antioxidant in food systems. Huang *et al.* (1999) showed that the LF decreased prooxidant effect of iron in corn oil emulsions while Satue-Garcia *et al.* (2000) demonstrated that the LF caused concentration-dependent inhibition of oxidation in whey-based and casein-predominant infant formulas. The antioxidant activity of LF is related to its strong iron-binding capacity that is mediated by its two globular lobes that bind one iron atom each (Drobni *et al.*, 2004). The LF-iron complex is very stable between pH 4.0 and 7.0 (Ward *et al.*, 1996). Moreover, the protein maintains its iron-binding ability and solubility after mild pasteurization and UHT treatments (Mata *et al.*, 1998).

The bovine LF is also used in nutraceuticals and functional foods due to its well-characterized health benefits on human health. The supplementation of infant formulas with bovine LF increases the *Bifidobacterium* in the fecal flora and ferritin in the serum of infants, while suppressing *Enterobacteriaceae, Streptococcus,* and *Clostridium* in the fecal flora (Wakabayashi *et al.*, 2006). The LF is also known as an anti-inflammatory protein (Conneely, 2001). Moreover, there is also significant evidence to show that LF suppresses or inhibits carcinogenesis (Tsuda *et al.*, 2010; Aboda *et al.*, 2020). In Japan, Korea, and Indonesia, the LF is used in infant formulas (Tomita *et al.*, 2009). In Japan, the bovine LF is also added in yogurt, skim milk, and milk-type drinks (Wakabayashi *et al.*, 2006). Thus, the use of LF in bioactive packaging might be considered in development of functional foods.

The commercial LF is obtained from skimmed bovine milk or whey (or clostrum whey) by using mainly cation-exchange chromatographic methods. Fast-flow cation-exchange chromatography following a two-step ultrafiltration (firstly from a 100 kDa and secondly from a 10 kDa membrane) (Lu *et al.*, 2007), or cation exchange membrane process conducted with multiple modules following ultrafiltration (Ulber *et al.*, 2001) are suggested for large-scale LF production. The enzymatic hydrolysis of purified LF is also used in production of a highly-potent bactericidal peptide, called lactoferricin (LFC). The LFC owes its antibacterial effect to its ability to cause depolarization of cytoplasmic membranes of Gram-positive and Gram-negative bacteria (Liu *et al.*, 2011). This peptide is also reported to have a strong antifungal activity (De Lucca and Walsh 1999). Liceaga-Gesualdo *et al.* (2001) found that LFC effectively inhibited both spore germination and mycelial growth of *Penicillium* spp. LFC corresponds to the 24 (17-41) or 25 (17-42) amino acid residues near the N terminus of lactoferrin (Recio and Visser, 1999). This location was reported as a region distinct from lactoferrin's iron-binding sites (De Lucca and Walsh, 1999). Therefore, it is thought that the antibacterial activity of LF, other than its iron-binding ability, originates from LFC. The list of bacteria susceptible to LFC (0.3 to 150 µg/ml) includes *E. coli, S. enterica* serovar Enteritidis, *K. pneumoniae, Proteus vulgaris, Yersinia enterocolitica, Pseudomonas aeruginosa, Campylobacter jejuni, Staphylococcus aureus, Streptococcus mutans, Corynebacterium diphtheriae, Listeria monocytogenes* and *Clostridium perfringens* while *P. fluorescens, Enterococcus faecalis,* and *Bifidobacterium bifidum* are highly resistant to LFC (Bellamy *et al.*, 1992). However, it should be noted that the antibacterial effectiveness of LFC reduces in the presence of Na^+, K^+, Mg^{2+} and Ca^{2+} ions (Bellamy *et al.*, 1992).

Due to their effectiveness on a wide range of microorganisms, LF and LFC have been tested in different edible packaging. In the USA, a commercial LF preparation (LF together with sodium bicarbonate, citric acid, sodium chloride, and carrageenan) designed to keep LF in active form 'activated LF' was approved by FDA as an antimicrobial spray on the surfaces of beef carcasses to prevent contamination during processing or as an antimicrobial for treatment of beef surface (65.2 mg/kg beef) before packaging to extend its shelf-life (Naidu, 2002). A solid-phase immunoassay based on a fluorescent-labelled LF-antibody was also developed to ensure the homogenous application and quantification of activated LF on the meat surface (Naidu, 2002). Thus, it appears that the activated LF combined with carrageenan is one of most carefully designed active edible coating applications. However, antimicrobial test results of LF in most edible films are less promising than those of soluble LF in microbiological media since edible film-forming materials and/or other ingredients used in film solution might contain metal contaminants that could neutralize the LF (transformation of apo form into halo form). Min and Krochta (2005) found that LF and LF pepsin hydrolysate incorporated into whey protein isolate (WPI) films lacked antifungal activity on *P. commune*. The WPI films loaded with LF also showed weak antibacterial activity on *S. enterica* and *E. coli* O157:H7 (Min *et al.*, 2005d). The LF incorporated into chitosan films did not also show considerable inactivation of *L. monocytogenes* and *E. coli* O157:H7 (Brown *et al.*, 2008). Moreover, the LF incorporated into paper containing carboxymethyl cellulose did not show antimicrobial effect on *L. innocua*, but it caused significant inactivation against *E. coli* (Barbiroli *et al.*, 2012). It was also found that the bacterial cellulose (BC) film loaded with LF caused significant reduction of *E. coli* when it was employed as sausage wraps, but this film showed limited antimicrobial activity against *S. aureus* (Padrao *et al.*, 2016). Al-Nabulsi *et al.* (2006) followed a different method than the other workers and employed LF in WPI films following encapsulation conducted by emulsification of LF within a lipid mixture (22% butter fat, 78% corn oil) using polyglycerol polyricinoleate as an emulsifier. It was determined that the WPI films loaded with encapsulated LF and chelating agents (e.g. Na-lactate and Na-bicarbonate) used to protect bioactive protein from neutralization caused significant reduction in *Carnobacterium viridans* in Bologna sausages. Further studies are needed to test encapsulated activated LF combined with chelating agents in different edible films.

4.3.2 Nisin

Nisin (NIS), a class I bacteriocin obtained commercially from some strains of *Lactococcus lactis* (formerly *Streptococcus lactis*), is one of the most extensively studied antimicrobial agents for development of active edible films and coatings. NIS is a cationic polypeptide composed of 34 amino acids and shows antimicrobial activity by interacting with the anionic phospholipids at the bacterial surfaces and forming pores and dissipating proton motive forces at the bacterial membrane (Sudağıdan and Yemenicioğlu, 2012). *L. lactis* strains could produce three different nisin variants – NIS A, NIS Z (contains asparagine at position 27 instead of histidine in NIS A), and NIS Q (differs from NIS A with four amino acid substitution) (Yoneyama *et al.*, 2008), but NIS Z (MW: 3330 Da) is the major NIS form observed in the food that contained commercial NIS preparations (Schneider *et al.*, 2011). In the USA, the FDA approved GRAS status (with GMP of 250 ppm in finished product) of NIS as an antimicrobial agent to inhibit the growth of *C. botulinum* spores and toxin formation in pasteurized cheese spreads, pasteurized processed cheese spreads, pasteurized cheese spread with fruits, vegetables, or meats (FDA, 1988, 2011). In the EU, the NIS is an authorized food additive (E 234) for use in cream (only clotted cream at 10 mg/kg), unripened cheese (only mascarpone cheese at 10 mg/kg), ripened cheese and cheese products (at 12 mg/kg), heat-treated meat products (at 25 mg/kg), pasteurized liquid egg (at 6.25 mg/kg), and some desserts (only semolina and tapioca puddings and similar products at 3 mg/kg) (EC, 2008; EFSA, 2017). In Japan, NIS is used in meat products, cheeses (except processed cheese), and wiped cream products at the levels of 12.5 mg/kg, in processed egg products, and miso at the level of 5 mg/kg, processed cheese confectionary at the level of 6.25 mg/kg, and in wet starch-based confectionery at the level of 3 mg/kg (JETRO, 2011).

The NIS is not effective against Gram-negative bacteria since it cannot overcome the protective LPS layer on their cell walls. However, it shows potent antimicrobial effect on critical pathogenic Gram-positive bacteria, such as *S. aureus, C. botulinum, L. monocytogenes*, and *Bacillus cereus* (Vukomanović *et al.*, 2017*)*. The effectiveness of NIS in inhibiting growth of *C. botulinum* spores is one of the most important attributes of this bacteriocin. It was reported that the *C. botulinum* spores of type A and C show more NIS resistance than type B and E that have moderate and low NIS resistance, respectively (Scott and Taylor, 1981). However, Mazzotta *et al.* (1997) found that the frequency of NIS resistance for *C. botulinum* type A and B vegetative cells and spores can be reduced considerably by reducing pH around 5.5 and increasing salt concentration to 3% in a medium. NIS is also effective on Gram-positive lactic acid spoilage bacteria (Sozbilen *et al.*, 2018) and *C. tyrobutyricum* – a spoilage bacterium that causes late blowing defect in ripening cheeses (Ávila *et al.*, 2014). The potency of NIS against *S. aureus* is also important since this bacterium is resistant against the action of LYS, an active agent frequently employed in ripening cheeses to prevent late blowing. Sudağıdan and Yemenicioglu (2012) showed that NIS at 25 µg/mL is sufficient to inactivate all of the 25 *S. aureus* strains isolated from raw milk and cheese samples. The breasts of dairy cows are mostly contaminated with enterotoxigenic strains of *S. aureus* that cause mastitis; thus, raw milk and traditional cheeses made from raw milk cause food poisoning very frequently (Oliver *et al.*, 2005; Pinto *et al.*, 2011). The *L. monocytogenes* also frequently colonizes on the breasts of dairy cows and causes contamination of milk (Muraoka *et al.*, 2003). Both *S. aureus* and *L. monocytogenes* are well known for their ability to form biofilms (Sudağıdan and Yemenicioglu, 2012). Therefore, these two bacteria cause risks even in cheeses obtained from heat-treated milk and whey when processing equipment is not properly decontaminated, and curd obtained at post-heating is not handled and stored in accordance with hygienic rules (Jakobsen *et al.*, 2011). Thus, the application of NIS in active packaging of cheeses could be very beneficial to improve the safety of these products. It was determined that the interaction of NIS with cheese fat globules causes loss of part of the NIS activity, but the remaining active NIS is mostly sufficient to show the desired preservation effects against major spoilage (e.g. *C. tyrobutyricum, C. butyricum, C. sporogenes*) and pathogenic bacteria (e.g. *C. botulinum, L. monocytogenes and S. aureus*) in cheeses (Somers and Taylor, 1987; Delves-Broughton *et al.*, 1996; Chollet *et al.*, 2008). In contrast, application of NIS in raw meat is not suggested since high amount of naturally occurring glutathione (GSH) in meat complexes with NIS and neutralizes the majority of its activity (Rose *et al.*, 1999). It was found that more than 50% of NIS in vacuum-packed minced raw beef lost within 10h at 8°C (Stergiou *et al.*, 2006). Rose *et al.* (2002) determined that the conjugation of GSH by NIS occurred through nonenzymatic mechanisms by binding of 1-3 GSH on to NIS, but the presence of enzyme glutathione-S-transferase in meat accelerated the conjugation reaction. Interestingly, the NIS maintains most of its activity in cooked or heat-treated processed meat products (Murray and Richard, 1997; Stergiou *et* *al.*, 2006). Stergiou *et al.* (2006) showed that loss of free sulfhydryl groups during cooking was a result of the reaction of GSH with meat proteins. As a result, no considerable free GSH remains in heat-treated meat to interact with the added NIS. Therefore, the use of NIS in active packaging of cooked meats is more suitable than that of raw meat. The neutralization of NIS caused by conjugation of GSH is less likely in raw fish due to its considerably lower levels of glutathione than fresh beef, pork, and poultry (Rose *et al.*, 1999). However, the action of inherent proteases in raw or smoked fish might also be a factor affecting the stability of NIS (Aasen *et al.*, 2003). Thus, initial levels of NIS used in active packaging and release rates of NIS from films must be adjusted not only according to the growth rates and NIS resistance of target bacteria, but also factors effective in NIS neutralization.

The NIS has been tested extensively in different edible films against mainly Gram-positive bacteria, but particular efforts have been spent to develop NIS containing active zein and whey protein films. The stability of NIS activity during classical solution-cast zein film-making process conducted by using ethanol as a solvent is low (NIS retains only ~12% of its added activity in films) (Dawson *et al.*, 2003). However, the solution-cast zein films still show effective antimicrobial activity on Gram-positive bacteria. Hoffman *et al.* (2001) incorporated NIS (37.5 µg/mL of film forming solution) into solution-cast zein films and tested these films against *L. monocytogenes* in broth media achieved 4 log reduction in this bacterium. These authors also determined that the combination of NIS with EDTA gave films having bacteriostatic activity against *S. enterica* serovar Enteritidis. Ku and Song (2007) obtained 1.4 log reduction of *L. monocytogenes* by using NIS (12000 IU/ml of film-forming solution) loaded solution-cast zein films in broth media. Padgett *et al.* (1998) developed NIS (6000 µg NIS/g film) loaded heat-pressed zein films effective on *L. plantarum* in zone inhibition assay. Padgett *et al.* (2000) also employed NIS (1500 µg/mL of film-forming solution) loaded solution-cast zein films against *L. plantarum* and achieved 2-3 log reduction of this bacteria in broth media. Different applications of NIS loaded zein films and coatings on refrigerated foods also proved the effectiveness of these films to control *L. monocytogenes* on ready-to-eat chicken (Janes *et al.*, 2002) and turkey frankfurters (Lungu and Johnson, 2005), and to suppress TVC in fish balls (Lin *et al.*, 2011).

The whey protein films have also been incorporated with NIS extensively. The pI of major whey protein fractions changes between 4.4 and 5.4 (Etzel, 2004). Therefore, the release of NIS from WPI films is highly pH-dependent. As pH approaches neutrality, the increased negative charges in WPI proteins start to bind the positively-charged NIS (pI: 8.8), while release of NIS from WPI films accelerates as pH gets lower than pH 4.4. and WPI proteins gain positive charges that repel NIS. Rossi-Márquez *et al.* (2009) investigated release profiles of NIS from WPI films at pH 4.0, 5.0, and 7.0 found that the highest NIS release occurred at pH 4.0. It was also found that the sensitivity of bacteria against NIS increases as pH at neutrality is reduced around pH 5.0 (Gänzle *et al.*, 1999). Therefore, the pH of the WPI solution during film preparation is highly effective in antimicrobial activity of NIS. Ko *et al.* (2001) showed that NIS loaded WPI films prepared at pH 3.0 caused greater reduction (2.4 log) of *L. monocytogenes* inoculated on the film surfaces than those prepared at pH 2.0 (0.8 log), 7.0 (1.2 log), and 8.0 (1.7 log). Pintado *et al.*

(2009) adjusted the pH of NIS loaded WPI films around pH 3.0 using lactic, malic and citric acids, and tested the zone inhibition potential of films against *L. innocua*. According to these authors, the most potent NIS loaded films are those acidified with malic acid, followed by citric acid, and lactic acid. Pintado *et al.* (2009) hypothesized that the higher effectiveness of malic and citric acids than lactic acid on bacterial inactivation was due to high metal-chelating capacity and ability of these two acids to dissociate following diffusion into cells. Gadang *et al.* (2008) showed that without addition of malic acid, NIS and NIS-grape seed extract mixture loaded WPI coatings were ineffective on *L. monocytogenes* inoculated on to turkey frankfurters. These authors proposed that the penetration of malic acid into *Listeria* cells and reduction of cell pH by this organic acid increased the susceptibility of bacteria against NIS and grape seed polyphenols. All these findings suggest the beneficial effects of using organic acids in combination with NIS in WPI films. On the other hand, Eswaranandam *et al.* (2004) did not determine any beneficial effect of using NIS (210 IU/g protein in film solution) in soy protein isolate (SPI) films (original film pH: 6.95) alone or in combination with organic acids, such as lactic, malic, tartaric and citric acids (film pHs: 3.35-4.5) against *L. monocytogenes*, *S. gamineria*, and *E. coli* O157:H7. These authors found that NIS-loaded SPI films were ineffective at the test concentration and antimicrobial effect of films originated from organic acids. In contrast, Ko *et al.* (2001) obtained NIS (6000 IU/ml of film solution) loaded SPI films effective on *L. monocytogenes* when they set the film pH to 3.0 using HCL. Padgett *et al.* (1998) used classical zone inhibition assay and determined minimum inhibitory concentration of NIS loaded heat-pressed SPI films against *L. plantarum* as 0.1 mg/g film, but these films did not show antimicrobial activity against *E. coli* even when they were incorporated with EDTA in addition to NIS. There is also some interest to use NIS in gelatin (GEL) films. For example, Luciano *et al.* (2020) developed NIS (56 mg/g of GEL) incorporated porcine GEL films effective on *L. monocytogenes* and *S. aureus*. NIS at 3% (w/w) incorporated into electrospun porcine GEL fiber mats also showed antimicrobial activity on *L. monocytogenes, S. aureus,* and *L. plantarum* (Dheraprasart *et al.*, 2009). NIS-loaded GEL films were also used in food applications to control *L. monocytogenes* in refrigerated Rainbow trout fillet slices (Han *et al.*, 2013) and bologna (Min *et al.*, 2010). Batpho *et al.* (2017) conducted an innovative study by vacuum impregnating NIS into collagen casings that were found effective against *L. monocytogenes* in sausages. Other protein-based antimicrobial edible packaging loaded with NIS are Na-caseinate (Kristo *et al.*, 2008) and wheat gluten films (Teerakarn *et al.*, 2002), while some NIS loaded polysaccharide-based films or coatings include chitosan film and coating (Cé *et al.*, 2012), alginate coating (Cutter and Siragusa, 1996), tapioca starch film (Sanjurjo *et al.*, 2006), pullulan film (Pattanayaiying *et al.*, 2015), bacterial cellulose film (Nguyen *et al.*, 2008), pectin film (Jin *et al.*, 2009), methylcellulose/hydroxypropylmethylcellulose coating (Franklin *et al.*, 2004), hydroxypropyl methylcellulose film (Imran *et al.*, 2010), guar gum coating (Garavito *et al.*, 2020), konjac glucomannan-gellan blend film (Xu *et al.*, 2007), pectin, alginate, carrageenan, starch, and xanthan gum coatings (Juck *et al.*, 2010), and corn starch-halloysite nanocomposite film (Meira *et al.*, 2016).

4.3.3 Pediocin

Pediocins, a group of bacteriocins produced by *Pediococcus* spp., have a good potential as antimicrobial components of active packaging (Espitia *et al.*, 2016). Although the NIS is still the only bacteriocin approved as a food additive, Class II (subclass IIa) pediocins, like pediocin PA-1 (MW: 4624 Da) and pediocin AcH (MW: 4628 Da) have been produced commercially by fermentation from *Pediococcus acidilactici* PAC1 and *P. acidilactici* H, respectively (Biswas *et al.*, 1991; Marugg *et al.*, 1992). The pediocin PA-1 and AcH are both 44 amino acid antimicrobial peptides with two disulfide bonds. These pediocins contain a cationic N-terminal that is important for electrostatic binding of bacteriocin to the receptors of target bacterial membrane, and a hydrophobic/amphiphilic C-terminal that is associated with target cell recognition (specificity), membrane insertion, and resulting pore formation (Espitia *et al.*, 2016; Porto *et al.*, 2017). The pore formation on bacterial membranes by pediocin triggers bacterial inactivation by dissipating the transmembrane electrical potential and causing cellular efflux of salts and amino acids (Porto *et al.*, 2017). Similar to NIS, pediocins are not very effective on Gram-negative bacteria. The antimicrobial spectrum of pediocins against Gram-positive bacteria is also narrower than that of NIS, but pure or partially pure pediocins tested at the laboratory media have been found more effective on *L. monocytogenes* than commercial NIS. Coventry *et al.* (1995) found that partially purified pediocin PO2, a pediocin very identical to pediocin PA-1/AcH, showed lower MICs than commercial NIS against four different *L. monocytogenes* strains. Cintas *et al.* (1998) also showed that the purified pediocin PA-1 had lower MICs than purified NIS A against five different *L. monocytogenes* strains. Katla *et al.* (2003) proved the higher antilisterial activity of pediocin PA-1 than commercial NIS by conducting tests on two hundred *L. monocytogenes* strains isolated from food and food industry environment. However, some reports related to comparative food applications of pediocins and NIS are contradictory due possibly to differences in interactions and stability of these bacteriocins in the food. For example, Bari *et al.* (2005) reported that commercial NIS was more effective than pediocin purified from *P. acidilactici* LET 01 to inactivate *L. monocytogenes* on some fresh-cut vegetables (e.g. broccoli and cabbage). Murray and Richard (1997) also found NIS A more effective than pediocin AcH to reduce *Listeria* load in fresh, refrigerated ground pork meat since pediocin AcH lost its activity in meat rapidly, probably due to the action of proteolytic enzymes. El-Khateib *et al.* (1993) reported that commercial NIS was slightly more effective (caused 0.5 log greater inactivation) than pediocin PO1 obtained from *P. acidolactici PO1* when these bacteriocins were employed in raw meat cubes inoculated with *L. monocytogenes*. The results of these studies suggest that the pediocin might be instable when it is added directly into food. Degan and Luchansky (1992) showed that the recovery of pediocin in slurries of beef tallow and muscle could be increased partially by encapsulation of this bacteriocin into liposomes of phosphatidylcholine. Narsaiah *et al.* (2013) showed that hybrid capsules of alginate (2%)-guar gum (0.4%) incorporated with pediocin-loaded nanoliposomes of phosphatidylcholine were more effective than free pediocin to inactivate *L. innocua* in broth media. However, further studies are needed to prove these works in foods. Most of the pediocins are also less effective than NIS on *S. aureus, B. cereus, C. tyrobutrycum, Brochothrix thermosphacta, Lactococcus* spp., but more effective than NIS on *Enterococcus faecalis, C. perfringens,* and *C. botulinum* (Coventry *et al.*, 1995; Cintas *et al.*, 1998). The effects of NIS and pediocin on some bacterial spores also vary considerably. For example, commercial NIS accelerated thermal inactivation of *B. subtilis* spores without showing a considerable effect on thermal inactivation rates of *B. licheniformis* spores while pediocin obtained from *P. acidilactici* NRRL B-5627 acted on these spores just in the opposite way (Cabo *et al.*, 2009). Therefore, the combination of pediocins with NIS has been suggested to prevent spoilage of sous vide products (e.g. mushroom and shellfish salads) by the *Bacillus* spores (Cabo *et al.*, 2009). Bari *et al.* (2005) also combined pediocin with NIS to achieve greater decimal reductions on *L. monocytogenes*. The combination of pediocins with NIS might also contribute to overcoming bacterial-resistance problems originating from extensive application of NIS (Murray and Richard, 1997). One great advantage of pediocins over NIS is that unlike NIS, they lack a considerable antimicrobial effect on infant *Bifidobacteria* isolates and commercial strains of *Bifidobacteria* used extensively to obtain functional foods (Kheadr *et al.*, 2004; Blay *et al.*, 2007). Thus, development of antilisterial and bioactive edible films loaded with pediocin and *Bifidobacteria* might be possible in future.

In the literature, studies related to incorporation of pediocin into biodegradable polymeric materials, such as polylactic acid-based films (Woraprayote *et al.*, 2013), and cellulose-based polymeric films and casings (Ming *et al.*, 1997; Santiago-Silva *et al.*, 2009; Espitia *et al.*, 2013a) have shown the potential application of this bacteriocin for active packaging of meat products to inhibit *L. monocytogenes*. However, studies to incorporate pediocins into edible films and coatings are limited with alginate coatings (Narsaiah *et al.*, 2015), methyl cellulose-zinc oxide nanoparticle composite films (Espitia *et al.*, 2013b), and corn starch and corn starch-halloyside nanoclay composite films (Meira *et al.*, 2016). Meira *et al.* (2016) compared antimicrobial performance of corn starch films with commercial pediocin (ALTA™ 2345) at 1% (w/v) and nisin (Nisaplin®) at 0.4% (w/v) against *L. monocytogenes* and *C. perfringens*. According to these authors, the pediocin loaded starch films were less effective on *L. monocytogenes* than NIS loaded films. Moreover, pediocin loaded films also lacked antimicrobial activity against *C. perfringens* while NIS loaded films were effective against this bacterium (Meira *et al.*, 2016). Further studies are expected related to pediocin loaded edible films if future regulations allow direct use of this bacteriocin in food products.

4.3.4 Sakacin

Sakacins are also Class II (subclass IIa) bacteriocins that are more effective on *L. monocytogenes* than both pediocin and nisin. Sakacins, such as sakacin A, G, K and P obtained from *Lb. sakei* LB 706, *Lb. curvatus* ACU-1, *Lb. sake* CTC 494, and *Lb sakei* LTH 673 are some of the most extensively studied sakacins against *Listeria*, respectively. Katla *et al.* (2003) proved the antilisterial potency of sakacin A and P against two hundred *L. monocytogenes* strains isolated from food samples and food industry environments (Table 4.1). This study also proved the greater effectiveness of sakacins than NIS against *Listeria*. Trineta *et al*. (2010) also showed the effectiveness of sakacin-A on *L. monocytogenes* strains isolated from major outbreaks taking place in different parts of the world between 1972 and 2002. Most of the sakacins are not effective on Gram-negative bacteria, but sakacin C2, obtained from *Lb sakei* C2, is a unique sakacin that shows antimicrobial activity on critical Gram-negative bacteria, such as *E. coli, S. enterica* serovar Typhimurium and *Shigella flexneri* (Gao *et al*., 2010). It was also reported that sakacin C2 caused membrane permeabilization and depolarization of trans membrane electrical potential of *E. coli*, but further studies are needed to show how it overcomes the protective LPS layer of Gram-negative bacteria. According to the findings of Aasen *et al*. (2003), sakacin shows better stability in the heat-treated food than in the raw animal origin food that contains active proteases capable of destabilizing bacteriocins (e.g. chicken and fish products). It was also determined that the NIS recoveries from such protease active food were much higher than those of sakacin (Aasen *et al*., 2003).

Table 4.1: Comparative antilisterial activities of different bacteriocins against two hundred *L. monocytogenes* strains isolated from food samples and food industry environments (Katla *et al.,* 2003)

Bacteriocins	IC$_{50}$ (mg/mL)
Sakacin P	0.01-0.61
Sakacin A	0.16-44.2
Pediocin PA-1	0.1-7.34
Nisin	2.2-781

The sakacins are among the potential candidate antimicrobials that might be suitable for antimicrobial packaging. In the literature, studies related to the use of sakacin P and K in edible films are scarce, but different antilisterial packaging materials have been obtained by incorporating sakacin A into pullulan films (Trinetta *et al*., 2010), pullulan-xanthan-locust bean gum blend films (Trinetta *et al*., 2011), nanocellulose fiber mats (Mapelli *et al*., 2019), and gelatin coating cast onto surface of polyethylene coated paper (Barbiroli *et al*., 2017). Moreover, Rivas *et al*. (2018) impregnated Sakacin G into natural bovine, ovine and pork casings, and collagen and cellulosic films used for artificial casings.

4.3.5 Polylysine

ε-polylysine (PL) is an antimicrobial peptide produced by aerobic fermentation by *Streptomyces albulus* (Yoshida and Nagasawa, 2003). The PL formed by 25 to 35 L-lysine residues shows antimicrobial effect both on major Gram-positive and Gram-negative food pathogenic bacteria, such as *L. monocytogenes, E. coli* O157:H7, and *S. enterica* serovar Typhimurium (Geornaras and Sofos, 2005; Chang *et al*., 2010). Li *et al*. (2014) attributed the mechanism of antibacterial action of PL to its ability to change the cell membrane permeability, and to cause disappearance of soluble cellular proteins possibly since it interferes with protein synthesis or it causes protein aggregation and leakage. The study of Liu *et al.* (2015) suggested that the PL is also capable of interacting with the bacterial DNA when it is penetrated into cells. Lin *et al.* (2018a) reached similar findings with Li *et al.* (2014), but they also determined that the PL blocks the Emden-Meyerhof-Parnas pathway by inhibiting three key-enzyme in this pathway.

In Japan, PL is used in sliced fish and fish sushi at levels between 1000 and 5000 ppm, in different traditional Japanese daily dishes (Nimono) at levels up to 500 ppm, and in foods, such as boiled rice, noodle soup stocks, soups, noodles, cooked vegetables, sukiyaki (Japanese beef steak), potato salad, steamed cakes, and custard cream at levels between 10 and 500 ppm (Hiraki *et al.*, 2003). In the USA, FDA first recognized PL as GRAS for use in cooked rice or sushi rice (FDA, 2004), but then PL application was approved for a greater variety of foods, such as baked goods, beverages, cheeses, fresh and processed meat, poultry and fish, milk and milk products, etc. at the maximum level of 0.025% by weight (FDA, 2010a). This development turned PL into a popular antimicrobial agent for edible packaging applications. The high compatibility of PL with zein was demonstrated by Ünalan *et al.* (2011b), who also tested this natural antimicrobial in chitosan, whey and alginate films. Ünalan *et al.* (2011b) attributed the good antimicrobial potential of PL in zein films to the hydrophobic nature and limited charged groups of zein that prevent the complexation and immobilization of PL with the film matrix by charged-charge interactions. Garcia (2020) developed PL-zein nanoparticles and applied these as a coating to prevent softening and fungal spoilage of avocadoes by *Colletotrichum* spp. Cai *et al.* (2015) applied PL loaded alginate coatings to control microbial quality of sea bass. Tang *et al.* (2017) developed antimicrobial alginate nanobiocomposite films by incorporating these films with PL and cellulose nanocrystals. Recently, there has also been a great interest in developing antimicrobial films, using PL and chitosan. Different workers incorporated PL directly into chitosan films (Li *et al.*, 2018; Na *et al.*, 2018; Li *et al.*, 2019), while Xu *et al.* (2020) incorporated PL into chitosan-gelatin composite films. Wang *et al.* (2014) interacted PL with chitosan, using natural Maillard reaction and developed antimicrobial films from obtained conjugates, while Wu *et al.* (2019b) prepared composite antimicrobial films of PL-chitosan conjugate by using tripolyphosphate as a crosslinking agent. Lin *et al.* (2018b) developed electrospun PL-chitosan nanofiber mats and applied these mats to control *S. enterica* serovar Typhimurium and *S. enterica* serovar Enteritidis on raw chicken cubes, while Liu *et al.* (2020) developed antimicrobial mats from electrospun PL-gelatin-chitosan nanofibers. The PL is also employed to develop antimicrobial starch films intended for preservation of slices of bread against *A. parasiticus* and *P. expansum* (Luz *et al.*, 2018), while Sun *et al.* (2019) developed PL loaded distarch phosphate (a resistant starch form with GRAS status)-crystalline cellulose nanocomposite films. Several interesting innovative approaches have also been developed to use PL in active food packaging. For example, Weng *et al.* (2014) developed PL incorporated surimi edible films obtained by using surimi homogenate as film-forming solution (2% w/v surimi protein). Alemán *et al.* (2016) obtained antimicrobial packaging by encapsulating PL into shrimp peptide hydrolysate-phosphatidylcholine liposomes, and incorporating these encapsules into edible wraps prepared from cooked shrimp muscle homogenate. Mizielińska *et al.* (2018) employed antimicrobial cellulose carton boxes coated with methyl cellulose solutions containing PL for packaging of fish fillets. All these studies clearly showed the increased importance of PL as an emerging antimicrobial agent.

4.4 Phenolic compounds

The phenolic compounds in plants, secondary metabolites produced in shikimate/ phenylpropanoid, and polyketide pathways are considered the most important pool for bioactive compounds. The use of plant phenolic compounds in active edible food packaging has been boosted not only because of potent antimicrobial and antioxidant effects of these natural compounds in food, but also because of their different bioactive properties. The frequent intake of phenolic compounds reduces the risk of major diseases, such as cancer, cardiovascular disease, and diabetes and shows a number of important health benefits (e.g. antioxidant, anti-inflammation, antiaging, anticoagulant etc.) (Lopez-Rubio *et al.*, 2006; Wang *et al.*, 2012a; Bijak *et al.*, 2019).

4.4.1 Classification of phenolic compounds

Phenolic compounds in food are generally classified as flavonoids and non-flavonoids that include phenolic acids, xanthones, stilbenes, lignans, and tannins (Durazzo *et al.*, 2019). The details of molecular properties and characteristics of phenolic compounds have been well-documented in the literature (Papuc *et al.*, 2017; Singla *et al.*, 2019; Durazzo *et al.*, 2019). The most important sources of phenolic compound in human diet are fresh fruits and vegetables that are very rich in phenolic acids, flavonoids, and tannins. The phenolic acids contain a single aromatic ring (phenolic ring), and consist of hydroxycinnamic acids (C_6C_3, phenolic ring with —CH=CH—COOH) and hydroxyl benzoic acids (C_6C_1, phenolic ring with —COOH group). The flavonoids are polyphenols ($C_6C_3C_6$) that contain two phenolic rings (ring A and ring B) combined with a three carbon to form an oxygenated heterocycle (Ring C) (Cutrim and Cortez, 2018). Due to variations in $C_6C_3C_6$ substituents, different flavonoid subclasses are formed, such as flavones, flavonones, flavonols, flavanols (flavan-3-ols), isoflavones, flavanones, and anthocyanins. Tannins are also divided into two subclasses – hydrolysable tannins (formed by gallotannins, ellagitannins, and complex tannins) and condensed tannins (catechin tannins or proanthocyanidins) that are oligomers or polymers of flavonols ($C_6C_3C_6)_n$ linked mostly through C4 → C8 bonds (Smeriglio *et al.*, 2017).

4.4.2 Antioxidant mechanisms of phenolic compounds

Phenolic compounds extracted from different plant sources are the most extensively studied natural compounds for active edible packaging since they prevent oxidation of lipids and cause inhibition of spoilage/pathogenic microorganisms in food, owing to their antioxidant and antimicrobial capacities, respectively. The antioxidant activity of polyphenols originates mostly from their free radical scavenging activity that is mediated mainly by their ability to conduct H atom transfer to free radicals. The free radicals targeted by polyphenols in biological systems include superoxide anion radical ($O2^{\cdot-}$), hydroxyl radical ($^{\cdot}OH$ or $HO^{\cdot}$), peroxyl radical

(ROO$\cdot$), alcoxyl radical (RO$\cdot$), polyunsaturated fatty acid radical (PUFA$\cdot$), nitric oxide radical (NO$\cdot$), thiyl radical (RS$\cdot$), carbon centered radical (R$\cdot$), etc. (Di Meo *et al.*, 2013; Papuc *et al.*, 2017). It is thought that the 'H atom transfer' during free radical scavenging of polyphenols occurs by three different mechanisms. These mechanisms given below were described and illustrated using Eq. 1 to 6 according to Di Meo *et al.*, 2013. The first mechanism involves pure hydrogen atom transfer (HAT) and proton-coupled electron transfer (PCET). The HAT is a mechanism that the proton and electron of H atom are transferred to the same atomic orbital in the free radical. The main difference of PCET from HAT is that it takes place at several molecular orbitals, but in both mechanisms the proton and electron transfer occurs at one step as shown in reaction (Eq.1).

$$ArOH + R^{\cdot} \rightarrow ArO^{\cdot} + RH \qquad (1)$$

Where ArOH shows the –OH group of polyphenol, R$\cdot$ shows free radical, RH shows neutralized free radical and ArO$\cdot$ shows formed less-reactive antioxidant radical.

The second H atom transfer mechanism is ET-PT that is a two-step reaction, initiated by electron transfer (Eq. 2) and followed by proton transfer (Eq. 3)

$$ArOH + R^{\cdot} \rightarrow ArOH^{\cdot +} + R^{-} \qquad (2)$$

$$ArOH^{\cdot +} + R^{-} \rightarrow ArO^{\cdot} + RH \qquad (3)$$

where ArOH$^{+\cdot}$ and R^{-} are less reactive antioxidant cation radical and energetically stable species with an even number of electrons, respectively.

The third H atom transfer mechanism is sequential proton loss-electron transfer (SPLET). The SPLET is a three-step reaction that the first two steps progress in inverse order of ET-PT (Eqs. 4, 5 and 6).

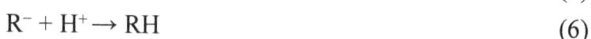

$$ArOH \rightarrow ArO^{-} + H^{+} \qquad (4)$$

$$ArO^{-} + R^{\cdot} \rightarrow ArO^{\cdot} + R^{-} \qquad (5)$$

$$R^{-} + H^{+} \rightarrow RH \qquad (6)$$

It is thought that the weight of these three H transfer mechanisms in overall free radical scavenging capacity might change, depending on antioxidant characteristics of polyphenols (e.g. molecular dissociation enthalpy, proton affinity, proton dissociation enthalpy, ionization potential, etc.) and the properties of food and environmental conditions (e.g. polarity/nonpolarity and pH of food) (Di Meo *et al.*, 2013; Olszowy, 2019). For example, studies conducted by using quercetin as a model antioxidant suggest that the PCET should be the major mechanism effective in non-polar media (e.g. inside lipid bilayer membrane or in lipid fractions found in food) and acidic pH (e.g. in stomach) media to inhibit chain reaction in lipid peroxidation (Di Meo *et al.*, 2013). Nakayama and Uno (2015) showed that the PCET was involved in scavenging of superoxide radical by polyphenols, such as (+)-catechin, quercetin, and rutin. However, Rajan and Muraleedharan (2017) identified HAT as the major free radical scavenging mechanism for gallic acid.

Another antioxidant mechanism of polyphenols is their chelating ability of metal atoms, such as Fe^{+3} and Cu^{+2} that participate in Fenton reactions to generate the most potent radical HO• in the presence of H_2O_2 (eq. 7) (Leopoldini *et al.*, 2011). The gallo tannins are good examples of phenolic compounds with iron-chelating capacity (Engels *et al.*, 2011). Moreover, hydroxycinnamic acids with o-diphenol structure, such as caffeic and chlorogenic acids, are well known with their strong iron-chelating capacity (Andjelković *et al.*, 2006). The polyphenols are also capable of decomposing H_2O_2 into water by donating protons and electrons to this nonradical reactive species that play important roles in the formation of HO• in food, especially in meat and meat products (Papuc *et al.*, 2017).

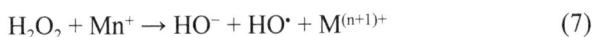

$$H_2O_2 + Mn^+ \rightarrow HO^- + HO^• + M^{(n+1)+} \tag{7}$$

The antioxidant capacity of phenolic compounds is highly related to their chemical structure (Kim and Lee, 2004). For example, it is well-known that phenolic acids, having a hydroxycinnamic acid structure (C_6C_3, phenolic ring with —CH=CH—COOH), generally show higher antioxidant activity than phenolic acids with hydroxyl benzoic acid structure (C_6C_1, phenolic ring with —COOH group) (Olszowy, 2019). It is also generally accepted that diphenols show higher antioxidant activity than monophenols, and the triphenols show higher antioxidant activity than diphenols (Brand-Williams *et al.*, 1995; Siquet *et al.*, 2006). The presence of o-diphenol structure in phenolic acids (e.g. catechol) and flavonoids (e.g. OH substitutes at 3′, 4′ positions of B ring as in epicatechin and epicatechin gallate) is mostly associated with high antioxidant capacity since such polyphenols have the ability to rapidly oxidize into o-quinones when they contact with free radicals (Rösch *et al.*, 2003) (Figs. 4.1a-4.1c). The vicinal substitution of three OH groups in phenolic acids (e.g. gallic acid) and flavonoids (e.g. —OH at 3′, 4′, 5′ positions of B ring as in epigallocatechin and epigallocatechin gallate), and gallic acid group (galloyl moiety) esterified to hydroxyl group at position 3 of C ring in flavonoids (e.g. epigallocatechin gallate and epicatechin gallate) also cause a strong antioxidant capacity (Figs. 4.1d and 4.1f) (Chu *et al.*, 2017). Other molecular properties associated with high antioxidant potential of flavonoids are the presence OH groups at the 3 and 5 positions, and the existence of 2,3-double bond in conjugation with a 4-oxo function of a carbonyl group in the C-ring (Yordi *et al.*, 2012).

4.4.3   Antimicrobial mechanisms of phenolic compounds

The phenolic compounds play a central role in the defense mechanisms of plants against microbial pathogens. It is well known that oxidation of polyphenols by polyphenol oxidase and peroxidase into quinons in damaged plant tissues is an important mechanism that improves the antimicrobial activity of polyphenols against plant pathogens (Bennett and Wallsgrove, 1994; Li and Steffens, 2002). Moreover, recent studies also show that the flavonoids in plants form a concealed antimicrobial mechanism against fungal pathogens. This mechanism was unveiled when it was determined that the flavonoids deglycosylated by extracellular fungal β-glycosidases showed considerably higher antimicrobial activity than their glycosylated forms (Sudheeran *et al.*, 2020).

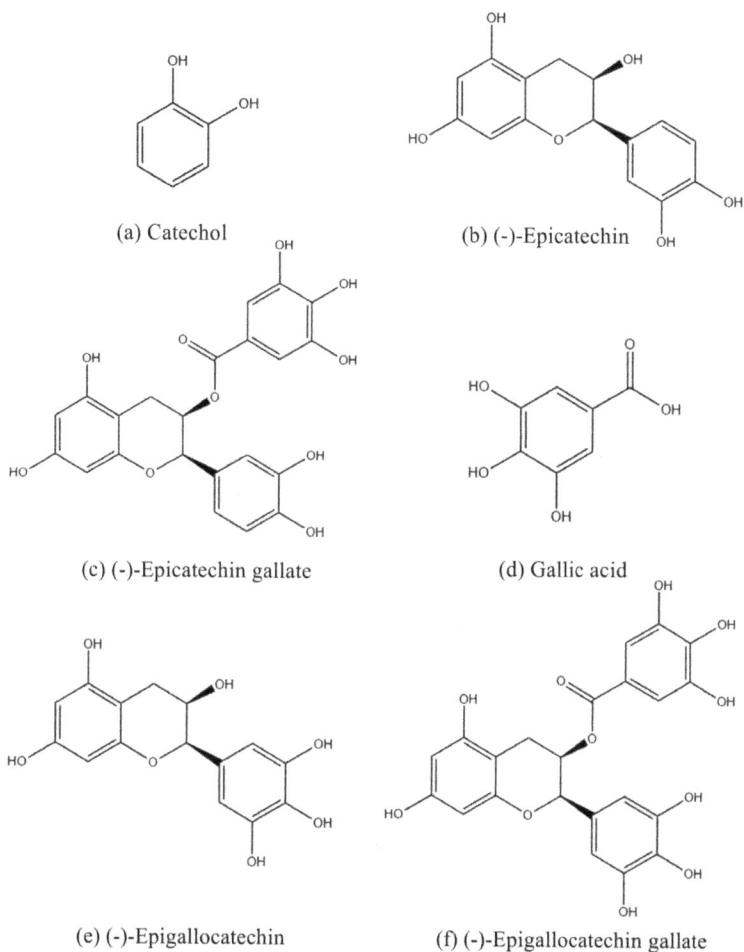

(a) Catechol

(b) (-)-Epicatechin

(c) (-)-Epicatechin gallate

(d) Gallic acid

(e) (-)-Epigallocatechin

(f) (-)-Epigallocatechin gallate

Fig. 4.1: Examples of polyphenols having critical antioxidant groups

The antimicrobial activity of phenolic compounds occurs by multiple mechanisms, including direct action on microbial metabolism, deprivation of critical substrates of microorganism (e.g. iron and zinc), complex formation with cell wall components, membrane disruption and/or permeabilization, denaturation of membrane proteins, inhibition of bacterial adhesion, biofilm formation and quorum-sensing, inhibition of energy metabolism and DNA synthesis, inactivation of critical bacterial enzyme systems (e.g. protein reductases (FabG/FabI), DNA gyrase) (Cowan, 1999; Zhang and Rock, 2004; Gradišar *et al.*, 2007; Engels *et al.*, 2011; Daglia, 2012; Camele *et al.*, 2019; Chibane *et al.*, 2019). Due to their different molecular properties, antimicrobial mechanisms and spectrums of phenolic compounds show significant variations. For example, Daglia (2012) reported that phenolic acids show mainly antibacterial activity while flavonoids, such as flavanols and flavonols, and hydrolysable tannins show antibacterial, antifungal, and antiviral activities, and condensed tannins show antibacterial and antiviral activities. In general, phenolic acids show greater antimicrobial activity on Gram-positive than Gram-negative bacteria as protective LPS forms a barrier, especially against hydrophilic phenolic acids (Cueva *et al.*, 2010). Sánchez-Maldonado *et al.*, (2011) reported that the antimicrobial activity of hydroxybenzoic acids on *E. coli* showed an inverse relationship with their number of hydroxyl groups (MICs of gallic acid (3 × —OH groups) and p-hydroxybenzoic acid (1 × —OH group) are 0.49 and 0.12 g. L^{-1}, respectively). However, these authors did not find the same correlation for hydroxycinnamic acids. Gutiérrez-Larraínzar *et al.* (2012) also found that the potent antioxidant gallic acid was not very effective on Gram-negative bacteria, such as *E. coli* and *P. fluorescens*, but it was highly effective on Gram-positive *S. aureus*. The studies to relate molecular structures of other polyphenols with their antimicrobial activity also revealed that phenolic compounds with pyrogallol moiety (a phenol ring having —OH substituents at position 3, 4, 5) show higher antimicrobial activity than those with catechol moiety (a phenol ring with o-diphenol structure) and resorcinol moiety (a phenol ring with m-diphenol structure) (Taguri *et al.*, 2006). It was also reported that the presence of galloyl moiety in some polyphenols (e.g. green tea catechins) makes a crucial contribution to their antimicrobial activity since this group increases affinity of polyphenols to bacterial cell walls and membranes (Kajiya *et al.*, 2001; Cui *et al.*, 2012), and it is highly effective in inhibiting bacterial enzymes by anchoring polyphenols into their active sites (e.g. bacterial DNA gyrase and protein reductases such as FabG and FabI) (Zhang *et al.*, 2004; Gradišar *et al.*, 2007). However, the LPS layer of Gram-negative bacteria (e.g. *Enterobacteriaceae*) might show barrier effect even against most active polyphenols, such as epigallocatechin (containing a pyrogallol group at the B ring) and epigallocatechin gallate (containing a pyrogallol at the B ring and a galloyl moiety at the C rings) (Taguri *et al.*, 2006). The resistance of Gram-negative bacteria was also observed against different phenolic extracts (Proestos *et al.*, 2006; Engels *et al.*, 2011). It is thought that the negative charges of LPS surrounding the PG layer of Gram-negative bacteria repel polyphenols while PG in Gram-positive bacteria generally absorb polyphenols; thus, this paves the way to bacterial inactivation (Patra, 2012). Cui *et al.* (2012) proposed that the epigallocatechin gallate acted on *S. aureus* first by binding to the PG layer to interfere with the proteins and peptides at this layer via H-bonding through —OH groups and galloyl moiety. According to these authors, the binding of this flavonoid then caused cell envelope's aggregation that led to subsequent lysis of *S. aureus* cells. However, it should be noted that the critically important Gram-positive pathogenic bacteria *L. monocytogenes* shows some significant resistance to most polyphenols, including epigallocatechin gallate. Taguri *et al.* (2006) found that *L. monocytogenes* was considerably less susceptible to a great variety of pure polyphenols than pathogenic Gram-positive bacteria, such as *B. cereus*, *C. perfringens*, and *S. aureus*. Tenore *et al.* (2011) tested fractionated phenolic compounds from three different wines and reported that *L. monocytogenes* showed lower susceptibility to these fractions than *B. cereus*, *S. aureus*, and *E. faecalis*. In contrast, Proestos *et al.* (2006) determined that a majority of phenolic extracts from 13 aromatic plants were effective on *L. monocytogenes*. Moreover, it was determined that the phenolic acid pyrogallol

(Taguri *et al.*, 2006) and the hydrolysable tannin ellagic acid (a dimeric derivative of gallic acid) (Shen *et al.*, 2014) were highly effective on *L. monocytogenes*. Therefore, a combination of different polyphenols with each other or with other natural antimicrobial agents (e.g. nisin, lysozyme, etc.) might be a strategy against Gram-positive bacteria targeted during food packaging applications. Overcoming the resistance of Gram-negative bacteria against hydrophilic polyphenols is more challenging and generally needs their application at high concentrations that mostly cause problems in aroma and flavor of foods. For example, it was determined that the increased concentration of epigallocatechin gallate induced oxidative stress on *E. coli,* causing inactivation of this bacterium due to formation of pore-like lessons (grooves) on its membrane (Cui *et al.*, 2012). Effective inhibition of Gram-negative bacteria with low concentrations of polyphenols needs application of non-polar polyphenols that might overcome the LPS barrier easily. Phenolic components of essential oils, such as eugenol, carvacrol, and thymol are well known due to their ability to inhibit Gram-negative bacteria, such as *E. coli* and *P. fluorescens* (Hemaiswarya and Doble, 2009; Gutiérrez-Larraínzar *et al.*, 2012). However, these polyphenols are applicable only to a limited number of foods due to their highly aromatic nature.

4.4.4 Major phenolic compounds used in active packaging

4.4.4.1 Catechins

The application of pure catechins and phenolic extracts, rich in catechins (mainly green tea extracts) in active packaging as antioxidant, antimicrobial or bioactive agent, has been continuously increasing since these compounds are well characterized for their active properties and health benefits, and they are already involved in human diet in large amounts (Yılmaz, 2006). The catechins belong to the flavanol group that includes some of the most potent antioxidant and antimicrobial polyphenols. According to the position of two hydrogen atoms at ring C (positions 2 and 3), catechins are classified as trans-type (e.g., (+)-catechin, (-)-gallocatechin, (-)-catechin gallate, (-)-gallocatechin gallate), and cis-type (e.g., (-)-epicatechin, (-)-epigallocatechin gallate, (-)-epicatechin gallate, (-)-epigallocatechin). However, green tea, the most common source of catechins, contains mainly cis-type catechins. The dominant and most active catechin in green tea is (-)-epigallocatechin gallate that forms almost 30% of total green tea phenolic compounds and 59% of total green tea catechins (Botten *et al.*, 2015; Rahardiyan, 2019). The catechins owe their antimicrobial properties to the position and number of their critical functional moieties, especially pyrogallol and galloyl moieties (Kajiya *et al.*, 2001; Zhang and Rock, 2004; Taguri *et al.*, 2006; Gradišar *et al.*, 2007; Cui *et al.*, 2012). Besides their direct antimicrobial effects, the main green tea catechin (-)-epigallocatechin gallate, could also effectively inhibit biofilm formation of both Gram-positive and Gram-negative bacteria (Jagani *et al.*, 2009). In contrast, although they show good antioxidant activity, group members lacking indicated critical moieties, such as (+)-catechin and (-)-epicatechin, show weak antimicrobial activity (Cueva *et al.*, 2010). In addition to the presence of necessary moieties, being in the cis- or trans-configuration might also affect the biological activities (e.g. affinity on cell membranes) of catechins. Studies by Kajiya *et al.* (2001) on lipid bilayers suggest that the catechins having cis-type show greater ability to penetrate from cell membranes than those having trans-types.

In the literature, (+)-catechin has been used extensively to achieve antimicrobial and antioxidant activity in solution-cast edible films and coatings. Although (+)-catechin is only a minor component of green tea catechins, it is a promising antioxidant since it can be produced in considerable amounts, using callus and suspension-culture cells, especially from hypocotyls of *Pelargonum hydropiper* seedlings (Ono *et al.*, 1998). Moreover, it was demonstrated that some traditional Asian medicinal plants, like gambir (*Uncaria gambir*) and bark of *Ulmus pumila* contain considerable amounts of (+)-catechin as the main phenolic constituent (Anggraini *et al.*, 2011; Cho *et al.*, 2016). The (+)-catechin alone or in combination with gallic acid and LYS were used in zein films to improve their antioxidant and antimicrobial activity (Arcan and Yemenicioglu, 2011; Arcan and Yemenicioglu, 2014; Ünalan *et al.*, 2013). According to Arcan and Yemenicioğlu (2011), (+)-catechin (at 3 mg per cm^2 of film) alone acted as an effective plasticizer for the zein films that suffer from classical brittleness problems, but showed no antimicrobial activity on *L. monocytogenes* and *C. jejuni*. Moreover, the solution-cast zein films incorporated with (+)-catechin-gallic acid mixture (each at 3 mg per cm^2 of film) and tested on cheese slices did not cause a significant antimicrobial activity on inoculated *L. monocytogenes*, but effectively prevented lipid oxidation in packed cheeses (Ünalan *et al.*, 2013). In contrast, application of solution-cast films of silk fibroin (Ku *et al.*, 2008a) and agar (*Gelidium corneum* extract) (Ku *et al.*, 2008b) incorporated with (+)-catechin (100 mg per mL of FFS) on sausages caused 0.83-1.83 log inactivation of inoculated *E. coli O157:H7* and *L. monocytogenes*. The gelatin films, incorporated with (+)-catechin-LYS mixture and characterized for their different properties, also inhibited lipid oxidation and reduced total microbial load of minced pork meat significantly (Rawdkuen *et al.*, 2012; Kaewprachu *et al.*, 2015). Some other studies also exist and are related to incorporation of (+)-catechin-Kradon (*Careya sphaerica Roxb.*) extract mixture into fish myofibrillar protein films (Kaewprachu *et al.*, 2017), (+)-catechin into soy protein-CMC blend films (Han *et al.*, 2015), (+)-catechin-nisin mixture into gelatin films (Kaewprachu *et al.*, 2018), (+)-catechin into potato starch films supported with waxy maize starch nanocrystals (Sessini *et al.*, 2016), (+)-catechin-lysozyme mixture into rice-flour-gelatin-nanoclay films (Pattarasiriroj *et al.*, 2020). The conjugation of (+)-catechin with alginate and inulin by exploitation of ascorbic acid/hydrogen peroxide redox pair was also conducted to obtain antioxidant coatings, suitable for food applications (Spizzirri *et al.*, 2010). However, further studies are needed to characterize antimicrobial and antioxidant performances of developed (+)-catechin incorporated films in different food applications.

(-)-Epigallocatechin gallate, purified mostly from green tea, is also used extensively in active edible films obtained by the classical solution-casting method. Promising studies have been conducted to develop (-)-epigallocatechin gallate loaded Na-alginate-carboxymethyl cellulose films and apply these films for coating of fresh pork meat to control its antimicrobial load and lipid oxidation (Ruan *et al.*, 2019a,b). Liang *et al.* (2017a,b) obtained edible films with sustained (-)-epigallocatechin gallate release by first encapsulating this polyphenol into chitosan nanoparticles coated with zein, and then by incorporating obtained chitosan/zein nanoparticles into chitosan hydrochloride films. Mittal *et al.* (2021) also applied a different approach and developed chitosan films incorporated with chitosan-(-)-epigallocatechin gallate conjugate (chitosan:chitosan-polyphenol conjugate ratio: 8:2) obtained by using ascorbic acid-hydrogen peroxide redox pair. Also (-)-epigallocatechin gallate was incorporated into pea starch-guar gum composite films (Saberi *et al.*, 2017), konjac glucomannan-carboxymethyl chitosan composite films (Sun *et al.*, 2020), chitosan films reinforced by bacterial cellulose (Wang *et al.*, 2018), alginate films (Biao *et al.*, 2019), and bovine (Wang *et al.*, 2019a) and fish gelatin films (Tammineni *et al.*, 2012; Nilsuwan *et al.*, 2018). It is also important to note that different from most other workers that employed (-)-epigallocatechin gallate in solution-cast films, Nilsuwan *et al.* (2019) incorporated (-)-epigallocatechin gallate into fish gelatin films obtained by thermo-compression molding.

4.4.4.2 Quercetin

Quercetin is a flavonol-group flavonoid that is abundant in fruits and vegetables (Fig 4.2). Some glycosylated derivatives of quercetin also exist, such as isoquercitrin (or isoquercetin) and rutin formed by attachment of glucose and rutinose (a disaccharide formed by rhamnose and glucose) on to 3-OH group of quercetin (Magar and Sohng, 2020), respectively. The onion waste is a rich source of quercetin and its glycosylated derivatives that could also be converted to quercetin by enzymatic hydrolysis using microbial β-glucosidases (Turner *et al.*, 2006; Ko *et al.*, 2011). In the USA, FDA approved the GRAS status of quercetin in beverages and beverage bases, grain products, and pastas, processed fruits and fruit juices, and soft candies at levels up to 500 mg/serving (FDA, 2010b). The quercetin shows low solubility in water, but it can be modified to increase its water solubility. For example, enzymatically-modified isoquercitrin (EMIQ), a water-soluble modified

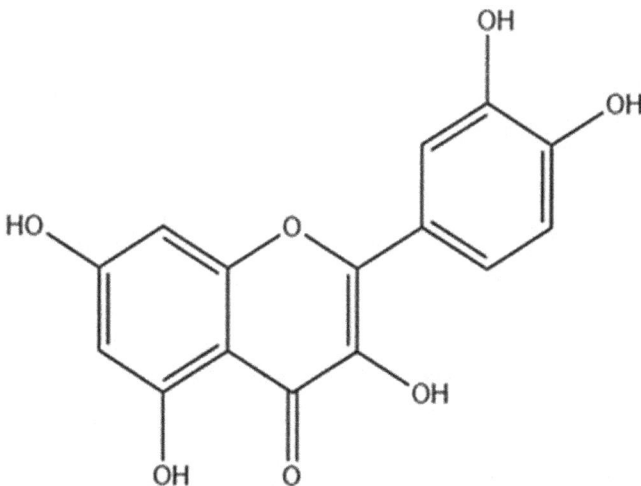

Fig. 4.2: Chemical structure of quercetin

quercetin derivative composed mainly of alpha-glucosylquercetin, has been used as an antioxidant food additive in USA (FDA, 2007) and Japan (JFCRF, 2014). The EMIQ has been approved as GRAS by the FDA for its application at 150 mg/kg in a wide range of food products (FDA, 2007). The EMIQ is generally produced from rutin extracted from *Sophora japonica.* During this process, the rutin hydrolysates are glycosylated via glycosyltransferase in the presence of dextrin (Jadeja and Devkar, 2014).

Due to its outstanding antioxidant and antimicrobial properties, quercetin and its derivatives (mainly rutin) have been incorporated into different edible films. It is important to note that a particular interest has been shown on using quercetin with chitosan-based films. Different workers obtained active food packaging materials by incorporating quercetin directly into chitosan films (Souza *et al.*, 2015; Rubini *et al.*, 2020; Zhang *et al.*, 2020a). Quercetin loaded edible chitosan films have also been developed by nutraceutical industry to enhance oral administration of this bioactive compound (Pandit *et al.*, 2020). Some interesting approaches have also been developed by incorporating quercetin-starch complex into chitosan-gelatin blend films (Yadav *et al.*, 2020) or by grafting enzymatically oxidized quercetin or its derivative rutin (quercetin or rutin quinones) on to chitosan (Torres *et al.*, 2012). Diao *et al.* (2020a) also achieved grafting of quercetin on to chitosan films, but these authors exploited ascorbic acid-hydrogen peroxide redox-pair for this purpose. Souza *et al.* (2018a) developed layer-by-layer coatings by using sequential application of quercetin loaded lecithin/chitosan nanoparticles and κ-carrageenan. Moreover, a mixture of quercetin with chitosan and hordein was used to obtain edible films from electrospun nanofibers (Li *et al.*, 2020). Some innovative applications have also been developed by using quercetin and carboxymethyl cellulose. For example, Bai *et al.* (2019) developed Al^{+3} sensing quercetin loaded carboxymethyl cellulose films by exploiting color change created by quercetin-Al^{+3} complex. These authors reported that such antioxidant films could provide an extra benefit of detecting food produced by using aluminum equipment, vessels, cups, etc. Silva-Weiss *et al.* (2018) achieved sustained release of quercetin and rutin by encapsulating these bioactive compounds by using liposomes based on dipalmitoyl lecithin, and then incorporating obtained liposomes into carboxymethyl cellulose films. Farrag *et al.* (2018) formed insoluble complexes of quercetin with corn or pea starch, and incorporated these microparticles into starch films. These authors reported that the films with quercetin-corn starch microparticles showed considerably faster quercetin release rates than those with quercetin-pea starch loaded films. Other studies with quercetin are related to incorporation of this bioactive compound into films of gelatin (Rubini *et al.*, 2020), gelatin-chitosan composite (Benbettaïeb *et al.*, 2016), kafirin (Giteru *et al.*, 2015), and cassia gum (Cao *et al.*, 2020).

4.4.4.3 Curcumin

Curcumin is the main active phenolic compound present in rhizome (underground stem) of turmeric plant (*Curcuma longa*) that belongs to the ginger family (*Zingiberaceae*) (Fig. 4.3). The curcumin solubilized properly (mostly in ethanol or methanol) is a broad-spectrum antimicrobial effective on bacteria, fungi, and

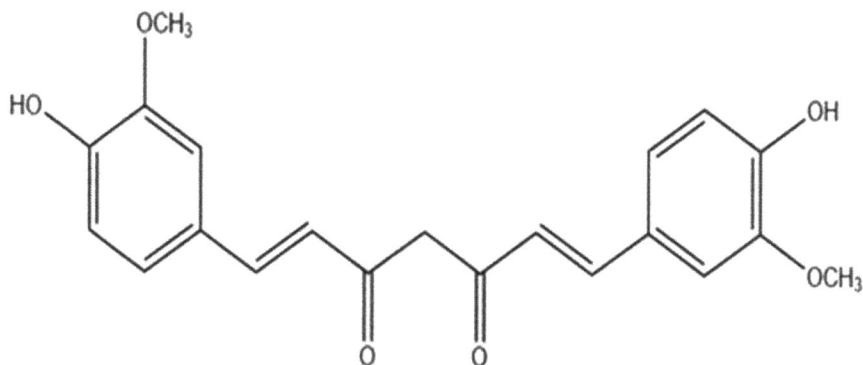

Fig. 4.3: Chemical structure of curcumin

viruses (Moghadamtousi *et al.*, 2014). However, although there is a good interest in using curcumin in active edible packaging, the limited water solubility of this phenolic compound reduces its effective use as an antimicrobial and antioxidant. Musso *et al.* (2017) added curcumin solubilized in ethanolic solution into gelatin film-forming solutions, but although the solution-cast gelatin films showed some antioxidant activity, they did not show antimicrobial activity in the classical zone of inhibition test. Roy and Rhim (2020a) developed curcumin loaded carrageenan, agar, and chitosan films by emulsifying film-forming solutions by using Tween-80 and ultrasonication. These authors found that the addition of curcumin into carbohydrate-based films caused significant increases in free radical scavenging and antimicrobial activity of these films against *L. monocytogenes* and *E. coli.* The highest curcumin release was observed from carrageenan films, followed by agar films that released almost 50% less curcumin than carrageenan films. In contrast, chitosan formed a tight complex with curcumin and prevented release of this bioactive compound from films (Roy and Rhim, 2020a). Liu *et al.* (2016) also developed antimicrobial chitosan films effective on *S. aureus* and *Rhizoctonia solani* by directly incorporating curcumin into chitosan films without using an emulsifier. These workers also studied molecular simulation of chitosan-curcumin interaction and proposed that the interaction occurred mainly by H-bonding. Da Silva *et al.* (2019) also incorporated curcumin into hydroxypropyl methylcellulose films, using Tween-80 and homogenization technique, but they also applied an additional strategy of modifying crystalline structure of curcumin via antisolvent precipitation method before it was added into films. The antisolvent precipitation applied by using ethanol as a solvent caused a dramatic reduction in the crystal size and geometry (turned to dendritic needles format) of curcumin, and this increased its solubility (~6.8-fold) during release tests of films conducted in ethanol used as a lipophilic food simulant (da Silva *et al.*, 2019). Details of nanonization of curcumin particles, using antisolvent precipitation procedure and stabilization of nanosized particles during storage, were described in the study of Yadav and Kumar (2014). Moreover, recent developments in film making (e.g. nonofiber mats via electrospinning) and encapsulation technologies that employ nanoparticles or nanovesicles (e.g. liposomes) have also provided new opportunities to improve active properties and solubility profiles of curcumin (Hussain *et al.*, 2017). For example, Wang *et al.* (2019b) obtained antioxidant curcumin-loaded films from nanodispersions of caseinate or caseinate/zein composite (ratio: 3:1 or 1:1) prepared by a pH-driven method (dispersion at pH 12 and then neutralization) without using ethanol or organic solvents. Nieto-Suaza *et al.* (2019) developed native and acetylated banana starch nanoparticles, loaded with curcumin and incorporated these nanoparticles into banana starch edible films that delivered curcumin in lipophilic food simulants (ethanol 50% v/v, and vegetable oil) (EU Commission Regulation No l0/2011). However, these edible films showed limited curcumin solubility in hydrophilic food simulants (ethanol 10% v/v, and acetic acid 3%, w/v) (Nieto-Suaza *et al.*, 2019).

Abdou *et al.* (2018) prepared nanoemulsions of curcumin with garlic and cinnamon essential oils, and sunflower oil, and then dispersed these nanoemulsions into pectin solutions (6% w/v) (pectin:nanoemulsion ratio = 3:2). It was reported that coating of chicken fillets with these nanoemulsion/pectin film solutions was beneficial in improving their microbial and oxidative quality. Wang *et al.* (2017, 2019c) developed curcumin-loaded konjac glucomannan/zein nanofiber mats and zein electrospun nanofibers that showed antimicrobial effect on *E. coli* and *S. aureus*. Alehosseini *et al.* (2019) encapsulated curcumin into phosphatidylcholine liposomes and prepared encapsulated curcumin-loaded electrospun gelatin nanofiber mats. These authors also developed electrospun mats using zein, but they solubilized curcumin directly in ethanolic zein film solutions without applying liposome encapsulation step. It was reported that the gelatin-based nanofiber mats showed faster and higher curcumin release (60 to 75% of added curcumin within one day) than zein-based ones (~45% of added curcumin within one day) in release tests conducted in 50% ethanol used to simulate lipophilic food. Moreover, it was only gelatin-based nanofibers that showed curcumin release (15-16% of added curcumin within four days) in vegetable oil used as a highly lipophilic food simulant (Alehosseini *et al.*, 2019).

4.4.4.4 *Tocopherols*

The tocopherols are naturally existing monophenolic antioxidants with vitamin E activity. Due to their lipid solubility, the tocopherols act as potent antioxidants both in edible fats and oils, but they naturally exist mainly in vegetable oils (Niki and Abe, 2019). Foods contain four different tocopherols (α-, β-, γ- and δ-tocopherol) that consist of a chromanol ring with a saturated phytyl side chain at C2, but differ in number and position of methyl groups in the chromanol ring (Niki and Abe, 2019). However, the α-tocopherol, that has three methyl groups at C5, C7 and C8 of its chromanol ring, is used extensively as an antioxidant in food and edible packaging since it is the most widely distributed and biologically active tocopherol form in terms of vitamin E activity (Viñas *et al.*, 2014) (Fig. 4.4).

In the literature, extensive studies have been conducted to employ α-tocopherol in development of antioxidant edible films and coatings suitable for food susceptible to lipid oxidation. Yıldırım and Barutçu Mazı (2017) found that incorporation of α-tocopherol improved the performance of zein coatings applied to control lipid oxidation of roasted hazelnuts. In contrast, Han *et al.* (2008) determined that similar

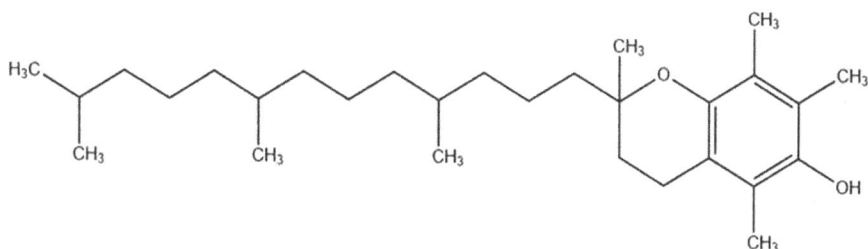

Fig. 4.4: The chemical structure of α-tocopherol

reductions in lipid oxidation occurred when roasted peanuts were coated with whey protein coating formulated with or without incorporating α-tocopherol-ascorbic palmitate mixture. The application of solution-cast α-tocopherol loaded edible films in preservation of meat products has also attracted some interest. Wu *et al.* (2001) developed starch-alginate blend and starch-alginate-stearic acid emulsion-based films incorporated or impregnated with a commercial tocopherol mixture (composed of α-, β-, γ- and δ-tocopherol at 14, 1, 62 and 23%, respectively). The application of developed tocopherol incorporated or impregnated films in active packaging retarded the lipid oxidation of ground beef patties, but all antioxidant edible films were less effective than polyester vacuum bags in preventing oxidative changes in the patties (Wu *et al.*, 2001). Abd Hamid *et al.* (2019) also applied α-tocopherol loaded carrageenan films for active packaging of beef patties, but they effectively prevented the lipid oxidation and metmyoglobin formation in patties. Agudelo-Cuartas *et al.* (2020) developed a different approach and used α-tocopherol in whey protein films after preparing oil-in-water nanoemulsion of this antioxidant compound by the microfluidization method. These authors developed antioxidant and antifungal edible films by incorporating nanoencapsulated α-tocopherol in combination with natamycin. The α-tocopherol was also incorporated into solution-cast films of chitosan (Martins *et al.,* 2012), zein-chitosan composite (Zhang *et al.*, 2020b), chitosan-montmorillonite nanocomposite (Dias *et al.*, 2014), Na-caseinate (Fabra *et al.*, 2011), gelatin (Jongjareonrak *et al.*, 2008), and carboxymethyl cellulose (Martelli *et al.*, 2017). Moreover, electrospinning has also been employed to obtain α-tocopherol loaded active packaging materials. For example, Fabra *et al.* (2016) encapsulated α-tocopherol, using different shell materials, such as zein, whey protein isolate, and soy protein isolate and electrospun/electrospray these structures on to gluten films to obtain active bilayer films.

4.4.4.5 Anthocyanins

The anthocyanins are flavonol group natural water-soluble color compounds that give red, violet, purple, and blue color to different fruits and vegetables (Khoo *et al.*, 2017). The anthocyanins are glycosides of different sugars (e.g. glucose, rhamnose, galactose, or arabinose) with anthocyanidins (e.g. cyanidin, delphinidin, pelargonidin, peonidin, petunidin, or malvidin) having the basic structure of flavylium ion ($C_{15}H_{11}O^+$) (Cemeroğlu *et al.*, 2014). Although anthocyanidins show antioxidant and antimicrobial activity, the increased interest to use these bioactive compounds in active packaging is related to their pH-dependent color-changing ability originating from complex multi-equilibrium states of flavylium structure (Roy and Rhim, 2020b). In general, anthocyanins have (1) a red color at highly acidic pH, (2) light red color/colorless at slightly to moderately acidic pH, (3) purple color around neutral pH, and (4) blue color at alkaline pH (Table 4.2). Therefore, color change of anthocyanin-loaded films could be employed as an indicator to detect acid formation (colorless anthocyanins turn into red as pH reduces) by microbial spoilage (e.g. action of lactic acid bacteria in food) and some enzymatic reactions (e.g. increased free fatty acid content of oils by lipase), or formation of volatile nitrogenous bases, as ammonia (red colored anthocyanins turn into blue as pH increases) (e.g. as in stored fresh meat and fish during spoilage).

Table 4.2: pKa values and color of different equilibrium species of an anthocyanin

Equilibrium species	pKa	Color
Flavylium cation	1.4-4.8[a]	Red
Quinoidal base	4-5[b]	Red
Chalcone	4-5[c]	Colorless
Quinoidal monoanion	7.5-8.0[d]	Purple
Quinoidal dianion	11.0[e]	Blue

[a] Amić *et al.* (1999)
[b-e] Asenstorfer *et al.* (2003)

Different solution-cast edible films have been obtained by using anthocyanins from various sources. These films were intended for active packaging of the food products or as indicator headspace stickers for packed food. For example, Wang *et al.* (2019d) nanoencapsulated cranberry anthocyanins using chitosan chloride and carboxymethyl cellulose, and then incorporated this material into gelatin films intended for olive oil antioxidant packaging. In another study, wine grape pomace anhocyanins encapsulated with maltodextrin or gum Arabic were incorporated into starch films intended for sun-flower oil antioxidant packaging (Stoll *et al.*, 2016). Zhang *et al.* (2019) conducted an interesting study by employing Roselle calyx anthocyanin loaded starch/polyvinyl alcohol composite films as headspace stickers to detect spoilage-induced total volatile bases formation by color change for packed fresh pork meat spoiled intentionally at 25°C (red to light red: 0-36 h, light blue to dark blue: 48-60 h, and orange: 72 h). Recently, *Lycium ruthenicum* Murr. anthocyanin loaded cassava starch films were also applied as smart headspace sticker by Qin *et al.* (2019) to detect volatile bases formation in fresh pork meat left for spoilage at 25°C. However, the indicator films loaded with *Lycium ruthenicum Murr.* anthocyanin films showed a different pH-color response than Roselle calyx anthocyanins (purple: 0-8 h, very dark purple: 16 h, dark purple: 24-32 h, green-yellow: 40-48 h). The headspace stickers developed by Chen *et al.* (2020a), who incorporated a combination of purple sweet potato anthocyanins with curcumin into starch/polyvinyl alcohol composite films are also very promising since these headspace stickers discriminate among fresh (light brown: zero to two days), semi-fresh (light green: four to six days) and spoiled fish (light greenish-yellowish: eight to 10 days) kept under refrigerated storage at 4°C. Moreover, the headspace stickers of natural edible bacterial cellulose films loaded with *Echium amoenum* flower anthocyanins were also highly effective in determining spoilage of shrimp kept under refrigerated storage at 4°C (Mohammadalinejhad *et al.*, 2020). Musso *et al.* (2019) also incorporated water and alcohol extracts of red cabbage anthocyanins into bovine gelatin films intended for food active packaging, but they found that the ethanolic extract gave better pH-color change and ammonia responses than water extracts. They also determined that the developed anthocyanin loaded bovine gelatin films had antioxidant activity, but they lacked in showing antimicrobial activity against test bacteria, such as *S. enterica* serovar Enteritidis, *E. coli, S. aureus* and *B. cereus*. Uranga *et al.* (2018) used water extracts of red cabbage anthocyanins to obtain compression-molded antioxidant fish gelatin films, but they did not show the pH-response capabilities of their films. Some other studies in this field were conducted by incorporation of red raspberry anthocyanins into soy protein isolate films (Wang *et al.*, 2012b), anthocyanin-rich haskap berry phenolic extract into fish gelatin films (Liu *et al.*, 2019b), red cabbage anthocyanins into *Artemisia sphaerocephala* Krasch. gum/carboxymethyl cellulose composite film (Liang *et al.*, 2019) and red cabbage anhocyanins into oxidized chitosan nanocrystal/konjac glucomannan films (Wu *et al.*, 2020). Further details of developing anthocyanin-loaded edible films and their food applications could be reached from the detailed review of Roy and Rhim (2020b).

4.4.4.6   Crude phenolic extracts

Although the aqueous or alcoholic crude phenolic extracts from a great variety of plant materials (from leaves or other edible parts of tea, herb or species, agro-industrial wastes, such as pomace and meal, or inedible parts of plants, such as peel, seed, stem or root) have been tested in different edible films and coatings, green tea (GTE) and grape seed extracts (GSE) have played a central role in development of active edible films. The GTE formed heavily by catechins, and GSE – a source of oligomeric proanthocyanidines, are GRAS substances that have been used in different foods and nutraceuticals as natural antioxidant and antimicrobial compounds (Perumalla and Hettiarachchy, 2011). These extracts owe their popularity mainly to their well-characterized health benefits that are particularly demanded by the consumers (Perumalla and Hettiarachchy, 2011). It is well known that frequent intake of antioxidant polyphenols supports anti-aging by maintaining the balance between the accumulation of reactive oxygen species (ROS) and antioxidant reserves in the human body (Yan *et al.*, 2020). Moreover, more specific current knowledge suggests that frequent intake of green tea polyphenols shows some major health benefits, especially against cardiovascular diseases and certain cancer types (Abe and Inoue, 2020), while grape seed proanthocyanidins contribute to human health mainly by showing preventive effects on different cancer types, cardiovascular diseases, neurodegenerative diseases, obesity and type 2 diabetes, and by supporting microbial ecology, and modulating gut microbiota (Unusan, 2020). Table 4.3 shows the major phenolic constituents in different plant extracts, and Table 4.4 gives uses of these phenolic extracts in different edible films and

Table 4.3: Major phenolic compounds determined in different phenolic extracts

Extract[a] (Extraction solvent)	Major phenolic compounds	References
GTE (Methanol or aqueous)	Aqueous extract: Caffeine and epigallocatechin gallate (two major forms) are followed by epigallocatechin and epicatechin gallate; methanol extract: Epigallocatechin gallate (the major form) is followed by epigallocatechin, caffeine, and epicatechin gallate.	Unachukwu *et al.* (2010)
GSE (Aqueous Me$_2$CO, 66%)[b]	Flavan-3-ol monomers (e.g. (+)-catechin, (-)-epicatechin, (-)-epicatechin-3-o-gallate)[b] and their oligomers (2-7 subunits) and polymers (8 to 24+ monomers)[c] called proanthocyanidine.	Waterhouse *et al.* (2000), Kennedy *et al.* (2000b)
GSE (Aqueous)	Catechin, epicatechin and proanthocyanidines (dimers and trimers mainly).	Chedea *et al.* (2011)
GFSE (Aqueous methanol, 80%)	Flavanones (major form is naringin while others were formed mainly by hesperidin and neohesperidin) dominate over phenolic acids (the dominating one is gallic acid).	Xi *et al.* (2015)
PPE (Aqueous ethanol or aqueous methanol, 80%)	Ellagitannins (punicalagins are dominant forms).	Nair *et al.* (2018), Gullon *et al.* (2016)
CLVE (Aqueous methanol, 80%)	Phenolic acids (dominant form is gallic acid), flavonol glucosides, phenolic volatile oils (eugenol, acetyl eugenol and tannins).	Shan *et al.* (2005)
RME (Aqueous methanol, 80%)	Phenolic acids (caffeic acid, rosmarinic acid, caffeoyl derivatives), phenolic diterpenes (carnosic acid, carnosol, epirosmanol), volatile compounds (carvacrol) and flavonoids.	
SAE (Aqueous methanol, 80%)	Phenolic acids (rosmarinic acid), phenolic diterpenes (carnosic acid), volatile compounds, flavonoids.	
RME (Aqueous)	Rosmarinic acid (dominant form), chlorogenic and caffeic acids.	Gómez-Estaca *et al.* (2009)
ORE (Aqueous)	Rosmarinic acid (dominant form), gallic acid and protocatechuic acid.	Gómez-Estaca *et al.* (2009)
THYE (Aqueous or aqueous methanol, 80%)	Aqueous extract: Luteolin 7-O-glucuronide, rosmarinic acid and luteolin 7-O-glucoside (three major forms) are followed by hesperetin-7-O-rutinoside; Aqueous methanol: Luteolin 7-O-glucoside and rosmarinic acid (two major forms) are followed by luteolin 7-O-glucuronide and hesperetin-7-O-rutinoside.	Martins *et al.* (2015)

GARE (aqueous ethanol, 30%)	Flvonoids (quercetin is the major form followed by kaempferol 3-O-glucoside, quercetin 3-O-galactoside, and kaemferol) dominate over phenolic acids (major forms are caffeic and ferulic acids).		Yang *et al.* (2020)
CINE (aqueous or aqueous ethanol, 69%)	Sinapic acid is the major form followed by cinnamic acid.		Dvorackova *et al.* (2015)
OLE (Aqueous ethanol, 80%)	Oleuropein is the major form followed by hydroxytyrosol, tyrosol, and caffeic acid.		Soleimanifar *et al.* (2020)

[a] GTE: Green tea extract, GSE: Grape seed extract, GFSE: Grapefruit seed extract, PPE: Pomegranate peel extract, CLVE: Clove extract, RME: Rosemary extract, ORE: Oregano extract, SAE: Sage extract, GARE: Garlic extract, CINE: Cinnamon extract, OLE: Olive leaf extract, THYE: Thyme extract

[b] Flavon-3-ol monomers determined by Kennedy *et al.* (2000b) in aqueous Me_2CO extracts

[c] Number of monomers in oligomers and polymers defined by Waterhouse *et al.* (2000)

Table 4.4: Antimicrobial activity of edible packaging incorporated with major phenolic extracts

Phenolic extract[a] (extraction solvent)	Edible material	Phenolic extract concentrations in film or coating solution/ Antimicrobial effect[b]	References
GTE (Not given)	Soy protein isolate film	Extract at 2% (w/w)[c]/*S. aureus* was inhibited, but *E. coli* O157:H7, *S. enterica* serovar Typhimurium and *P. aeruginosa* were not affected by extract at 4%.	Kim *et al.* (2006)
GTE (Water or aqueous ethanol, 40/80%)	Tapioca starch based coatings	Extract at 1-5% (w/v)[c]/*L. monocytogenes, S. aureus,* and *B. cereus* were inhibited, but *E. coli* and *S. enterica* showed no considerable inhibition.	Chiu and Lai (2010)
GTE (Water)	Agar or agar-gelatin (2:1) films	Extract (0.1 g/ml) diluted 1:1 with film making solution/26 different bacteria were inhibited including *S. aureus, B. cereus, L. monocytogenes, E. coli, C. perfringens, P. aeruginosa, V. parahaemolyticus,* and *Y. enterocolitica.*	Giménez *et al.* (2013)
GTE (Water)	Furcellaran-gelatin (1:2) film	Extract at 5% and 10% (w/w)[c]/*S. aureus, E.coli* and *Henseniaspora uvarum* were inhibited.	Jamróz *et al.* (2019)

(Contd.)

Phenolic extract[a] (extraction solvent)	Edible material	Phenolic extract concentrations in film or coating solution/ Antimicrobial effect[b]	References
GTE (Not given)	Chitosan film	Extract at 5-15% (w/v)[c]/ Murine norovirus (MNV) titer decreased by 1.6 and 4.5 logs at 5 and 10% of GTE, respectively. GTE at 15% decreased MNV titer to undetectable levels. Number of *L. innocua* and *E. coli* decreased by ~7 log at 5-15% of GTE (test method: Films were incubated for 24 h with virus at 23°C or with bacteria at 37°C).	Amankwaah *et al.* (2020)
GTE or GSE (Not given)	Na-alginate-oleic acid-soybean oil (1:0.25:0.25) film	Extracts at 0.75% (w/w)[f]/ GTE and GSE decreased murine norovirus titers by 1.92 and 1.67 log, and hepatitis A titers by 1.92 and 1.50 log, respectively (test method: Films were incubated overnight at 37°C with virus suspension).	Fabra *et al.* (2018)
GTE or GFSE (Not given)	Gelatin-*Gelidium corneum* (An algae rich in agarose) (1:0.15) film	GFSE at 0.1% (w/v)[e] or GTE at 4.2%/Number of *E. coli* O157:H7 and *L. monocytogenes* decreased by 2.08 and 3.30 log CFU/g with GFSE, and by 0.77 and 0.91 log CFU/g with GTE loaded films, respectively (test method: Survivors in inoculated films were counted).	Hong *et al.* (2009)
GSE (Water)	Soy protein isolate film	Extract at 1% (w/w)[c]/ Number of *L. monocytogenes* showed a limited decrease while number of *E. coli* O157:H7 and *S. enterica* serovar Typhimurium did not decrease considerably (test method: Survivors in inoculated films were counted).	Sivarooban *et al.* (2008)

GSE (Ethanol)	Gelatin and chitosan films	Extract at 1% (v/w)[d]/*L. monocytogenes* and *S. aureus* were inhibited while *B. cereus* and *B. subtilis* showed limited, and *E. coli* O157:H7 and *S. enterica* serovar Typhimurium showed no inhibition.	Shahbazi (2017)
GARE (water)	CMC chitosan coating	Extract at 50% (w/v)[e]/*S. aureus* and *E. coli* were inhibited.	Diao *et al.* (2020b)
Phenolic extract[a] (extraction solvent)	Edible material	Phenolic extract concentrations in film or coating solution/ Antimicrobial effect[b]	References
GFSE (Not given)	Alginate-carrageenan-CMC (1:1:1) coating	Extract at 6.7% (w/w)[f]/ Number of *E. coli* and *L. monocytogenes* decreased by ~6 log within 9 and 3 h, respectively (test method: Dried coating was mixed (4 mg/ml) with cultures in broth).	Shankar and Rhim (2018)
GFSE (Not given)	CMC or CMC-chitin nanocrystal film	Extract at 5% (w/w)[f]/*E. coli* and *S. aureus* were inhibited (test method: Survivors in inoculated films were counted).	Oun and Rhim (2020)
GFSE (Not given)	Rape seed protein-gelatin (1:1) film	Extract at 0.5-1.5% (w/v)[e]/ *L. monocytogenes* and *E. coli* O157:H7 were inhibited.	Jang *et al.* (2011)
PPE (Water or methanol)	Chitosan or locust bean gum films	Extracts at 0.18 and 0.36 g/ml of chitosan and locust bean gum film forming solution, respectively/*P. digitatum* was inhibited.	Kharchoufi *et al.* (2018)
PPE (Aqueous ethanol, 80%)	Chitosan or alginate coating solution	Extract at 1% (w/v)[e]/ *Colletotrichum gloeosporioides* was inhibited (test method: Coating solution with extract at 1% was mixed with melt agar at 1% and then solidified agar was inoculated with fungi).	Nair *et al.* (2018)
PPE (Water)	Zein film	Extract at 25-75 mg/g of film forming solution/*E. coli*, *P. perfringens*, *Enterococci*	Mushtaq *et al.* (2018)

(Contd.)

Phenolic extract[a] (extraction solvent)	Edible material	Phenolic extract concentrations in film or coating solution/ Antimicrobial effect[b]	References
		faecalis, S. aureus, Proteus vulgaris and *S.* ser. Typhi were inhibited.	
RME or CINE (Ethanol)	Gelatin-chitosan (1:1) film	Extracts at 1% (w/w)[c]/*E. coli* and *S. aureus* were inhibited.	Bonilla and Sobral (2016)
THYE (Aqueous ethanol)	Whey protein film	Extract at 5-15% (w/w) [f]/*E. coli* and *S. aureus* were inhibited.	Aziz and Almasi (2018)
CLVE (Aqueous ethanol, 60%)	Zein film	Extract at 4-6 mg/ cm^2/*X. vesicatoria* and *E. amylovora* were inhibited at 4 mg/cm^2, but *E. carotovora* was inhibited at 6 mg/cm^2.	Alkan and Yemenicioğlu (2016)
Apple peel phenolic extract (Not given)	Apple puree-pectin film	Extract at 1.5-10% (w/w)[c]/*L. monocytogenes* was inhibited.	Du *et al.* (2011)
RME (Ethanol)	Whey protein concentrate film	Extract at 1% (w/w)[c]/*L. monocytogenes* and *S. aureus* were decreased minimum by 2.5 log (test method: Survivors in inoculated films were counted).	Andrade *et al.* (2018)
CITE (Aqueous)	Gelatin or methyl cellulose film	Extract at 1% (w/v)[e]/ *L. innocua* and *A. hydrophila* and *A. caviae* were inhibited, but *P. fluorescens* showed no considerable inhibition.	Iturriaga *et al.* (2012)

[a] GTE: Green tea extract, GSE: Grape seed extract, GFSE: Grapefruit seed extract, PPE: Pomegranate peel extract, CLVE: Clove extract, RME: Rosemary extract, ORE: Oregano extract, SAE: Sage extract, GARE: Garlic extract, CINE: Cinnamon extract, THYE: Thyme extract; CITE: Citrus extract

[b] Antimicrobial effect was tested by the zone of inhibition method unless otherwise was indicated

[c] w/w film forming solution

[d] v/w film forming solution

[e] w/v film forming solution

[f] w/w hydrocolloid

coatings. In general, most film development studies focus mainly on determination of effective microbial inhibition concentrations of plant extracts. Different studies have shown the antimicrobial potentials and limitations (e.g. bacterial resistance) of GTE and GSE loaded edible films (Kim *et al.*, 2006; Sivarooban *et al.*, 2008; Hong *et al.*, 2009; Chiu and Lai, 2010; Giménez *et al.*, 2013; Theivendran *et al.*, 2006;

Shahbazi, 2017). In contrast, there are a limited number of studies about antiviral activity of edible films and coatings. Recent studies by Fabra *et al.* (2018) and Amankwaah *et al.* (2020) showed antiviral activity of GTE and GSE loaded edible films against some major enteric viruses (e.g. norovirus and hepatitis A). Another recent study conducted by screening of different green tea catechins for their ability to inhibit the main protease (Mpro) of SARS CoV-2 (Covid-19) revealed that some GTE polyphenols, such as EGCG, ECG, and GCG (gallocatechin-3-gallate) could be effective Covid-19 protease inhibitors (Ghosh *et al.*, 2020). Further studies are needed to discover the potential of phenolic extract loaded edible films and coatings to eliminate SARS-CoV-2 in contaminated food and food packaging materials.

It is also important to note that the use of crude phenolic extracts, such as pomegranate peel extract and grapefruit seed extract in active edible packaging, has gained a significant interest since these value-added extracts obtained from important agro-industrial wastes show potent antimicrobial activity. The antibacterial and antifungal activity of films loaded with pomegranate peel extracts (Kharchoufi *et al.*, 2018; Nair *et al.*, 2018; Mushtaq *et al.*, 2018) and antibacterial activity of films loaded with grapefruit seed extract (Hong *et al.*, 2009; Jang *et al.*, 2011; Shankar and Rhim, 2018; Oun and Rhim, 2020) are quite promising in controlling microbial safety and spoilage in susceptible food products. Furthermore, the effectiveness of clove extract loaded edible films on major plant pathogenic bacteria provides a good opportunity to control pre- and post-harvest losses in fresh fruits and vegetables (Alkan and Yemenicioğlu, 2016).

4.4.4.7 Essential oils

The essential oils (EOs) are complex mixtures of many different active components (up to 60 components) that contain several major (e.g. terpenes, terpenoids, and aromatic compounds) and many minor fractions (Bakkali *et al.*, 2018). The EOs and their main components have been incorporated into edible films extensively (Sánchez-González *et al.*, 2011; Sivakumar and Bautista-Baños, 2014; Shahidi and Hossain, 2020). Some of the most frequently tested EO components in active edible packaging and their source EOs are thymol and carvacrol of oregano EO, thymol of thyme EO, cinnamaldehyde of cinnamon EO, and anethole of anise EO, eugenol of clove EO, limonene of orange peel EO, geraniol and geranyl acetate of palmarosa (rosa grass) EO, and citral of lemongrass EO, respectively (Hosseini *et al.*, 2009; Bahram *et al.*, 2014; Smitha and Rana, 2015; Alkan and Yemenicioğlu, 2016; Chen *et al.*, 2016; Randazzo *et al.*, 2016; Yao *et al.*, 2017; Yeddes *et al.*, 2020). The EOs or their components have different advantages, such as broad antimicrobial spectrum on both fungi and bacteria (from food-borne pathogens to plant pathogens), anti-biofilm and anti-quorum sensing activity (Camele *et al.*, 2019), formation of antimicrobial and antioxidant vapors at the headspace of packages, as well as diffusion into depths of foods (Kurita *et al.*, 1981; Alvarez *et al.*, 2014; Alkan and Yemenicioğlu, 2016; Boyaci *et al.*, 2019). Therefore, EOs could be employed not only in food-contact packaging, such as edible films and coatings, but also in non-food contact packaging, such as antimicrobial stickers applied locally on the internal surface of packaging to release essential oil vapors into food headspace (Lopes *et al.*, 2018; Tracz *et al.*, 2018). In the USA, FDA considers thymol and carvacrol among food additives permitted for direct addition to food (FDA, 2020a), while major essential oil components, such as limonene, geraniol, geranyl acetate, citral, cinnamaldehyde (FDA, 2020b), and eugenol (FDA, 2020c) are considered GRAS (Fig. 4.5). Some examples of using pure major GRAS EO components in different types of edible packaging materials are listed in Table 4.5. However, the crude EOs of oregano, citrus, thyme, anise, garlic, ginger, clove, sage, rosemary, pepper, etc. have found a more widespread application than pure essential oil components in active edible packaging (Table 4.6). In particular, oregano EO plays a central role in antimicrobial edible film studies since it is a highly potent antibacterial and antifungal, and it is one of the rare essential oils having sensory properties compatible with different food products (Seydim and Sarikus, 2006; Avila-Sosa *et al.*, 2010; Alvarez *et al.*, 2014; Cattelan *et al.*, 2015; Camele *et al.*, 2019).

Since the EOs and their components are volatile substances showing limited water solubility, the main challenge of obtaining edible films loaded with these hydrophobic substances is to form a good emulsion in the film-forming solution with the aid of suitable emulsifying agents (e.g. lecithin, Tween-20, Tween-80). Otherwise, coarse EO droplets show coalescence at or close to film surface and loss rapidly by evaporation (Çavdaroğlu *et al.*, 2019). The application of surface active proteins, such as Na-caseinates and whey proteins (Yemenicioğlu *et al.*, 2020) in

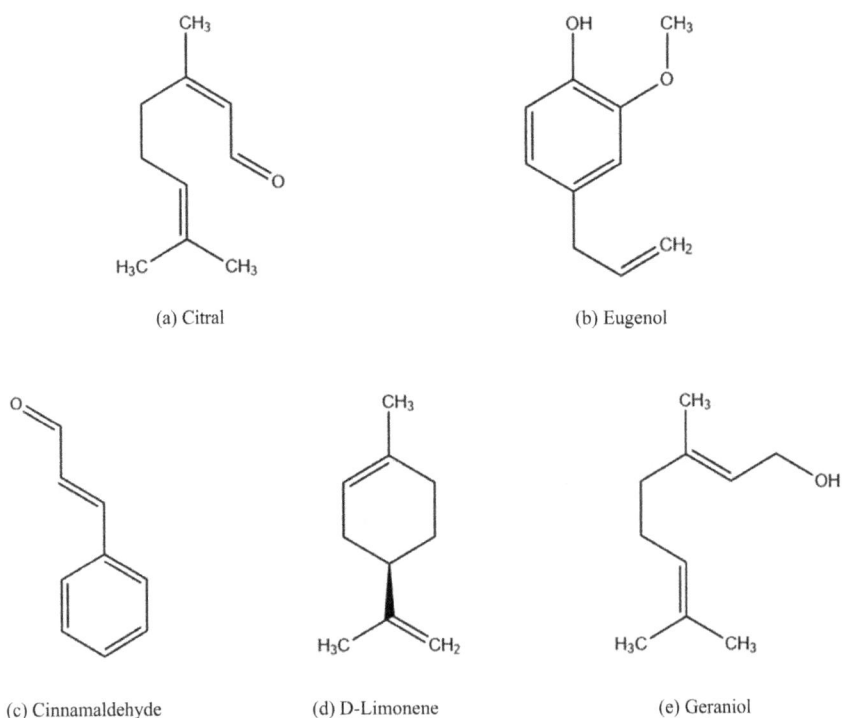

(a) Citral   (b) Eugenol

(c) Cinnamaldehyde   (d) D-Limonene   (e) Geraniol

Fig. 4.5: Essential oil components having Generally Recognized as Safe (GRAS) status

Table 4.5: Antimicrobial activity of edible films and coatings incorporated with major GRAS essential oil components (FDA, 2020b)

Essential Oil (EO)[a]	Edible material	Effective EO concentrations in film or coating solution/ Antimicrobial effect[b]	References
CIT	Kafirin films	EO at 2.5% (w/w)[c] /*C. jejuni, L. monocytogenes* and *P. fluorescens* were inhibited.	Giteru *et al.* (2015)
CIT	Alginate-apple puree film	EO at 0.5% (g/g)[c]/*E. coli* was inhibited.	Rojas-Graü *et al.* (2007)
CIT or EUG	Zein film	EO at 1-4 mg/cm²/*E. amylovora, E. carotovora, X. vesicatoria* were inhibited, but *P. syringae* did not show considerable inhibition.	Alkan and Yemenicioğlu (2016)
EUG	Zein film	EO ≥3%/*L. innocua* and *E. coli* were inhibited.	Boyacı *et al.* (2019)
EUG	Fig or citrus pectin film	EO at 1-2% (w/w)[c]/*L. innocua* was inhibited.	Çavdaroğlu *et al.* (2019)
EUG	Yam starch film	EO at 1-5% (g/g)[d]/ *E. coli, S. aureus,* and *L. monocytogenes* were inhibited.	Cheng *et al.* (2019)
EUG	Chitosan and pectin films	EO at 2% (g/g)[c]/*C. jejuni* was inhibited.	Wagle *et al.* (2019).
CIN	Gliadin film	EO at 3% (g/g)[d] /*P. expansum* and *A. niger* were inhibited.	Balaguer *et al.* (2013)
CIN	Paper coated with soy protein-CIN emulsion	EO at 5 mg/l of air in a jar/*E. coli* was inhibited. *B. cinerea* growth was delayed for 2 weeks (test method: Media inoculated with microorganisms were kept in a closed jar with coated paper at the lid).	Ben Arfa *et al.* (2007)
CIN	Electrospun fish gelatin nanofibers	EO at 5-30% (w/w)[d]/ *E. coli* O157:H7, *S. enterica serovar* Typhimurium and *L. monocytogenes* were inhibited.	Liu *et al.* (2018)
LO	Chitosan-gelatin blend film (1:1)	EO at 0.5-1% (w/w)[c]/ *E. coli* was inhibited.	Yao *et al.* (2017)

(Contd.)

Table 4.5: *(Contd.)*

Essential Oil (EO)[a]	Edible material	Effective EO concentrations in film or coating solution/ Antimicrobial effect[b]	References
LO	Whey protein-maltodextrin (1:2) conjugate emulsion coating	Nanoemulsion with EO at 5% (w/w)[c]/ MIC of nanoemulsion causing inhibitory zones on *Bacillus cereus*, *E. faecalis*, *E. coli* and *S.* ser. Typhi was 12.5 µL (test method: Nanoemulsion was added into wells opened on agar surface).	Sonu *et al.* (2018)
GER	Chitosan-starch blend film (1:1)	EO at 0.5% (w/w) [c]/*Agrobacterium tumefaciens*, *E. carotovora*, *Corynebacterium fascians*, and *P. solanacearum* were inhibited.	Badawy *et al.* (2016)
GER	Gum Arabic emulsion coating	EO at 2.5% (v/v)[e]/*B. cereus* and *E. coli* were inhibited.	Syed *et al.* (2020)

[a] CIT: Citral, EUG: Eugenol, CIN: Cinnamaldehyde, LO: D-Limonene, GER: Geraniol
[b] Antimicrobial effect was tested by the classical zone of inhibition method (disc diffusion method) unless otherwise was indicated
[c] w/w film forming solution
[d] w/w hydrocolloid
[e] v/v film forming solution

film making might be beneficial to obtain homogenous EO loaded films. However, the formation of homogenously distributed nanosized EO emulsion droplets showing sufficient emulsion stability needs an effective homogenization. The combination of mechanical homogenization with ultrasonication or microfluidization has been suggested to obtain nanosized emulsions and maximal film homogeneity (Acevedo-Fani *et al.*, 2015; Arnon-Rips *et al.*, 2019). It was also found that the formation of double-layer emulsions (layer-by-layer assembly) is more effective in increasing the stability of essential oil emulsions than mono-layer emulsions. For example, Li *et al.* (2020) reported that double emulsion of thyme EO (a chitosan layer surrounding whey protein layer around EO globule) was much more stable, and showed higher antimicrobial activity than its mono-layer emulsion (whey protein layer around EO globule). The nanoencapsulation of essential oils (e.g. formation of nanoparticles, nanofibers, or nanoliposomes) also helps not only to obtain a more homogenous distribution of EOs within films, but also to improve the controlled release and resulting antimicrobial efficiency of packaging materials loaded with nanoencapsules (Esmaeili *et al.*, 2020). Finally, additional measures, such as increasing concentrations of hydrocolloids and sugars in film-forming solution

Table 4.6: Antimicrobial activity of edible films and coatings incorporated with major essential oils

Essential Oil (EO)	Edible material	EO concentrations in film or coating solution/ Antimicrobial effect[a]	References
Oregano EO	Starch and chitosan film	EO at 0.5-2% (v/v)[b]/ Both film inhibited *A. niger* at 0.75%. EO at 2% and 0.5% were needed for chitosan and starch films to inhibit *Pencicillium* spp., respectively.	Avila-Sosa *et al.* (2010)
Oregano EO	Alginate film	EO at 1-1.5% (w/v)[c]/ *S. aureus*, *L. monocytogenes*, *E. coli*, and *S. enterica* serovar Enteritidis were inhibited.	Benavides *et al.* (2012)
Oregano EO	Pectin film	EO at 15.7 mg/ml of film forming solution/ *E. coli* O157:H7, *S. aureus* and *L. monocytogenes* were inhibited. Moreover, oregano EO showed anti-quorum sensing activity based on violacein production by *C. violaceum*.	Alvarez *et al.* (2014)
Oregano EO	Chitosan film	EO at 1-4% (w/w)[d]/ *L. monocytogenes* was inhibited more effectively than *E. coli* O157:H7.	Zivanovic *et al.* (2005)
Oregano EO	Fish gelatin-chitosan blend film	EO at 2-3% (v/v)[b]/*E. coli* and *S. aureus* were inhibited by 2% of EO. *B. subtilis*, *S. enterica* serovar Enteritidis, *Shiga bacillus* were inhibited by 3% of EO.	Wu *et al.* (2014)
Oregano EO and sweet potato anthocyanins	Nanocellulose pH-indicator film	EO at 4% (w/w)[e]/*E. coli* and *L. monocytogenes* were inhibited (sweet potato anthocyanin did not show antibacterial activity, but served as pH-indicator).	Chen *et al.* (2020b)
Oregano, garlic, or rosemary EO	Whey protein isolate film	EO at 2-4% (w/w)[d]/Oregano EO at 2% and garlic EO at 3% inhibited *S. aureus*, *S. enterica* serovar Enteritidis, *L. monocytogenes*, *L. plantarum*, and *E. coli* O157:H7. Rosemary EO caused no inhibition even at 4%.	Seydim and Sarikus (2006)

(Contd.)

Table 4.6: (*Contd.*)

Essential Oil (EO)	Edible material	EO concentrations in film or coating solution/ Antimicrobial effect[a]	References
Oregano or sage EO	Whey protein isolate film	EO at 1-4% (w/w)[d]/*S. enterica* serovar Enteritidis, *S. aureus* and *L. innocua* were inhibited by oregano EO (*L. innocua* showed the highest susceptibility to oregano EO). Films with Sage EO caused no inhibition at this concentration range.	Royo *et al.* (2010)
Oregano or thyme EO	Soy protein film	EOs at 1-5% (v/v)[b]/*S. aureus, E. coli* O157:H7, *P. aeruginosa* and *L. plantarum* were inhibited.	Emiroğlu *et al.* (2010)
Garlic EO	Alginate film	EO at 0.2-0.4% (v/v)[b]/*S. aureus* and *B. cereus* were inhibited. Films with Garlic EO caused no inhibition on *E. coli* and *S. enterica* serovar Typhimurium.	Pranoto *et al.* (2005)
Oregano EO or cinnamon EO	Chitosan film	EO at 0.25-4% (v/v)[b]/*A. niger* was inhibited with EO ≥ 0.25%. *P. digitatum* was inhibited by cinnamon and oregano EO at 0.5 and 0.25%, respectively (Test method: Films were cast onto Petri dish lids to cover agars inoculated with fungi)	Avila-Sosa *et al.* (2012)
Cinnamon EO	Whey protein film	EO at 0.8% (v/v)[b]/*B. subtilis, L. lactis, E. coli,* and *L. monocytogenes* were inhibited.	Bahram *et al.* (2014)
Cinnamon EO	Pickering emulsion coating of pectin-zein nanoparticles	EO at 0.12-0.16 μL/ml (EO volume/air space)/*B. cinerea* and *A. alternate* were inhibited at 0.12 and 0.16 μL/ml, respectively (test method: Media inoculated with fungi were incubated 96 h in a closed Petri dish with emulsion impregnated paper at the lid).	Jiang *et al.* (2020)
Anise EO	Whey protein film	EO at 4-6% (v/v)[b]/*A. flavus, Penicillium* spp. and *S. aureus* were inhibited.	Matan (2012)
Lemon EO	Chitosan film	EO at 0.5% (w/w)[d]/Delayed the growth of *L. monocytogenes*	Randazzo *et al.* (2016)

		(test method: Films were placed at the entire surface of inoculated Petri dishes and then survivors were enumerated by plate counting)	
Lemon EO	Alginate-montmorillonite composite film	EO at 0.5-1.5% (v/w)[f]/*S. aureus, B. cereus, E. coli,* and *S. enterica* serovar Enteritidis were inhibited	Hammoudi *et al.* (2020)
Lemon or grapefruit EO	Chitosan-corn starch composite film	EO at 1 or 3% (w/w)[d]/ *L. innocua* was inhibited by lemon EO, but not by grapefruit EO. Films of both EOs did not cause inhibition on *R. stolonifer, P. expansum,* and *Escherichia coli.*	Bof *et al.* (2016)
Orange EO	Corn starch film	EO at 0.3-0.7 μL/g of film forming solution/ *L. monocytogenes* and *S. aureus* were inhibited considerably by EO at ≥ 0.3 and 0.7 μL/g, respectively (test method: Films were placed at the entire surface of inoculated Petri dishes and then growth colonies were counted).	Do Evangelho *et al.* (2019)
Orange EO	Gelatin coating	EO at 1 and 2% (v/w)[g]/*B. subtilis, S. aureus, E. coli, P. aeruginosa,* and *C. albicans* were inhibited.	Alparslan *et al.* (2016)
Ginger EO	Na-caseinate	EO at 5% (v/v)[b]/*L. monocytogenes* and *S. enterica* serovar Typhimurium were inhibited.	Noori *et al.* (2018)
Ginger EO	Chitosan-montmorillonite composite film	EO at 0.5-1% (v/v)[b]/*S. aureus* and *B cereus* were inhibited with EO at 0.5 and 1%, respectively. *L. monocytogenes* caused no considerable inhibition even at EO of 2%.	Souza *et al.* (2018b)
Lemongrass EO	Chitosan film	EO at 3-9% (w/w)[e]/*B. cereus, E. coli, S.* ser. Typhi, and *L. monocytogenes* were inhibited.	Lyn and Hanani (2020)
Lemongrass EO	Chitosan coating	EO at 1% (v/v)[b]/Reduction occurred in growth of *Colletotrichum capsici* (test method: Coating solutions were mixed with agar medium inoculated with fungi).	Ali *et al.* (2015)

(Contd.)

Table 4.6: (*Contd.*)

Essential Oil (EO)	Edible material	EO concentrations in film or coating solution/ Antimicrobial effect[a]	References
Rosemary EO	Chitosan film	EO at 1.5% (v/v)[b]/*L. monocytogenes* was inhibited, but *E. coli* and *S. agalactiae* showed no considerable inhibition.	Abdollahi *et al.* (2012)
Rosemary EO	Gelatin film	EO at 0.5-2% (w/v)[f]/*E. coli, P. aeruginosa,* and *E. faecalis* were inhibited (*E. coli* showed less susceptibility than the other bacteria).	Yeddes *et al.* (2019)
Thyme EO	Chitosan-starch composite film	EO at 1-2% (w/w)[b] / *L. monocytogenes* was inhibited by EO at 1%, but *E. coli* O157:H7, *S. enterica serovar* Typhimurium, and *S. aureus* were inhibited by EO at 2%.	Mehdizadeh *et al.* (2012)
Clove EO	Gelatin or gelatin-chitosan composite film	EO at 0.75 ml/g of hydrocolloid/*E. coli, P. fluorescens, L. acidophilus,* and *L. innocua* were inhibited.	Gómez-Estaca *et al.* (2010)
Oregano, clove, tea tree, coriander, mastic thyme, laurel, rosemary or sage EOs	Whey protein film	EO at 2-8% (w/w)[d]/Oregano EO at 2-8% showed more inhibition than clove EO at 3-8% against *L. innocua, S. aureus,* and *S. enterica* serovar Enteritidis, but films of both EO caused no considerable inhibition against *P. fragi.* Tea tree, coriander, mastic thyme, laurel, rosemary, and sage EOs showed no considerable inhibition on test bacteria.	Fernández-Pan *et al.* (2012)
Orange, thyme, sage, pepper, or clove EO	Gelatin coating	EO at 2-10% (v/w)[f]/Thyme, orange, sage, pepper EOs at 5% and clove EO at 10% inhibited *S. aureus.* Thyme and orange EOs at 2%, sage and clove EOs at 5%, and pepper EO at 10% inhibited *E. coli.*	Alparslan (2018)
Orange, anise or cinnamon EO	Chitosan-zein composite film	EOs at 250 ppm (in film forming solution)/Anise and cinnamon EO inhibited	Escamilla-García *et al.* (2017)

		Rhizopus sp. and *Penicillium* sp, but orange EO showed limited inhibition.	
Thyme, sage, or lemongrass EO	Alginate film	Thyme EO at 1% (v/v)[b]/*E. coli* was inhibited, but other EOs showed no inhibition at this concentration.	Acevedo-Fani *et al.* (2015)
Bergamot EO	Whey protein film	EO at 1.4-5.6% (w/w)[d]/*E. coli* was inhibited, but *S. aureus* showed limited inhibition.	Çakmak *et al.* (2020)
Daphne or rosemary EO	Zein nanofibers	MICs / MICs for nanofiber were 0.42 and 0.36 mg/ml for daphne EO, and 0.64 and 0.50 mg/ml for rosemary EO against *S. aureus* and *L. monocytogenes*, respectively (test method: nanofibers solubilized in aq. ethanol were mixed with cultures, and microbial growth was evaluated by turbidity measurement).	Göksen *et al.* (2020)
Basil EO	Chitosan coating	EO at 0.5% (w/w)[d]/*A. niger, A. flavus, Fusarium* sp. and *Penicillium* sp. growth was inhibited by coating solution (test method: coating solutions were mixed with agar medium inoculated with fungi).	Hemalatha *et al.* (2017)
Black cumin EO	Milk protein concentrate film	EO AT 1.27% (w/v)[e]/*S. aureus* and *E. coli* were inhibited.	Ghamari *et al.* (2021)

[a] Antimicrobial effect was assayed by the classical disc diffusion test (zone of inhibition test) unless otherwise was indicated
[b] v/v film forming solution
[c] w/v film forming solution
[d] w/w film forming solution
[e] w/w hydrocolloid
[f] v/w film forming solution
[g] v/w hydrocolloid

might also be helpful since this increases the viscosity and helps in controlling coalescence of essential oil droplets during film solution preparation and drying (Alarcón-Moyano *et al.*, 2017).

References

Aasen, I.M., S. Markussen, T. Møretrø, T. Katla, L. Axelsson and K. Naterstad (2003). Interactions of the bacteriocins sakacin P and nisin with food constituents, *Int. J. Food Microbiol.*, 87: 35-43.

Abd Hamid, K.H., W.A. Wan Yahaya, N.A.Z. Mohd Saupy, M.P. Almajano and N.A. Mohd Azman (2019). Semi-refined carrageenan film incorporated with α-tocopherol: Application in food model, *J. Food Process. Pres.*, 43: e13937.

Abdollahi, M., M. Rezaei and G. Farzi (2012). Improvement of active chitosan film properties with rosemary essential oil for food packaging, *Int. J. Food Sci. Technol.*, 47: 847-853.

Abdou, E.S., G.F. Galhoum and E.N. Mohamed (2018). Curcumin loaded nanoemulsions/ pectin coatings for refrigerated chicken fillets, *Food Hydrocoll.*, 83: 445-453.

Abe, S.K. and M. Inoue (2020). Green tea and cancer and cardiometabolic diseases: A review of the current epidemiological evidence, *Eur. J. Clin. Nutr.*, 75: 865-876.

Aboda, A., W. Taha, I. Attia, A. Gad, M.M. Mostafa, M.A. Abdelwadod *et al.* (2020). Iron bond bovine lactoferrin for the treatment of cancers and anemia associated with cancer cachexia, pp. 243-254. *In:* M.R. Singh, D. Singh, J. Kanwar and N.S. Chauhan (Eds.). *Advances and Avenues in the Development of Novel Carriers for Bioactives and Biological Agents.* Academic Press, Cambridge, USA.

Acevedo-Fani, A., L. Salvia-Trujillo, M.A. Rojas-Graü and O. Martín-Belloso (2015). Edible films from essential-oil-loaded nanoemulsions: Physicochemical characterization and antimicrobial properties, *Food Hydrocoll.*, 47: 168-177.

Agudelo-Cuartas, C., D. Granda-Restrepo, P.J. Sobral, H. Hernandez and W. Castro (2020). Characterization of whey protein-based films incorporated with natamycin and nanoemulsion of α-tocopherol, *Heliyon*, 6: e03809.

Alarcón-Moyano, J.K., R.O. Bustos, M.L. Herrera and S.B Matiacevich (2017). Alginate edible films containing microencapsulated lemongrass oil or citral: Effect of encapsulating agent and storage time on physical and antimicrobial properties, *J. Food Sci. Technol.*, 54: 2878-2889.

Alehosseini, A., L.G. Gómez-Mascaraque, M. Martínez-Sanz and A. López-Rubio (2019). Electrospun curcumin-loaded protein nanofiber mats as active/bioactive coatings for food packaging applications, *Food Hydrocoll.*, 87: 758-771.

Alemán, A., I. Mastrogiacomo, M.E. López-Caballero, B. Ferrari, M.P. Montero and M.C. Gómez-Guillén (2016). A novel functional wrapping design by complexation of ε-polylysine with liposomes entrapping bioactive peptides, *Food Bioproc. Tech.*, 9: 1113-1124.

Ali, A., N.M. Noh and M.A. Mustafa (2015). Antimicrobial activity of chitosan enriched with lemongrass oil against anthracnose of bell pepper, *Food Packag. Shelf-Life*, 3: 56-61.

Alkan, D. and A. Yemenicioğlu (2016). Potential application of natural phenolic antimicrobials and edible film technology against bacterial plant pathogens, *Food Hydrocoll.*, 55: 1-10.

Al-Nabulsi, A.A., J.H. Han, Z. Liu, E.T. Rodrigues-Vieira and R.A. Holley (2006). Temperature-sensitive microcapsules containing lactoferrin and their action against *Carnobacterium viridans* on bologna, *J. Food Sci.*, 71: M208-M214.

Alparslan, Y. (2018). Antimicrobial and antioxidant capacity of biodegradable gelatin film forming solutions incorporated with different essential oils, *J. Food Meas. Charact.*, 12: 317-322.

Alparslan, Y., H.H. Yapıcı, C. Metin, T. Baygar, A. Günlü and T. Baygar (2016). Quality assessment of shrimps preserved with orange leaf essential oil incorporated gelatin, *LWT-Food Sci. Technol.*, 72: 457-466.

Alvarez, M.V., L.A. Ortega-Ramirez, M.M. Gutierrez-Pacheco, A.T. Bernal-Mercado, I. Rodriguez-Garcia, G.A. Gonzalez-Aguilar *et al.* (2014). Oregano essential oil-pectin edible films as anti-quorum sensing and food antimicrobial agents, *Front. Microbiol.*, 5: 699.

Amankwaah, C., J. Li, J. Lee and M.A. Pascall (2020). Antimicrobial activity of chitosan-based films enriched with green tea extracts on murine norovirus, *Escherichia coli*, and *Listeria innocua*, *Int. J. Food Sci.*, 2020: 3941924

Amić, D., D. Davidović-Amić, D. Bešlo, B. Lučić and N. Trinajstić (1999). Prediction of pK values, half-lives, and electronic spectra of flavylium salts from molecular structure, *J. Chem. Inf. Comput. Sci.*, 39: 967-973.

Andersson, Y., S. Lindquist, C. Lagerqvist and O. Hernell (2000). Lactoferrin is responsible for the fungistatic effect of human milk, *Early Hum. Dev.*, 59: 95-105.

Andjelković, M., J. Van Camp, B. De Meulenaer, G. Depaemelaere, C. Socaciu, M. Verloo and R. Verhe (2006). Iron-chelation properties of phenolic acids bearing catechol and galloyl groups, *Food Chem.*, 98: 23-31.

Andrade, M.A., R. Ribeiro-Santos, M.C.C. Bonito, M. Saraiva and A. Sanches-Silva (2018). Characterization of rosemary and thyme extracts for incorporation into a whey protein based film, *LWT-Food Sci. Technol.*, 92: 497-508.

Anggraini, T., A. Tai, T. Yoshino and T. Itani (2011). Antioxidative activity and catechin content of four kinds of Uncaria gambir extracts from West Sumatra, Indonesia, *Afr. J. Biochem. Res.*, 5: 33-38.

Arcan, I. and A. Yemenicioğlu (2011). Incorporating phenolic compounds opens a new perspective to use zein films as flexible bioactive packaging materials, *Food Res. Int.*, 44: 550-556.

Arcan, I. and A. Yemenicioğlu (2013). Development of flexible zein–wax composite and zein–fatty acid blend films for controlled release of lysozyme, *Food Res. Int.*, 51: 208-216.

Arcan, I. and A. Yemenicioglu (2014). Controlled release properties of zein–fatty acid blend films for multiple bioactive compounds, *J. Agric. Food Chem.*, 62: 8238-8246.

Arıca, M.Y., M. Yılmaz, E. Yalçın and G. Bayramoğlu (2004). Affinity membrane chromatography: Relationship of dye-ligand type to surface polarity and their effect on lysozyme separation and purification, *J. Chromatogr. B.*, 805: 315-323.

Arnon-Rips, H., R. Porat and E. Poverenov (2019). Enhancement of agricultural produce quality and storability using citral-based edible coatings; the valuable effect of nano-emulsification in a solid-state delivery on fresh-cut melons model, *Food Chem.*, 277: 205-212.

Asenstorfer, R.E., P.G. Iland, M.E. Tate and G.P. Jones (2003). Charge equilibria and pKa of malvidin-3-glucoside by electrophoresis, *Anal. Biochem.*, 318: 291-299.

Australia and New Zealand Act (1991). Schedule 18, *Processing Aids*, Section S18.9. No 148. https://studylib.net/doc/7019014/normal--food-standards-australia-new-zealand

Ávila, M., N. Gómez-Torres, M. Hernández and S. Garde (2014). Inhibitory activity of reuterin, nisin, lysozyme and nitrite against vegetative cells and spores of dairy-related Clostridium species, *Int. J. Food Microbiol.*, 172: 70-75.

Avila-Sosa, R., E. Hernández-Zamoran, I. López-Mendoza, E. Palou, M.T.J. Munguía, G.V Nevárez-Moorillón and A. López-Malo (2010). Fungal inactivation by Mexican oregano (*Lippia berlandieri Schauer*) essential oil added to amaranth, chitosan, or starch edible films, *J. Food Sci.*, 75: M127-M133.

Avila-Sosa, R., E. Palou, M.T.J. Munguía, G.V. Nevárez-Moorillón, A.R.N. Cruz and A. López-Malo (2012). Antifungal activity by vapor contact of essential oils added to amaranth, chitosan, or starch edible films, *Int. J. Food Microbiol.*, 153: 66-72.

Aziz, S.G.G. and H. Almasi (2018). Physical characteristics, release properties, and antioxidant and antimicrobial activities of whey protein isolate films incorporated with thyme (*Thymus vulgaris* L.) extract-loaded nanoliposomes, *Food Bioprocess Tech.*, 11: 1552-1565.

Badawy, M.E., E.I. Rabea, N.E. Taktak and M.A. El-Nouby (2016). The antibacterial activity of chitosan products blended with monoterpenes and their biofilms against plant pathogenic bacteria, *Sci.*, 2016: 1796256.

Bahram, S., M. Rezaei, M. Soltani, A. Kamali, S.M. Ojagh and M. Abdollahi (2014). Whey protein concentrate edible film activated with cinnamon essential oil, *J. Food Process. Preserv.*, 38: 1251-1258.

Bai, R., X. Zhang, H. Yong, X. Wang, Y. Liu and J. Liu (2019). Development and characterization of antioxidant active packaging and intelligent Al3+-sensing films based on carboxymethyl chitosan and quercetin, *Int. J. Biol. Macromol.*, 126: 1074-1084.

Bakkali, F., S. Averbeck, D. Averbeck and M. Idaomar (2008). Biological effects of essential oils – A review, *Food Chem. Toxicol.*, 46: 446-475.

Balaguer, M.P., G. Lopez-Carballo, R. Catala, R. Gavara and P. Hernandez-Munoz (2013). Antifungal properties of gliadin films incorporating cinnamaldehyde and application in active food packaging of bread and cheese spread foodstuffs, *Int. J. Food Microbiol.*, 166: 369-377.

Banani, H., D. Spadaro, D. Zhang, S. Matic, A. Garibaldi and M.L. Gullino (2015). Postharvest application of a novel chitinase cloned from *Metschnikowia fructicola* and overexpressed in *Pichia pastoris* to control brown rot of peaches, *Int. J. Food Microbiol.*, 199: 54-61.

Bankar, S.B., M.V. Bule, R.S. Singhal and L. Ananthanarayan (2009). Glucose oxidase – An overview, *Biotechnol. Adv.*, 27: 489-501.

Barbiroli, A., F. Bonomi, G. Capretti, S. Iametti, M. Manzoni, L. Piergiovanni and M. Rollini (2012). Antimicrobial activity of lysozyme and lactoferrin incorporated in cellulose-based food packaging, *Food Control*, 26: 387-392.

Barbiroli, A., A. Musatti, G. Capretti, S. Iametti and M. Rollini (2017). Sakacin – A antimicrobial packaging for decreasing Listeria contamination in thin-cut meat: Preliminary assessment, *J. Sci. Food and Agri.*, 97: 1042-1047.

Bari, M.L., D.O. Ukuku, T. Kawasaki, Y. Inatsu, K. Isshiki and S. Kawamoto (2005). Combined efficacy of nisin and pediocin with sodium lactate, citric acid, phytic acid, and potassium sorbate and EDTA in reducing the *Listeria monocytogenes* population of inoculated fresh-cut produce, *J. Food Prot.*, 68: 1381-1387.

Batpho, K., W. Boonsupthip and C. Rachtanapun (2017). Antimicrobial activity of collagen casing impregnated with nisin against foodborne microorganisms associated with ready-to-eat sausage, *Food Control*, 73: 1342-1352.

Bayarri, M., N. Oulahal, P. Degraeve and A. Gharsallaoui (2014). Properties of lysozyme/ low methoxyl (LM) pectin complexes for antimicrobial edible food packaging, *J. Food Eng.*, 131: 18-25.

Bellamy, W., M. Takase, H. Wakabayashi, K. Kawase and M. Tomita (1992). Antibacterial spectrum of lactoferricin B, a potent bactericidal peptide derived from the N-terminal region of bovine lactoferrin, *J. Appl. Bacteriol.*, 73: 472-479.

Ben Arfa, A., L. Preziosi-Belloy, P. Chalier and N. Gontard (2007). Antimicrobial paper based on a soy protein isolate or modified starch coating including carvacrol and cinnamaldehyde, *J. Agric. Food Chem.*, 55: 2155-2162.

Benavides, S., R. Villalobos-Carvajal and J.E. Reyes (2012). Physical, mechanical and antibacterial properties of alginate film: Effect of the crosslinking degree and oregano essential oil concentration, *J. Food Eng.*, 110: 232-239.

Benbettaïeb, N., O. Chambin, T. Karbowiak and F. Debeaufort (2016). Release behavior of quercetin from chitosan-fish gelatin edible films influenced by electron beam irradiation, *Food Control*, 66: 315-319.

Bennett, R.N. and R.M. Wallsgrove (1994). Secondary metabolites in plant defense mechanisms, *New Phytol.*, 127: 617-633.

Bhatia, S. and A. Bharti (2015). Evaluating the antimicrobial activity of Nisin, Lysozyme and Ethylenediaminetetraacetate incorporated in starch based active food packaging film, *J. Food Sci. Technol.*, 52: 3504-3512.

Biao, Y., C. Yuxuan, T. Qi, Y. Ziqi, Z. Yourong, D.J. McClements and C. Chongjiang (2019). Enhanced performance and functionality of active edible films by incorporating tea polyphenols into thin calcium alginate hydrogels, *Food Hydrocoll.*, 97: 105197.

Bijak, M., A. Sut, A. Kosiorek, J. Saluk-Bijak and J. Golanski (2019). Dual anticoagulant/ antiplatelet activity of polyphenolic grape seeds extract, *Nutrients*, 11: 93.

Biswas, S.R., P. Ray, M.C. Johnson and B. Ray (1991). Influence of growth conditions on the production of a bacteriocin, pediocin AcH, by *Pediococcus acidilactici* H, *Appl. Environ. Microbiol.*, 57: 1265-1267.

Blay, G.L., C. Lacroix, A. Zihler and I. Fliss (2007). *In vitro* inhibition activity of nisin A, nisin Z, pediocin PA-1 and antibiotics against common intestinal bacteria, *Lett. Appl. Microbiol.*, 45: 252-257.

Boeckx, T., A.L. Winters, K.J. Webb and A.H. Kingston-Smith (2015). Polyphenol oxidase in leaves: Is there any significance to the chloroplastic localization? *J. Exp. Bot.*, 66: 3571-3579.

Bof, M.J., A. Jiménez, D.E. Locaso, M.A. Garcia and A. Chiralt (2016). Grapefruit seed extract and lemon essential oil as active agents in corn starch-chitosan blend films, *Food Bioprocess Tech.*, 9: 2033-2045.

Bonilla, J. and P.J. Sobral (2016). Investigation of the physicochemical, antimicrobial and antioxidant properties of gelatin-chitosan edible film mixed with plant ethanolic extracts, *Food Biosci.*, 16: 17-25.

Botten, D., G. Fugallo, F. Fraternali and C. Molteni (2015). Structural properties of green tea catechins, *J. Phys. Chem. B*, 119: 12860-12867.

Boyacı, D., F. Korel and A. Yemenicioğlu (2016). Development of activate-at-home-type edible antimicrobial films: An example pH-triggering mechanism formed for smoked salmon slices using lysozyme in whey protein films, *Food Hydrocoll.*, 60: 170-178.

Boyacı, D. and A. Yemenicioğlu (2018). Expanding horizons of active packaging: Design of consumer-controlled release systems helps risk management of susceptible individuals, *Food Hydrocoll.*, 79: 291-300.

Boyacı, D., G. Iorio, G.S. Sozbilen, D. Alkan, S. Trabattoni, F. Pucillo *et al.* (2019). Development of flexible antimicrobial zein coatings with essential oils for the inhibition of critical pathogens on the surface of whole fruits: Test of coatings on inoculated melons, *Food Packag. Shelf-Life*, 20: 100316.

Brand-Williams, W., M.E. Cuvelier and C.L.W.T. Berset (1995). Use of a free radical method to evaluate antioxidant activity, *LWT-Food Sci. Technol.*, 28: 25-30.

Brown, C.A., B. Wang and J.H. Oh (2008). Antimicrobial activity of lactoferrin against foodborne pathogenic bacteria incorporated into edible chitosan film, *J. Food Prot.*, 71: 319-324.

Cabo, M.L., B. Torres, J.R. Herrera, M. Bernardez and L. Pastoriza (2009). Application of nisin and pediocin against resistance and germination of Bacillus spores in sous vide products, *J. Food Prot.*, 72: 515-523.

Cai, L., A. Cao, F. Bai and J. Li (2015). Effect of ε-polylysine in combination with alginate coating treatment on physicochemical and microbial characteristics of Japanese sea bass (*Lateolabrax japonicas*) during refrigerated storage, *LWT-Food Sci. Technol.*, 62: 1053-1059.

Çakmak, H., Y. Özselek, O.Y. Turan, E. Fıratlıgil and F. Karbancioğlu-Güler (2020). Whey protein isolate edible films incorporated with essential oils: Antimicrobial activity and barrier properties, *Polym. Degrad. Stab.*, 179: 109285.

Camele, I., H.S. Elshafie, L. Caputo and V. De Feo (2019). Anti-quorum sensing and antimicrobial effect of mediterranean plant essential oils against phytopathogenic bacteria, *Front. Microbiol.*, 10: 2619.

Cao, L., H. Feng, F. Meng, J. Li and L. Wang (2020). Fabrication of a high tensile and antioxidative film via a green strategy of self-growing needle-like quercetin crystals in cassia gum for lipid preservation, *J. Clean. Prod.*, 266: 121885.

Cattelan, M.G., M.B. M. de Castilhos, D.C.M.N. da Silva, A.C. Conti-Silva and F.L. Hoffmann (2015). Oregano essential oil: Effect on sensory acceptability, *Nutr. Food Sci.*, 45: 574-582.

Çavdaroğlu, E., S. Farris and A. Yemenicioğlu (2020). Development of pectin–eugenol emulsion coatings for inhibition of Listeria on webbed-rind melons: A comparative study with fig and citrus pectins, *Int. J. Food Sci. Tech.*, 55: 1448-1457.

Cé, N., C.P. Noreña and A. Brandelli (2012). Antimicrobial activity of chitosan films containing nisin, peptide P34, and natamycin, *CyTA-J. Food*, 10: 21-26.

Cemeroglu, B., A. Yemenicioğlu and M. Özkan (2014). Meyve sebzelerin bileşimi, pp. 95-107. *In:* B. Cemeroğlu. (Ed.). *Meyve ve sebze işleme teknolojisi*, Başkent Klişe Matbaacılık, Ankara, TR.

Cha, D.S., J.H. Choi, M.S. Chinnan and H.J. Park (2002). Antimicrobial films based on Na-alginate and κ-carrageenan, *LWT-Food Sci. Technol.*, 35: 715-719.

Chang, H., C. Yang and Y. Chang (2000). Rapid separation of lysozyme from chicken egg white by reductants and thermal treatment, *J. Agric. Food Chem.*, 48: 161-164.

Chang, S.S., W.Y.W. Lu, S.H. Park and D.H. Kang (2010). Control of foodborne pathogens on ready-to-eat roast beef slurry by ε-polylysine, *Int. J. Food Microbiol.*, 141: 236-241.

Chedea, V.S., C. Echim, C. Braicu, M. Andjelkovic, R. Verhe and C. Socaciu (2011). Composition in polyphenols and stability of the aqueous grape seed extract from the Romanian variety 'Merlot Recas', *J. Food Biochem.*, 35: 92-108.

Chen, H., X. Hu, E. Chen, S. Wu, D.J. McClements, S. Liu *et al.* (2016). Preparation, characterization, and properties of chitosan films with cinnamaldehyde nanoemulsions, *Food Hydrocoll.*, 61: 662-671.

Chen, H.Z., M. Zhang, B. Bhandari and C.H. Yang (2020a). Novel pH-sensitive films containing curcumin and anthocyanins to monitor fish freshness, *Food Hydrocoll.*, 100: 105438.

Chen, S., M. Wu, P. Lu, L. Gao, S. Yan and S. Wang (2020b). Development of pH indicator and antimicrobial cellulose nanofibre packaging film based on purple sweet potato anthocyanin and oregano essential oil, *Int. J. Biol. Macromol.*, 149: 271-280.

Cheng, J., H. Wang, S. Kang, L. Xia, S. Jiang, M. Chen and S. Jiang (2019). An active packaging film based on yam starch with eugenol and its application for pork preservation, *Food Hydrocoll.*, 96: 546-554.

Chibane, L.B., P. Degraeve, H. Ferhout, J. Bouajila and N. Oulahal (2019). Plant antimicrobial polyphenols as potential natural food preservatives, *J. Sci. Food Agric.*, 99: 1457-1474.

Chiu, P.E. and L.S. Lai (2010). Antimicrobial activities of tapioca starch/decolorized hsian-tsao leaf gum coatings containing green tea extracts in fruit-based salads, romaine hearts and pork slices, *Int. J. Food Microbiol.*, 139: 23-30.

Cho, M., S.B. Ko, J.M. Kim, O.H. Lee, D.W. Lee and J.Y. Kim (2016). Influence of extraction conditions on antioxidant activities and catechin content from bark of *Ulmus pumila* L., *Appl. Biol. Chem.*, 59: 329-336.

Chollet, E., I. Sebti, A. Martial-Gros and P. Degraeve (2008). Nisin preliminary study as a potential preservative for sliced ripened cheese: NaCl, fat and enzymes influence on nisin concentration and its antimicrobial activity, *Food Control.*, 19: 982-989.

Chu, C., J. Deng, Y. Man and Y. Qu (2017). Green tea extracts epigallocatechin-3-gallate for different treatments, *Biomed. Res. Int.*, 2017: 5615647.

Chung, W. and R.E. Hancock (2000). Action of lysozyme and nisin mixtures against lactic acid bacteria, *Int. J. Food Microbiol.*, 60: 25-32.

Cintas, L.M., P. Casaus, M.F. Fernández and P.E. Hernández (1998). Comparative antimicrobial activity of enterocin L50, pediocin PA-1, nisin A and lactocin S against spoilage and foodborne pathogenic bacteria, *Food Microbiol.*, 15: 289-298.

Cissé, M., J. Polidori, D. Montet, G. Loiseau and M.N. Ducamp-Collin (2015). Preservation of mango quality by using functional chitosan-lactoperoxidase systems coatings, *Postharvest Biol. Technol.*, 101: 10-14.

Conneely, O.M. (2001). Antiinflammatory activities of lactoferrin, *J. Am. Coll. Nutr.*, 20: 389S-395S.

Coventry, M.J., K. Muirhead and M.W. Hickey (1995). Partial characterisation of pediocin PO2 and comparison with nisin for biopreservation of meat products, *Int. J. Food Microbiol.*, 26: 133-145.

Cueva, C., M.V. Moreno-Arribas, P.J. Martín-Álvarez, G. Bills, M.F. Vicente, A. Basilio *et al.* (2010). Antimicrobial activity of phenolic acids against commensal, probiotic and pathogenic bacteria, *Res. Microbiol.*, 161: 372-382.

Cui, Y., Y.J. Oh, J. Lim, M. Youn, I. Lee, H.K. Pak *et al.* (2012). AFM study of the differential inhibitory effects of the green tea polyphenol (−)-epigallocatechin-3-gallate (EGCG) against Gram-positive and Gram-negative bacteria, *Food Microbiol.*, 29: 80-87.

Cutrim, C.S. and M.A.S. Cortez (2018). A review on polyphenols: Classification, beneficial effects and their application in dairy products, *Int. J. Dairy Technol.*, 71: 564-578.

Cutter, C.N. and G.R. Siragusa (1996). Reduction of *Brochothrix thermosphacta* on beef surfaces following immobilization of nisin in calcium alginate gels, *Lett. Appl. Microbiol.*, 23: 9-12.

Da Silva, M.N., J. de Matos Fonseca, H.K. Feldhaus, L.S. Soares, G.A. Valencia, C.E. de Campos *et al.* (2019). Physical and morphological properties of hydroxypropyl methylcellulose films with curcumin polymorphs, *Food Hydrocoll.*, 97: 105217.

Daglia, M. (2012). Polyphenols as antimicrobial agents, *Curr. Opin. Biotechnol.*, 23: 174-181.

Das, S.N., C. Neeraja, P.V.S.R.N. Sarma, J.M. Prakash, P. Purushotham, M. Kaur *et al.* (2012). Microbial chitinases for chitin waste management, pp. 135-150. *In:* T. Satyanarayana, B.V. Johri and A. Prakash (Ed.). *Microorganisms in Environmental Management*, Springer, Dordrecht., NL.

Datta, S., M.E. Janes, Q.G. Xue, J. Losso and J.F. La Peyre (2008). Control of *Listeria monocytogenes* and Salmonella anatum on the surface of smoked salmon coated with calcium alginate coating containing oyster lysozyme and nisin, *J. Food Sci.*, 73: M67-M71.

Davidson, P.M., T.M. Taylor and S.E. Schmidt (2012). Chemical preservatives and natural antimicrobial compounds, pp. 765-801. *In:* M.P. Doyle and R.L. Buchanan (Eds.), *Food Microbiology: Fundamentals and Frontiers*, ASM Press, Washigton D.C., USA.

Dawson, P.L., D.E. Hirt, J.R. Rieck, J.C. Acton and A. Sotthibandhu (2003). Nisin release from films is affected by both protein type and film-forming method, *Food Res. Int.*, 36: 959-968.

Degan, A.J. and J.B. Luchansky (1992). Influence of beef tallow and muscle on the antilisterial activity of pediocin AcH and liposome-encapsulated pediocin AcH, *J. Food Prot.*, 55: 552-554.

De Lucca, A.J. and T.J. Walsh (1999). Antifungal peptides: Novel therapeutic compounds against emerging pathogens, *Antimicrob. Agents Chemother.*, 43: 1-11.

Delves-Broughton, J., P. Blackburn, R.J. Evans and J. Hugenholtz (1996). Applications of the bacteriocin, nisin, *Antonie Leeuwenhoek*, 69: 93-202.

De Roos, A.L., P. Walstra and T.J. Geurts (1998). The association of lysozyme with casein, *Int. Dairy J.*, 8: 319-324.

Dheraprasart, C., S. Rengpipat, P. Supaphol and J. Tattiyakul (2009). Morphology, release characteristics, and antimicrobial effect of nisin-loaded electrospun gelatin fiber mat, *J. Food Prot.*, 72: 2293-2300.

Diao, Y., X. Yu, C. Zhang and Y. Jing (2020a). Quercetin-grafted chitosan prepared by free radical grafting: Characterization and evaluation of antioxidant and antibacterial properties, *J. Food Sci. Technol.*, 7: 2259-2268.

Diao, X., Y. Huan and B. Chitrakar (2020b). Extending the shelf-life of ready-to-eat spiced chicken meat: Garlic aqueous extracts-carboxymethyl chitosan ultrasonicated coating solution, *Food Bioprocess. Tech.*, 13: 786-796.

Dias, M.V., V.M. Azevedo, S.V. Borges, N.D.F.F. Soares, R.V. de Barros Fernandes, J.J. Marques and É.A.A. Medeiros (2014). Development of chitosan/montmorillonite nanocomposites with encapsulated α-tocopherol, *Food Chem.*, 165: 323-329.

Di Meo, F., V. Lemaur, J. Cornil, R. Lazzaroni, J.L. Duroux, Y. Olivier and P. Trouillas (2013). Free radical scavenging by natural polyphenols: Atom versus electron transfer, *J. Phys. Chem. A*, 117: 2082-2092.

Do Evangelho, J.A., G. da Silva Dannenberg, B. Biduski, S.L.M. El Halal, D.H. Kringel, M.A. Gularte *et al.* (2019). Antibacterial activity, optical, mechanical, and barrier properties of corn starch films containing orange essential oil, *Carbohydr. Polym.*, 222: 114981.

Drobni, P., J. Näslund and M. Evander (2004). Lactoferrin inhibits human papillomavirus binding and uptake *in vitro*, *Antiviral Res.*, 64: 63-68.

Du, W.X., C.W. Olsen, R.J. Avena-Bustillos, M. Friedman and T.H. McHugh (2011). Physical and antibacterial properties of edible films formulated with apple skin polyphenols, *J. Food Sci.*, 76: M149-M155.

Duan, J., S.L. Park, M.A. Daeschel and Y. Zhao (2007). Antimicrobial chitosan-lysozyme (CL) films and coatings for enhancing microbial safety of Mozzarella cheese, *J. Food Sci.*, 72: 355-362.

Dubey, M.K., A. Zehra, M. Aamir, M. Meena, L. Ahirwal, S. Singh *et al.* (2017). Improvement strategies, cost effective production, and potential applications of fungal glucose oxidase (GOD): Current updates, *Front. Microbiol.*, 8: 1032.

Durán, N., M.A. Rosa, A. D'Annibale and L. Gianfreda (2002). Applications of laccases and tyrosinases (phenoloxidases) immobilized on different supports: A review, *Enzyme Microb. Technol.*, 31: 907-931.

Durazzo, A., M. Lucarini, E.B. Souto, C. Cicala, E. Caiazzo, A.A. Izzo *et al.* (2019). Polyphenols: A concise overview on the chemistry, occurrence, and human health, *Phytother. Res.*, 33: 2221-2243.

Dvorackova, E., M. Snoblova, L. Chromcova and P. Hrdlicka (2015). Effects of extraction methods on the phenolic compounds contents and antioxidant capacities of cinnamon extracts, *Food Sci. Biotechnol.*, 24: 1201-1207.

EC (European Commission) (2001). No 2066/2001, Amending Regulation (EC) No 1622/2000 as regards the use of lysozyme in wine products, *Official Journal of the European Communities*, 278: 9-10.

EC (European Commission) (2008). No 1333/2008. Regulation of the European Parliament and of the council of 16 December on food additives, *Official Journal of the European Communities*, 354: 16-33.

EFSA (European Food Safety Authority) (2017). Safety of nisin (E 234) as a food additive in the light of new toxicological data and the proposed extension of use, *EFSA Journal*, 15(12): 5063.

Ehsani, A., M. Hashemi, M. Aminzare, M. Raeisi, A. Afshari, A.M. Alizadeh and M. Rezaeigolestani (2019). Comparative evaluation of edible films impregnated with sage essential oil or lactoperoxidase system: Impact on chemical and sensory quality of carp burgers, *Journal of Food Process. Preserv.*, 43: e14070.

Ehsani, A., M. Hashemi, A. Afshari, M. Aminzare, M. Raeisi and T. Zeinali (2020). Effect of different types of active biodegradable films containing lactoperoxidase system or sage essential oil on the shelf-life of fish burger during refrigerated storage, *LWT-Food Sci.Technol.*, 117: 108633.

Elegir, G., A. Kindl, P. Sadocco and M. Orlandi (2008). Development of antimicrobial cellulose packaging through laccase-mediated grafting of phenolic compounds, *Enzyme Microb. Technol.*, 43: 84-92.

El-Fakharany, E.M., V.N. Uversky and E.M. Redwan (2017). Comparative analysis of the antiviral activity of camel, bovine, and human lactoperoxidases against herpes simplex virus type 1, *Appl. Biochem.*, 182: 294-310.

El-Khateib, T.A.L.A.A.T., A.E. Yousef and H.W. Ockerman (1993). Inactivation and attachment of *Listeria monocytogenes* on beef muscle treated with lactic acid and selected bacteriocins, *J. Food Prot.*, 56: 29-33.

Elliot, R.M., J.C. McLay, M.J. Kennedy and R.S. Simmons (2004). Inhibition of foodborne bacteria by the lactoperoxidase system in a beef cube system, *Int. J. Food Microbiol.*, 91: 73-81.

Ellison, R., T.J. Giehl and F.M. La Force (1988). Damage of the outer membrane of enteric Gram-negative bacteria by lactoferrin and transferrin, *Infect. Immun.*, 56: 2774-2781.

Emiroğlu, Z.K., G.P. Yemiş, B.K. Coşkun and K. Candoğan (2010). Antimicrobial activity of soy edible films incorporated with thyme and oregano essential oils on fresh ground beef patties, *Meat Sci.*, 86: 283-288.

Engels, C., A. Schieber and M.G. Gänzle (2011). Inhibitory spectra and modes of antimicrobial action of gallotannins from mango kernels (*Mangifera indica* L.), *Appl. Environ. Microbiol.*, 77: 2215-2223.

EPCD (European Parliament and Council Directive) (1995). pp. 1-63. No 95/2/EC. *Food Additives, Other than Colours and Sweeteners.*

Escamilla-García, M., G. Calderón-Domínguez, J.J. Chanona-Pérez, A.G. Mendoza-Madrigal, P. Di Pierro, B.E. García-Almendárez *et al.* (2017). Physical, structural, barrier, and antifungal characterization of chitosan–zein edible films with added essential oils, *Int. J. Mol. Sci.*, 18: 2370.

Esmaeili, H., N. Cheraghi, A. Khanjari, M. Rezaeigolestani, A.A. Basti, A. Kamkar and E.M. Aghaee (2020). Incorporation of nanoencapsulated garlic essential oil into edible films: A novel approach for extending shelf-life of vacuum-packed sausages, *Meat Sci.*, 166: 108135.

Espitia, P.J.P., J.J.R. Pacheco, N.R.D. Melo, N.D.F.F. Soares and A.M. Durango (2013a). Packaging properties and control of *Listeria monocytogenes* in bologna by cellulosic films incorporated with pediocin, *Brazilian J. Food Technol.*, 16: 226-235.

Espitia, P.J.P., N.D.F.F. Soares, R.F. Teófilo, J.S. dos Reis Coimbra, D.M. Vitor, R.A. Batista *et al.* (2013b). Physical–mechanical and antimicrobial properties of nanocomposite films with pediocin and ZnO nanoparticles, *Carbohydr. Polym.*, 94: 199-208.

Espitia, P.J.P., C.G. Otoni and N.F.F. Soares (2016). Pediocin applications in antimicrobial food packaging systems, pp. 445-454. *In:* J. Barros-Velázquez (Ed.). *Antimicrobial Food Packaging*. Academic Press, New York, USA.

Eswaranandam, S., N.S. Hettiarachchy and M.G Johnson (2004). Antimicrobial activity of citric, lactic, malic, or tartaric acids and nisin-incorporated soy protein film against *Listeria monocytogenes, Escherichia coli* O157: H7, and *Salmonella gaminara, J. Food Sci.,* 69: FMS79-FMS84.

Etzel, M.R. (2004). Manufacture and use of dairy protein fractions, *J. Nutr.,* 134: 996S-1002S.

Fabra, M.J., A. Hambleton, P. Talens, F. Debeaufort and A. Chiralt (2011). Effect of ferulic acid and α-tocopherol antioxidants on properties of sodium caseinate edible films, *Food Hydrocoll.,* 25: 1441-1447.

Fabra, M.J., L. Sánchez-González and A. Chiralt (2014). Lysozyme release from isolate pea protein and starch based films and their antimicrobial properties, *LWT-Food Sci. Technol.,* 55: 22-26.

Fabra, M.J., A. López-Rubio and J.M. Lagaron (2016). Use of the electrohydrodynamic process to develop active/bioactive bilayer films for food packaging applications, *Food Hydrocoll.,* 55: 11-18.

Fabra, M.J., I. Falcó, W. Randazzo, G. Sánchez and A. López-Rubio (2018). Antiviral and antioxidant properties of active alginate edible films containing phenolic extracts, *Food Hydrocoll.,* 81: 96-103.

FAO (Food and Agriculture Organization of the United Nations) (1999). *Manual on the Use of the LP-System in Milk Handling and Preservation,* Animal Production Service, Animal Production and Health Division, Global Lactoperoxidase Programme, Rome, 28pp.

Farnaud, S. and R.W. Evans (2003). Lactoferrin – A multifunctional protein with antimicrobial properties, *Mol. Immunol.,* 40: 395-405.

Farrag, Y., W. Ide, B. Montero, M. Rico, S. Rodríguez-Llamazares, L. Barral, and R. Bouza (2018). Starch films loaded with donut-shaped starch-quercetin microparticles: Characterization and release kinetics, *Int. J. Biol. Macromol.,* 118: 2201-2207.

FDA (US Food and Drug Administration) (1988). Nisin preparation: Affirmation of GRAS status as direct human food ingredient, *Fed. Register,* 53: 11247. 21 CFR 184.

FDA (1998). 63: 12421-12426. 21 CFR 184.

FDA (2004). GRAS Notice No. GRN 000135.

FDA (2007). Gras Notice No. GRN 220.

FDA (2010a). Gras Notice No. GRN 336.

FDA (2010b). Gras Notice No. GRN 000341.

FDA (2011). 21 CFR 184. 1538.

FDA (2020a). 21CFR172.515.

FDA (2020b). 21CFR182.60

FDA (2020c). 21CFR184.1257

Felse, P.A. and T. Panda (2000). Production of microbial chitinases – A revisit, *Bioprocess. Eng.,* 23: 127-134.

Fernández-Pan, I., M. Royo and J. Ignacio Mate (2012). Antimicrobial activity of whey protein isolate edible films with essential oils against food spoilers and foodborne pathogens, *J. Food Sci.,* 77: M383-M390.

Ferri, S., K. Kojima and K. Sode (2011). Review of glucose oxidases and glucose dehydrogenases: A bird's eye view of glucose sensing enzymes, *J. Diabetes Sci. Technol.,* 5: 1068-1076.

Franklin, N.B., K.D. Cooksey and K.J. Getty (2004). Inhibition of *Listeria monocytogenes* on the surface of individually packaged hot dogs with a packaging film coating containing nisin, *J. Food Prot.,* 67: 480-485.

Gadang, V.P., N.S. Hettiarachchy, M.G. Johnson and C. Owens (2008). Evaluation of antibacterial activity of whey protein isolate coating incorporated with nisin, grape seed extract, malic acid, and EDTA on a turkey frankfurter system, *J. Food Sci.*, 73: M389-M394.

Gänzle, M.G., S. Weber and W.P. Hammes (1999). Effect of ecological factors on the inhibitory spectrum and activity of bacteriocins, *Int. J. Food Microbiol.*, 46: 207-217.

Gao, Y., S. Jia, Q. Gao and Z. Tan (2010). A novel bacteriocin with a broad inhibitory spectrum produced by Lactobacillus sake C2, isolated from traditional Chinese fermented cabbage, *Food Control*, 21: 76-81.

Garavito, J., D. Moncayo-Martínez and D.A. Castellanos (2020). Evaluation of antimicrobial coatings on preservation and shelf-life of fresh chicken breast fillets under cold storage, *Foods*, 9: 1203.

Garcia, F. (2020). *A Study of Zein and e-Polylysine Hydrocolloidic Coatings and Their Effect on Hass Avocado (Persea Americana) Shelf-Life*, doctoral dissertation, California State Polytechnic University, Pomona, USA.

Geornaras, I. and J.N. Sofos (2005). Activity of ε–polylysine against *Escherichia coli* O157: H7, *Salmonella typhimurium*, and *Listeria monocytogenes*, *J. Food Sci.*, 70: M404-M408.

Ghamari, M.A., S. Amiri, M. Rezazadeh-Bari and L. Rezazad-Bari (2021). Physical, mechanical, and antimicrobial properties of active edible film based on milk proteins incorporated with *Nigella sativa* essential oil, *Polym. Bull.*, 79: 1097-1117.

Ghosh, R. and Z.F. Cui (2000). Purification of lysozyme using ultrafiltration, *Biotechnol. Bioeng.*, 68: 191-203.

Ghosh, R., A. Chakraborty, A. Biswas and S. Chowdhuri (2020). Evaluation of green tea polyphenols as novel corona virus (SARS CoV-2) main protease (Mpro) inhibitors – An in silico docking and molecular dynamics simulation study, *J. Biomol. Struct. Dyn.*, 39: 4362-4374.

Gialamas, H., K.G. Zinoviadou, C.G. Biliaderis and K.P. Koutsoumanis (2010). Development of a novel bioactive packaging based on the incorporation of *Lactobacillus sakei* into sodium-caseinate films for controlling *Listeria monocytogenes* in foods, *Food Res. Int.*, 43: 2402-2408.

Gill, A.O. and R.A. Holley (2000). Inhibition of bacterial growth on ham and bologna by lysozyme, nisin and EDTA, *Food Res. Int.*, 33: 83-90.

Giménez, B., A.L. De Lacey, E. Pérez-Santín, M.E. López-Caballero and P. Montero (2013). Release of active compounds from agar and agar–gelatin films with green tea extract, *Food Hydrocoll.*, 30: 264-271.

Gingerich, A., U. Patel, J. Hanson, B. Rada and R.A. Tripp (2017). Antiviral activity of cell free hypothiocyanite against various subtypes of Influenza virus, 198: 148.17.

Giteru, S.G., R. Coorey, D. Bertolatti, E. Watkin, S. Johnson and Z. Fang (2015). Physicochemical and antimicrobial properties of citral and quercetin incorporated kafirin-based bioactive films, *Food Chem.*, 168: 341-347.

Göksen, G., M.J. Fabra, H.I. Ekiz and A. López-Rubio (2020). Phytochemical-loaded electrospun nanofibers as novel active edible films: Characterization and antibacterial efficiency in cheese slices, *Food Control.*, 112: 107133.

Gómez-Estaca, J., L. Bravo, M.C. Gómez-Guillén, A. Alemán and P. Montero (2009). Antioxidant properties of tuna-skin and bovine-hide gelatin films induced by the addition of oregano and rosemary extracts, *Food Chem.*, 112: 18-25.

Gómez-Estaca, J., A.L. De Lacey, M.E., López-Caballero, M.C. Gómez-Guillén and P. Montero (2010). Biodegradable gelatin–chitosan films incorporated with essential oils as antimicrobial agents for fish preservation, *Food Microbiol.*, 27: 889-896.

Gradišar, H., P. Pristovšek, A. Plaper and R. Jerala (2007). Green tea catechins inhibit bacterial DNA gyrase by interaction with its ATP binding site, *J. Med. Chem.*, 50: 264-271.

Grasselli, M., S.A. Camperi, A.A.N. del Carizo and O. Cascone (1999). Direct lysozyme separation from egg white by dye membrane affinity chromatography, *J. Sci. Food Agric.*, 79: 333-339.

Gullon, B., M.E. Pintado, J.A. Pérez-Álvarez and M. Viuda-Martos (2016). Assessment of polyphenolic profile and antibacterial activity of pomegranate peel (*Punica granatum*) flour obtained from co-product of juice extraction, *Food Control.*, 59: 94-98.

Gunalan, G., D. Sadhana, and P. Ramya (2012). Production, optimization of chitinase using *Aspergillus flavus* and its biocontrol of phytopathogenic fungi, *J. Pharm. Res.*, 5: 3151-3154.

Gutiérrez-Larraínzar, M., J. Rúa, I. Caro, C. de Castro, D. de Arriaga, M.R. García-Armesto and P. del Valle (2012). Evaluation of antimicrobial and antioxidant activities of natural phenolic compounds against foodborne pathogens and spoilage bacteria, *Food Control.*, 26: 555-563.

Hamid, R., M.A. Khan, M. Ahmad, M.M. Ahmad, M.Z. Abdin, J. Musarrat and S. Javed (2013). Chitinases: An update, *Journal of Pharm. Bioallied Sci.*, 5: 21-29.

Hammoudi, N., H. Ziani Cherif, F. Borsali, K. Benmansour and A. Meghezzi (2020). Preparation of active antimicrobial and antifungal alginate-montmorillonite/lemon essential oil nanocomposite films, *Mater. Technol.*, 35: 383-394.

Han, J.H., H.M. Hwang, S. Min and J.M. Krochta (2008). Coating of peanuts with edible whey protein film containing α-tocopherol and ascorbyl palmitate, *J. Food Sci.*, 73: E349-E355.

Han, Y., N. Tammineni, G. Ünlü, B. Rasco and C. Nindo (2013). Inhibition of *Listeria monocytogenes* on rainbow trout (*Oncorhynchus mykiss)* using trout skin gelatin edible films containing nisin, *J. Food Chem. Nutr.*, 1: 06-15.

Han, J., S.H. Shin, K.M. Park and K.M. Kim (2015). Characterization of physical, mechanical, and antioxidant properties of soy protein-based bioplastic films containing carboxymethylcellulose and catechin, *Food Sci. Biotechnol.*, 24: 939-945.

Hanušová, K., L. Vápenka, J. Dobiáš and L. Mišková (2013). Development of antimicrobial packaging materials with immobilized glucose oxidase and lysozyme, *Open Chem.*, 11: 1066-1078.

Hemaiswarya, S. and M. Doble (2009). Synergistic interaction of eugenol with antibiotics against Gram negative bacteria, *Phytomedicine*, 16: 997-1005.

Hemalatha, T., T. Uma Maheswari, R. Senthil, G. Krithiga and K. Anbukkarasi (2017). Efficacy of chitosan films with basil essential oil: Perspectives in food packaging, *J. Food Meas. Charact.*, 11: 2160-2170.

Hiraki, J., T. Ichikawa, S.I. Ninomiya, H. Seki, K. Uohama, H. Seki et al. (2003). Use of ADME studies to confirm the safety of ε-polylysine as a preservative in food, *Regul. Toxicol. Pharmacol.*, 37: 328-340.

Hoffman, K.L., I.Y. Han and P.L. Dawson (2001). Antimicrobial effects of corn zein films impregnated with nisin, lauric acid, and EDTA, *J. Food Prot.*, 64: 885-889.

Hong, Y.H., G.O. Lim and K.B. Song (2009). Physical properties of Gelidium corneum–gelatin blend films containing grapefruit seed extract or green tea extract and its application in the packaging of pork loins, *J. Food Sci.*, 74: C6-C10.

Hosseini, M.H., S.H. Razavi and M.A. Mousavi (2009). Antimicrobial, physical and mechanical properties of chitosan-based films incorporated with thyme, clove and cinnamon essential oils, *J. Food Process. Preserv.*, 33: 727-743.

Hou, W. and Y. Lin (1997). Egg white lysozyme purification with sweet potato [*Ipomoea batatas* (L.) Lam] leaf preparations, *J. Agric. Food Chem.*, 45: 4487-4489.

Huang, S.W., M.T. Satué-Gracia, E.N. Frankel and J.B. German (1999). Effect of lactoferrin on oxidative stability of corn oil emulsions and liposomes, *J. Agric. Food Chem.*, 47: 1356-1361.

Huang, W., H. Xu, Y. Xue, R. Huang, H. Deng and S. Pan (2012). Layer-by-layer immobilization of lysozyme–chitosan–organic rectorite composites on electrospun nanofibrous mats for pork preservation, *Food Res. Int.*, 48: 784-791.

Huang, W., X. Li, Y. Xue, R. Huang, H. Deng and Z. Ma (2013). Antibacterial multilayer films fabricated by LBL immobilizing lysozyme and HTCC on nanofibrous mats, *Int. J. Biol. Macromol.*, 53: 26-31.

Hussain, Z., H.E. Thu, M.W. Amjad, F. Hussain, T.A. Ahmed and S. Khan (2017). Exploring recent developments to improve antioxidant, anti-inflammatory and antimicrobial efficacy of curcumin: A review of new trends and future perspectives, *Mater. Sci. Eng.*, 77: 1316-1326.

Ibrahim, H.R., K. Kobayashi and A. Kato (1993). Length of hydrocarbon chain and antimicrobial action to gram-negative bacteria of fatty acylated lysozyme, *J. Agric. Food Chem.*, 41: 1164-1168.

Ibrahim, H.R., S. Higashiguchi, L.R. Juneja, M. Kim and T. Yamamoto (1996a). A structural phase of heat-denatured lysozyme with novel antimicrobial action, *J. Agric. Food Chem.*, 44: 1416-1423.

Ibrahim, H.R., S. Higashiguchi, M. Koketsu, L.R. Juneja, M. Kim, T. Yamamoto *et al.* (1996b). Partially unfolded lysozyme at neutral pH agglutinates and kills gram-negative and gram-positive bacteria through membrane damage mechanism, *J. Agric. Food Chem.*, 44: 3799-3806.

Imran, M., S. El-Fahmy, A.M. Revol-Junelles and S. Desobry (2010). Cellulose derivative based active coatings: Effects of nisin and plasticizer on physico-chemical and antimicrobial properties of hydroxypropyl methylcellulose films, *Carbohydr. Polym.*, 81: 219-225.

Iturriaga, L., O.I. Labarrieta and I.M. de Marañón (2012). Antimicrobial assays of natural extracts and their inhibitory effect against *Listeria innocua* and fish spoilage bacteria, after incorporation into hydrocolloid edible films, *Int. J. Food Microbiol.*, 158: 58-64.

Jadeja, R.N. and R.V. Devkar (2014). Polyphenols in chronic diseases and their mechanisms of action, pp. 615-623. *In:* R.R. Watson, V.R. Preedy and S. Zibadi (Eds.). *Polyphenols in Human Health and Disease*, Academic Press. Amsterdam, NL.

Jagani, S., R. Chelikani and D.S. Kim (2009). Effects of phenol and natural phenolic compounds on biofilm formation by *Pseudomonas aeruginosa*, *Biofouling*, 25: 321-324.

Jahani, S., A. Shakiba and L. Jahani (2015). The antimicrobial effect of lactoferrin on gram-negative and gram-positive bacteria, *Int. J. Infect.*, 2: e27954.

Jakobsen, R.A., R. Heggebø, E.B. Sunde and M. Skjervheim (2011). *Staphylococcus aureus* and *Listeria monocytogenes* in Norwegian raw milk cheese production, *Food Microbiol.*, 28: 492-496.

Jamróz, E., P. Kulawik, P. Krzyściak, K. Talaga-Ćwiertnia and L. Juszczak (2019). Intelligent and active furcellaran-gelatin films containing green or pu-erh tea extracts: Characterization, antioxidant and antimicrobial potential, *Int. J. Biol. Macromol.*, 122: 745-757.

Janes, M.E., S. Kooshesh and M.G. Johnson (2002). Control of *Listeria monocytogenes* on the surface of refrigerated, ready-to-eat chicken coated with edible zein film coatings containing nisin and/or calcium propionate, *J. Food Sci.*, 67: 2754-2757.

Jang, S.A., Y.J. Shin and K.B. Song (2011). Effect of rapeseed protein–gelatin film containing grapefruit seed extract on 'Maehyang' strawberry quality, *Int. J. Food Sci. Technol.*, 46: 620-625.

Jasour, M.S., A. Ehsani, L. Mehryar and S.S. Naghibi (2015). Chitosan coating incorporated with the lactoperoxidase system: An active edible coating for fish preservation, *J. Sci. Food Agric.*, 95: 1373-1378.

Jenssen, H. and R.E. Hancock (2009). Antimicrobial properties of lactoferrin, *Biochimie*, 91: 19-29.

JETRO (Japanese Trade Organization) (2011). Specifications and standards for foods, food additives, etc. under the Food Sanitation Act. https://www.jetro.go.jp/en/reports/regulations/pdf/foodext201112e.pdf

JFCRF (The Japan Food Chemical Research Foundation) (2014). List of existing food additives (Effective from January 30, 2014). https://www.ffcr.or.jp/en/tenka/list-of-existing-food-additives/list-of-existing-food-additives.html_

Jiang, C.M., M.C. Wang, W.H. Chang and H.M. Chang (2001). Isolation of lysozyme from hen egg albumen by alcohol-insoluble cross-linked pea pod solid ion-exchange chromatography, *J. Food Sci.*, 66: 1089-1092.

Jiang, Y., D. Wang, F. Li, D. Li and Q. Huang (2020). Cinnamon essential oil pickering emulsion stabilized by zein-pectin composite nanoparticles: Characterization, antimicrobial effect and advantages in storage application, *Int. J. Biol. Macromol.*, 148: 1280-1289.

Jin, T., L. Liu, C.H. Sommers, G. Boyd and H. Zhang (2009). Radiation sensitization and postirradiation proliferation of *Listeria monocytogenes* on ready-to-eat deli meat in the presence of pectin-nisin films, *J. Food Prot.*, 72: 644-649.

Jongjareonrak, A., S. Benjakul, W. Visessanguan and M. Tanaka (2008). Antioxidative activity and properties of fish skin gelatin films incorporated with BHT and α-tocopherol, *Food Hydrocoll.*, 22: 449-458.

Jooyandeh, H., A. Aberoumand and B. Nasehi (2011). Application of lactoperoxidase system in fish and food products: A review, *Am. Eurasian J. Agric. Environ. Sci.*, 10: 89-96.

Juck, G., H. Neetoo and H. Chen (2010). Application of an active alginate coating to control the growth of *Listeria monocytogenes* on poached and deli turkey products, *Int. J. Food Microbiol.*, 142: 302-308.

Kaewprachu, P., K. Osako, S. Benjakul and S. Rawdkuen (2015). Quality attributes of minced pork wrapped with catechin–lysozyme incorporated gelatin film, *Food Packag. Shelf-Life*, 3: 88-96.

Kaewprachu, P., N. Rungraeng, K. Osako and S. Rawdkuen (2017). Properties of fish myofibrillar protein film incorporated with catechin-Kradon extract, *Food Packag. Shelf-Life*, 13: 56-65.

Kaewprachu, P., C.B. Amara, N. Oulahal, A. Gharsallaoui, C. Joly, W. Tongdeesoontorn *et al.* (2018). Gelatin films with nisin and catechin for minced pork preservation, *Food Packag. Shelf-Life*, 18: 173-183.

Kajiya, K., S. Kumazawa and T. Nakayama (2001). Steric effects on interaction of tea catechins with lipid bilayers, *Biosci. Biotechnol. Biochem.*, 65: 2638-2643.

Kandemir, N., A. Yemenicioglu, Ç. Mecitoglu, Z.S. Elmaci, A. Arslanoglu, Y. Göksungur and T. Baysal (2005). Production of antimicrobial films by incorporation of partially purified lysozyme into biodegradable films of crude exopolysaccharides obtained from *Aureobasidium pullulans* fermentation, *Food Technol. Biotechnol.*, 43: 343-350.

Katla, T., K. Naterstad, M. Vancanneyt, J. Swings and L. Axelsson (2003). Differences in susceptibility of *Listeria monocytogenes* strains to sakacin P, sakacin A, pediocin PA-1, and nisin, *Appl. Environmen. Microbiol.*, 69: 4431-4437.

Kennedy, M., A.L. O'Rourke, J. McLay and R. Simmonds (2000a). Use of ground beef model to assess the effect of the lactoperoxidase system on the growth of *Escherichia coli O157:H7*, *Listeria monocytogenes* and *Staphylococcus aureus* in red meat, *Int. J. Food Microbiol.*, 57: 147-158.

Kennedy, J.A., M.A. Matthews and A.L. Waterhouse (2000b). Changes in grape seed polyphenols during fruit ripening, *Phytochemistry*, 55: 77-85.

Kharchoufi, S., L. Parafati, F. Licciardello, G. Muratore, M. Hamdi, G. Cirvilleri and C. Restuccia (2018). Edible coatings incorporating pomegranate peel extract and biocontrol yeast to reduce *Penicillium digitatum* postharvest decay of oranges, *Food Microbiol.*, 74: 107-112.

Kheadr, E., N. Bernoussi, C. Lacroix and I. Fliss (2004). Comparison of the sensitivity of commercial strains and infant isolates of bifidobacteria to antibiotics and bacteriocins, *Int. Dairy J.*, 14: 1041-1053.

Khoo, H.E., A. Azlan, S.T. Tang and S.M. Lim (2017). Anthocyanidins and anthocyanins: Colored pigments as food, pharmaceutical ingredients, and the potential health benefits, *Food Nutr. Res.*, 61: 1361779.

Kim, D.O. and C.Y. Lee (2004). Comprehensive study on vitamin C equivalent antioxidant capacity (VCEAC) of various polyphenolics in scavenging a free radical and its structural relationship, *Crit. Rev. Food Sci. Nutr.*, 44: 253-273.

Kim, K.M., B.Y. Lee, Y.T. Kim, S.G. Choi, J.S. Lee, S.Y. Cho and W.S. Choi (2006). Development of antimicrobial edible film incorporated with green tea extract, *Food Sci. Biotechnol.*, 15: 478-481.

Ko, S., M.E. Janes, N.S. Hettiarachchy and M.G. Johnson (2001). Physical and chemical properties of edible films containing nisin and their action against *Listeria monocytogenes*, *J. Food Sci.*, 66: 1006-1011.

Ko, M.J., C.I. Cheigh, S.W. Cho and M.S. Chung (2011). Subcritical water extraction of flavonol quercetin from onion skin, *J. Food Eng.*, 102: 327-333.

Kramer, K.J. and S. Muthukrishnan (1997). Insect chitinases: Molecular biology and potential use as biopesticides, *Insect Biochem. Mol. Biol.*, 27: 887-900.

Kristo, E., K.P. Koutsoumanis and C.G. Biliaderis (2008). Thermal, mechanical and water vapor barrier properties of sodium caseinate films containing antimicrobials and their inhibitory action on *Listeria monocytogenes*, *Food Hydrocoll.*, 22: 373-386.

Ku, K.J. and K.B. Song (2007). Physical properties of nisin-incorporated gelatin and corn zein films and antimicrobial activity against *Listeria monocytogenes*, *J. Microbiol. Biotechnol.*, 17: 520-523.

Ku, K.J., Y.H. Hong and K.B. Song (2008a). Preparation of a silk fibroin film containing catechin and its application, *Food Sci. Biotechnol.*, 17: 1203-1206.

Ku, K.J., Y.H. Hong and K.B. Song (2008b). Mechanical properties of a *Gelidium corneum* edible film containing catechin and its application in sausages, *J. Food Sci.*, 73: C217-C221.

Kurita, N., M. Miyaji, R. Kurane and Y. Takahara (1981). Antifungal activity of components of essential oils, *Agr. Biol. Chem.*, 45: 945-952.

Lee, H. and S.C. Min (2013). Antimicrobial edible defatted soybean meal-based films incorporating the lactoperoxidase system, *LWT-Food Sci. Technol.*, 54: 42-50.

Leopoldini, M., N. Russo and M. Toscano (2011). The molecular basis of working mechanism of natural polyphenolic antioxidants, *Food Chem.*, 125: 288-306.

Li, S., J. Sun, J. Yan, S. Zhang, C. Shi, D.J. McClements and F. Liu (2020). Development of antibacterial nanoemulsions incorporating thyme oil: Layer-by-layer self-assembly of whey protein isolate and chitosan hydrochloride, *Food Chem.*, 339: 128016.

Liang, J., H. Yan, X. Wang, Y. Zhou, X. Gao, P. Puligundla and X. Wan (2017a). Encapsulation of epigallocatechin gallate in zein/chitosan nanoparticles for controlled applications in food systems, *Food Chem.*, 231: 19-24.

Liang, J., H. Yan, J. Zhang, W. Dai, X. Gao, Y. Zhou et al. (2017b). Preparation and characterization of antioxidant edible chitosan films incorporated with epigallocatechin gallate nanocapsules, *Carbohydr. Polym.*, 171: 300-306.

Liang, T., G. Sun, L. Cao, J. Li and L. Wang (2019). A pH and NH$_3$ sensing intelligent film based on *Artemisia sphaerocephala* Krasch. gum and red cabbage anthocyanins anchored by carboxymethyl cellulose sodium added as a host complex, *Food Hydrocoll.*, 87: 858-868.

Li, L. and J.C. Steffens (2002). Overexpression of polyphenol oxidase in transgenic tomato plants results in enhanced bacterial disease resistance, *Planta*, 215: 239-247.

Li, Y.Q., Q. Han, J.L. Feng, W.L. Tian and H.Z. Mo (2014). Antibacterial characteristics and mechanisms of ε-poly-lysine against *Escherichia coli* and *Staphylococcus aureus*, *Food Control*, 43: 22-27.

Li, X., H. Tu, M. Huang, J. Chen, X. Shi, H. Deng *et al.* (2017). Incorporation of lysozyme-rectorite composites into chitosan films for antibacterial properties enhancement, *Int. J. Biol. Macromol.*, 102: 789-795.

Li, Y.N., Q.Q. Ye, W.F. Hou and G.Q. Zhang (2018). Development of antibacterial ε-polylysine/chitosan hybrid films and the effect on citrus, *Int. J. Biol. Macromol.*, 118: 2051-2056.

Li, N.A., W. Liu, Y. Shen, J. Mei and J. Xie (2019). Coating effects of ε-polylysine and rosmarinic acid combined with chitosan on the storage quality of fresh half-smooth tongue sole (*Cynoglossus semilaevis Günther*) fillets, *Coatings*, 9: 273.

Li, S., Y. Yan, X. Guan and K. Huang (2020). Preparation of a hordein-quercetin-chitosan antioxidant electrospun nanofibre film for food packaging and improvement of the film hydrophobic properties by heat treatment, *Food Packag. Shelf-Life*, 23: 100466.

Liburdi, K., I. Benucci and M. Esti (2014). Lysozyme in wine: An overview of current and future applications, *Compr. Rev. Food Sci. F.*, 13: 1062-1073.

Liceaga-Gesualdo, A., E.C.Y. Li-Chan and B.J. Skura (2001). Antimicrobial effect of lactoferrin digest on spores of a *Penicillium* sp. isolated from bottled water, *Food Res. Int.*, 34: 501-506.

Lin, L.S., B.J. Wang and Y.M. Weng (2011). Quality preservation of commercial fish balls with antimicrobial zein coatings, *J. Food Qual.*, 34: 81-87.

Lin, L., Y. Gu, C. Li, S. Vittayapadung and H. Cui (2018a). Antibacterial mechanism of ε-poly-lysine against *Listeria monocytogenes* and its application on cheese, *Food Control*, 91: 76-84.

Lin, L., L. Xue, S. Duraiarasan and C. Haiying (2018b). Preparation of ε-polylysine/chitosan nanofibers for food packaging against *Salmonella* on chicken, *Food Packag. Shelf-Life*, 17: 134-141.

Liu, S., H. Azakami and A. Kato (2000a). Improvement in the yield of lipophilized lysozyme by the combination with Maillard-type glycosylation, *Food/Nahrung*, 44: 407-410.

Liu, S.T., T. Sugimoto, H. Azakami and A. Kato (2000b). Lipophilization of lysozyme by short and middle chain fatty acids, *J. Agric. Food Chem.*, 48: 265-269.

Liu, Y., F. Han, Y. Xie and Y. Wang (2011). Comparative antimicrobial activity and mechanism of action of bovine lactoferricin-derived synthetic peptides, *Biometals*, 24: 1069-1078.

Liu, H., H. Pei, Z. Han, G. Feng and D. Li (2015). The antimicrobial effects and synergistic antibacterial mechanism of the combination of ε-polylysine and nisin against *Bacillus subtilis*, Food Control, 47: 444-450.

Liu, Y., Y. Cai, X. Jiang, J. Wu and X. Le (2016). Molecular interactions, characterization and antimicrobial activity of curcumin–chitosan blend films, *Food Hydrocoll.*, 52: 564-572.

Liu, F., F. Türker Saricaoglu, R.J. Avena-Bustillos, D.F. Bridges, G.R. Takeoka, V.C. Wu *et al.* (2018). Preparation of fish skin gelatin-based nanofibers incorporating cinnamaldehyde by solution blow spinning, *Int. J. Mol. Sci.*, 19: 618.

Liu, D., S. Meng, Z. Xiang, N. He and G. Yang (2019a). Antimicrobial mechanism of reaction products of *Morus notabilis* (mulberry) polyphenol oxidases and chlorogenic acid, *Phytochemistry*, 163: 1-10.

Liu, J., H. Yong, Y. Liu, Y. Qin, J. Kan and J. Liu (2019b). Preparation and characterization of active and intelligent films based on fish gelatin and haskap berries (*Lonicera caerulea* L.) extract, *Food Packag. Shelf-Life*, 22: 100417.

Liu, F., Y. Liu, Z. Sun, D. Wang, H. Wu, L. Du and D. Wang (2020). Preparation and antibacterial properties of ε-polylysine-containing gelatin/chitosan nanofiber films, *Int. J. Biol. Macromol.*, 164: 3376-3387.

Lopes, M.I., J.T. Martins, L.P. Fonseca and A.A. Vicente (2011). Effect of glucose oxidase incorporation in chitosan edible films properties, *Proceedings of the 6th CIGR Section VI International Symposium 'Towards a Sustainable Food Chain' Food Process, Bioprocessing and Food Quality Management*, Nantes, France.

Lopes, L.F., G. Meca, K.C. Bocate, T.M. Nazareth, K. Bordin and F.B. Luciano (2018). Development of food packaging system containing allyl isothiocyanate against Penicillium nordicum in chilled pizza: Preliminary study, *J. Food Process. Preserv.*, 42: e13436.

Lopez-Rubio, A., R. Gavara and J.M. Lagaron (2006). Bioactive packaging: Turning foods into healthier foods through biomaterials, *Trends in Food Sci. Technol.*, 17: 567-575.

Lu, R.R., S.Y. Xu, Z. Wang and R.J. Yang (2007). Isolation of lactoferrin from bovine colostrum by ultrafiltration coupled with strong cation exchange chromatography on a production scale, *J. Membrane Sci.*, 297: 152-161.

Luciano, C.G., L. Tessaro, R.V. Lourenço, A.M.Q.B. Bittante, A.M. Fernandes, I.C.F. Moraes, and P.J. do Amaral Sobral (2020). Effects of nisin concentration on properties of gelatin film-forming solutions and their films, *Int. J. Food Sci. Tech.*, 56: 587-599.

Lungu, B. and M.G. Johnson (2005). Fate of *Listeria monocytogenes* inoculated onto the surface of model turkey frankfurter pieces treated with zein coatings containing nisin, sodium diacetate, and sodium lactate at 4 C, *J. Food Prot.*, 68: 855-859.

Luz, C., J. Calpe, F. Saladino, F.B. Luciano, M. Fernandez-Franzón, J. Mañes and G. Meca (2018). Antimicrobial packaging based on ε-polylysine bioactive film for the control of mycotoxigenic fungi in vitro and in bread, *J. Food Process. Preserv.*, 42: e13370.

Lyn, F.H. and Z.N. Hanani (2020). Effect of Lemongrass (*Cymbopogon citratus*) essential oil on the properties of chitosan films for active packaging, *J. Packag. Technol. Res.*, 4: 33-44.

Magar, R.T. and J.K. Sohng (2020). A review on structure, modifications and structure-activity relation of quercetin and its derivatives, *J. Microbiol. Biotechnol.*, 30: 11-20.

Makki, F. and T.D. Durance (1996). Thermal inactivation of lysozyme as influenced by pH, sucrose and sodium chloride and inactivation and preservative effect in beer, *Food Res. Int.*, 29: 635-645.

Mangalassary, S., I. Han, J. Rieck, J. Acton and P. Dawson (2008). Effect of combining nisin and/or lysozyme with in-package pasteurization for control of *Listeria monocytogenes* in ready-to-eat turkey bologna during refrigerated storage, *Food Microbiol.*, 25: 866-870.

Mapelli, C., A. Musatti, A. Barbiroli, S. Saini, J. Bras, D. Cavicchioli and M. Rollini (2019). Cellulose nanofiber (CNF)–sakacin-A active material: Production, characterization and application in storage trials of smoked salmon, *J. Sci. Food and Agric.*, 99: 4731-4738.

Marchetti, M., F. Superti, M.G. Ammendolia, P. Rossi, P. Valenti and L. Seganti (1999). Inhibition of poliovirus type 1 infection by iron-, manganese- and zinc-saturated lactoferrin, *Med. Microbiol. Immunol.*, 187: 199-204.

Martelli, S.M., C. Motta, T. Caon, J. Alberton, I.C. Bellettini, A.C.P do Prado *et al.* (2017). Edible carboxymethyl cellulose films containing natural antioxidant and surfactants: α-tocopherol stability, *in vitro* release and film properties, *LWT-Food Sci. Technol.*, 77: 21-29.

Martínez-Zavala, S.A., U.E. Barboza-Pérez, G. Hernández-Guzmán, D.K. Bideshi and J.E. Barboza-Corona (2020). Chitinases of *Bacillus thuringiensis*: Phylogeny, modular structure, and applied potentials, *Front. Microbiol.*, 10: 3032.

Martins, J.T., M.A. Cerqueira and A.A. Vicente (2012). Influence of α-tocopherol on physicochemical properties of chitosan-based films, *Food Hydrocoll.*, 27: 220-227.

Martins, N., B.L. Arros, C. Santos-Buelga, S. Silva, M. Henriques and I.C. Ferreira (2015). Decoction, infusion and hydroalcoholic extract of cultivated thyme: Antioxidant and antibacterial activities, and phenolic characterization, *Food Chem.*, 167: 131-137.

Marugg, J.D., C.F. Gonzalez, B.S. Kunka, A.M. Ledeboer, M.J. Pucci, M.Y. Toonen *et al.* (1992). Cloning, expression, and nucleotide sequence of genes involved in production of pediocin PA-1, and bacteriocin from Pediococcus acidilactici PAC1.0, *Appl. Environ. Microbiol.*, 58: 2360-2367.

Masuda, T., Y. Ueno and N. Kitabatake (2001). Sweetness and enzymatic activity of lysozyme, *J. Agric. Food Chem.*, 49: 4937-4941.

Mata, L., L. Sánchez, D.R. Headon and M. Calvo (1998). Thermal denaturation of human lactoferrin and its effect on the ability to bind iron, *J. Agric. Food Chem.*, 46: 3964-3970.

Matan, N. (2012). Antimicrobial activity of edible film incorporated with essential oils to preserve dried fish (*Decapterus maruadsi*), *Int. Food Res. J.*, 19: 1733-1738.

Mayer, A.M. (1986). Polyphenol oxidases in plants – Recent progress, *Phytochemistry*, 26: 11-20.

Mazzotta, A.S., A.D. Crandall and T.J. Montville (1997). Nisin resistance in *Clostridium botulinum* spores and vegetative cells, *Appl. Environ. Microbiol.*, 63: 2654-2659.

Mecitoglu, Ç., A. Yemenicioglu, A. Arslanoglu, Z.S. Elmacı, F. Korel and A.E. Çetin (2006). Incorporation of partially purified hen egg white lysozyme into zein films for antimicrobial food packaging, *Food Res. Int.*, 39: 12-21.

Mecitoğlu, Ç. and A. Yemenicioğlu (2007). Partial purification and preparation of bovine lactoperoxidase and characterization of kinetic properties of its immobilized form incorporated into cross-linked alginate films, *Food Chem.*, 104: 726-733.

Mehdizadeh, T., H. Tajik, S.M.R. Rohani and A.R. Oromiehie (2012). Antibacterial, antioxidant and optical properties of edible starch-chitosan composite film containing *Thymus kotschyanus* essential oil, *In. Vet. Res. Forum*, 3: 167-173.

Meira, S.M.M., G. Zehetmeyer, J.M. Scheibel, J.O. Werner and A. Brandelli (2016). Starch-halloysite nanocomposites containing nisin: Characterization and inhibition of *Listeria monocytogenes* in soft cheese, *LWT-Food Sci. Technol.*, 68: 226-234.

Meschini, R., D. D'Eliseo, S. Filippi, L. Bertini, B.M. Bizzarri, L. Botta *et al.* (2018). Tyrosinase-treated hydroxytyrosol-enriched olive vegetation waste with increased antioxidant activity promotes autophagy and inhibits the inflammatory response in human THP-1 monocytes, *J. Agric. Food Chem.*, 66: 12274-12284.

Min, S. and J.M. Krochta (2005). Inhibition of *Penicillium commune* by edible whey protein films incorporating lactoferrin, lacto-ferrin hydrolysate, and lactoperoxidase systems, *J. Food Sci.*, 70: M87-M94.

Min, S., L.J. Harris, J.H. Han and J.M. Krochta (2005a). *Listeria monocytogenes* inhibition by whey protein films and coatings incorporating lysozyme, *J. Food Prot.*, 68: 2317-2325.

Min, S., L.J. Harris and J.M. Krochta (2005b). Antimicrobial effects of lactoferrin, lysozyme, and the lactoperoxidase system and edible whey protein films incorporating the lactoperoxidase system against *Salmonella enterica* and *Escherichia coli* O157: H7, *J. Food Sci.*, 70: M332-M338.

Min, S., L.J. Harris and J.M. Krochta (2005c). *Listeria monocytogenes* inhibition by whey protein films and coatings incorporating the lactoperoxidase system, *J. Food Sci.*, 70: M317-M324.

Min, S., L.J. Harris and J.M. Krochta (2005d). Antimicrobial effects of lactoferrin, lysozyme, and the lactoperoxidase system and edible whey protein films incorporating the lactoperoxidase system against *Salmonella enterica* and *Escherichia coli* O157: H7, *J. Food Sci.*, 70: M332-M338.

Min, S., L.J. Harris and J.M. Krochta (2006). Inhibition of *Salmonella enterica* and *Escherichia coli* O157: H7 on roasted turkey by edible whey protein coatings incorporating the lactoperoxidase system, *J. Food Prot.*, 69: 784-793.

Min, B.J., I.Y. Han and P.L. Dawson (2010). Antimicrobial gelatin films reduce *Listeria monocytogenes* on turkey bologna, *Poult. Sci.*, 89: 1307-1314.

Ming, X., G.H. Weber, J.W. Ayres and W.E. Sandine (1997). Bacteriocins applied to food packaging materials to inhibit *Listeria monocytogenes* on meats, *J. Food Sci.*, 62: 413-415.

Mittal, A., A. Singh, S. Benjakul, T. Prodpran, K. Nilsuwan, N. Huda and K. de la Caba (2021). Composite films based on chitosan and epigallocatechin gallate grafted chitosan: Characterization, antioxidant and antimicrobial activities, *Food Hydrocoll.*, 111: 106384.

Mizielińska, M., U. Kowalska, M. Jarosz and P. Sumińska (2018). A comparison of the effects of packaging containing nano ZnO or polylysine on the microbial purity and texture of Cod (*gadus morhua*) fillets, *Nanomaterials*, 8: 158.

Moghadamtousi, S., H.Z. Abdul Kadir, P. Hassandarvish, H. Tajik, S. Abubakar and K. Zandi (2014). A review on antibacterial, antiviral, and antifungal activity of curcumin, *BioMed Res. Int.*, 2014: 186864.

Mohamed, C., K.A. Clementine, M. Didier, L. Gérard and D.C.M. Noëlle (2013). Antimicrobial and physical properties of edible chitosan films enhanced by lactoperoxidase system, *Food Hydrocoll.*, 30: 576-580.

Mohammadalinejhad, S., H. Almasi and M. Moradi (2020). Immobilization of *Echium amoenum* anthocyanins into bacterial cellulose film: A novel colorimetric pH indicator for freshness/spoilage monitoring of shrimp, *Food Control*, 113: 107169.

Molayi, R., A. Ehsani and M. Yousefi (2018). The antibacterial effect of whey protein–alginate coating incorporated with the lactoperoxidase system on chicken thigh meat, *Food Sci. Nutr.*, 6: 878-883.

Muraoka, W., C. Gay, D. Knowles and M. Borucki (2003). Prevalence of *Listeria monocytogenes* subtypes in bulk milk of the Pacific northwest, *J. Food Prot.*, 66: 1413-1419.

Murillo-Martínez, M.M., S.R. Tello-Solís, M.A. García-Sánchez and E. Ponce-Alquicira (2013). Antimicrobial activity and hydrophobicity of edible whey protein isolate films formulated with nisin and/or glucose oxidase, *J. Food Sci.*, 78: M560-M566.

Murray, M. and J.A. Richard (1997). Comparative study of the antilisterial activity of nisin A and pediocin AcH in fresh ground pork stored aerobically at 5 C, *J. Food Prot.*, 60: 1534-1540.

Mushtaq, M., A. Gani, A. Gani, H.A. Punoo and F.A. Masoodi (2018). Use of pomegranate peel extract incorporated zein film with improved properties for prolonged shelf-life of fresh Himalayan cheese (Kalari/kradi), *Innov. Food Sci. Emerg. Technol.*, 48: 25-32.

Musso, Y.S., P.R. Salgado and A.N. Mauri (2017). Smart edible films based on gelatin and curcumin, *Food Hydrocoll.*, 66: 8-15.

Musso, Y.S., P.R. Salgado and A.N. Mauri (2019). Smart gelatin films prepared using red cabbage (*Brassica oleracea* L.) extracts as solvent, *Food Hydrocoll.*, 89: 674-681.

Na, S., J.H. Kim, H.J. Jang, H.J. Park and S.W. Oh (2018). Shelf-life extension of Pacific white shrimp (*Litopenaeus vannamei*) using chitosan and ε-polylysine during cold storage, *Int. J. Biol. Macromol.*, 115: 1103-1108.

Naidu, A.S. (2002). Activated lactoferrin – A new approach to meat safety, *Food Technology*, 56: 40-46.

Nair, M.S., A. Saxena and C. Kaur (2018). Characterization and antifungal activity of pomegranate peel extract and its use in polysaccharide-based edible coatings to extend the shelf-life of capsicum (*Capsicum annuum* L.), *Food Bioprocess Tech.*, 11: 1317-1327.

Nakamura, S., A. Kato and K. Kobayashi (1991). New antimicrobial characteristics of lysozyme-dextran conjugate, *J. Agric. Food Chem.*, 39: 647-650.

Nakamura, S., A. Kato and K. Kobayashi (1992). Bifunctional lysozyme-galactomannan conjugate having excellent emulsifying properties and bactericidal effect, *J. Agric. Food Chem.*, 40: 735-739.

Nakayama, T. and B. Uno (2015). Importance of proton-coupled electron transfer from natural phenolic compounds in superoxide scavenging, *Chem. Pharm. Bull.*, 63: 967-973.

Narsaiah, K., S.N. Jha, R.A. Wilson, H.M. Mandge, M.R. Manikantan, R.K. Malik and S. Vij (2013). Pediocin-loaded nanoliposomes and hybrid alginate–nanoliposome delivery systems for slow release of pediocin, *Bionanoscience*, 3: 37-42.

Narsaiah, K., R.A. Wilson, K. Gokul, H.M. Mandge, S.N. Jha, S. Bhadwal *et al.* (2015). Effect of bacteriocin-incorporated alginate coating on shelf-life of minimally processed papaya (*Carica papaya* L.), *Postharvest Biol. Technol.*, 100: 212-218.

Neri-Numa, I.A., H.S. Arruda, M.V. Geraldi, M.R.M. Júnior and G.M. Pastore (2020). Natural prebiotic carbohydrates, carotenoids and flavonoids as ingredients in food systems, *Curr. Opin. Food Sci.*, 33: 98-107.

Nguyen, V.T., M.J. Gidley and G.A. Dykes (2008). Potential of a nisin-containing bacterial cellulose film to inhibit *Listeria monocytogenes* on processed meats, *Food Microbiol.*, 25: 471-478.

Niaz, B., F. Saeed, A. Ahmed, M. Imran, A.A. Maan, M.K.I. Khan *et al.* (2019). Lactoferrin (LF): A natural antimicrobial protein, *Int. J. Food Prop.*, 22: 1626-1641.

Nieto-Suaza, L., L. Acevedo-Guevara, L.T. Sánchez, M.I. Pinzón and C.C. Villa (2019). Characterization of Aloe vera-banana starch composite films reinforced with curcumin-loaded starch nanoparticles, *Food Struct.*, 22: 100131.

Niki, E. and K. Abe (2019). Vitamin E: Structure, properties and functions, pp. 1-11. *In:* E. Niki (Ed.). *Vitamin E: Chemistry and Nutritional Benefits,* vol. 11, Royal Society of Chemistry, Croydon, UK.

Nilsuwan, K., S. Benjakul and T. Prodpran (2018). Properties and antioxidative activity of fish gelatin-based film incorporated with epigallocatechin gallate, *Food Hydrocoll.*, 80: 212-221.

Nilsuwan, K., P. Guerrero, K. de la Caba, S. Benjakul and T. Prodpran (2019). Properties of fish gelatin films containing epigallocatechin gallate fabricated by thermo-compression molding, *Food Hydrocoll.*, 97: 105236.

Noori, S., F. Zeynali and H. Almasi (2018). Antimicrobial and antioxidant efficiency of nanoemulsion-based edible coating containing ginger (*Zingiber officinale*) essential oil and its effect on safety and quality attributes of chicken breast fillets, *Food Control*, 84: 312-320.

Oliver, S.P., B.M. Jayarao and R.A. Almeida (2005). Foodborne pathogens in milk and the dairy farm environment: Food safety and public health implications, *Food-borne Pathog. Dis.*, 2: 115-129.

Olszowy, M. (2019). What is responsible for antioxidant properties of polyphenolic compounds from plants? *Plant Physiol. Biochem.*, 144: 135-143.

Ono, K., M. Nakao, M. Toyota, Y. Terashi, M. Yamada, T. Kohno and Y. Asakawa (1998). Catechin production in cultured Polygonum hydropiper cells, *Phytochemistry*, 49: 1935-1939.

Oun, A.A. and J.W. Rhim (2020). Preparation of multifunctional carboxymethyl cellulose-based films incorporated with chitin nanocrystal and grapefruit seed extract, *Int. J. Biol. Macromol.*, 152: 1038-1046.

Padgett, T., I.Y. Han and P.L. Dawson (1998). Incorporation of food-grade antimicrobial compounds into biodegradable packaging films, *J. Food Prot.*, 61: 1330-1335.

Padgett, T., Y. Han and P.L. Dawson (2000). Effect of lauric acid addition on the antimicrobial efficacy and water permeability of corn zein films containing nisin, *J. Food Process. Preserv.*, 24: 423-432.

Padrao, J., S. Gonçalves, J.P. Silva, V. Sencadas, S. Lanceros-Méndez, A.C. Pinheiro *et al.* (2016). Bacterial cellulose-lactoferrin as an antimicrobial edible packaging, *Food Hydrocoll.*, 58: 126-140.

Pandit, A.P., S.B. Omase and V.M. Mute (2020). A chitosan film containing quercetin-loaded transfersomes for treatment of secondary osteoporosis, *Drug Deliv. Transl. Res.*, 10: 1495-1506.

Papuc, C., G.V. Goran, C.N. Predescu, V. Nicorescu and G. Stefan (2017). Plant polyphenols as antioxidant and antibacterial agents for shelf-life extension of meat and meat products: Classification, structures, sources, and action mechanisms, *Compr. Rev. Food Sci. Food Saf.*, 16: 1243-1268.

Park, S.I., M.A. Daeschel and Y. Zhao (2004). Functional properties of antimicrobial lysozyme-chitosan composite films, *J. Food Sci.*, 69: M215-M221.

Patra, A.K. (2012). An overview of antimicrobial properties of different classes of phytochemicals, pp. 1-32. *In:* A.K. Patra (Ed.). *Dietary Phytochemicals and Microbes*, Springer, Dordrecht, NL.

Pattanayaiying, R., H. Aran and C.N. Cutter (2015). Incorporation of nisin Z and lauric arginate into pullulan films to inhibit food-borne pathogens associated with fresh and ready-to-eat muscle foods, *Int. J. Food Microbiol.*, 207: 77-82.

Pattarasiriroj, K., P. Kaewprachu and S. Rawdkuen (2020). Properties of rice flour-gelatin-nanoclay film with catechin-lysozyme and its use for pork belly wrapping, *Food Hydrocoll.*, 107: 105951.

Pavli, F., C. Tassou, G.J.E. Nychas and N. Chorianopoulos (2018). Probiotic incorporation in edible films and coatings: Bioactive solution for functional foods, *Int. J. Mol. Sci.*, 19: 150.

Perumalla, A.V.S. and N.S. Hettiarachchy (2011). Green tea and grape seed extracts – Potential applications in food safety and quality, *Food Res. Int.*, 44: 827-839.

Pintado, C.M., M.A. Ferreira and I. Sousa (2009). Properties of whey protein–based films containing organic acids and nisin to control *Listeria monocytogenes*, *J. Food Prot.*, 72: 1891-1896.

Pinto, M.S., A.F. de Carvalho, A.C. dos Santos Pires, A.A.C. Souza, P.H.F. da Silva, D. Sobral *et al.* (2011). The effects of nisin on *Staphylococcus aureus* count and the physicochemical properties of traditional Minas Serro cheese, *Int. Dairy J.*, 21: 90-96.

Popper, L. and D. Knorr (1997). Inactivation of yeast and filamentous fungi by the lactoperoxidase-hydrogen peroxide-thiocyanate-system, *Food/Nahrung*, 41: 29-33.

Porto, M.C.W., T.M. Kuniyoshi, P.O.S. Azevedo, M. Vitolo and R.S. Oliveira (2017). *Pediococcus* spp. an important genus of lactic acid bacteria and pediocin producers, *Biotechnol. Adv.*, 35: 361-374.

Pranoto, Y., V.M. Salokhe and S.K. Rakshit (2005). Physical and antibacterial properties of alginate-based edible film incorporated with garlic oil, *Food Res. Int.*, 38: 267-272.

Proestos, C., I.S. Boziaris, G.J. Nychas and M. Komaitis (2006). Analysis of flavonoids and phenolic acids in Greek aromatic plants: Investigation of their antioxidant capacity and antimicrobial activity, *Food Chem.*, 95: 664-671.

Pruitt, K.M., J. Tenovuo, R.W. Andrews and T. McKane (1982). Lactoperoxidase-catalyzed oxidation of thiocyanate: A polarographic study of the oxidation products, *Biochemistry*, 21: 562-567.

Qin, Y., Y. Liu, H. Yong, J. Liu, X. Zhang and J. Liu (2019). Preparation and characterization of active and intelligent packaging films based on cassava starch and anthocyanins from *Lycium ruthenicum* Murr., *Int. J. Biol. Macromol.*, 134: 80-90.

Rahardiyan, D. (2019). Antibacterial potential of catechin of tea (*Camellia sinensis*) and its applications, *Food Res.*, 3: 1-6.

Rahman, A.N.F., M. Ohta, K. Nakatani, N. Hayashi and S. Fujita (2012). Purification and characterization of polyphenol oxidase from cauliflower (*Brassica oleracea* L.), *J. Agric. Food Chem.*, 60: 3673-3678.

Rajan, V.K. and K. Muraleedharan (2017). A computational investigation on the structure, global parameters and antioxidant capacity of a polyphenol, gallic acid, *Food Chem.*, 220: 93-99.

Randazzo, W., A. Jiménez-Belenguer, L. Settanni, A. Perdones, M. Moschetti, E. Palazzolo *et al.* (2016). Antilisterial effect of citrus essential oils and their performance in edible film formulations, *Food Control*, 59: 750-758.

Rawdkuen, S., P. Suthiluk, D. Kamhangwong and S. Benjakul (2012). Mechanical, physico-chemical, and antimicrobial properties of gelatin-based film incorporated with catechin-lysozyme, *Chem. Cent. J.*, 6: 1-10.

Recio, I. and S. Visser (1999). Two ion-exchange chromatographic methods for the isolation of antibacterial peptides from lactoferrin: *In situ* enzymatic hydrolysis on an ion-exchange membrane, *J. Chromatogr. A*, 831: 191-201.

Redwan, E.M., H.A. Almehdar, E.M. El-Fakharany, A.W.K Baig and V.N. Uversky (2015). Potential antiviral activities of camel, bovine, and human lactoperoxidases against hepatitis C virus genotype 4, *RSC Adv.*, 5: 60441-60452.

Rivas, F.P., M.E. Cayré, C.A. Campos and M.P. Castro (2018). Natural and artificial casings as bacteriocin carriers for the biopreservation of meats products, *J. Food Saf.*, 38: e12419.

Rojas-Graü, M.A., R.J. Avena-Bustillos, C. Olsen, M. Friedman, P.R. Henika, O. Martín-Belloso *et al.* (2007). Effects of plant essential oils and oil compounds on mechanical, barrier and antimicrobial properties of alginate–apple puree edible films, *J. Food Eng.*, 81: 634-641.

Rösch, D., M. Bergmann, D. Knorr and L.W. Kroh (2003). Structure-antioxidant efficiency relationships of phenolic compounds and their contribution to the antioxidant activity of sea buckthorn juice, *J. Agric. Food Chem.*, 51: 4233-4239.

Rose, N.L., P. Sporns, M.E. Stiles and L.M. McMullen (1999). Inactivation of nisin by glutathione in fresh meat, *J. Food Sci.*, 64: 759-762.

Rose, N.L., M.M. Palcic, P. Sporns and L.M. McMullen (2002). Nisin: A novel substrate for glutathione S-transferase isolated from fresh beef, *J. Food Sci.*, 67: 2288-2293.

Rossi-Márquez, G., J.H. Han, B. García-Almendárez, E. Castaño-Tostado and C. Regalado-González (2009). Effect of temperature, pH and film thickness on nisin release from antimicrobial whey protein isolate edible films, *J. Sci. Food Agric.*, 89: 2492-2497.

Rostami, H., S. Abbaszadeh and S. Shokri (2017). Combined effects of lactoperoxidase system-whey protein coating and modified atmosphere packaging on the microbiological, chemical and sensory attributes of Pike-Perch fillets, *J. Food Sci. Tech. Mys.*, 54: 3243-3250.

Roy, S. and J.W. Rhim (2020a). Preparation of carbohydrate-based functional composite films incorporated with curcumin, *Food Hydrocoll.*, 98: 105302.

Roy, S. and J.W. Rhim (2020b). Anthocyanin food colorant and its application in pH-responsive color change indicator films, *Crit. Rev. Food Sci. Nutr.*, 61: 2297-2325.

Royo, M., I. Fernández-Pan and J.I. Maté (2010). Antimicrobial effectiveness of oregano and sage essential oils incorporated into whey protein films or cellulose-based filter paper, *J. Sci. Food Agric.*, 90: 1513-1519.

Ruan, C., Y. Zhang, J. Wang, Y. Sun, X. Gao, G. Xiong and J. Liang (2019a). Preparation and antioxidant activity of sodium alginate and carboxymethyl cellulose edible films with epigallocatechin gallate, *Int. J. Biol. Macromol.*, 134: 1038-1044.

Ruan, C., Y. Zhang, Y. Sun, X. Gao, G. Xiong and J. Liang (2019b). Effect of sodium alginate and carboxymethyl cellulose edible coating with epigallocatechin gallate on quality and shelf-life of fresh pork, *Int. J. Biol. Macromol.*, 141: 178-184.

Rubini, K., E. Boanini, A. Menichetti, F. Bonvicini, G.A. Gentilomi, M. Montalti and A. Bigi (2020). Quercetin loaded gelatin films with modulated release and tailored anti-oxidant, mechanical and swelling properties, *Food Hydrocoll.*, 109: 106089.

Saberi, B., Q.V. Vuong, S. Chockchaisawasdee, J.B. Golding, C.J. Scarlett and C.E. Stathopoulos (2017). Physical, barrier, and antioxidant properties of pea starch-guar gum biocomposite edible films by incorporation of natural plant extracts, *Food Bioproc. Tech.*, 10: 2240-2250.

Sanchez, L., M. Calvo and J.H. Brock (1992). Biological role of lactoferrin, *Arch. Dis. Child.*, 67: 657-661.

Sánchez-Maldonado, A.F., A. Schieber and M.G. Gänzle (2011). Structure-function relationships of the antibacterial activity of phenolic acids and their metabolism by lactic acid bacteria, *J. Appl. Microbiol.*, 111: 1176-1184.

Sánchez-González, L., M. Vargas, C. González-Martínez, A. Chiralt and M. Chafer (2011). Use of essential oils in bioactive edible coatings: A review, *Food Eng. Rev.*, 3: 1-16.

Sandhya, C., L.K. Adapa, K.M. Nampoothiri, P. Binod, G. Szakacs and A. Pandey (2004). Extracellular chitinase production by Trichoderma harzianum in submerged fermentation, *J. Basic Microbiol.*, 44: 49-58.

Sanjurjo, K., S. Flores, L. Gerschenson and R. Jagus (2006). Study of the performance of nisin supported in edible films, *Food Res. Int.*, 39: 749-754.

Santiago-Silva, P., N.F. Soares, J.E. Nóbrega, M.A. Júnior, K.B. Barbosa, A.C.P. Volp *et al.* (2009). Antimicrobial efficiency of film incorporated with pediocin (ALTA® 2351) on preservation of sliced ham, *Food Control*, 20: 85-89.

Santos, J.C., R.C. Sousa, C.G. Otoni, A.R. Moraes, V.G. Souza, E.A.A. Medeiros *et al.* (2018). Nisin and other antimicrobial peptides: Production, mechanisms of action, and application in active food packaging, *Innov. Food Sci. Emerg.*, 48: 179-194.

Satué-Gracia, M.T., E.N. Frankel, N. Rangavajhyala and J.B. German (2000). Lactoferrin in infant formulas: Efect on oxidation, *J. Agric. Food Chem.*, 48: 4984-4990.

Schneider, N., K. Werkmeister and M. Pischetsrieder (2011). Analysis of nisin A, nisin Z and their degradation products by LCMS/MS, *Food Chem.*, 127: 847-854.

Scott, V.N. and S.L. Taylor (1981). Effect of nisin on the outgrowth of *Clostridium botulinum* spores, *J. Food Sci.*, 46: 117-126.

Seifu, E., E.M. Buys and E.F. Donkin (2004). Quality aspects of gouda cheese made from goat milk preserved by the lactoperoxidase system, *Int. Dairy J.*, 14: 581-589.

Seifu, E., E.M. Buys and E.F. Donkin (2005). Significance of the lactoperoxidase system in the dairy industry and its potential applications: A review, *Trends Food Sci. Technol.*, 16: 137-154.

Sessini, V., M.P. Arrieta, J.M. Kenny and L. Peponi (2016). Processing of edible films based on nanoreinforced gelatinized starch, *Polym. Degrad. Stab.*, 132: 157-168.

Seydim, A.C. and G. Sarikus (2006). Antimicrobial activity of whey protein based edible films incorporated with oregano, rosemary and garlic essential oils, *Food Res. Int.*, 39: 639-644.

Shahbazi, Y. (2017). The properties of chitosan and gelatin films incorporated with ethanolic red grape seed extract and *Ziziphora clinopodioides* essential oil as biodegradable materials for active food packaging, *Int. J. Biol. Macromol.*, 99: 746-753.

Shahidi, F. and A. Hossain (2020). Preservation of aquatic food using edible films and coatings containing essential oils: A review, *Crit. Rev. Food Sci. Nutr.*, 62: 66-105.

Shan, B., Y.Z. Cai, M. Sun and H. Corke (2005). Antioxidant capacity of 26 spice extracts and characterization of their phenolic constituents, *J. Agric. Food Chem.*, 53: 7749-7759.

Shankar, S. and J.W. Rhim (2018). Antimicrobial wrapping paper coated with a ternary blend of carbohydrates (alginate, carboxymethyl cellulose, carrageenan) and grapefruit seed extract, *Carbohydr. Polym.*, 196: 92-101.

Shen, X., X. Sun, Q. Xie, H. Liu, Y. Zhao, Y. Pan *et al.* (2014). Antimicrobial effect of blueberry (*Vaccinium corymbosum* L.) extracts against the growth of *Listeria monocytogenes* and *Salmonella enteritidis*, *Food Control*, 35: 159-165.

Shin, Y.O., E. Rodil and J.H. Vera (2003). Selective preparation of lysozyme from egg white using AOT, *J. Food Sci.*, 68: 595-599.

Shivalee, A., K. Lingappa and D. Mahesh (2018). Influence of bioprocess variables on the production of extracellular chitinase under submerged fermentation by *Streptomyces pratensis* strain KLSL55, *J. Genet. Eng. Biotechnol.*, 16: 421-426.

Shokri, S., A. Ehsani and M.S. Jasour (2015). Efficacy of lactoperoxidase system-whey protein coating on shelf-life extension of rainbow trout fillets during cold storage (4 C), *Food Bioprocess Tech.*, 8: 54-62.

Shokri, S. and A. Ehsani (2017). Efficacy of whey protein coating incorporated with lactoperoxidase and α-tocopherol in shelf-life extension of Pike-Perch fillets during refrigeration, *LWT-Food Sci. Technol.*, 85: 225-231.

Silva, B.D.S., C.J. Ulhoa, K.A. Batista, F. Yamashita and K.F. Fernandes (2011). Potential fungal inhibition by immobilized hydrolytic enzymes from *Trichoderma asperellum*, *J. Agric. Food Chem.*, 59: 8148-8154.

Silva, B.D.S., C.J. Ulhoa, K.A. Batista, M.C. Di Medeiros, R.R. da Silva Filho, F. Yamashita and K.F. Fernandes (2012). Biodegradable and bioactive CGP/PVA film for fungal growth inhibition, *Carbohydr. Polym.*, 89: 964-970.

Silva-Weiss, A., M. Quilaqueo, O. Venegas, M. Ahumada, W. Silva, F. Osorio and B. Giménez (2018). Design of dipalmitoyl lecithin liposomes loaded with quercetin and rutin and their release kinetics from carboxymethyl cellulose edible films, *J. Food Eng.*, 224: 165-173.

Şimşek, Ş. and A. Yemenicioğlu (2007). Partial purification and kinetic characterization of mushroom stem polyphenoloxidase and determination of its storage stability in different lyophilized forms, *Process Biochemistry*, 42: 943-950.

Singla, R.K., A.K. Dubey, A. Garg, R.K. Sharma, M. Fiorino, S.M. Ameen *et al.* (2019). Natural polyphenols: Chemical classification, definition of classes, subcategories, and structures, *J. AOAC Int.*, 102: 1397-1400.

Siquet, C., F. Paiva-Martins, J.L. Lima, S. Reis and F. Borges (2006). Antioxidant profile of dihydroxy- and trihydroxyphenolic acids – A structure-activity relationship study, *Free Radic. Res.*, 40: 433-442.

Sivakumar, D. and S. Bautista-Baños (2014). A review on the use of essential oils for postharvest decay control and maintenance of fruit quality during storage, *Crop Prot.*, 64: 27-37.

Sivarooban, T., N.S. Hettiarachchy and M.G. Johnson (2008). Physical and antimicrobial properties of grape seed extract, nisin, and EDTA incorporated soy protein edible films, *Food Res. Int.*, 41: 781-785.

Smeriglio, A., D. Barreca, E. Bellocco and D. Trombetta (2017). Proanthocyanidins and hydrolysable tannins: Occurrence, dietary intake and pharmacological effects, *Br. J. Pharmacol.*, 174: 1244-1262.

Smitha, G.R. and V.S. Rana (2015). Variations in essential oil yield, geraniol and geranyl acetate contents in palmarosa (*Cymbopogon martinii,* Roxb. Wats. var. motia) influenced by inflorescence development, *Ind. Crops Prod.*, 66: 150-160.

Soleimanifar, M., S.M. Jafari and E. Assadpour (2020). Encapsulation of olive leaf phenolics within electrosprayed whey protein nanoparticles; production and characterization, *Food Hydrocoll.*, 101: 105572.

Somers, E.B. and S.L. Taylor (1987). Antibotulinal effectiveness of nisin in pasteurized process cheese spreads, *J. Food Prot.*, 50: 842-848.

Sonu, K.S., B. Mann, R. Sharma, R. Kumar and R. Singh (2018). Physico-chemical and antimicrobial properties of d-limonene oil nanoemulsion stabilized by whey protein–maltodextrin conjugates, *J. Food Sci. Tech.*, 55: 2749-2757.

Souza, M.P., A.F. Vaz, H.D. Silva, M.A. Cerqueira, A.A. Vicente and M.G. Carneiro-da-Cunha (2015). Development and characterization of an active chitosan-based film containing quercetin, *Food Bioprocess Tech.*, 8: 2183-2191.

Souza, M.P., A.F. Vaz, T.B. Costa, M.A. Cerqueira, C.M. De Castro, A.A. Vicente and M.G. Carneiro-da-Cunha (2018a). Construction of a biocompatible and antioxidant multilayer coating by layer-by-layer assembly of κ-carrageenan and quercetin nanoparticles, *Food Bioprocess Tech.*, 11: 1050-1060.

Souza, V.G., J.R. Pires, É.T. Vieira, I.M. Coelhoso, M.P. Duarte and A.L. Fernando (2018b). Shelf-Life assessment of fresh poultry meat packaged in novel bionanocomposite of chitosan/montmorillonite incorporated with ginger essential oil, *Coatings*, 8: 177.

Sozbilen, G.S., F. Korel and A. Yemenicioğlu (2018). Control of lactic acid bacteria in fermented beverages using lysozyme and nisin: Test of traditional beverage boza as a model food system, *Int. J. Food Sci.Tech.*, 53: 2357-2368.

Sozbilen, G.S. and A. Yemenicioğlu (2020). Decontamination of seeds destined for edible sprout production from Listeria by using chitosan coating with synergetic lysozyme-nisin mixture, *Carbohydr. Polym.*, 235: 115968.

Spizzirri, U.G., O. Parisi, I.F. Iemma, G. Cirillo, F. Puoci, M. Curcio and N. Picci (2010). Antioxidant-polysaccharide conjugates for food application by eco-friendly grafting procedure, *Carbohydr. Polym.*, 79: 333-340.

Stergiou, V.A., L.V. Thomas and M.R. Adams (2006). Interactions of nisin with glutathione in a model protein system and meat, *J. Food Prot.*, 69: 951-956.

Stoll, L., T.M.H. Costa, A. Jablonski, S.H. Flôres and A. de Oliveira Rios (2016). Microencapsulation of anthocyanins with different wall materials and its application in active biodegradable films, *Food Bioprocess Tech.*, 9: 172-181.

Su, C. and B.H. Chiang (2006). Partitioning and purification of lysozyme from chicken egg white using aqueous two-phase system, *Process. Biochem.*, 41: 257-263.

Sudagidan, M. and A. Yemenicioğlu (2012). Effects of nisin and lysozyme on growth inhibition and biofilm formation capacity of *Staphylococcus aureus* strains isolated from raw milk and cheese samples, *J. Food Prot.*, 75: 1627-1633.

Sudheeran, P.K., R. Ovadia, O. Galsarker, I. Maoz, N. Sela, D. Maurer *et al.* (2020). Glycosylated flavonoids: Fruit's concealed antifungal arsenal, *New Phytol.*, 225: 1788-1798.

Sun, H., X. Shao, M. Zhang, Z. Wang, J. Dong and D. Yu (2019). Mechanical, barrier and antimicrobial properties of corn distarch phosphate/nanocrystalline cellulose films incorporated with nisin and ε-polylysine, *Int. J. Biol. Macromol.*, 136: 839-846.

Sun, J., H. Jiang, M. Li, Y. Lu, Y. Du, C. Tong *et al.* (2020). Preparation and characterization of multifunctional konjac glucomannan/carboxymethyl chitosan biocomposite films incorporated with epigallocatechin gallate, *Food Hydrocoll.*, 105: 105756.

Superti, F., M.G. Ammendolia, P. Valenti and L. Seganti (1997). Antirotaviral activity of milk proteins: Lactoferrin prevents rotavirus infection in the enterocyte-like cell line HT-29, *Med. Microbiol. Immunol.*, 186: 83-91.

Syed, I., P. Banerjee and P. Sarkar (2020). Oil-in-water emulsions of geraniol and carvacrol improve the antibacterial activity of these compounds on raw goat meat surface during extended storage at 4°C, *Food Control*, 107: 106757.

Taguri, T., T. Tanaka and I. Kouno (2006). Antibacterial spectrum of plant polyphenols and extracts depending upon hydroxyphenyl structure, *Biol. Pharm. Bull.*, 29: 2226-2235.

Takahashi, M., Y. Okakura, H. Takahashi, M. Imamura, A. Takeuchi, H. Shidara *et al.* (2018). Heat-denatured lysozyme could be a novel disinfectant for reducing hepatitis A virus and murine norovirus on berry fruit, *Int. J. Food Microbiol.*, 266: 104-108.

Takaichi, M. and K. Oeda (2000). Transgenic carrots with enhanced resistance against two major pathogens, *Erysiphe heraclei* and *Alternaria dauci*, *Plant Sci.*, 153: 135-144.

Tammineni, N., G. Ünlü, B. Rasco, J. Powers, S. Sablani and C. Nindo (2012). Trout-skin gelatin-based edible films containing phenolic antioxidants: Effect on physical properties and oxidative stability of cod-liver oil model food, *J. Food Science*, 77: E342-E347.

Tang, Q., D. Pan, Y. Sun, J. Cao and Y. Guo (2017). Preparation, characterization and antimicrobial activity of sodium alginate nanobiocomposite films incorporated with ε-polylysine and cellulose nanocrystals, *J. Food Process. Preserv.*, 41: e13120.

Teerakarn, A., D.E. Hirt, J.C. Acton, J.R. Rieck and P.L. Dawson (2002). Nisin diffusion in protein films: Effects of film type and temperature, *J. Food Sci.*, 67: 3019-3025.

Tenore, G.C., A. Basile and E. Novellino (2011). Antioxidant and antimicrobial properties of polyphenolic fractions from selected Moroccan red wines, *J. Food Sci.*, 76: C1342-C1348.

Theivendran, S., N.S. Hettiarachchy and M.G. Johnson (2006). Inhibition of *Listeria monocytogenes* by nisin combined with grape seed extract or green tea extract in soy protein film coated on turkey frankfurters, *J. Food Sci.*, 71: M39-M44.

Tomita, M., H. Wakabayashi, K. Shin, K. Yamauchi, T. Yaeshima and K. Iwatsuki (2009). Twenty-five years of research on bovine lactoferrin applications, *Biochimie*, 91: 52-57.

Torres, E., V. Marín, J. Aburto, H.I. Beltrán, K. Shirai, S. Villanueva and G. Sandoval (2012). Enzymatic modification of chitosan with quercetin and its application as antioxidant edible films, *Appl. Biochem. Microbiol.*, 48: 151-158.

Touch, V., S. Hayakawa, S. Yamada and S. Kaneko (2004). Effects of a lactoperoxidase-thiocyanate-hydrogen peroxide system on *Salmonella enteritidis* in animal or vegetable foods, *Int. J. Food Microbiol.*, 93: 175-183.

Tracz, B.L., K. Bordin, K.C.P. Bocate, R.V. Hara, C. Luz, R.E.F. Macedo *et al.* (2018). Devices containing allyl isothiocyanate against the growth of spoilage and mycotoxigenic fungi in mozzarella cheese, *J. Food Process Preserv.*, 42: e13779.

Trinetta, V., J.D. Floros and C.N. Cutter (2010). Sakacin a-containing pullulan film: An active packaging system to control epidemic clones of *Listeria monocytogenes* in ready-to-eat foods, *J. Food Saf.*, 30: 366-381.

Trinetta, V., C.N. Cutter and J.D. Floros (2011). Effects of ingredient composition on optical and mechanical properties of pullulan film for food-packaging applications, *LWT-Food Sci. Technol.*, 44: 2296-2301.

Tripathi, V. and B. Vashishtha (2006). Bioactive compounds of colostrum and its application, *Food Rev. Int.*, 22: 225-244.

Tsuda, H., T. Kozu, G. Iinuma, Y. Ohashi, Y. Saito, D. Saito *et al.* (2010). Cancer prevention by bovine lactoferrin: From animal studies to human trial, *Biometals*, 23: 399-409.

Turner, C., P. Turner, G. Jacobson, K. Almgren, M. Waldebäck, P. Sjöberg *et al.* (2006). Subcritical water extraction and β-glucosidase-catalyzed hydrolysis of quercetin glycosides in onion waste, *Green Chem.*, 8: 949-959.

Ulber, R., K. Plate, T. Weiss, W. Demmer, H. Buchholz and T. Scheper (2001). Downstream processing of bovine lactoferrin from sweet whey, *Acta Biotechnol.*, 21: 27-34.

Unachukwu, U.J., S. Ahmed, A. Kavalier, J.T. Lyles and E.J. Kennelly (2010). White and green teas (*Camellia sinensis* var. *sinensis*): Variation in phenolic, methylxanthine, and antioxidant profiles, *J. Food Sci.*, 75: C541-C548.

Unalan, İ.U., F. Korel and A. Yemenicioğlu (2011a). Active packaging of ground beef patties by edible zein films incorporated with partially purified lysozyme and Na$_2$EDTA, *Int. J. Food Sci. Technol.*, 46: 1289-1295.

Ünalan, İ.U., K.D.A. Ucar, I. Arcan, F. Korel and A. Yemenicioğlu (2011b). Antimicrobial potential of polylysine in edible films, *Food Sci. Technol. Res.*, 17: 375-380.

Ünalan, İ.U., I. Arcan, F. Korel and A. Yemenicioğlu (2013). Application of active zein-based films with controlled release properties to control *Listeria monocytogenes* growth and lipid oxidation in fresh Kashar cheese, *Innov. Food Sci. Emerg.*, 20: 208-214.

Unusan, N. (2020). Proanthocyanidins in grape seeds: An updated review of their health benefits and potential uses in the food industry, *J. Funct. Foods*, 67: 103861.

Uranga, J., A. Etxabide, P. Guerrero and K. de la Caba (2018). Development of active fish gelatin films with anthocyanins by compression molding, *Food Hydrocoll.*, 84: 313-320.

Vámos-Vigyázó, L. and N.F. Haard (1981). Polyphenol oxidases and peroxidases in fruits and vegetables, *Crit. Rev. Food Sci. Nutr.*, 15: 49-127.

Viñas, P., M. Bravo-Bravo, I. López-García, M. Pastor-Belda and M. Hernández-Córdoba (2014). Pressurized liquid extraction and dispersive liquid-liquid microextraction for determination of tocopherols and tocotrienols in plant foods by liquid chromatography with fluorescence and atmospheric pressure chemical ionization-mass spectrometry detection, *Talanta*, 119: 98-104.

Vukomanović, M., V. Žunič, Š. Kunej, B. Jančar, S. Jeverica and D. Suvorov (2017). Nano-engineering the antimicrobial spectrum of lantibiotics: Activity of nisin against gram negative bacteria, *Sci. Rep.*, 7: 1-13.

Wagle, B.R., S. Shrestha, K. Arsi, I. Upadhyaya, A.M. Donoghue and D.J. Donoghue (2019). Pectin or chitosan coating fortified with eugenol reduces *Campylobacter jejuni* on chicken wingettes and modulates expression of critical survival genes, *Poult. Sci.*, 98: 1461-1471.

Wakabayashi, H., K. Yamauchi and M. Takase (2006). Lactoferrin research, technology and applications, *Int. Dairy J.*, 16: 1241-1251.

Wang, S., M.F. Marcone, S. Barbut and L.T. Lim (2012a). Fortification of dietary hydrocolloids-based packaging material with bioactive plant extracts, *Food Res. Int.*, 49: 80-91.

Wang, S., M. Marcone, S. Barbut and L.T. Lim (2012b). The impact of anthocyanin-rich red raspberry extract (ARRE) on the properties of edible soy protein isolate (SPI) films, *J. Food Sci.*, 77: C497-C505.

Wang, Y., F. Liu, C. Liang, F. Yuan and Y. Gao (2014). Effect of Maillard reaction products on the physical and antimicrobial properties of edible films based on ε-polylysine and chitosan, *J. Sci. Food Agric.*, 94: 2986-2991.

Wang, H., L. Hao, P. Wang, M. Chen, S. Jiang and S. Jiang (2017). Release kinetics and antibacterial activity of curcumin loaded zein fibers, *Food Hydrocoll.*, 63: 437-446.

Wang, X., Y. Xie, H. Ge, L. Chen, J. Wang, S. Zhang *et al.* (2018). Physical properties and antioxidant capacity of chitosan/epigallocatechin-3-gallate films reinforced with nano-bacterial cellulose, *Carbohydr. Polym.*, 179: 207-220.

Wang, Q., J. Cao, H. Yu, J. Zhang, Y. Yuan, X. Shen and C. Li (2019a). The effects of EGCG on the mechanical, bioactivities, cross-linking and release properties of gelatin film, *Food Chem.*, 271: 204-210.

Wang, L., J. Xue and Y. Zhang (2019b). Preparation and characterization of curcumin loaded caseinate/zein nanocomposite film using pH-driven method, *Ind. Crops Prod.*, 130: 71-80.

Wang, L., R.J. Mu, Y. Li, L. Lin, Z. Lin and J. Pang (2019c). Characterization and antibacterial activity evaluation of curcumin loaded konjac glucomannan and zein nanofibril films, *LWT-Food Sci. Technol.*, 113: 108293.

Wang, S., P. Xia, S. Wang, J. Liang, Y. Sun, P. Yue and X. Gao (2019d). Packaging films formulated with gelatin and anthocyanins nanocomplexes: Physical properties, antioxidant activity and its application for olive oil protection, *Food Hydrocoll.*, 96: 617-624.

Ward, P.P., X. Zhou and O.M. Conneely (1996). Cooperative interactions between the amino- and carboxyl-terminal lobes contribute to the unique iron-binding stability of lactoferrin, *J. Biol. Chem.*, 27: 12790-12794.

Waterhouse, A.L., S. Ignelzi and J.R. Shirley (2000). A comparison of methods for quantifying oligomeric proanthocyanidins from grape seed extracts, *Am. J. Enol. Vitic.*, 51: 383-389.

Wei, L., C. Cai, J. Lin, L. Wang and X. Zhang (2011). Degradation controllable biomaterials constructed from lysozyme-loaded Ca-alginate microparticle/chitosan composites, *Polymer*, 52: 5139-5148.

Weng, W.Y., Z. Tao, G.M. Liu, W.J. Su, K. Osako, M. Tanaka and M.J. Cao (2014). Mechanical, barrier, optical properties and antimicrobial activity of edible films prepared from silver carp surimi incorporated with ε-polylysine, *Packaging Technol. Sci.*, 27: 37-47.

Wilcox, C.P., A.K. Weissinger, R.C. Long, L.C. Fitzmaurice, T.E. Mirkov and H.E. Swaisgood (1997). Production and purification of an active bovine lysozyme in tobacco (*Nicotiana tabacum*): Utilization of value-added crop plants traditionally grown under intensive agriculture, *J. Agric. Food Chem.*, 45: 2793-2797.

Woraprayote, W., Y. Kingcha, P. Amonphanpokin, J. Kruenate, T. Zendo, K. Sonomoto *et al.* (2013). Anti-listeria activity of poly (lactic acid)/sawdust particle biocomposite film impregnated with pediocin PA-1/AcH and its use in raw sliced pork, *Int. J. Food Microbiol.*, 167: 229-235.

Wu, Y., C.L. Weller, F. Hamouz, S. Cuppett and M. Schnepf (2001). Moisture loss and lipid oxidation for precooked ground-beef patties packaged in edible starch-alginate-based composite films, *J. Food Sci*, 66: 486-493.

Wu, J., S. Ge, H. Liu, S. Wang, S. Chen, J. Wang *et al.* (2014). Properties and antimicrobial activity of silver carp (*Hypophthalmichthys molitrix*) skin gelatin-chitosan films incorporated with oregano essential oil for fish preservation, *Food Packag. Shelf-Life*, 2: 7-16.

Wu, T., Q. Jiang, D. Wu, Y. Hu, S. Chen, T. Ding *et al.* (2019a). What is new in lysozyme research and its application in food industry? A review, *Food Chem.*, 274: 698-709.

Wu, C., J. Sun, Y. Lu, T. Wu, J. Pang and Y. Hu (2019b). *In situ* self-assembly chitosan/ε-polylysine bionanocomposite film with enhanced antimicrobial properties for food packaging, *Int. J. Biol. Macromol.*, 132: 385-392.

Wu, C., Y. Li, J. Sun, Y. Lu, C. Tong, L. Wang *et al.* (2020). Novel konjac glucomannan films with oxidized chitin nanocrystals immobilized red cabbage anthocyanins for intelligent food packaging, *Food Hydrocoll.*, 98: 105245.

Xi, W., G. Zhang, D. Jiang and Z. Zhou (2015). Phenolic compositions and antioxidant activities of grapefruit (*Citrus paradisi Macfadyen*) varieties cultivated in China, *Int. J. Food Sci. Nutr.*, 66: 858-866.

Xu, X., B. Li, J.F. Kennedy, B.J. Xie and M. Huang (2007). Characterization of konjac glucomannan–gellan gum blend films and their suitability for release of nisin incorporated therein, *Carbohydr. Polym.*, 70: 192-197.

Xu, J., R. Wei, Z. Jia and R. Song (2020). Characteristics and bioactive functions of chitosan/gelatin-based film incorporated with ε-polylysine and astaxanthin extracts derived from by-products of shrimp (*Litopenaeus vannamei*), *Food Hydrocoll.*, 100: 105436.

Yadav, D. and N. Kumar (2014). Nanonization of curcumin by antisolvent precipitation: Process development, characterization, freeze drying and stability performance, *Int. J Pharm.*, 477: 564-577.

Yadav, S., G.K. Mehrotra, P. Bhartiya, A. Singh and P.K. Dutta (2020). Preparation, physicochemical and biological evaluation of quercetin based chitosan-gelatin film for food packaging, *Carbohydr. Polym.*, 227: 115348.

Yan, Z., Y. Zhong, Y. Duan, Q. Chen and F. Li (2020). Antioxidant mechanism of tea polyphenols and its impact on health benefits, *Anim. Nutr.*, 6: 115-123.

Yang, D., F.R. Dunshea and H.A. Suleria (2020). LC-ESI-QTOF/MS characterization of Australian herb and spices (garlic, ginger, and onion) and potential antioxidant activity, *J. Food Process. Preserv.*, 44: e14497.

Yao, Y., D. Ding, H. Shao, Q. Peng and Y. Huang (2017). Antibacterial activity and physical properties of fish gelatin-chitosan edible films supplemented with D-limonene, *Int. J. Polym. Sci.*, 2017: 1837171.

Yeddes, W., M. Nowacka, K. Rybak, I. Younes, M. Hammami, M. Saidani-Tounsi and D. Witrowa-Rajchert (2019). Evaluation of the antioxidant and antimicrobial activity of rosemary essential oils as gelatin edible film component, *Food Sci. Technol. Res.*, 25: 321-329.

Yeddes, W., K. Djebali, W.A. Wannes, K. Horchani-Naifer, M. Hammami, I. Younes and M.S. Tounsi (2020). Gelatin-chitosan-pectin films incorporated with rosemary essential oil: Optimized formulation using mixture design and response surface methodology, *Int. J. Biol. Macromol.*, 154: 92-103.

Yemenicioğlu, A. (2016a). Zein and its composites and blends with natural active compounds: Development of antimicrobial films for food packaging, pp. 503-513. *In:* J. Barros-Velazquez (Ed.). *Antimicrobial Food Packaging.* Academic Press, London, UK.

Yemenicioglu, A. (2016b). Strategies for controlling major enzymatic reactions in fresh and processed vegetables, pp. 377-388. *In:* Y.H. Hui and Ö. Evranuz (Eds.). *Handbook of Vegetable Preservation and Processing,* 2nd ed. CRC Press, New York, USA.

Yener, F.Y., F. Korel and A. Yemenicioğlu (2009). Antimicrobial activity of lactoperoxidase system incorporated into cross-linked alginate films, *J. Food Sci.*, 74: M73-M79.

Yıldırım, E. and I. Barutçu Mazı (2017). Effect of zein coating enriched by addition of functional constituents on the lipid oxidation of roasted hazelnuts, *J. Food Process Eng.*, 40: e12515.

Yilmaz, Y. (2006). Novel uses of catechins in foods, *Trends Food Sci. Technol.*, 17: 64-71.

Yoneyama, F., M. Fukao, T. Zendo, J. Nakayama and K. Sonomoto (2008). Biosynthetic characterization and biochemical features of the third natural nisin variant, nisin Q, produced by *Lactococcus lactis* 61-14, *J. Appl. Microbiol.*, 105: 1982-1990.

Yordi, E.G., E.M. Pérez, M.J. Matos and E.U. Villares (2012). Antioxidant and pro-oxidant effects of polyphenolic compounds and structure-activity relationship evidence, pp. 24-48. *In:* J. Bouayed and T. Bohn (Eds.). *Nutrition, Well-being and Health.* IntechOpen. Rijeca, HR.

Yoshida, S. (1991). Isolation of lactoperoxidase and lactoferrins from bovine milk acid whey by carboxymethyl cation exchange chromatography, *J. Dairy Sci.*, 74: 1439-1444.

Yoshida, T. and T. Nagasawa (2003). ε-Poly-L-lysine: microbial production, biodegradation and application potential, *Appl. Microbiol. Biotechnol.*, 62: 21-26.

Zdarta, J., A.S. Meyer, T. Jesionowski and M. Pinelo (2018). Developments in support materials for immobilization of oxidoreductases: A comprehensive review, *Adv. Colloid Interfac.*, 258: 1-20.

Zhang, Y.M. and C.O. Rock (2004). Evaluation of epigallocatechin gallate and related plant polyphenols as inhibitors of the FabG and FabI reductases of bacterial type II fatty-acid synthase, *J. Biol. Chem.*, 279: 30994-31001.

Zhang, T., P. Zhou, Y. Zhan, X. Shi, J. Lin, Y. Du *et al.* (2015). Pectin/lysozyme bilayers layer-by-layer deposited cellulose nanofibrous mats for antibacterial application, *Carbohydr. Polym.*, 117: 687-693.

Zhang, J., X. Zou, X. Zhai, X. Huang, C. Jiang and M. Holmes (2019). Preparation of an intelligent pH film based on biodegradable polymers and roselle anthocyanins for monitoring pork freshness, *Food Chem.*, 272: 306-312.

Zhang, N., F. Bi, F. Xu, H. Yong, Y. Bao, C. Jin and J. Liu (2020a). Structure and functional properties of active packaging films prepared by incorporating different flavonols into chitosan based matrix, *Int. J. Biol. Macromol.*, 165: 625-634.

Zhang, L., Z. Liu, Y. Sun, X. Wang and L. Li (2020b). Effect of α-tocopherol antioxidant on rheological and physicochemical properties of chitosan/zein edible films, *LWT-Food Sci. Technol.*, 118: 108799.

Zhou, B., Y. Li, H. Deng, Y. Hu and B. Li (2014). Antibacterial multilayer films fabricated by layer-by-layer immobilizing lysozyme and gold nanoparticles on nanofibers, *Colloid. Surface B*, 116: 432-438.

Zivanovic, S., S. Chi and A.F. Draughon (2005). Antimicrobial activity of chitosan films enriched with essential oils, *J. Food Sci.*, 70: M45-M51.

CHAPTER

5

Strategies and Methods of Enhancing the Performance of Active Edible Packaging

5.1 Introduction

The development of active edible packaging effective in food applications is not possible without a detailed knowledge about design strategies and methods used in current packaging concepts (e.g. antimicrobial, antioxidant, bioactive or flavor-release packaging). To increase the shelf-life of food, most active edible packaging materials are designed to show antimicrobial activity and/or to inhibit lipid oxidation in the food. Some active edible packaging materials prevent browning or discoloration of food by inhibiting oxidative enzymes or chelating metal atoms. In some other cases, the active edible packaging delivers texture enhancers, flavoring substances, bioactive polyphenols, probiotic and/or prebiotics, protective cultures or vitamins on to the food surface (Lopez-Rubio *et al.*, 2006; Rojas-Graü *et al.*, 2009; Yemenicioğlu *et al.*, 2020). The design of packaging materials helps to choose the most suitable active components for the target food, and to determine how long the active components will maintain their functionality in the food system without losing their active properties. In general, active edible packaging starts delivering active agents immediately after it contacts the food surface, but packaging incorporated with volatile active agents might show its active properties without contacting the food surface. In advanced packaging materials the rates of delivering active agents on to the food surface or at food headspace could be controlled (controlled release) depending on the shelf-life of food. Moreover, the initiation of delivery (timing) could also be controlled by designing triggering mechanisms for active packaging (Boyacı *et al.*, 2016). There are also different methods to enhance the active properties of edible packaging, such as application of synergetic mixtures, nanoencapsulation, coencapsulation, and combinational preservation methods (hurdles). This chapter helps to understand major strategies and methods used to enhance the performance of different packaging concepts during food applications.

5.2 Enhancing the performance of antimicrobial packaging

The performance of antimicrobial packaging could be enhanced by choosing effective natural antimicrobials against targeted pathogenic and/or spoilage microorganisms, applying synergistic mixtures of antimicrobials and controlled release strategies, and using supporting combinational preservation methods.

5.2.1 Choosing suitable natural antimicrobials

The primary objective of most antimicrobial edible packaging is not the substantial reduction of total microbial load of food as in chemical disinfection or thermal processing since none of the natural antimicrobials or their mixtures show such a potency or broad antimicrobial spectrum. Instead, the antimicrobial edible packaging mostly targets the inactivation of most critical pathogenic and/or spoilage microorganisms in the food. The incorporation of bacteriophages (e.g. *Salmonella*, *Listeria* or *E. coli* phages) into edible films also initiated the trend of achieving selective inhibition of a single type of critical pathogen in food (Cui *et al.*, 2017a; Radford *et al.*, 2017). For example, antimicrobial chitosan films incorporated by liposome-encapsulated bacteriophage and applied against *E. coli* O157:H7 in beef is a good example about the selective inactivation of a single pathogen in food (Cui *et al.*, 2017a). The most important advantage of a selective inactivation of critical pathogen is that this prevents inactivation of desired microorganism in the food, such as probiotic bacteria, protective cultures, starter cultures, and native flora that contribute to the health benefits of food or formation of its main characteristics during ripening or fermentation (e.g. starter cultures or native flora in ripening cheese or fermented sausages). A less specific example for targeting a critical pathogen in antimicrobial packaging is the use of enzyme lysozyme against *Listeria monocytogenes*. The lysozyme owes its popularity to its high stability in packaging materials and in food during storage as well as lack of negative effects on aroma and taste of food. However, the major limitation of lysozyme as an antimicrobial compound is lack of its antimicrobial effect on Gram-negative bacteria. Moreover, Gram-positive bacteria, such as *Staphylococcus aureus* could show a great resistance against lysozyme (Sudagidan and Yemenicioglu, 2012). Therefore, the lysozyme is frequently combined with other antimicrobials, such as phenolic compounds, chelating agents, or inherently antimicrobial polymers like chitosan to support its antimicrobial activity. In contrast, the use of an essential oil in packaging might provide a more potent antimicrobial effect and a broader antimicrobial spectrum, but most essential oils cause substantial modifications in mechanical and barrier properties of edible films (Alkan and Yemenicioğlu, 2016) and show undesired effects on aroma and taste of food (Gutierrez *et al.*, 2009). Therefore, a combination of essential oils with other antimicrobials or preservation methods (e.g. modified atmosphere packaging) is frequently applied to minimize their concentration in antimicrobial packaging. Thus, it is clear that the potency and antimicrobial spectrum are not the only parameters during choosing of antimicrobial agents in edible packaging. A suitable antimicrobial agent is the one that shows the following properties: (1) antimicrobial activity against critical pathogens and/ or spoilage microorganism in the food, (2) stability in edible packaging during its manufacturing and storage, and in packaged food during storage following its release, (3) minimum effects on desired properties (e.g. homogeneity, mechanical and barrier properties) of edible packaging and on sensory attributes of food, and (4) ability to show synergetic interactions with other antimicrobials and/or preservation methods.

Choosing suitable antimicrobial(s) needs some basic knowledge about specific microbiological risks of food (e.g. being the vector of a pathogen) and spoilage microorganisms associated with this food. In a food product that is a frequent vector for one or several critical pathogenic bacteria, the reduction or elimination of risks from these pathogens might help minimization of food poisoning risks considerably. Moreover, this also reduces recalls and great economic losses. The *E. coli* O157:H7 in beef burgers (Ansay *et al.*, 1999) and cured smoked sausages (Sartz *et al.*, 2008), *Campylobacter jejuni* in raw chicken meat (Alkan *et al.*, 2011), *S. enterica* serovar Typhimurium and *S. enterica* serovar Enteritidis in eggs (Whiley and Ross, 2015), *Listeria monocytogenes* in raw and cold smoked salmon (Hoffman *et al.*, 2003) and ready to eat salmon roe and minced tuna (Miya *et al.*, 2010), *L. monocytogenes* and *Salmonella* in cooked ham (Aymerich *et al.*, 2005) are some of the example target pathogenic bacteria in food that could be targeted by antimicrobial packaging. On the other hand, *Penicillium commune* in cheese (Kure *et al.*, 2001), *Penicillium* species in bread (Suhr and Nielsen, 2004), *Botrytis cinerea* in table grapes, strawberry, tomato (Zoffoli *et al.*, 1999; You *et al.*, 2016; Petrasch *et al.*, 2019), *Penicillium expansum* and *P. digitatum* in apples and oranges (Vilanova *et al.*, 2014), *Colletotrichum gloeosporioides* in mango, papaya and avocado (Ochoa-Velasco *et al.*, 2021), *Fusarium* spp. in potato tubers (Stefańczyk *et al.*, 2016), *Erwinia amylovora, E. carotovora, Xanthomonas vesicatoria*, and *P. syringae* in some fresh fruits and vegetables (Alkan and Yemenicioğlu, 2016) are only some of the example spoilage fungi and bacteria that might be targeted specifically by antimicrobial packaging to reduce major economic losses.

Examples of natural antimicrobial compounds tested in different packaging materials and found effective against six major pathogenic bacteria frequently targeted in antimicrobial packaging are listed in Table 5.1. It is clear that a majority of the natural antimicrobials are essential oils or their major active components. Carvacrol is one of the most effective natural antimicrobial essential oil components. Therefore, many reports exist related to antimicrobial activity of carvacrol-loaded packaging materials against food pathogens. However, this natural compound is classified as a food additive (flavoring agents and related substances) by the U.S. Food and Drug Administration (FDA, 2021a). Thus, for minimization of carvacrol used in edible films, it is important to combine this essential oil with other natural antimicrobials. In contrast, some other potent essential oil components, such as eugenol, limonene, cinnamaldehyde, geraniol, and citral are having a GRAS (Generally Recognized As Safe) status for food applications (FDA, 2021b; FDA, 2021c). Other essential oils highly effective on bacterial pathogens are *Zataria multiflora* Boiss, oregano, and tea tree essential oils. The use of lactoperoxidase system that is effective on both Gram-positive and Gram-negative bacteria, and combination of nisin with other natural antimicrobials (e.g. lysozyme, green tea, or grape seed phenolic extract, and cationic surfactant lauric arginate) against Gram-positive pathogenic bacteria are some effective strategies that could be employed in antimicrobial packaging.

Table 5.1: Some natural antimicrobials frequently used in edible packaging against major food pathogens

Pathogen	Natural antimicrobials[a,b]
E. coli O157:H7	Allyl isothiocyanate (Nadarajah *et al.*, 2005), carvacrol* (Rojas-Graü *et al.*, 2007), *Artemisia annua* EO (Cui *et al.*, 2017b), catechin (Ku *et al.*, 2008), *E. coli* bacteriophage (Cui *et al.*, 2017a), *Zataria multiflora* Boiss EO (ZEO)* (Moradi *et al.*, 2016), lysozyme-disodium EDTA (Ünalan *et al.*, 2011), lactoperoxidase system (Min *et al.*, 2005b), geraniol (Yegin *et al.*, 2016), pomegranate peel extract (Surendhiran *et al.*, 2020).
L. monocytogenes	Lysozyme-nisin (Sozbilen and Yemicioğlu, 2020), nisin-grape seed extract or nisin-green tea extract (Theivendran *et al.*, 2006), lactoperoxidase system (Min *et al.*, 2005a), lemon EO (Randazzo *et al.*, 2016), *Listeria* bacteriophage (Radford *et al.*, 2017), nisin-lauric arginate (Pattanayaiying *et al.*, 2015), oregano EO* (Pavli *et al.*, 2019), carvacrol* (Lopresti *et al.*, 2021), ZEO* (Moradi *et al.*, 2016), apricot kernel EO (Wang *et al.*, 2020a), grape seed extract (Jang *et al.*, 2011), *Lactobacillus plantarum* postbiotics (Yordshahi *et al.*, 2020), cinnamaldehyde* (Otoni *et al.*, 2014), tea tree EO* (Sánchez-González *et al.*,2011), eugenol (Guerreiro *et al.*, 2015).
S. aureus	Nisin (Millette *et al.*, 2007), nisin-lauric arginate (Pattanayaiying *et al.*, 2015), oregano EO* (Seydim and Sarikus, 2006), Ginger EO (Kavas *et al.*, 2016), Anise EO (Matan *et al.*, 2012), Garlic oil (Pranoto *et al.*, 2005), carvacrol* (Lopresti *et al.*, 2021), ZEO* (Shojaee-Aliabadi *et al.*, 2014), cinnamaldehyde* (Otoni *et al.*, 2014), tea tree EO* (Fernández-Pan *et al.*, 2012), bergamot EO (Sánchez-González *et al.*, 2011), eugenol* (Li *et al.*, 2021).
S. enterica serovar Typhimurium	Carvacrol* (Severino *et al.*, 2015), oregano EO* (Hashemi and Khaneghah, 2017), thyme EO (Hu *et al.*, 2019), *S. enterica* serovar Typhimurium bacteriophage (Radford *et al.*, 2017), ZEO* (Shojaee-Aliabadi *et al.*, 2014), geraniol (Yegin *et al.*, 2016), lactoperoxidase system (Lee and Min, 2013), tea tree EO* (Cui *et al.*, 2018), *Pediococcus acidilactici* postbiotics (İncili *et al.*, 2021), eugenol* (Guerreiro *et al.*, 2015).
S. enterica serovar Enteritidis	Oregano EO* (Ehivet *et al.*, 2011), citral (Prakash *et al.*, 2020), cinnamaldehyde* (Otoni *et al.*, 2014), cinnamon leaf EO (Raybaudi-Massilia *et al.*, 2008), tea tree EO* (Fernández-Pan *et al.*, 2012), carvacrol* (Wang *et al.*, 2020b).
C. jejuni	Cinnamaldehyde* (Mild *et al.*, 2011), gallic acid (Alkan *et al.*, 2011), allyl isothiocyanate (Olaimat *et al.*, 2014), citral (Giteru *et al.*, 2015), carvacrol* (Shrestha *et al.*, 2019), resveratrol (Silva *et al.*, 2016), eugenol* (Wagle *et al.*, 2019).

[a] Natural antimicrobials effective on minimum half of the given 6 pathogens were underlined. Those effective on minimum 4 of the 6 pathogens were also indicated with a star.
[b] EO: Essential oil

The examples of natural antimicrobials used effectively in packaging against spoilage fungi and bacteria are also listed in Table 5.2. The essential oils and their major components (alone or in combination) have also been used effectively in edible coatings against spoilage fungi and bacterial plant pathogens. The use of

Table 5.2: Some natural antimicrobials used in edible packaging against spoilage microorganisms

Spoilage fungi/bacteria	Natural antimicrobials[a]
Penicillium commune	Natamycin (Torrijos *et al.*, 2021), lactoperoxidase system (Min and Krochta, 2005c), cinnamon EO, clove EO (Souza *et al.*, 2013).
Colletotrichum gloeosporioides	Carvacrol-thymol mixture (Ochoa-Velasco *et al.*, 2021), Rue herb (*Ruta graveolens*) EO (Peralta-Ruiz *et al.*, 2020), cinnamaldehyde (Permana *et al.*, 2021), *Meyerozyma caribbica* (biocontrol yeast) (Iñiguez-Moreno *et al.*, 2020).
C. gloeosporioides, Rhizopus stolonifer	Thyme EO (Bosquez-Molina *et al.*, 2010), cinnamon-thyme EO combination (Monzón-Ortega *et al.*, 2018).
Botrytis cinerea	Thymol (Robledo *et al.*, 2018), grapefruit seed extract (Xu *et al.*, 2007), tea seed oil (Tran *et al.*, 2021), *Candida sake* (biocontrol yeast) (Marín *et al.*, 2016).
P. Italicum, P. digitatum	Lemongrass EO, citral (El-Mohamedy *et al.*, 2015), cress (*Lepidium sativum*) extract, pomegranate peel extract (Tayel *et al.*, 2016).
P. Italicum	Bergamot EO (Sánchez-González *et al.*, 2010), *Debaryomyces hansenii* (biocontrol yeast) (González-Estrada *et al.*, 2017).
Aspergillus niger, P. digitatum	Cinnamon EO, oregano EO (Avila-Sosa *et al.*, 2012).
P. expansum	*Metschnikowia pulcherrima* (biocontrol yeast) (Settier-Ramírez *et al.*, 2022), garlic extract (Heras-Mozos *et al.*, 2019).
P. expansum, A. flavus	Cinnamon EO (Xing *et al.*, 2010a).
P. digitatum	Pomegranate peel extract (PPE), PPE-*Wickerhamomyces anomalus* (biocontrol yeast) combination (Kharchoufi *et al.*, 2018), grapefruit seed extract (Aloui *et al.*, 2014).
Erwinia amylovora, E. carotovora, Xanthomonas vesicatoria, Pseudomonas syringae	Gallic acid (effective on all bacterial plant pathogens), citral (effective on all plant pathogens except *P. syringae*), eugenol (effective on *P. syringae*, but not on other plant pathogens). Clove extract (effective on all plant pathogens except *P. syringae*) (Alkan and Yemenicioğlu, 2016).
A. flavus, A. niger, Pennicillium spp., *Fusarium* spp.	Basil EO (Hemalatha *et al.*, 2017).

[a] EO: Essential oil

phenolic extracts from pomegranate peel and grapefruit seed in packaging has also attracted a significant interest since this provides an opportunity to utilize agro-industrial wastes, especially fruit processing wastes into value-added phenolic extracts. Moreover, it is a rapidly emerging trend to use biocontrol yeast incorporated coatings against major spoilage fungi, such as *B. cinerea, C. gloeosporioides, P. Italicum,* and *P. expansum.*

5.2.2 Application of synergistic mixtures of antimicrobials

The combinational use of synergetic antimicrobials is another powerful strategy that could be employed to enhance performance of antimicrobial packaging. The mechanism of synergy for different antimicrobial mixtures and example packaging materials employing these mixtures are given in Table 5.3. Combination of lysozyme and nisin is a typical example for natural antimicrobials having a synergy. These two natural antimicrobials are also extensively used separately or in combination with other antimicrobial agents (e.g. essential oils, phenolic extracts, chelating agents, antimicrobial peptides, and proteins) to develop alternative synergetic mixtures. The main objective of combining lysozyme and nisin is to enhance the effectiveness of antimicrobial packaging against Gram-positive bacteria, such as *L. monocytogenes, Clostridium* spp., lactic acid spoilage bacteria, etc. On the other hand, since the main limitation of lysozyme and nisin is lack of their capacity to overcome lipopolysaccharide (LPS) layer in Gram-negative bacteria, these two antimicrobials are also combined with those capable of penetrating or eliminating LPS layer (e.g. EDTA, lactoferrin, and ε-polylysine) (Ellison and Giehl, 1991; Murdock *et al.*, 2007; Ünalan *et al.*, 2011; Liu *et al.*, 2015). An interesting example for developing synergy formulation came from Sadiq *et al.* (2016) who encapsulated nisin within monolaurin nano-micelles. These authors determined that the nisin released from nano-micelles and its encapsulant monolaurin showed synergetic antimicrobial activity on *S. aureus.*

The synergetic mixtures of some essential oils (e.g. eugenol-citral, carvacrol-citral) are also effectively used to enhance their antimicrobial potency against fungi and bacteria (Chueca *et al.*, 2014; Ju *et al.*, 2020a; Ju *et al.*, 2020b; Cao *et al.*, 2021). In contrast, some other essential oils, such as thymol and carvacrol showed only an additive effect against bacteria, such as *Pseudomonas aeruginosa* and *Staphylococcus aureus* (Lambert *et al.* (2001). A synergetic effect was also observed for the following antimicrobials: nisin and reuterin against *Listeria monocytogenes* (Arqués *et al.*, 2004); lactoperoxidase and reuterin against *Escherichia coli* O157:H7 and *Salmonella enterica* (Arqués *et al.*, 2008), nisin and lactoferrin against *L. monocytogenes* and *E. coli* O157:H7 (Murdock *et al.*, 2007). Moreover, according to Mokhtar *et al.* (2017), the binary mixtures of quercetin, kaempferol, and caffeic acid also showed synergetic antimicrobial effect against *Staphylococcus aureus* and *Pseudomonas aeruginosa*. They also reported that the most effective combination for *S. aureus* is quercetin-caffeic acid mixture while *Pseudomonas aeruginosa* showed the highest inactivation with quercetin-kaempferol mixture. However, further studies are needed to clarify exact mechanism of synergy for these combinations and to show their effectiveness in food packaging applications.

Table 5.3: Examples of different synergetic mixtures and their edible packaging applications

Synergetic mixture	Example packaging application	Potential synergy mechanism/ Test bacteria[a]
Lysozyme-nisin	Chitosan film and coating (Sozbilen and Yemenicioğlu, 2021)	Lysozyme and nisin enhance action of each other to damage bacterial cell wall and membrane/Lactic acid bacteria and *Listeria monocytogenes* (Chung and Hancock, 2000).
Lysozyme-Na$_2$ EDTA	Zein film (Ünalan *et al.*, 2011), Pullulan film (Kandemir *et al.*, 2005) Whey protein isolate coating (Abdul-Rahman, 2019)	EDTA destabilizes lipopolysaccharide outer layer at Gram-negative cell walls and exposes peptidoglycan layer to lysozyme/ *S. enterica* serovar Typhimurium and *E. coli* O157:H7 (Ünalan *et al.*, 2011).
Lysozyme-eugenol	Casein nanoparticles (Wang *et al.*, 2021)	Lysozyme hydrolyzes the cell walls of bacteria and paves the way for eugenol diffusion into cells/*S. aureus* and *Bacillus* sp. (Wang *et al.*, 2021).
Lysozyme-lactoferrin	Cellulose-based packaging material (Barbiroli *et al.*, 2012), chitosan film (Brown *et al.*, 2008)	Lactoferrin destabilizes lipopolysaccharide outer layer at Gram-negative cell walls and exposes peptidoglycan layer to lysozyme/ *Vibrio cholerae*, *S. enterica* serovar Typhimurium, and *E. coli* (Ellison and Giehl, 1991).
Nisin-ε-polylysine	Chitosan coating (Song *et al.*, 2017)	Cell wall damage caused by nisin promotes the cellular uptake of polylysine and its interaction with DNA/*B. subtilis*, *E. coli*, *S. aureus* (Liu *et al.*, 2015).
Nisin-thymol	Zein capsules (Xiao *et al.*, 2011)	Thymol induced destabilization of bacterial membrane structure results in increased permeability for nisin/*L. monocytogenes* and *B. subtilis* (Ettayebi *et al.*, 2000).
Nisin-grape seed extract	Soy protein coating (Theivendran *et al.*, 2006), Soy protein films (Sivarooban *et al.*, 2008)	Cell wall and membrane damage caused by nisin promotes the cellular uptake of grape seed extract and this blocks some vital metabolic pathways, such as TCA cycle, energy metabolism, and amino acid biosynthesis/*L. monocytogenes* (Zhao *et al.*, 2020).
Eugenol-citral	Corn starch granules impregnated with essential oils in perforated sachets (Ju *et al.*, 2020a)	Eugenol mainly damages and permeabilizes cell membrane while citral mainly causes membrane lipid peroxidation and generation of reactive oxygen species/*A. niger* and *P. roqueforti* (Ju *et al.*, 2020a; Ju *et al.*, 2020b).

(Contd.)

Table 5.3: (*Contd.*)

Synergetic mixture	Example packaging application	Potential synergy mechanism/ Test bacteria[a]
Carvacrol-citral	Guar gum/sago starch/ whey protein isolate composite film (Dhumal *et al.*, 2019a). Guar gum/chitosan/whey protein isolate film (Dhumal *et al.*, 2019b)	The mechanism of synergy is unclear, but both growth and biofilm formation are inhibited. Apparent cell-wall damage occurs/*Cronobacter sakazakii*, (Cao *et al.*, 2021). The exact mechanism of synergy is unclear. Reactive oxygen species were generated (no involvement of TCA cycle and Fenton reaction) possibly by oxidation of particular cell components. DNA damage was evident/*E. coli* (Chueca *et al.*, 2014).

[a] Some of the test bacteria given might not be the target of given antimicrobial edible film application. These packaging examples aim to make readers aware that synergetic mixtures could be employed in indicated edible films.

5.2.3 Application of controlled release strategies

The controlled release strategies applied by employing different methods, such as encapsulation and nanotechnology-based approaches or by modification of film properties could be effective in enhancing the performance of antimicrobials used in active packaging. However, the design and application of controlled release strategies need overcoming of a number of challenges originating from the complexity of food systems. The food surface is the most susceptible part for microbial contamination and spoilage (Ünalan *et al.*, 2013). Therefore, a sufficient amount of antimicrobial should be released continuously from the packaging on to the food surface (or into food headspace) to maintain at least its minimum inhibitory concentration against the spoilage flora or pathogenic microbial contaminants during the storage of food. A too slow a release rate might be insufficient to control microbial spoilage of food and to inactivate pathogenic contaminants (Han, 2005). Similarly, a too fast a release of antimicrobials from the packaging is also mostly undesirable since this exhausts the antimicrobial reserves in the packaging rapidly and leaves the food surface unprotected after diffusion of released antimicrobials through depths of food. The release rate of the antimicrobial from the packaging on to the food surface must be compatible with diffusion rate of released antimicrobial through depths of food (Appendini and Hotchkiss, 2002). Moreover, the release rate of antimicrobial from films should also be much greater than its rate of neutralization in the food due to interactions with the food components (e.g. hydrocolloids, lipids, enzymes, etc.). The release rate and the total amount of antimicrobial incorporated into packaging should also be compatible with the shelf-life of packed food and microbiological parameters, such as initial load, resistance, growth rate, inactivation rate, etc. A packed food that will be stored for a long period needs incorporation of high amounts of antimicrobial into its packaging as well as an effective sustained release mechanism for packaging. However, even optimized antimicrobial packaging is not effective on microbial inactivation if initial microbial load of food is already too high. Moreover, a rarely found resistant pathogen, a newly emerging pathogen, or a pathogen from an unusual route of transmission might survive from antimicrobial packaging. All these parameters show that it is very difficult to form a fully mechanistic antimicrobial packaging system for a given food. However, antimicrobial packaging is a preservation method that reduces the risk of food spoilage or poisoning by most potential contaminants when designed properly and applied to food having high initial microbial quality.

A too fast a release profile is a frequently observed problem with most hydrocolloid-based films since their hydrophilic nature allows rapid swelling in aqueous environment and resulting increase of antimicrobial release rates (Buonocore *et al.*, 2005; Mastromatteo *et al.*, 2010; Almasi *et al.*, 2020). Therefore, sustaining of antimicrobial release is the most frequent controlled release problem in antimicrobial packaging. In contrast, the acceleration of the release rate is a rarely targeted strategy in antimicrobial packaging (e.g. used in food highly susceptible to microbial spoilage or antimicrobials having low solubility). Different but frequently used mechanisms employed to achieve sustained release of antimicrobials from edible packaging include (1) encapsulation, (2) modification of film properties (e.g. porosity, hydrophobicity, tortuosity, etc.), and (3) increasing degree of film cross-linking, but less frequent methods, such as (4) application of multilayer film and coatings, and (5) pH-response controlled release are also applied for some packaging (Table 5.4). These strategies could be employed in combination, depending on the challenges related to antimicrobial agent (e.g. high solubility, volatility, low molecular weight) and/or packaging material (e.g. high porosity, solubility or rapid swelling). Moreover, controlled release of multiple antimicrobials having different molecular weights and chemical structures might need a combination of several controlled release mechanisms in the same packaging material (Arcan and Yemenicioglu, 2014).

5.2.3.1 Application of encapsulation

One of the most applicable and effective controlled release strategies is incorporation of micro- or nano-encapsulated antimicrobials into edible packaging (Liu *et al.*, 2017; Amor *et al.*, 2021). The nanoencapsulation of phenolic compounds and essential oils by using inclusion complexes, electrospun nanofibers, solid lipid nanoparticles, nanoliposomes, nanomicelles, nanoemulsions, etc. before incorporation into packaging is the most powerful tool to control release rates of these natural compounds, and to enhance their antimicrobial performance during food applications (Severino *et al.*, 2014; Barba *et al.*, 2015; Donsì *et al.*, 2015; Liu *et al.*, 2017; He *et al.*, 2019; Lee *et al.*, 2020; Xiong *et al.*, 2020; Esmaeili *et al.*, 2020; Pirozzi *et al.*, 2020; Dini *et al.*, 2020; Jiang *et al.*, 2021). The encapsulation also helps in reducing loss of volatile essential oils from packaging materials and suppression of their undesired odors. The application of nanoliposomes in antimicrobial packaging might be very useful since this method provides multiple advantages, such as encapsulation of both hydrophilic and hydrophobic compounds (Bahrami *et al.*, 2020), reduction of evaporation for volatile compounds (Sebaaly

Table 5.4. Different mechanisms used for controlled release of natural antimicrobials from edible packaging

Material	Antimicrobial	Controlled release mechanism	References
Zein film	Lysozyme	Increased film hydrophobicity and tortuosity by incorporated carnauba wax (5% w/w of zein), and noncovalent cross-linking and reduced porosity of films by incorporated catechin (6 mg/cm^2) caused reduction in release rate of lysozyme.	Arcan and Yemenicioğlu (2013)
Na-caseinate film	Lysozyme	The release of lysozyme was controlled by modifying pH, but increasing degree of film crosslinking by CaCl$_2$ or transglutaminase caused immobilization of lysozyme.	De Souza *et al.* (2010)
Whey protein-oleic acid film	Lysozyme	Acidification of films by lemon juice triggered lysozyme release due to repulsive charge-charge interactions formed between lysozyme and whey protein film.	Boyacı *et al.* (2016)
Whey protein film	Nisin	The pH dropped below pI of whey proteins triggered nisin release due to repulsive charge-charge interactions formed between nisin and whey protein film.	Rossi-Márquez *et al.* (2009)
Zein film	Thymol	Addition of spelt bran (25 to 75% w/w of zein) into zein monolayer film increased thymol release rate. Zein multilayer films (three layers of cast film with thymol at the inner layer) showed sustained thymol release rates.	Mastromatteo *et al.* (2009)
Soy protein coated paper	Carvacrol	Temperature (5 to 20°C or 30°C) and moisture increase (60 to 80 or 100%) triggered carvacrol release due to glass transition phenomenon (glassy to rubbery state change) and increased chain protein mobility and carvacrol diffusivity.	Chalier *et al.* (2009)
Na-alginate-chitosan-Na-alginate coating	Cinnamon EO	Multilayered coating showed sustained release for cinnamon EO.	Zhang *et al.* (2019)

Chitosan-gelatin (1:3) film	Tyrosol, quercetin	Electron beam irradiation-induced cross-linking of dried films sustained the release rates of tyrosol, but not very effective in reducing quercetin release rates.	Benbettaïeb *et al.* (2016a,c)
Alginate breads	Thyme EO	Incorporation of cellulose nanocrystals (10-40% w/w of alginate) into alginate breads sustained the release rates of thyme EO.	Criado *et al.* (2019)
Gelatin-pullulan (5:1) film	Clove EO	Incorporation of clove EO pickering emulsion (2-6% w/w of film forming solution) obtained using tween 80-whey protein isolate-inulin mixture into films sustained release of clove EO.	Shen *et al.* (2021)
Whey protein film	Eugenol-carvacrol	Films incorporated with eugenol-carvacrol/β-cyclodextrin inclusion complex showed sustained release rates for eugenol and carvacrol.	Barba *et al.* (2015)
Gelatin film	Tea polyphenols	Tea polyphenols encapsulated into chitosan nanoparticles were incorporated into gelatin films to achieve sustained release of tea polyphenols.	Liu *et al.* (2017)
Zein film	Catechin	Films incorporated with catechin/β-cyclodextrin inclusion complex showed sustained release rates for catechin.	Jiang *et al.* (2021)
Agar based film	*Artemisia annua* EO	Film incorporated with EO loaded nanoliposomes (formed by mixing EO with soy lecithin-cholesterol mixture, 5:1, w/w) showed sustained release rate.	Cui *et al.* (2017b)
Carboxymethyl cellulose film	Quercetin, rutin	Film incorporated with quercetin or rutin loaded dipalmitoyl lecithin liposomes showed sustained release rate.	Silva-Weiss *et al.* (2018)
Dextran/zein nanofibers	Curcumin	The increased zein content in the electrospun nanofibers enhanced intermolecular interaction between dextran and zein through H-bonding. Change of zein content from 15 to 25% (w/v) at 50% (w/v) dextran concentration enhanced sustained release of curcumin.	Luo *et al.* (2021)

(Contd.)

Table 5.4. (*Contd.*)

Material	Antimicrobial	Controlled release mechanism	References
Gelatin nanofibers	Curcumin	Solubilization and release of curcumin from electrospun nanofibers was achieved by incorporating surfactant tween-80 into fibers.	Deng *et al.* (2017)
Alginate microbreads	Nisin	Ca^{+2} cross-linked micro-breads obtained from w/o emulsion of alginate-resistant starch (RS) blend were more effective to sustain release rates of nisin than cross-linked w/o emulsion of alginate microbreads. The alginate microbreads with RS were more packed and crackles than those without RS.	Hosseini *et al.* (2014)
Alginate-pectin microcapsules	Carvacrol	Alginate-pectin microcapsules loaded with carvacrol (11%, w/w) and kept in perforated miracloth pouches showed sustained release properties	Sun *et al.* (2021)
Carboxymethyl cellulose film	Curcumin	Film incorporated with curcumin-loaded nanohydrogel (obtained by lactoferrin-glycomacropeptide electrostatic interaction and thermal gelation) showed increased cumulative release of hydrophobic curcumin in 10% (v/v) solution of ethanol that simulate hydrophilic food. In contrast, free curcumin interacted with film matrix and showed considerably lower cumulative release.	Bourbon *et al.* (2021)
Na-alginate Pickering emulsion breads coated with $CaCO_3$ microparticles	Lysozyme	The release of lysozyme encapsulated into the breads could be controlled by pH change (rapid at pH 5.0, but slow at pH 7.0).	Xu et al. (2019)
Zein microparticles	Nisin and thymol	The zein capsules controlled release rates of nisin depending on change of pH, thymol content (turned porous capsules into dense ones) and glycerol content (nisin release rate showed a parallel increase to glycerol	Xiao *et al.* (2011)

		content). In contrast, the thymol release showed a parallel reduction to glycerol content.	
Fish gelatin film	Cinnamon EO	The incorporation of nanoliposome encapsulated cinnamon EO into films sustained the release rate of EO.	Wu *et al.* (2015)
Solid lipid nanoparticles	Carvacrol	Incorporating carvacrol into solid lipid nanoparticles of propylene glycol monopalmitate and glyceryl monostearate controlled its release rate and enhanced its antimicrobial activity.	He *et al.* (2019)

EO: Essential oil

et al., 2015; Wu *et al.*, 2015), and improvement of antimicrobial properties due to ability of nanoliposomes to penetrate membranes and to release their contents directly into the cell (Liolios *et al.*, 2009). These advantages explain the potency of edible films, coatings, and nanofibers incorporated with essential oil-loaded nanoliposomes during food applications (Cui *et al.*, 2017b; Cui *et al.*, 2018; Esmaeili *et al.*, 2020; Mehdizadeh *et al.*, 2021). The use of nanoemulsions of essential oils in edible packaging is also highly beneficial since most essential oils, normally insoluble in aqueous film-forming solutions, can be solubilized (dispersed) following emulsification. Moreover, the incorporation of essential oil loaded nanoemulsions into edible films mostly gave better antimicrobial performance in food applications than edible films loaded with free essential oils (Noori *et al.*, 2018; Xiong *et al.*, 2020). Therefore, the nano-based encapsulation methods are currently playing the central role in enhancing controlled release properties and antimicrobial performance of edible packaging.

5.2.3.2 Modification of film properties

The modification of film hydrophobicity and tortuosity by incorporated lipids is a classical strategy applied to improve sustained release properties of packaging materials. The increased hydrophobicity of edible films helps to control their swelling rate that is a major factor affecting the release properties. Moreover, incorporated lipids also entrap (or encapsulate) hydrophobic antimicrobials or increase film tortuosity due to the dispersed micro/nano lipid droplets within the film matrix. This method, applied by incorporating lipids into edible films by effective homogenization to form a composite film structure, is also beneficial in improving water vapor barrier properties of protein- and polysaccharide-based edible films (Shellhammer and Krochta, 1997; Aguirre-Joya *et al.*, 2019). Waxes or fatty acids have long been used to reduce the release rates of sorbates from whey protein films (Ozdemir and Floros, 2003) and methylcellulose and hydroxypropyl methylcellulose films (Vojdani and Torres, 1990). The controlled release of lysozyme from zein films was also achieved by forming composite films with different waxes and fatty acids (Arcan and Yemenicioğlu, 2013). It was reported that the incorporation of waxes with high melting point (e.g. carnauba wax) into zein films by hot homogenization was more effective in reducing lysozyme release rates than incorporating waxes with low melting point (e.g. candelilla and beeswax) (Arcan and Yemenicioğlu, 2013). Similarly, the incorporation of long chained (C_{18}) fatty acids (e.g. linoleic or oleic acid) by homogenization in the presence of lecithin was found more effective than incorporation of short chained (C_{12}) fatty acids (e.g. lauric acid) to reduce lysozyme release rates of zein films (Arcan and Yemenicioğlu, 2014). The melted lipids can also be employed as a coating layer on the surface of edible films, mainly to improve their water vapor barrier properties (Weller *et al.*, 1998), but this is not a frequently used method to enhance the sustained release properties of antimicrobial films since formed impermeable lipid layer might also prevent diffusion of hydrophilic antimicrobials.

5.2.3.3 Increasing degree of film cross-linking

Increasing the degree of film cross-linking is another frequently used controlled release strategy that could be applied both to protein and polysaccharide-based edible films. The cross-linking increases the network within the film matrix due to enhanced intra- and inter-molecular interactions among hydrocolloids in film matrix. However, this method needs careful monitoring of free/immobilized antimicrobial ratios in the film and creating a balance to provide the necessary amount of antimicrobial release on the food surface. The edible films could be cross-linked covalently or non-covalently by using physical, chemical, or enzymatic methods. The major physical methods used for cross-linking hydrocolloid based edible films are heating and irradiation (UV irradiation, γ-irradiation or electron beam irradiation) (Fathi *et al.*, 2018; Lee *et al.*, 2005; Benbettaïeb *et al.*, 2016a). The heating causes unfolding of proteins and this exposes buried thiol groups that participate in covalent-crosslinking of proteins via intermolecular disulfide bond formation. The heating could be applied for cross-linking of films based on whey, casein, and soy proteins that contain sufficient cysteine residues, but this method is not effective for cross-linking films of gelatin due to lack of cysteine residues in this protein (Benbettaïeb *et al.*, 2016b). The irradiation methods also cause covalent cross-linking of protein-based films, but this occurs by generation of free radicals that induce formation of intermolecular covalent bonds mainly in the form of bityrosine bridges (Le Tien *et al.*, 2000; Benbettaïeb *et al.*, 2016b). Therefore, irradiation methods, such as UV irradiation, γ-irradiation, or electron beam irradiation might be employed for cross-linking of proteins, such as gelatin that could not be cross-linked effectively by heating (Otoni *et al.*, 2012; Benbettaïeb *et al.*, 2016d; Bessho *et al.*, 2007).

The ability of Ca^{+2} ions to cross-link hydrocolloids by creating ionic interactions has also been exploited to achieve controlled release of antimicrobial compounds. For example, the Ca^{+2} induced crosslinking of films, coatings or macro/micro breads of polysaccharides, such as pectin, alginate, gellan, etc. can be used for controlled delivery of different phenolic compounds and essential oils (Jantrawut *et al.*, 2013; Dey *et al.*, 2020; Lan *et al.*, 2020; Dadwal *et al.*, 2021; Mostaghimi *et al.*, 2021). The Ca^{+2} induced crosslinking is applied by immersing packaging (e.g. self-standing film) or coated food into $CaCl_2$ solution, or by injecting hydrocolloid solution (or emulsion) into $CaCl_2$ solution to form breads. The spraying of $CaCl_2$ solution on to self-standing films or coated food surface is also preferred when immersion causes significant loss of incorporated active compound. The extrusion technology employing $CaCl_2$ cross-linked alginate-based casings is also currently used commercially to obtain sausages (Hilbig *et al.*, 2020), but further studies are needed to obtain antimicrobial casings with controlled release properties. The release rates of such cross-linked delivery systems might be modified by change of hydrocolloid concentration or concentration of $CaCl_2$ used in cross-linking. Moreover, it was also reported that blending of alginate or pectin with other hydrocolloids (e.g. starch, chitosan) might be employed to control the release rates, or to increase encapsulation efficiency and cumulative release of natural antimicrobial compounds from Ca^{+2} cross-linked breads (Torpol *et al.*, 2019; Kurozawa *et al.*, 2017). The incorporation of cross-linked alginate or pectin macro/micro breads into edible films, coatings, pads or gels might also be very useful for controlled release of natural antimicrobials. The edible films of sodium caseinate can also be cross-linked by Ca^{+2} ions that bind mainly to the anionic clusters of casein phosphoseryl residues, but such cross-lined films might not be suitable for delivery of macromolecules like lysozyme due to extensive trapping or immobilization (De Souza *et al.*, 2010).

The cross-linking of edible packaging can also be conducted by adding intact (catechin, vanillin, ellagitannins, etc.) (Peng *et al.*, 2010; Arcan and Yemenicioğlu, 2013; Chen *et al.*, 2020) or chemically (non-enzymatic) oxidized phenolic compounds (ferulic, caffeic, tannic acids, green tea extract, etc.) into hydrocolloid solutions (Choi *et al.*, 2018; Leite *et al.*, 2021). The intact phenolic compounds cause non-covalent cross-linking mainly by forming hydrophobic interactions and extensive intra- and intermolecular hydrogen bonding among natural polymers. The hydrophobic interactions formed between aromatic rings (e.g. A and C rings of flavonoids) of polyphenols and hydrophobic residues/groups in hydrocolloids (e.g. aliphatic side chains of proteins or methyl groups of pectin) are accepted as the major mechanisms for polyphenol-hydrocolloid complexation (Tang *et al.*, 2003; He *et al.*, 2006; Liu *et al.*, 2020a), while extensive hydrogen bonds that occur following complexation stabilize this structure. In proteins, the hydrogen bonding forms mainly between hydroxyl groups of polyphenols and carbonyl groups of protein (Arcan and Yemenicioğlu, 2013), while hydrogen bonding occurs mainly between hydroxyl groups of polyphenols and oxygen atoms in different groups/linkages (e.g. carboxyl/carboxylic acid and hydroxyl groups, oxygen atom of glycosidic linkages) of polysaccharides (Wu *et al.*, 2009; Jakobek, 2015; Liu *et al.*, 2020a). The increased molecular weight (as in tannins and polymerized procyanidines) of polyphenols means greater capacity to interact with hydrocolloids (Liu *et al.*, 2020a). In addition, attractive charge-charge interactions between oppositely charged groups in polyphenols (e.g. positively charged flavylium ions and negative charges of carboxylate groups in hydroxybenzoic acids) and hydrocolloids (e.g. basic or acidic side chains of protein, carboxylate of pectin, protonated amine group of chitosan) also contribute to stabilization effect. The oxidation of polyphenols conducted by treating them with molecular oxygen or hydrogen peroxide also creates reactive semiquinones or quinones that could cause covalent cross-linking of proteins by interacting with their nucleophilic groups (e.g. side chains of cysteine, lysine, histidine) (Rohn, 2014). However, control of protein cross-linking in this method is challenging due to non-specific modifications in reactive groups of protein.

The enzymatic cross-linking conducted by transglutaminase or oxidoreductase group enzyme (e.g. peroxidase, laccase, tyrosinase) is another promising method that is used to cross-link proteins, polysaccharides, and protein-polysaccharide mixtures. The enzymatic method is frequently used for stabilization of protein-polysaccharide complex structures (e.g. hydrogels, nanoparticles, emulsions, electrostatic complexes) and to control release of drugs or bioactive compounds (e.g. cinnamaldehyde) (Littoz and McClements, 2008; Yin *et al.*, 2012; Tormos *et al.*, 2015; Huber *et al.*, 2017; Liu *et al.*, 2021). Thus, incorporation of antimicrobial loaded enzyme cross-linked nano- or micro-structures (encapsules) into edible films might be an alternative controlled release strategy. The application of enzymatic cross-linking is very attractive since this method is an environmentally friendly approach that might be employed not only to enhance controlled release properties of films, but also to improve film characteristics, such as thermal, mechanical, and barrier properties (Quan *et al.*, 2018). The mechanisms of cross-linking for proteins and polysaccharides by different enzymes are given in Table 5.5. The main challenge to enzymatic cross-linking of some hydrocolloids is that the reactive groups (substrates of enzyme) of globular proteins might be inaccessible to the enzyme. This is really a challenge for the transglutaminase, but the use of some innovative methods (e.g. application of ultrasound, high pressure processing, UV irradiation, microwave heating) that enhance accessibility of proteins to an enzyme, increase the enzyme activity, or modify proteins (e.g. increased cross-linking and polymerization, and reduced solubility) were found beneficial to obtain material structures that could be exploited for controlled release of natural antimicrobials (Qin *et al.*, 2016; Gharibzahedi *et al.*, 2018). The combination of transglutaminase with ultrasound treatment is particularly suggested for controlled release of natural agents since this method might be employed to convert porous gels into denser and more homogenous alveolar protein gels (Gharibzahedi *et al.*, 2018). Too much cross-linking might also cause undesirable effects during application of transglutaminase in controlled release studies. For example, De Souza *et al.* (2010) tried to control release rates of lysozyme with transglutaminase cross-linking of sodium caseinate films, but extensive cross-linking of film proteins blocked the antimicrobial enzyme release from films. Thus, it is essential to control the degree of cross-linking and to choose suitable antimicrobials (e.g. low molecular weight ones having minimum interaction with the cross-linked film matrix). The accessibility problem observed for proteins during transglutaminase cross-linking could be overcome for oxidoreductases by involving low molecular weight phenolic substrates (mainly phenolic acids) in reaction mixtures. The enzyme oxidized free phenolic compounds more easily penetrate into depths of hydrocolloids to interact with reactive groups and to induce their covalent cross-linking. However, all methods depending on oxidation (enzymatic or non-enzymatic) of free phenolic compounds carry the disadvantage of browning that might cause undesired changes in the color of developed packaging material.

Table 5.5: The mechanisms of cross-linking proteins and polysaccharides
by different enzymes

Cross-linker	Cross-linking mechanism(s)	Result
Transglutaminase	Transglutaminase causes inter- and intra-molecular covalent cross-linking of proteins by catalyzing acyl-transfer reactions between γ-carboxyamide groups of glutamine residues and ε-amino groups of lysine residues of proteins. The covalent bond formed is called an isopeptide bond (DeJong and Koppelman, 2002).	Protein-protein cross-linking
Tyrosinase	The enzyme is capable to catalyze hydroxylation of monophenols to diphenols, and oxidation of o-diphenols to o-quinones. Thus, it is capable to form o-quinones from phenolic side chains of tyrosine residues of proteins. It was proposed that the o-quinones react spontaneously mainly via 1,4-additions with the side chains of lysine, tyrosine, histidine, and cysteine residues of proteins and form covalent protein–protein cross-links (Heck *et al.*, 2013).	Protein-protein cross-linking
Laccase	Laccase catalyzes oxidation of its substrates by removing one electron. Thus, the oxidation of tyrosine residues of proteins by this enzyme forms free radicals that cause non-enzymatic formation of covalent bond (ether bond) between oxidized tyrosine residues (Mattinen *et al.*, 2005; Mattinen *et al.*, 2006).	Protein-protein cross-linking
Peroxidase	Peroxidases catalyze oxidation of their substrates by removing one or two electrons. The oxidation of tyrosine residues of protein by peroxidase mainly via removing one electron forms free radicals that cause non-enzymatic formation of covalent bonds between oxidized tyrosine residues of proteins (McCormick *et al.*, 1998).	Protein-protein cross-linking
Tyrosinase, laccase or peroxidase	The enzyme catalyzed oxidation of phenolic moieties (e.g. esterified ferulic acid) in polysaccharides causes covalent cross-linking of polysaccharides with each other or with protein (Oudgenoeg *et al.*, 2001; Littoz and McClements, 2008).	Polysaccharide-polysaccharide or protein-polysaccharide cross-linking

(Contd.)

Table 5.5: (*Contd.*)

Cross-linker	Cross-linking mechanism(s)	Result
Tyrosinase, laccase or peroxidase	The use of free low molecular weight phenolic compounds (ferulic, chlorogenic, vanillic, caffeic, sinapic, *p*-coumaric acids, catechin, eugenol, etc.) in reaction mixtures generates quinones or free radicals that interact with reactive groups of proteins and polysaccharides and cause their non-enzymatic covalent cross-linking (Oudgenoeg *et al.*, 2001; Selinheimo *et al.*, 2007; Ma *et al.*, 2011; Huber *et al.*, 2017; Li *et al.*, 2020).	Protein-protein, protein-polysaccharide, polysaccharide-polysaccharide cross-linking
Lysyl oxidase	Lysyl oxidase catalyzes oxidation of primary amine groups of lysyl side chains of proteins, such as gelatin and elastin, to corresponding aldehydes. The formed reactive aldehydes undergo further reactions (aldol condensation and Schiff base product formation) to form covalent cross-links (Bakota *et al.*, 2011).	Protein-protein cross-linking

5.2.4 Application of combinational preservation methods

The performance of edible antimicrobial packaging can also be enhanced by combining it with other preservation methods such as vacuum packaging (Yingyuad *et al.*, 2006) or modified atmosphere packaging (passive or active) (Carrión-Granda *et al.*, 2018; Tabassum and Khan, 2020), γ-irradiation (Severino *et al.*, 2014), pulsed light (Koh *et al.*, 2017), and high pressure processing (Pavli *et al.*, 2019). However, this approach has some limitations, such as high costs and the need for using plastic packaging (especially in VP and MAP). Further details and examples of combinational approaches with active edible packaging have been discussed at Chapter 7.

5.3 Enhancing the performance of antioxidant packaging

The success of antioxidant packaging is highly related to the use of suitable antioxidant compounds considering the potential oxidative changes (might be chemical and/or enzymatic) in food. The naturally existing antioxidants in food, such as ascorbic acid, phenolic compounds, carotenoids or tocopherols could be employed in antioxidant packaging alone or in combination to control chemical and/or enzymatic oxidative changes in food. The pure polyphenols, phenolic-rich plant extracts, and essential oils have been increasingly preferred in antioxidant packaging since they also provide additional benefits, such as antimicrobial activity and bioactivity. The selection of suitable antioxidants is a highly challenging process since they should be compatible both with edible packaging and ongoing oxidative mechanisms in the target food system. Moreover, the form of lipid (e.g. bulk oil or emulsion), lipid content, and fatty acid profile of lipids also affect the selection of antioxidants. According to the 'polar paradox' theory, non-polar lipophilic antioxidants are more active in oil-in-water emulsion systems than hydrophilic antioxidants, while hydrophilic antioxidants are more effective in bulk oil systems than lipophilic antioxidants (Khan and Shahidi, 2000). However, different and more developed approaches are needed to understand the complex roles of micellar structures in emulsions and bulk oils in lipid oxidation (Villeneuve *et al.*, 2021). The combinational use of different antioxidants is also one of the strategies used to enhance effectiveness of antioxidant packaging. This strategy enables exploiting of potential synergies between different antioxidants and catching a broad antioxidant spectrum (e.g. iron chelation, free radical scavenging, and singlet oxygen quenching capacity in the same formulation) effective on different lipid oxidation mechanisms in the food. It is also highly beneficial to apply encapsulation of antioxidants incorporated into packaging since this increases stability of antioxidants by preventing their interactions with film components and improves their sustained release properties. The sustained release helps to provide only the necessary amount of fresh (unreacted) antioxidant on to the food surface continuously, throughout the storage period. This limits undesired reactions between antioxidant and food components and prevents potential prooxidant effects due to local concentration increase of released antioxidant on the food surface. Finally, antioxidant edible packaging can be supported by other packaging methods, such as vacuum or modified atmosphere packaging that helps in elimination of oxygen and in more effective control of oxidative changes in food.

5.3.1 Choosing suitable natural antioxidants

Choosing suitable antioxidants for edible packaging is a major factor affecting the performance of food application. The selection of antioxidant(s) is done according to the possible mechanisms of oxidation in the target food. The major oxidative changes in different types of food and suitable natural antioxidants are discussed below.

5.3.1.1 *Choosing suitable natural antioxidants for beef, pork, chicken and fish*

The use of suitable antioxidants for the target food is the most important parameter determining the performance of antioxidant packaging. Thus, it is essential to know the factors affecting the potential oxidative mechanisms in target food. For example, ferric hemochromogen and denatured ferric hemochromogen are major pro-oxidants that catalyze rapid lipid oxidation in raw and cooked meats, respectively (Love and Pearson, 1971; King and Whyte, 2006). However, the pink pigment of cured meat is converted into ferrous nitric oxide hemochromogen that cannot catalyze lipid oxidation rapidly (Love and Pearson, 1971). The non-heme iron in meat is also effective on the formation of lipid hydroperoxides (Amaral *et al.*, 2018). It is thought that NaCl used during curing of meet liberates iron from myoglobin; thus, free iron becomes an important factor accelerating lipid oxidation (Min *et al.*, 2010). The increased lipid oxidation in meat by addition of NaCl is also attributed to its disruption of membrane integrity and reduction of antioxidant enzyme activity (Mariutti and Bragagnolo, 2017). Moreover, lipoxygenase in raw meat is also involved in lipid oxidation since it catalyzes the oxidation of polyunsaturated free fatty acids and forms hydroperoxides (Min *et al.*, 2010).

The lipid oxidation in meat begins with oxidation of highly susceptible phospholipids that form almost 1% of meat tissue, and then it progresses with oxidation of neutral lipids that form a majority of meat lipids (Love and Pearson, 1971). The color changes in meat are linked to lipid oxidation since reactive products of lipid oxidation and myoglobin pigment oxidation (metmyoglobin) enhance the oxidative reactions of each other (Faustman *et al.*, 2010). Due to their lower myoglobin contents, refrigerated raw pork and chicken tissues are less susceptible to lipid oxidation than refrigerated raw beef tissue (Rhee and Ziprin, 1987). It was reported that the total iron and heme iron contents of beef are almost 1.5 and 1.6, and 3.8 and 2.8-fold higher than those of pork and chicken, respectively (Tang *et al.*, 2001). However, all types of meat become more susceptible to lipid oxidation after thermal processing or heat treatments applied during cooking. The heating increases lipid oxidation of meat by reducing the activation energy needed for lipid oxidation, breaking down hydroperoxides into free radicals, increasing free iron liberated from myoglobin, and inactivating antioxidant enzymes (Mei *et al.*, 1994; Amaral *et al.*, 2018). Although fish contains significantly lower free iron and heme-iron than beef, the hemoglobin is an important catalyst in lipid oxidation and peroxidation of fishes since their blood cannot be removed prior to processing (Maqsood *et al.*, 2012). Moreover, some fish (e.g. mackerel) are very susceptible to lipid oxidation due to their high polyunsaturated fatty acid content (Tang *et al.*, 2001).

The natural phenolic compounds are capable of controlling oxidative changes in meat and meat products since they can function not only as primary antioxidants by donating hydrogen atoms and quenching reactive free radicals, but also by acting as secondary antioxidants by chelating metal atoms (Oswell *et al.*, 2018). However, most of the phenolic antioxidants cannot be applied to meat and meat products due to their negative effects on sensory properties of these products. Some of the commercially available natural phenolic antioxidants tested for meat and meat products include grape seed or skin extract, green tea extract, pine bark extract, olive leaf extract, oregano extract, and rosemary extract (Shah *et al.*, 2014). However, it is the rosemary extract that has found most widespread application in the meat industry (Oswell *et al.*, 2018). In the literature, numerous studies exist related to effectiveness of natural phenolic loaded edible packaging materials on oxidative changes in meat and meat products. However, only minority of these studies are supported with sensory analysis. For example, oregano essential oil (EO), rosemary EO, apricot kernel essential oil (EO), cumin EO, and pomegranate peel extract-thyme EO mixture were employed for antioxidant edible packaging of raw beef, while garlic essential oil and *Terminalia arjuna* extract loaded edible film were used in cooked beef sausages to suppress lipid oxidation without causing adverse sensory effects (Vital *et al.*, 2016; Kalem *et al.*, 2018; Wang *et al.*, 2020a;

Behbahani *et al.*, 2020; Mehdizadeh *et al.*, 2020; Esmaeili *et al.*, 2020). The green tea extract and *Saturej khuzestanica* EO loaded edible coatings were also used to reduce lipid oxidation of vacuum packed cooked pork patties (Kang *et al.*, 2007) and lamb steaks (Pabast *et al.*, 2018) without adverse sensory effects, respectively. Some natural phenolic compounds approved with sensory analysis and antioxidant potential in edible packing applications of chicken include *Artemisia fragrans* EO, curcumin-cinnamon EO mixture, anise EO, *Zataria multiflora* EO-propolis extract mixture (Vital *et al.*, 2016; Abdou *et al.*, 2018; Mehdizadeh and Langroodi, 2019; Yaghoubi *et al.*, 2021; Fathi-Achachlouei *et al.*, 2021). Due to its well-known positive contribution in sensory properties of meat, Huang *et al.* (2020) did not conduct sensory tests for chicken meat treated with rosemary extract loaded edible coatings, but they showed the effectiveness of their antioxidant coatings to reduce lipid oxidation rates in coated samples. In one of the interesting studies, epigallocatechin gallate was employed in antioxidant edible packaging successfully to reduce lipid oxidation in fried salmon skins that contain high amounts of polyunsaturated lipids (Nilsuwan *et al.*, 2021). Some other examples of natural compounds employed for antioxidant edible packaging of fresh fish without sensory problems are lemon or orange seed extract, lime peel extract, pomegranate peel extract, propolis extract, and *Pulicaria gnaphalodes* (Vent.) Boiss. extract (Yuan *et al.*, 2016; Ucak *et al.*, 2021; Çoban, 2021; Mehdizadeh *et al.*, 2021; Khaledian *et al.*, 2021).

5.3.1.2 *Choosing suitable natural antioxidants for cheese*

Lipid oxidation is a major oxidative change that should be controlled during antioxidant packaging of cheese. The lipid oxidation rates of dairy products are affected by many different factors originating from milk composition, processing methods (e.g. homogenization, heat treatments and pasteurization), ripening and storage conditions (temperature, oxygen and light etc.). The major milk compositional factors affecting lipid oxidation are the degree of lipid unsaturation that might change depending on feeding management of cows (low- or high-fat diet), amounts of naturally existing milk antioxidants, such as vitamin C and tocopherols (mainly α-tocopherol), retinol, carotenoids (mainly β-carotene), caseins and amino acids, and amounts of pro-oxidants, such as metal atoms and oxidative enzymes (e.g. lactoperoxidase) (O'Connor and O'Brien, 2006; Revilla *et al.*, 2016; Rotondo *et al.*, 2021). The application of phenolic extracts and essential oils in antioxidant packaging of ripening cheeses should be conducted carefully since these natural antioxidants show undesired effects on sensory properties (odor and taste) of cheese and inhibit lactic acid starter cultures essential for cheese ripening. For example, it was determined that the lemon balm EO caused considerable inhibition of lactic acid starter culture in cheese while basil and thyme EOs did not cause considerable inhibition of starter culture (Licon *et al.*, 2020). However, the application of natural bioactive polyphenols in fresh and ripening cheeses' preservation is still highly attractive since this is a method to convert traditional cheese into a functional food. Moreover, recent trend of incorporating polyunsaturated fatty acids (PUFAs) into cheese to obtain functional products suffers from limited shelf-life due to increased sensitivity against lipid oxidation. Thus, edible antioxidant packaging might be a suitable tool for preservation of PUFA-enriched cheeses. The rosemary extract and essential oil are used frequently in cheese and other dairy products as alternatives to synthetic antioxidants since they can inhibit lipid oxidation and show different potential health benefits (Gad and Sayd, 2015; Ribeiro *et al.*, 2016). Some workers found that edible films loaded with rosemary essential oil and extract gave acceptable sensory properties (taste and odor) for some cheese (Al Mousawi and Khair, 2019; Pieretti *et al.*, 2019). However, it was determined that the sensory properties of cheese treated with oregano essential oil loaded edible films were superior than those treated with rosemary EO loaded coatings (Cano Embuena *et al.*, 2017; Pieretti *et al.*, 2019). Some other examples of natural active compounds employed in antioxidant packaging of cheese with acceptable sensory properties are black cumin (*Bunium persicum*) EO, *Pimpinella saxifrage* EO, oregano EO, basil EO, pine needle EO, and pomegranate peel extract (Mei *et al.*, 2015; Saravani *et al.*, 2019; Mushtaq *et al.*, 2018; Ksouda *et al.*, 2019; Mahcene *et al.*, 2021; Karunamay *et al.*, 2020). The double cream cheese treated with edible coatings loaded with garlic essential oil also gave satisfactory antimicrobial and sensory properties, highly preferred by the panelists (Molina-Hernández *et al.*, 2020).

5.3.1.3   *Choosing suitable natural antioxidants and browning inhibitors for fresh fruits and vegetables*

Coatings containing antioxidants and enzyme inhibitors are the most frequently used edible packaging materials for minimally processed fruits and vegetables suffering mainly from enzymatic browning. The enzymatic browning catalyzed by PPOs, such as tyrosinase, catecholoxidase, and laccase is a major problem during processing and refrigerated storage of minimally processed fruits and vegetables. The PPO in most fruits and vegetables is catechol oxidase that catalyzes oxidation of *o*-diphenols to o-quinones, but root vegetables (e.g., potatoes) and mushrooms contain mainly tyrosinase that is capable of catalyzing both hydroxylation of monophenols to o-diphenols (monophenolase activity) and oxidation of o-diphenols to o-quinones (Yemenicioğlu, 2016). The laccase catalyzes oxidation of both o-diphenols and p-diphenols to corresponding quinones, but this enzyme exists only in several products (e.g. some peach cultivars, mushrooms, and tomatoes) (Vámos-Vigyázó, 1981). The quinons formed by enzymatic oxidation of phenolic compounds undergo non-enzymatic polymerization reactions to form melanins that are responsible for the dark color formed on minimally processed fruit and vegetable surfaces. Some of the most challenging industrial problems caused by PPOs are observed on the surfaces of peeled and sliced fruits (e.g. apples, bananas, and eggplant) and root vegetables (e.g. potatoes and carrots), whole or sliced mushrooms and artichoke heads, and on the harvesting cut in the stem tissue of lettuce (Yemenicioğlu, 2016; Rizzo *et al.*, 2019).

Different naturally existing antioxidant agents and PPO inhibitors that could be applied for antioxidant packaging of minimally processed fruits and vegetables include ascorbic acid, organic acids (e.g. citric, malic and kojic acids), and L-cysteine (Table 5.6). The mixtures of ascorbic acid with organic acids and $CaCl_2$

are most frequently used browning inhibitors in coating solutions of fresh-cut fruits and vegetables (Yemenicioglu, 2016; Sanchís *et al.*, 2017). The CaCl$_2$ is not a PPO inhibitor, but it increases the cellular rigidity and integrity of plant tissues and helps in reducing tissue damage that causes contact of PPO in the cytoplasm with its phenolic substrates released from the disturbed vacuoles. The inhibition of enzymatic browning in fruits and vegetables by natural phenolic compounds is also very promising since these natural compounds not only inhibit PPO, but also reduce PPO formed o-quinones back to *o*-diphenols. Moreover, most of the polyphenols show antimicrobial activity that is important to control spoilage and pathogenic microorganisms. The inhibition of enzymatic browning in fresh-cut apples by green tea extract (Soysal, 2009), pumpkin seed extract, and hibiscus flower extract (Wessels *et al.*, 2014); in fresh-cut potatoes by green tea extract (Bobo *et al.*, 2022) and onion extract (Lee *et al.*, 2002); and in pineapple slices by citral (Prakash *et al.*, 2020) are some examples that show applicability of natural phenolic compounds against enzymatic browning. However, further studies are needed to test the effects of these polyphenols on aroma and taste of fruits and vegetables. A recent work by Kim *et al.* (2021) is also important since it applied different phenolic extract incorporated carboxymethyl cellulose (CMC) coatings to inhibit browning of whole bananas. The active coating was employed on the inedible peel surface of banana; thus, this minimized the potential effects of polyphenols on the taste of fruits. The developed antioxidant CMC coatings incorporated with *Morus alba* (mulberry) root extract prevented browning of banana peel surface for 12 days, while coatings incorporated with green tea extract did not show a considerable antibrowning effect at the coated banana surface (Kim *et al.*, 2021). The coatings of some hydrocolloids also show inherent anti-browning activity possibly due to the presence of phenolic residues having PPO inhibitory activity or reducing effect on PPO formed quinones. For example, the gum Arabic coating alone or enriched with Roselle (*Hibiscus*

Table 5.6: Mechanisms of enzymatic browning inhibition by some natural antioxidants and organic acids

Agent	Mechanism(s)	References
Ascorbic acid	Reduction of PPO formed quinones back to diphenols, competitive inhibition of PPO.	Mishra *et al.* (2012)
Citric acid	Mixed-type inhibition of PPO.	
Malic and citric acids	Reduction of pH below optimal activity of PPO, chelating of copper atom at the active site of PPO.	Vámos-Vigyázó (1981); Eskin (1990)
L-cysteine	Inhibition of melanin formation by forming conjugates with quinones, irreversible inhibition of PPO.	Valero *et al.* (1991)
Kojic acid	Reduction of PPO formed quinones back to diphenols, competitive inhibition of PPO.	Chen *et al.* (1991a), Chen *et al.* (1991b)
Green tea extract	Competitive inhibition of PPO.	Soysal (2009)
Onion extract	Non-competitive inhibition of PPO.	Lee *et al.* (2002)

sabdariffa L.) extract has been used for coating of blueberries to inhibit PPO enzyme responsible for loss of anthocyanins in this product (Yang *et al.*, 2019). Moreover, the strawberries coated with gum Arabic showed considerably lower PPO activity, but better color, appearance, and gloss during sensory tests (Tahir *et al.*, 2018). The Aloe vera gel, that contains acetylated mannan (acemannan) as a major hydrocolloid, also showed some PPO inhibitory and anti-browning effects when applied in coating of different products, such as sliced lotus roots (Ali *et al.*, 2020), mushrooms (Mirshekari *et al.*, 2019), litchi fruits (Ali *et al.*, 2019), but further studies are needed to clarify the exact mechanisms behind inherent anti-browning activity of Aloe vera gel.

5.3.2 Combination of suitable antioxidants

The combination of antioxidants is one of the major strategies to enhance effectiveness of antioxidant packaging since this enables exploiting synergetic effects of antioxidants. For example, it was reported that the combination of vitamin C (Niki *et al.*, 1995), lycopene (Shi *et al.*, 2007), or γ-terpinene (Mollica *et al.*, 2022) with vitamin E exhibited synergetic effects against lipid oxidation. Salminen and Russotti (2017) exploited synergistic interaction of vitamin C and green tea extract to reduce browning of fresh cut apple slices. It was also determined that free radical scavenging-based antioxidant activity for mixture of lycopene, vitamin E, vitamin C, and β-carotene at proper concentrations was substantially superior than the sum of their individual antioxidant activity (Liu *et al.*, 2008). The combination of antioxidants helps regeneration of exhausted antioxidants and inhibition of multiple oxidation catalysts and pro-oxidants (e.g. metal atoms, reactive oxygen species, enzymes, etc.) in food at the same time. For example, the mixture of sinapic acid with ascorbic acid and citric acid was employed against lipid oxidation to attain (1) free radical scavenging-based antioxidant activity (mainly by sinapic acid), (2) regeneration of sinapic acid (by vitamin C), and (3) chelation of pro-oxidant metal atoms (by citric acid) (Roschel *et al.*, 2019). The sufficient control of enzymatic browning using sulfite alternatives also needs use of proper combinations of antioxidants with each other and with other additives. For example, the mixture of vitamin C (or its derivatives) with citric acid and $CaCl_2$ provides (1) reduction of PPO formed quinones back to diphenols (by vitamin C), (2) reduction of pH below optimal activity of PPO (by vitamin C and citric acid), (3) chelation of copper atom at the active site of PPO (by citric acid), (4) competitive (by vitamin C) and mixed-type (by citric acid) inhibition of PPO, and (5) increased tissue integrity of plant material (by $CaCl_2$) (Yemenicioglu, 2016).

5.3.3 Application of controlled release strategies

The controlled release strategies used for antimicrobials (e.g. encapsulation and modification of film morphology, hydrophobicity, tortuosity, degree of cross-linking, etc.) could also be applied for antioxidants employed in edible packaging. The application of controlled release technology enhances the effectiveness of antioxidant packaging since it could help: (1) achieving continuous release of antioxidants from packaging material onto food surface during long food storage periods, (2) reducing risk of pro-oxidant effect at the food surface due to too much increase of released antioxidant at the food surface, (3) protecting antioxidants from neutralizing interactions with food components, and (4) reducing rapid loss of volatile antioxidants (e.g. essential oils) from edible packaging materials.

5.3.4 Application of combinational packaging methods

The effectiveness of antioxidant edible packaging (mostly coating) might be enhanced by combining this procedure with other packaging methods, such as vacuum (VP) or modified atmosphere packaging (MAP). Although the edible antioxidant films and coatings could be used for delivery of antioxidants, enzyme inhibitors, acids, chelating agents, and divalent ions (e.g. Ca^{+2}), the effective removal of oxygen from the packaging needs use of alternative packaging methods. In such combinational applications, susceptible products, such as meat and meat products, or cheese, are mostly treated first with an antioxidant coating, and then packed with impermeable (except respiring fruits and vegetables) plastic films using VP or MAP (Sayadi *et al.*, 2021; Catarino *et al.*, 2017; Mileriene *et al.*, 2021; Esmaeili *et al.*, 2020). To suppress the lipid oxidation (and also microbial growth), the O_2 is simply excluded from MAP gas mixtures. For example, the retail meat could be packed, using a mixture of 60-70% CO_2, 30-40% N_2, and < 0.5% CO (Sørheim *et al.*, 1997), while some cheese could be packed using 100% CO_2 (Alam and Goyal, 2011). In minimally processed fresh fruits and vegetables, the antioxidant coating is mostly combined with passive MAP using a semi-permeable (or perforated) plastic film that enables reaching of the equilibrium modified atmosphere conditions and ensures presence of sufficient residual O_2 in the package to prevent anaerobic respiration. For example, Sanchís *et al.* (2017) reduced browning in fresh-cut persimmon by combining passive MAP applied with perforated plastic film and pectin coating loaded with nisin (antimicrobial), citric acid (PPO inhibitor), and $CaCl_2$ (firming agent). Xing *et al.* (2010b) also successfully reduced enzymatic browning in fresh-cut lotus roots by combining ascorbic acid and citric acid loaded chitosan coating with passive MAP conducted by perforated plastic films. Thus, antioxidant edible coating combined with VP or MAP is a powerful tool that could be used to minimize oxidative changes in food. However, the sustainability of this combinational strategy might be questioned due to the need of using plastic (barrier or semi-permeable) packaging materials.

5.4 Enhancing the performance of flavor-release packaging

Encapsulation is a primary method to be used to enhance the performance of flavor-release edible packaging since increased stability and sustained release of flavor compounds are the key points for a successful application. The flavor compounds encapsulated previously could be incorporated into edible films or coatings, or free flavor compounds could be encapsulated during preparation of the packaging by the edible film matrix itself.

5.4.1 Application of encapsulation

5.4.1.1 Incorporation of encapsulated flavor compounds into packaging

The incorporation of encapsulated flavor compounds into packaging materials is the most important strategy employed to enhance the performance of flavor-release packaging. In general, maltodextrins, cyclodextrins, carrageenan, pectin, sodium-alginate, gelatin, gum Arabic, whey proteins, and zein were used alone or in proper combinations to encapsulate flavor compounds (Madene *et al.*, 2006; Chakraborty, 2017; Saifullah *et al.*, 2019). Although the classical encapsulation methods, such as spray drying and extrusion are used extensively for encapsulation of flavor compounds, methods, such as spray chilling and spray freeze drying, are more beneficial than these methods since they prevent loss of heat sensitive flavor compounds (Madene *et al.*, 2006; Saifullah *et al.*, 2019). Moreover, coaxial electrospinning and electrospraying are also promising novel methods employed in encapsulation of flavor compounds (Koo *et al.*, 2014; Wen *et al.*, 2017; Saifullah *et al.*, 2019). Electrospraying is a promising method for flavor-release coating since electrosprayed nanoencapsulated flavor agents could be dropped directly on to the food surface (Dhiman *et al.*, 2021). Moreover, the innovative idea of electrospraying flavor-loaded nanoparticles directly onto fabric surfaces could easily be adapted for coating of flavor compounds on to self-standing edible film surfaces (Ye *et al.*, 2019). The main benefits of employing nanotechnology in flavor encapsulation are better stability and sustained release profile that are beneficial to maintain desired flavor during long storage periods (Saifullah *et al.*, 2019). Moreover, the flavor-loaded nanoencapsules are mostly having a hydrophilic surface that helps their easy suspension in film and coating solutions. Therefore, application of nanoencapsulation also contributes to homogenous distribution of flavor agents within final dried packaging.

5.4.1.2 Use of film matrix as an encapsulant

The use of film matrix as an encapsulant is also sometimes employed as a strategy to enhance performance of flavor-release packaging. In this method, the flavor compound is mostly encapsulated within an emulsion-based film or coating. For example, Marcuzzo *et al.* (2010) encapsulated a mixture of different fruit flavor compounds into an emulsion film formed by homogenizing ι-carrageenan and fully hydrogenated vegetable oil. Similarly, Hambleton *et al.* (2009) showed encapsulation of different typical food flavors in emulsion-based films obtained by homogenizing Na-alginate with a lipid mixture (mixture of acetic acid ester of mono and diglycerides with beeswax). The complexes formed between flavor compounds and some hydrocolloids due to noncovalent bonds and interactions (e.g. hydrogen bonding, Van der Waals forces, ionic interactions, and hydrophobic interactions) could also be exploited to obtain flavor-release coatings. The starches are frequently employed for flavor-binding purposes since their amylose and amylopectin factions form complex with flavor compounds without interfering their aroma and taste perception (Arvisenet *et al.*, 2002). It was reported that the starch-flavor interaction occurs on the porous surfaces of starch via hydrogen bonds that intensify, depending on the polarity of flavor compounds. Different starch forms show variations in their flavor retention capacities. For example, maltodextrins, short chained oligosaccharides obtained by hydrolysis of starch, were reported to show the best flavor retention properties followed by pre-gelatinized starch and native granular starch (Boutboul *et al.*, 2002). The study by Laohakunjit and Kerdchoechuen (2007) is a typical example of applying starch coating in flavor-release packaging. These workers applied rice starch-pandan flavor (*Pandanus amaryllifolius* Roxb. leaf extract) complex-based suspension for coating of rice and improving its jasmine-like aromatic profile. Some of the other hydrocolloids that are capable of interacting and retaining flavor compounds in food formulations include xanthan gum (Xu *et al.*, 2017), acacia gum (Savary *et al.*, 2014), gelatin (Zafeiropoulou *et al.*, 2012), and pectin (Boland *et al.*, 2004). Thus, these hydrocolloids can also be used to develop flavor-release coating materials, depending on the type and strength of their interactions with the target flavor compounds.

5.5 Enhancing the performance of bioactive packaging

Bioactive packaging is the most challenging active packaging concept that should be supported by different technologies, methods, and strategies to increase solubility, stability, and bioavailability of bioactive susbstances employed in this method. The encapsulation and co-encapsulation of bioactive compounds (applied before incorporation into packaging materials) are the most important technologies used to enhance performance of bioactive packaging. However, it is also important to choose bioactive agents compatible with food composition and sensory properties (aroma and taste).

5.5.1 Application of encapsulation

The nano/micro encapsulation of bioactive agents mostly before incorporation into packaging materials provides one or more of the following benefits: (1) enabling suspension of water-insoluble bioactive agents in aqueous edible film-forming solutions, (2) protection of bioactive agents from undesired interactions and reactions (e.g. oxidation, complexation, degradation, etc.) in packaging and food during processing and storage, (4) prevention of volatile bioactive agent loss, (5) suppression of undesired aroma and taste of some bioactive agents, (6) stabilization of bioactive agents in gastric and intestinal fluids during passing from digestion system, and (7) enhancing bioavailability (mainly by nanoencapsulation) of bioactive agents (McClements *et al.*, 2015; Aboalnaja *et al.*, 2016; Chawda *et al.*, 2017; Rezaei *et al.*, 2019; Grgić *et al.*, 2020). Different methods used in encapsulation of bioactive compounds have been well documented in the literature (Đorđević *et al.*, 2015; Shishir *et al.*, 2018; Zhang *et al.*, 2020).

Due to their bioactive properties, such as antioxidant, antidiabetic, anticarcinogenic, antitumoral, anti-inflammatory, etc., polyphenols show different health benefits and protective effects against some chronic diseases (e.g. cancer, cardiovascular diseases, type-2 diabetes, neurodegenerative diseases, obesity).

Thus, extensive studies have been conducted to form hydrocolloid-based delivery systems for polyphenols (Zhang *et al.*, 2020). The solubility of polyphenols is an important factor for their homogenous distribution in edible packaging materials and sufficient release on to the food surface. Moreover, it is a truth that bioactive agents having low solubility cannot be absorbed efficiently in the gut and this causes low bioaccessibility. Therefore, increasing solubility of delivered bioactive compounds including polyphenols by use of nanotechnological methods (e.g. nanosuspensions, nanoemulsions, inclusion complexes, etc.) have become a primary tool in functional food development studies (Recharla *et al.*, 2017). However, a high solubility alone does not always mean high bioaccessibility. For example, resveratrol shows low water solubility, but high membrane bioaccessibility while green tea catechins are highly water soluble, but show low membrane bioaccessibility (Grgić *et al.*, 2020). Therefore, suitable delivery strategies should be selected for each bioactive compound, considering their solubility and bioaccessibility. For example, the use of cyclodextrins, cone-shaped cyclic oligosaccharides, having a hydrophobic core and hydrophilic shell, as nano carriers is a frequently used strategy to solubilize hydrophobic phenolic compounds (e.g. ellagic acid, quercetin, essential oils, etc.) and to improve their stability and bioavailability (Esfanjani *et al.*, 2016; Rezaei *et al.*, 2019). Non-viable yeast cells (mostly with *Saccharomyces cerevisiae*) are also applied as an encapsulant to increase solubility and stability of phenolic compounds, such as curcumin, resveratrol, and chlorogenic acid (Paramera *et al.*, 2011; Rezaei *et al.*, 2019). The hydrocolloid-based encapsulants employed for encapsulation of phenolic compounds include chitosan, starch, gum Arabic, soy protein isolate, gelatin, and caseins (Grgić *et al.*, 2020). The lipid-based delivery systems, such as micro- and nano-sized emulsions or double emulsions, liposomes, sphingosomes, solid lipid micro or nano particles, nanostructured lipid carriers, lipid-hydrocolloid hybrid nano particles are also used to encapsulate and improve stability, biocompatibility, and bioavailability of hydrophobic phenolic compounds (double emulsions are used both for hydrophilic and hydrophobic ones), and other hydrophobic bioactive agents, such as carotenoids, tocopherols, phytosterols, vitamins (A, D, E, K), and essential fatty acids (Lopez *et al.*, 2020; Barroso *et al.*, 2020; McClements and Öztürk, 2021; Tessaro *et al.*, 2021).

The encapsulation is also particularly important for stability of probiotics that are easily inactivated at the food surface due to processing and storage conditions as well as competition with native microbiota. Encapsulation also causes a significant improvement in gut colonization of probiotics (Corrêa-Filho *et al.*, 2019; Del Piano *et al.*, 2011). The alginate is one of the most versatile encapsulants for the probiotics. The alginates, alginate-milk protein or alginate-whey protein blends, and layer-by-layer coatings of alginate with chitosan are effectively used as an encapsulant to improve viability and gut colonization of probiotics (Shori, 2017).

5.5.2 Application of co-encapsulation

The co-encapsulation applied by combining different bioactive compounds within the same encapsule is a rising trend that is used to increase their bioavailability and stability, and to uncover their synergetic effects (Chawda *et al.*, 2017; Misra *et al.*,

2021). A frequently applied strategy is co-encapsulation of a mixture of hydrophilic and hydrophobic natural antioxidants. For example, Tavano *et al.* (2014) obtained synergetic free radical scavenging-based antioxidant activity between gallic acid and curcumin, and ascorbic acid and quercetin by co-encapsulating these agents into niosomal carriers. The co-encapsulation of epigallocatechin gallate with quercetin in liposomes also resulted in synergetic free radical scavenging-based antioxidant activity of these compounds (Chen *et al.*, 2019). The synergetic antioxidant effects of vitamin E with some carotenoids (e.g. β-carotene, lycopene, and lutein) (Shixian *et al.*, 2005) or with some polyphenols (e.g. quercetin, epicatechin gallate, and epigallocatechin gallate) (Murakami *et al.*, 2003) determined by *in-vitro* LDL oxidation and liposomal oxidation models, respectively, could also be used to prepare co-encapsulated antioxidant mixtures suitable for use in functional foods. Aditya *et al.* (2015) showed that co-encapsulation of curcumin and catechin using a W/O/W double emulsion increased *in-vitro* bioaccessibility and stability of these bioactive compounds in simulated gastrointestinal fluids. Moreover, Liu *et al.* (2020b) improved the stability of liposome encapsulated β-carotene by co-encapsulating this agent with vitamin C.

The co-encapsulation strategy is also applied to oils and fatty acids. The omega-3 fatty acids, such as alpha-linolenic acid (ALA) found in plant seeds, eicosapentaenoic acid (EPA) and docosahexaenoic acid (DHA) found in fish oils, and omega-6 fatty acid found mainly in vegetable oils are frequently delivered to obtain functional foods. The natural antioxidants (e.g. carotenoids, tocopherols, phenolic antioxidants, etc.) co-encapsulated with bioactive lipids prevent their oxidative changes during processing and storage (Chawda *et al.*, 2017). The co-encapsulation of fish oils with flavor compounds (e.g. vanillin-apple flavor mixture, limonene, etc.) is also applied to suppress the undesired odor formed in these oils during storage (Serfert *et al.*, 2010; Chen *et al.*, 2013). The use of nanoencapsulation strategies is also particularly beneficial to improve compatibility of bioactive lipids with food and to enhance their bioaccessibility (McClements and Öztürk, 2021).

The co-encapsulation is also conducted by encapsulation of probiotics with other bioactive compounds. For example, the presence of prebiotics, such as inulin, resistant starch, and oligofructose during encapsulation of probiotic bacteria with alginate increased the stability of these bacteria in simulated gastrointestinal conditions (And and Kailasapathy, 2005; Atia *et al.*, 2017). Okuro *et al.* (2013) also improved the viability of probiotic bacteria during storage by co-encapsulating bacteria with prebiotic polydextrose in solid lipid microparticles. The co-encapsulation of probiotic bacteria in the presence of some antioxidant polyphenols (e.g. green tea polyphenols) and omega-3 fatty acid source fish oils also increased the viability (Eratte *et al.*, 2015, 2018) and potential beneficial actions of probiotic bacteria (Misra *et al.*, 2021; Das, 2002).

5.5.3 Optimization of food composition

The composition of food is highly effective in the stability and bioavailability of bioactive agents. For example, the presence of suitable levels of vitamin C in food might be beneficial to increase the oxidative stability of phenolic compounds

(Peters *et al.*, 2010). The digestible carbohydrates and dietary lipids could also support the bioaccessibility of polyphenols, but food rich in dietary fiber (especially hemicellulose) and divalent minerals, as well as viscous and protein-rich food might limit the bioaccessibility of polyphenols (Palafox-Carlos *et al.*, 2011; Bohn, 2014). Some hydrocolloids used as functional additives could also adversely affect the bioaccessibility of polyphenols. For example, it was reported that the low methoxyl pectin used to form polyphenol enriched oil-in-water emulsions reduced bioaccessibility of flavonoids (Velderrain-Rodríguez *et al.*, 2021). However, it was reported that the polyphenols non-absorbed in the small intestines due to reduced bioaccessibility caused by dietary fibers (mostly bind polyphenols) reach large intestine and might help in creating a healthy antioxidant environment (Quirós-Sauceda *et al.*, 2014). Thus, further studies are needed to understand the exact health benefits of polyphenols (absorbed or non-absorbed), depending on food composition.

5.5.4 Reduction of undesired aroma and taste of bioactive agents

One of the most important problems of using bioactive agents in edible packaging is their undesired taste and aroma. Some examples include fishy taste of omega-3 fatty acids, characteristic taste and odor of essential oils, and bitter and astringent tastes of polyphenols and protein hydrolysates. The encapsulation technologies can be applied effectively to mask the undesired aroma and taste problems originating from bioactive agents. For example, Linde *et al.* (2009) reduced the bitter taste of soy protein hydrolysates considerably by employing α-cyclodextin as an encapsulant. The α-cyclodextin reduces bitterness both by encapsulating part of the protein hydrolysate at its internal site or by creating interactions between its external surface and amino acids/amino acid residues in the hydrolysate. Moreover, the sweet taste of α-cyclodextrin also contributes to the reduced perception of bitter taste (Linde *et al.*, 2009). The undesired tastes formed in foods due to bioactive compounds can also be eliminated or reduced by using some bitterness-inhibitor or flavor-modifying agents. For example, neohesperidin dihydrochalcone (NHDC) is a Generally Recognized as Safe (GRAS) agent obtained by hydrogenating neohesperidin, a flavonoid found in citrus fruits. Due to its pleasant sweet taste, the NHDC is used as a sweetener, but the most important attribute of this compound is its ability to reduce the perception of bitterness, saltiness, and sharp and spicy tastes (Gloria, 2003). Another approach is that strong flavoring agents are incorporated into the food to mask (reduce perception) the undesired taste or aroma originating from bioactive agent (Ley, 2008). For example, it was reported that an acceptable organoleptic profile could be obtained by combining essential oil carvacrol with vanilla aroma, and thymol with strawberry or vanilla aroma (Gutierrez *et al.*, 2009). Mukai *et al.* (2007) also determined that the image of the sweetness and sourness evoked by strawberry aroma is highly effective in increasing the threshold of bitterness in nutritional products originating form branched-chain amino acids, such as L-leucine, L-isoleucine, and L-valine (Mukai *et al.*, 2007). A different approach to suppress the undesired taste of bioactive agents is to employ congruence that is defined as the extent to which the taste and smell fits together (Amsellem and Ohla, 2016). Some congruent pairs, such as sucrose-citrus odor (Lim *et al.*, 2014) and sucrose-strawberry odor (Frank and Byram, 1988) could be employed to reduce the perception of an undesired taste originating from the bioactive agents.

References

Abdou, E.S., G.F. Galhoum and E.N. Mohamed (2018). Curcumin loaded nanoemulsions/ pectin coatings for refrigerated chicken fillets, *Food Hydrocoll.*, 83: 445-453.

Abdul-Rahman, S.M. (2019). Antimicrobial activity of whey protein isolate coating incorporated with partially purified duck egg white lysozyme and Na_2-EDTA and its use in chicken breast fillets preservation, *Biochem. Cell. Arch.*, 19: 2857-2863.

Aboalnaja, K.O., S. Yaghmoor, T.A. Kumosani and D.J. McClements (2016). Utilization of nanoemulsions to enhance bioactivity of pharmaceuticals, supplements, and nutraceuticals: Nanoemulsion delivery systems and nanoemulsion excipient systems, *Expert Opin. Drug Deliv.*, 13: 1327-1336.

Aditya, N.P., S. Aditya, H. Yang, H.W. Kim, S.O. Park and S. Ko (2015). Co-delivery of hydrophobic curcumin and hydrophilic catechin by a water-in-oil-in-water double emulsion, *Food Chem.*, 173: 7-13.

Aguirre-Joya, J.A., M.A. Cerqueira, J. Ventura-Sobrevilla, M.A. Aguilar-Gonzalez, E. Carbó-Argibay, L.P. Castro and C.N. Aguilar (2019). Candelilla wax-based coatings and films: Functional and physicochemical characterization, *Food Bioproc. Tech.*, 12: 1787-1797.

Alam, T. and G.K. Goyal (2011). Effect of MAP on microbiological quality of Mozzarella cheese stored in different packages at 7±1°C, *J. Food Sci. Technol.*, 48: 120-123.

Ali, S., A.S. Khan, A. Nawaz, M.A. Anjum, S. Naz, S. Ejaz and S. Hussain (2019). Aloe vera gel coating delays postharvest browning and maintains quality of harvested litchi fruit, *Postharvest Biol. Tech.*, 157: 110960.

Ali, S., M.A. Anjum, A. Nawaz, S. Naz, S. Hussain, S. Ejaz and H. Sardar (2020). Effect of pre-storage ascorbic acid and Aloe vera gel coating application on enzymatic browning and quality of lotus root slices, *J. Food Biochem.*, 44: e13136.

Alkan, D., L.Y. Aydemir, I. Arcan, H. Yavuzdurmaz, H.I. Atabay, C. Ceylan and A. Yemenicioğlu (2011). Development of flexible antimicrobial packaging materials against *Campylobacter jejuni* by incorporation of gallic acid into zein-based films, *J. Agric. Food Chem.*, 59: 11003-11010.

Alkan, D. and A. Yemenicioğlu (2016). Potential application of natural phenolic antimicrobials and edible film technology against bacterial plant pathogens, *Food Hydrocoll.*, 55: 1-10.

Almasi, H., M. Jahanbakhsh Oskouie and A. Saleh (2020). A review on techniques utilized for design of controlled release food active packaging, *Crit. Rev. Food Sci. Nutr.*, 61: 2601-2621.

Al Mousawi, A.J. and S.R. Khair (2019). The use of macro lament of alginate and rosemary in Monterey cheese coating, *Plant Arch.*, 19: 4369-4378.

Aloui, H., K. Khwaldia, L. Sánchez-González, L. Muneret, C. Jeandel, M. Hamdi and S. Desobry (2014). Alginate coatings containing grapefruit essential oil or grapefruit seed extract for grapes preservation, *Int. J. Food Sci. Technol.*, 49: 952-959.

Amaral, A.B., M.V.D. Silva and S.C.D.S. Lannes (2018). Lipid oxidation in meat: Mechanisms and protective factors – A review, *Food Sci. Technol.*, 38: 1-15.

Amor, G., M. Sabbah, L. Caputo, M. Idbella, V. De Feo, R. Porta *et al.* (2021). Basil essential oil: Composition, antimicrobial properties, and microencapsulation to produce active chitosan films for food packaging, *Foods*, 10: 121.

Amsellem, S. and K. Ohla (2016). Perceived odor-taste congruence influences intensity and pleasantness differently, *Chem. Senses*, 41: 677-684.

And, C.I. and K. Kailasapathy (2005). Effect of co-encapsulation of probiotics with prebiotics on increasing the viability of encapsulated bacteria under *in vitro* acidic and bile salt conditions and in yogurt, *J. Food Sci.*, 70: M18-M23.

Ansay, S.E., K.A. Darling and C.W. Kaspar (1999). Survival of *Escherichia coli* O157: H7 in ground-beef patties during storage at 2, –2, 15 and then –2°C, and –20°C, *J. Food Prot.*, 62: 1243-1247.

Appendini, P. and J.H. Hotchkiss (2002). Review of antimicrobial food packaging, *Innov. Food Sci. Emerg. Technol.*, 3: 113-126.

Arcan, I. and A. Yemenicioğlu (2013). Development of flexible zein–wax composite and zein–fatty acid blend films for controlled release of lysozyme, *Food Res. Int.*, 51: 208-216.

Arcan, I. and A. Yemenicioglu (2014). Controlled release properties of zein–fatty acid blend films for multiple bioactive compounds, *J. Agric. Food Chem.*, 62: 8238-8246.

Arqués, J.L., J. Fernández, P. Gaya, M. Nuñez, E. Rodríguez and M. Medina (2004). Antimicrobial activity of reuterin in combination with nisin against food-borne pathogens, *Int. J. Food Microbiol.*, 95: 225-229.

Arqués, J.L., E. Rodríguez, M. Nuñez and M. Medina (2008). Inactivation of gram-negative pathogens in refrigerated milk by reuterin in combination with nisin or the lactoperoxidase system, *Eur. Food Res. Technol.*, 227: 77-82.

Arvisenet, G., A. Voilley and N. Cayot (2002). Retention of aroma compounds in starch matrices: Competitions between aroma compounds toward amylose and amylopectin, *J. Agric. Food Chem.*, 50: 7345-7349.

Atia, A., A.I. Gomma, I. Fliss, E. Beyssac, G. Garrait and M. Subirade (2017). Molecular and biopharmaceutical investigation of alginate–inulin synbiotic coencapsulation of probiotic to target the colon, *J. Microencapsul.*, 34: 171-184.

Avila-Sosa, R., E. Palou, M.T.J. Munguía, G.V. Nevárez-Moorillón, A.R.N. Cruz and A. López-Malo (2012). Antifungal activity by vapor contact of essential oils added to amaranth, chitosan, or starch edible films, *Int. J. Food Microbiol.*, 153: 66-72.

Aymerich, T., A. Jofre, M. Garriga and M. Hugas (2005). Inhibition of *Listeria monocytogenes* and *Salmonella* by natural antimicrobials and high hydrostatic pressure in sliced cooked ham, *J. Food Prot.*, 68: 173-177.

Bahrami, A., R. Delshadi, E. Assadpour, S.M. Jafari and L. Williams (2020). Antimicrobial-loaded nanocarriers for food packaging applications, *Adv. Colloid Interface Sci.*, 278: 102140.

Bakota, E.L., L. Aulisa, K.M. Galler and J.D. Hartgerink (2011). Enzymatic cross-linking of a nanofibrous peptide hydrogel, *Biomacromolecules*, 12: 82-87.

Barba, C., A. Eguinoa and J.I. Maté (2015). Preparation and characterization of β-cyclodextrin inclusion complexes as a tool of a controlled antimicrobial release in whey protein edible films, *LWT-Food Sci. Technol.*, 64: 1362-1369.

Barbiroli, A., F. Bonomi, G. Capretti, S. Iametti, M. Manzoni, L. Piergiovanni and M. Rollini (2012). Antimicrobial activity of lysozyme and lactoferrin incorporated in cellulose-based food packaging, *Food Control*, 26: 387-392.

Barroso, L., C. Viegas, J. Vieira, C. Pego, J. Costa and P. Fonte (2020). Lipid-based carriers for food ingredients delivery, *J. Food Eng.*, 295: 110451.

Behbahani, B.A., M. Noshad and H. Jooyandeh (2020). Improving oxidative and microbial stability of beef using Shahri Balangu seed mucilage loaded with cumin essential oil as a bioactive edible coating, *Biocatal. Agric. Biotechnol.*, 24: 101563.

Benbettaïeb, N., A. Assifaoui, T. Karbowiak, F. Debeaufort and O. Chambin (2016a). Controlled release of tyrosol and ferulic acid encapsulated in chitosan–gelatin films after electron beam irradiation, *Radiat. Phys. Chem.*, 118: 81-86.

Benbettaïeb, N., J.P. Gay, T. Karbowiak and F. Debeaufort (2016b). Tuning the functional properties of polysaccharide–protein bio-based edible films by chemical, enzymatic, and physical cross-linking, *Compr Rev. Food Sci. Food Saf.*, 15: 739-752.

Benbettaïeb, N., O. Chambin, T. Karbowiak and F. Debeaufort (2016c). Release behavior of quercetin from chitosan-fish gelatin edible films influenced by electron beam irradiation, *Food Control,* 66: 315-319.

Benbettaïeb, N., T. Karbowiak, C.H. Brachais and F. Debeaufort (2016d). Impact of electron beam irradiation on fish gelatin film properties, *Food Chem.*, 195: 11-18.

Bessho, M., T. Kojima, S. Okuda and M. Hara (2007). Radiation-induced cross-linking of gelatin by using γ-rays: Insoluble gelatin hydrogel formation, *Bull. Chem. Soc. Jpn.*, 80: 979-985.

Bobo, G., C. Arroqui and P. Virseda (2022). Natural plant extracts as inhibitors of potato polyphenol oxidase: The green tea case study, *LWT-Food Sci. Technol.*, 153: 112467.

Bohn, T. (2014). Dietary factors affecting polyphenol bioavailability, *Nutr. Rev.*, 72: 429-452.

Boland, A.B., K. Buhr, P. Giannouli and S.M. van Ruth (2004). Influence of gelatin, starch, pectin and artificial saliva on the release of 11 flavour compounds from model gel systems, *Food Chem.*, 86: 401-411.

Bosquez-Molina, E., E. Ronquillo-de Jesús, S. Bautista-Baños, J.R. Verde-Calvo and J. Morales-López (2010). Inhibitory effect of essential oils against *Colletotrichum gloeosporioides* and *Rhizopus stolonifer* in stored papaya fruit and their possible application in coatings, *Postharvest Biol. Technol.*, 57: 132-137.

Bourbon, A.I., M.J. Costa, L.C. Maciel, L. Pastrana, A.A. Vicente and M.A. Cerqueira (2021). Active carboxymethylcellulose-based edible films: Influence of free and encapsulated curcumin on films' properties, *Foods*, 10: 1512.

Boutboul, A., P. Giampaoli, A. Feigenbaum and V. Ducruet (2002). Influence of the nature and treatment of starch on aroma retention, *Carbohydr. Polym.*, 47: 73-82.

Boyacı, D., F. Korel and A. Yemenicioğlu (2016). Development of activate-at-home-type edible antimicrobial films: An example pH-triggering mechanism formed for smoked salmon slices using lysozyme in whey protein films, *Food Hydrocoll.*, 60: 170-178.

Brown, C.A., B. Wang, and J.H. Oh (2008). Antimicrobial activity of lactoferrin against foodborne pathogenic bacteria incorporated into edible chitosan film, *J. Food Prot.*, 71: 319-324.

Buonocore, G.G., A. Conte, M.R. Corbo, M. Sinigaglia and M.A. Del Nobile (2005). Mono and multilayer active films containing lysozyme as antimicrobial agent, *Innov. Food Sci. Emerg. Technol.*, 6: 459-464.

Cano Embuena, A.I., M. Cháfer Nácher, A. Chiralt Boix, M.P. Molina Pons, M. Borrás Llopis, M.C. Beltran Martínez and C. González Martínez (2017). Quality of goat's milk cheese as affected by coating with edible chitosan-essential oil films, *Int. J. Dairy Technol.*, 70: 68-76.

Cao, Y., D. Zhou, X. Zhang, X. Xiao, Y. Yu and X. Li (2021). Synergistic effect of citral and carvacrol and their combination with mild heat against *Cronobacter sakazakii* CICC 21544 in reconstituted infant formula, *LWT-Food Sci. Technol.*, 138: 110617.

Carrión-Granda, X., I. Fernández-Pan, J. Rovira and J.I. Maté (2018). Effect of antimicrobial edible coatings and modified atmosphere packaging on the microbiological quality of cold stored hake (*Merluccius merluccius*) fillets, *J. Food Qual.*, 2018: 6194906.

Catarino, M.D., J.M. Alves-Silva, R.P. Fernandes, M.J. Gonçalves, L.R. Salgueiro, M.F. Henriques and S.M. Cardoso (2017). Development and performance of whey protein active coatings with *Origanum virens* essential oils in the quality and shelf-life improvement of processed meat products, *Food Control.*, 80: 273-280.

Chakraborty, S. (2017). Carrageenan for encapsulation and immobilization of flavor, fragrance, probiotics, and enzymes: A review, *J. Carbohydr. Chem.*, 36: 1-19.

Chalier, P., A. Ben Arfa, V. Guillard and N. Gontard (2009). Moisture and temperature triggered release of a volatile active agent from soy protein coated paper: Effect of glass transition phenomena on carvacrol diffusion coefficient, *J. Agric. Food Chem.*, 57: 658-665.

Chawda, P.J., J. Shi, S. Xue and S. Young Quek (2017). Co-encapsulation of bioactives for food applications, *Food Qual. Saf.*, 1: 302-309.

Chen, J.S., C.I. Wei and M.R. Marshall (1991a). Inhibition mechanism of kojic acid on polyphenol oxidase, *J. Agric. Food Chem.*, 39: 1897-1901.

Chen, J.S., C.I. Wei, R.S. Rolle, W.S. Otwell, M.O. Balaban and M.R. Marshall (1991b). Inhibitory effect of kojic acid on some plant and crustacean polyphenol oxidases, *J. Agric. Food Chem.*, 39: 1396-1401.

Chen, Q., D. McGillivray, J. Wen, F. Zhong and S.Y. Quek (2013). Co-encapsulation of fish oil with phytosterol esters and limonene by milk proteins, *J. Food Eng.*, 117: 505-512.

Chen, W., M. Zou, X. Ma, R. Lv, T. Ding and D. Liu (2019). Co-encapsulation of EGCG and quercetin in liposomes for optimum antioxidant activity, *J. Food Sci.*, 84: 111-120.

Chen, Y., L. Xu, Y. Wang, Z. Chen, M. Zhang and H. Chen (2020). Characterization and functional properties of a pectin/tara gum based edible film with ellagitannins from the unripe fruits of *Rubus chingii* Hu, *Food Chem.*, 325: 126964.

Chen, W., S. Ma, Q. Wang, D.J. McClements, X. Liu, T. Ngai and F. Liu (2021). Fortification of edible films with bioactive agents: A review of their formation, properties, and application in food preservation, *Crit. Rev. Food Sci. Nutr.* Doi: 10.1080/10408398.2021.1881435.

Choi, I., S.E. Lee, Y. Chang, M. Lacroix and J. Han (2018). Effect of oxidized phenolic compounds on cross-linking and properties of biodegradable active packaging film composed of turmeric and gelatin, *LWT-Food Sci. Technol.*, 93: 427-433.

Chueca, B., R. Pagán and D. García-Gonzalo (2014). Oxygenated monoterpenes citral and carvacrol cause oxidative damage in *Escherichia coli* without the involvement of tricarboxylic acid cycle and Fenton reaction, *Int. J. Food Microbiol.*, 189: 126-131.

Chung, W. and R.E. Hancock (2000). Action of lysozyme and nisin mixtures against lactic acid bacteria, *Int. J. Food Microbiol.*, 60: 25-32.

Çoban, M.Z. (2021). Effectiveness of chitosan/propolis extract emulsion coating on refrigerated storage quality of crayfish meat (*Astacus leptodactylus*) CyTA-J, *Food*, 19: 212-219.

Corrêa-Filho, L.C., M. Moldão-Martins and V.D. Alves (2019). Advances in the application of microcapsules as carriers of functional compounds for food products, *Appl. Sci.*, 9: 571.

Criado, P., C. Fraschini, M. Jamshidian, S. Salmieri, N. Desjardins, A. Sahraoui and M. Lacroix (2019). Effect of cellulose nanocrystals on thyme essential oil release from alginate beads: Study of antimicrobial activity against *Listeria innocua* and ground meat shelf-life in combination with gamma irradiation, *Cellulose*, 6: 5247-5265.

Cui, H., L. Yuan and L. Lin (2017a). Novel chitosan film embedded with liposome-encapsulated phage for biocontrol of *Escherichia coli* O157: H7 in beef, *Carbohydr. Polym.*, 177: 156-164.

Cui, H., L. Yuan, W. Li and L. Lin (2017b). Edible film incorporated with chitosan and *Artemisia annua* oil nanoliposomes for inactivation of *Escherichia coli* O157: H7 on cherry tomato, *Int. J. Food Sci. Tech.*, 52: 687-698.

Cui, H., M. Bai, C. Li, R. Liu and L. Lin (2018). Fabrication of chitosan nanofibers containing tea tree oil liposomes against *Salmonella* spp. in chicken, *LWT-Food Sci. Technol.*, 96: 671-678.

Dadwal, V., R. Joshi and M. Gupta (2021). Formulation, characterization and *in vitro* digestion of polysaccharide reinforced Ca-alginate microbeads encapsulating *Citrus medica* L. phenolics, *LWT-Food Sci. and Technol.*, 152: 112290.

Das, U.N. (2002). Essential fatty acids as possible enhancers of the beneficial actions of probiotics, *Nutrition*, 18: 786-789.

DeJong, G.A.H. and S.J. Koppelman (2002). Transglutaminase catalyzed reactions: Impact on food applications, *J. Food Sci.*, 67: 2798-2806.

Del Piano, M., S. Carmagnola, M. Ballarè, M. Sartori, M. Orsello, M. Balzarini and L. Mogna (2011). Is microencapsulation the future of probiotic preparations? The increased efficacy of gastro-protected probiotics, *Gut Microbes*, 2: 120-123.

Deng, L., X. Kang, Y. Liu, F. Feng and H. Zhang (2017). Effects of surfactants on the formation of gelatin nanofibres for controlled release of curcumin, *Food Chem.*, 231: 70-77.

De Souza, P.M., A. Fernández, G. López-Carballo, R. Gavara and P. Hernández-Muñoz (2010). Modified sodium caseinate films as releasing carriers of lysozyme, *Food Hydrocoll.*, 24: 300-306.

Dey, M., B. Ghosh and T.K. Giri (2020). Enhanced intestinal stability and pH sensitive release of quercetin in GIT through gellan gum hydrogels, *Colloid. Surface B.*, 196: 111341.

Dhiman, A., R. Suhag, A. Singh and P.K. Prabhakar (2021). Mechanistic understanding and potential application of electrospraying in food processing: A review, *Crit. Rev. Food Sci. Nutr.* https://doi.org/10.1080/10408398.2021.1926907

Dhumal, C.V., K. Pal and P. Sarkar (2019a). Synthesis, characterization, and antimicrobial efficacy of composite films from guar gum/sago starch/whey protein isolate loaded with carvacrol, citral and carvacrol-citral mixture, *J. Mater. Sci. Mater. Med.*, 30: 1-14.

Dhumal, C.V., K. Pal and P. Sarkar (2019b). Characterization of tri-phasic edible films from chitosan, guar gum, and whey protein isolate loaded with plant-based antimicrobial compounds, *Polym. Plast. Tech. Mat.*, 58: 255-269.

Dini, H., A.A. Fallah, M. Bonyadian, M. Abbasvali and M. Soleimani (2020). Effect of edible composite film based on chitosan and cumin essential oil-loaded nanoemulsion combined with low-dose gamma irradiation on microbiological safety and quality of beef loins during refrigerated storage, *Int. J. Biol. Macromol.*, 164: 1501-1509.

Donsì, F., E. Marchese, P. Maresca, G. Pataro, K.D. Vu, S. Salmieri *et al.* (2015). Green beans preservation by combination of a modified chitosan based-coating containing nanoemulsion of mandarin essential oil with high pressure or pulsed light processing, *Postharvest Biol. Technol.*, 106: 21-32.

Đorđević, V., B. Balanč, A. Belščak-Cvitanović, S. Lević, K. Trifković, A. Kalušević *et al.* (2015). Trends in encapsulation technologies for delivery of food bioactive compounds, *Food Eng. Rev.*, 7: 452-490.

Ehivet, F.E., B. Min, M.K. Park and J.H. Oh (2011). Characterization and antimicrobial activity of sweetpotato starch-based edible film containing origanum (*Thymus capitatus*) oil, *J. Food Sci.*, 76: C178-C184.

Ellison, R. and T.J. Giehl (1991). Killing of gram-negative bacteria by lactoferrin and lysozyme, *J. Clin. Invest.*, 88: 1080-1091.

El-Mohamedy, R.S., N.G. El-Gamal and A.R.T. Bakeer (2015). Application of chitosan and essential oils as alternatives fungicides to control green and blue moulds of citrus fruits, *Int. J. Curr. Microbiol. Appl. Sci.*, 4: 629-643.

Eratte, D., S. McKnight, T.R. Gengenbach, K. Dowling, C.J. Barrow and B.P. Adhikari (2015). Co-encapsulation and characterisation of omega-3 fatty acids and probiotic bacteria in whey protein isolate – Gum Arabic complex coacervates, *J. Funct. Foods*, 19: 882-892.

Eratte, D., K. Dowling, C.J. Barrow and B. Adhikari (2018). Recent advances in the microencapsulation of omega-3 oil and probiotic bacteria through complex coacervation: A review, *Trends in Food Sci. Technol.*, 71: 121-131.

Esfanjani, A.F. and S.M. Jafari (2016). Hydrocolloid nano-particles and natural nano-carriers for nano-encapsulation of phenolic compounds, *Colloid Surface B*, 146: 532-543.

Eskin, N.A.M. (1990). *Biochemistry of Foods*, 2nd ed. Academic Press, San Diego, USA, p. 539.

Esmaeili, H., N. Cheraghi, A. Khanjari, M. Rezaeigolestani, A.A. Basti, A. Kamkar and E.M. Aghaee (2020). Incorporation of nanoencapsulated garlic essential oil into edible films: A novel approach for extending shelf-life of vacuum-packed sausages, *Meat Sci.*, 166: 108135.

Ettayebi, K., J. El Yamani and B.D. Rossi-Hassani (2000). Synergistic effects of nisin and thymol on antimicrobial activities in *Listeria monocytogenes* and *Bacillus subtilis*, *FEMS Microbiol. Lett.*, 183: 191-195.

Fathi-Achachlouei, B., N. Babolanimogadam and Y. Zahedi (2021). Influence of anise (*Pimpinella anisum* L.) essential oil on the microbial, chemical, and sensory properties of chicken fillets wrapped with gelatin film, *Food Sci. Technol. Int.*, 27: 123-134.

Fathi, N., H. Almasi and M.K. Pirouzifard (2018). Effect of ultraviolet radiation on morphological and physicochemical properties of sesame protein isolate based edible films, *Food Hydrocoll.*, 85: 136-143.

Faustman, C., Q. Sun, R. Mancini and S.P. Suman (2010). Myoglobin and lipid oxidation interactions: Mechanistic bases and control, *Meat Sci.*, 86: 86-94.

Fernández-Pan, I., M. Royo and J. Ignacio Mate (2012). Antimicrobial activity of whey protein isolate edible films with essential oils against food spoilers and food-borne pathogens, *J. Food Sci.*, 77: M383-M390.

FDA (US Food and Drug Administration) (2021a). 21CFR172.515, Title 21, vol. 3.

FDA (US Food and Drug Administration) (2021b). 21CFR184.1257, Title 21, vol. 3.

FDA (US Food and Drug Administration) (2021c). 21CFR182.60, Title 21, vol. 3.

Frank, R.A. and J. Byram (1988). Taste-smell interactions are tastant and odorant dependent, *Chem. Senses*, 13: 445-455.

Gad, A.S. and A.F. Sayd (2015). Antioxidant properties of rosemary and its potential uses as natural antioxidant in dairy products – A review, *Food Nutr. Sci.*, 6: 179-193.

Gharibzahedi, S.M.T., S. Roohinejad, S. George, F.J. Barba, R. Greiner, G.V. Barbosa-Cánovas, and K. Mallikarjunan (2018). Innovative food processing technologies on the transglutaminase functionality in protein-based food products: Trends, opportunities and drawbacks, *Trends in Food Sci. Technol.*, 75: 194-205.

Giteru, S.G., R. Coorey, D. Bertolatti, E. Watkin, S. Johnson and Z. Fang (2015). Physicochemical and antimicrobial properties of citral and quercetin incorporated kafirin-based bioactive films, *Food Chem.*, 168: 341-347.

Gloria, M.B.A. (2003). Sweeteners, pp. 5695-5702. *In:* B. Caballero, L.C. Trugo and P.M. Finglas (Eds.). *Encyclopedia of Food Sciences and Nutrition*, 2nd ed. Academic Press, London, UK.

González-Estrada, R.R., E. Carvajal-Millán, J.A. Ragazzo-Sánchez, P.U. Bautista-Rosales and M. Calderón-Santoyo (2017). Control of blue mold decay on Persian lime: Application of covalently cross-linked arabinoxylans bioactive coatings with antagonistic yeast entrapped, *LWT-Food Sci. Technol.*, 85: 187-196.

Grgić, J., G. Šelo, M. Planinić, M. Tišma and A. Bucić-Kojić (2020). Role of the encapsulation in bioavailability of phenolic compounds, *Antioxidants*, 9: 923.

Guerreiro, A.C., C.M. Gago, M.L. Faleiro, M.G. Miguel and M.D. Antunes. 2015. The effect of alginate-based edible coatings enriched with essential oils constituents on *Arbutus unedo* L. fresh fruit storage, *Postharvest Biol. Technol.*, 100: 226-233.

Gutierrez, L., A. Escudero, R. Batlle and C. Nerin (2009). Effect of mixed antimicrobial agents and flavours in active packaging films, *J. Agric. Food Chem.*, 57: 8564-8571.

Hambleton, A., F. Debeaufort, A. Bonnotte and A. Voilley (2009). Influence of alginate emulsion-based films structure on its barrier properties and on the protection of microencapsulated aroma compound, *Food Hydrocoll.*, 23: 2116-2124.

Han, J.H. (2005). Antimicrobial packaging systems, pp. 80-108. *In:* J.H. Han (Ed.). *Innovations in Food Packaging*. Academic Press, London, UK.

Hashemi, S.M.B. and A.M. Khaneghah (2017). Characterization of novel basil-seed gum active edible films and coatings containing oregano essential oil, *Prog. Org. Coat.*, 110: 35-41.

He, Q., B. Shi and K. Yao (2006). Interactions of gallotannins with proteins, amino acids, phospholipids and sugars, *Food Chem.*, 95: 250-254.

He, J., S. Huang, X. Sun, L. Han, C. Chang, W. Zhang and Q. Zhong (2019). Carvacrol loaded solid lipid nanoparticles of propylene glycol monopalmitate and glyceryl monostearate: Preparation, characterization, and synergistic antimicrobial activity, *Nanomaterials*, 9: 1162.

Heck, T., G. Faccio, M. Richter and L. Thöny-Meyer (2013). Enzyme-catalyzed protein crosslinking, *Appl. Microbiol. Biotechnol.*, 97: 461-475.

Hemalatha, T., T. Uma Maheswari, R. Senthil, G. Krithiga and K. Anbukkarasi (2017). Efficacy of chitosan films with basil essential oil: Perspectives in food packaging, *J. Food Meas. Charact.*, 11: 2160-2170.

Heras-Mozos, R., V. Muriel-Galet, G. López-Carballo, R. Catalá, P. Hernández-Muñoz and R. Gavara (2019). Development and optimization of antifungal packaging for sliced pan loaf based on garlic as active agent and bread aroma as aroma corrector, *Int. J. Food Microbiol.*, 290: 42-48.

Hilbig, J., K. Hartlieb, K. Herrmann, J. Weiss and M. Gibis (2020). Influence of calcium on white efflorescence formation on dry fermented sausages with co-extruded alginate casings, *Food Res. Int.*, 131: 109012.

Hoffman, A.D., K.L. Gall, D.M. Norton and M. Wiedmann (2003). *Listeria monocytogenes* contamination patterns for the smoked fish processing environment and for raw fish, *J. Food Prot.*, 66: 52-60.

Hosseini, S.M., H. Hosseini, M.A. Mohammadifar, J.B. German, A.M. Mortazavian, A. Mohammadi *et al.* (2014). Preparation and characterization of alginate and alginate-resistant starch microparticles containing nisin, *Carbohydr. Polym.*, 103: 573-580.

Hu, W., K. Feng, Z. Xiu, A. Jiang and Y. Lao (2019). Thyme oil alginate-based edible coatings inhibit growth of pathogenic microorganisms spoiling fresh-cut cantaloupe, *Food Biosci.*, 32: 100467.

Jakobek, L. (2015). Interactions of polyphenols with carbohydrates, lipids and proteins, *Food Chem.*, 175: 556-567.

Huang, M., H. Wang, X. Xu, X. Lu, X. Song and G. Zhou (2020). Effects of nanoemulsion-based edible coatings with composite mixture of rosemary extract and ε-poly-L-lysine on the shelf-life of ready-to-eat carbonado chicken, *Food Hydrocoll.*, 102: 105576.

Huber, D., G. Tegl, M. Baumann, E. Sommer, E.G. Gorji, N. Borth *et al.* (2017). Chitosan hydrogel formation using laccase activated phenolics as cross-linkers, *Carbohydr. Polym.*, 157: 814-822.

İncili, G.K., P. Karatepe, M. Akgöl, B. Kaya, H. Kanmaz and A.A. Hayaloğlu (2021). Characterization of *Pediococcus acidilactici* postbiotic and impact of postbiotic-fortified chitosan coating on the microbial and chemical quality of chicken breast fillets, *Int. J. Biol. Macromol.*, 184: 429-437.

Iñiguez-Moreno, M., J.A. Ragazzo-Sánchez, J.C. Barros-Castillo, T. Sandoval-Contreras and M. Calderón-Santoyo (2020). Sodium alginate coatings added with Meyerozyma caribbica: Postharvest biocontrol of *Colletotrichum gloeosporioides* in avocado (*Persea americana* Mill. cv. Hass), *Postharvest Biol. Technol.*, 163: 111123.

Jang, S.A., Y.J. Shin and K.B. Song (2011). Effect of rapeseed protein–gelatin film containing grapefruit seed extract on 'Maehyang' strawberry quality, *Int. J. Food Sci. Technol.*, 46: 620-625.

Jantrawut, P., A. Assifaoui and O. Chambin (2013). Influence of low methoxyl pectin gel textures and in vitro release of rutin from calcium pectinate beads, *Carbohydr. Polym.*, 97: 335-342.

Jiang, L., F. Jia, Y. Han, X. Meng, Y. Xiao and S. Bai (2021). Development and characterization of zein edible films incorporated with catechin/β-cyclodextrin inclusion complex nanoparticles, *Carbohydr. Polym.*, 261: 117877.

Ju, J., Y. Xie, H. Yu, Y. Guo, Y. Cheng, R. Zhang and W. Yao (2020a). Synergistic inhibition effect of citral and eugenol against *Aspergillus niger* and their application in bread preservation, *Food Chem.*, 310: 125974.

Ju, J., Y. Xie, H. Yu, Y. Guo, Y. Cheng, H. Qian and W. Yao (2020b). Analysis of the synergistic antifungal mechanism of eugenol and citral, *LWT-Food Sci. Technol.*, 123: 109128.

Kalem, I.K., Z.F. Bhat, S. Kumar, S. Noor and A. Desai (2018). The effects of bioactive edible film containing *Terminalia arjuna* on the stability of some quality attributes of chevon sausages, *Meat Sci.*, 140: 38-43.

Kandemir, N., A. Yemenicioglu, Ç. Mecitoglu, Z.S. Elmaci, A. Arslanoglu, Y. Göksungur and T. Baysal (2005). Production of antimicrobial films by incorporation of partially purified lysozyme into biodegradable films of crude exopolysaccharides obtained from *Aureobasidium pullulans* fermentation, *Food Technol. Biotechnol.*, 43: 343-350.

Kang, H.J., C. Jo, J.H. Kwon, J.H. Kim, H.J. Chung and M.W. Byun (2007). Effect of a pectin-based edible coating containing green tea powder on the quality of irradiated pork patty, *Food Control*, 18: 430-435.

Karunamay, S., S.R. Badhe and V. Shulka (2020). Comparative study of essential oil of clove and oregano treated edible film in extending shelf-life of paneer, *Pharm. Innov. J.,* 9: 312-316.

Kavas, N., G. Kavas and D. Saygili (2016). Use of ginger essential oil-fortified edible coatings in Kashar cheese and its effects on *Escherichia coli* O157: H7 and *Staphylococcus aureus* CyTA, *J. Food.*, 14: 317-323.

Khaledian, S., S. Basiri and S.S. Shekarforoush (2021). Shelf-life extension of pacific white shrimp using tragacanth gum-based coatings containing Persian lime peel (*Citrus latifolia*) extract, *LWT-Food Sci. Technol.*, 141: 110937.

Khan, M.A. and F. Shahidi (2000). Oxidative stability of stripped and nonstripped borage and evening primrose oils and their emulsions in water, *J. Am. Oil Chem. Soc.*, 77: 963-969.

Kharchoufi, S., L. Parafati, F. Licciardello, G. Muratore, M. Hamdi, G. Cirvilleri and C. Restuccia (2018). Edible coatings incorporating pomegranate peel extract and biocontrol yeast to reduce *Penicillium digitatum* postharvest decay of oranges, *Food Microbiol.*, 74: 107-112.

Kim, J., J.Y. Choi, J. Kim and K.D. Moon (2021). Effects of edible coatings with various natural browning inhibitors on the qualitative characteristics of banana (*Musa acuminata* Cavendish Subgroup) during storage, *Korean J. Food Preserv.*, 28: 13-22.

King, N.J. and R. Whyte (2006). Does it look cooked? A review of factors that influence cooked meat color, *J. Food Sci.*, 71: R31-R40.

Koh, P.C., M.A. Noranizan, Z.A.N. Hanani, R. Karim and S.Z. Rosli (2017). Application of edible coatings and repetitive pulsed light for shelf-life extension of fresh-cut cantaloupe (*Cucumis melo* L. *reticulatus* cv. Glamour), *Postharvest Biol. Technol.*, 129: 64-78.

Koo, S.Y., K.H. Cha, D.G. Song, D. Chung and C.H. Pan (2014). Microencapsulation of peppermint oil in an alginate–pectin matrix using a coaxial electrospray system, *Int. J. Food Sci. Technol.*, 49: 733-739.

Ksouda, G., S. Sellimi, F. Merlier, A. Falcimaigne-Cordin, B. Thomasset, M. Nasri and M. Hajji (2019). Composition, antibacterial and antioxidant activities of *Pimpinella saxifraga* essential oil and application to cheese preservation as coating additive, *Food Chem.*, 288: 47-56.

Ku, K.J., Y.H. Hong and K.B. Song (2008). Mechanical properties of a *Gelidium corneum* edible film containing catechin and its application in sausages, *J. Food Sci.*, 73: C217-C221.

Kure, C.F., Y. Wasteson, J. Brendehaug and I. Skaar (2001). Mould contaminants on Jarlsberg and Norvegia cheese blocks from four factories, *Int. J. Food Microbiol.*, 70: 21-27.

Kurozawa, L.E. and M.D. Hubinger (2017). Hydrophilic food compounds encapsulation by ionic gelation, *Curr. Opin. Food Sci.*, 15: 50-55.

Lambert, R.J.W., P.N. Skandamis, P.J. Coote and G.J. Nychas (2001). A study of the minimum inhibitory concentration and mode of action of oregano essential oil, thymol and carvacrol, *J. Appl. Microbiol.*, 91: 453-462.

Lan, W., S. Li, S. Shama, Y. Zhao, D.E. Sameen, L. He and Y. Liu (2020). Investigation of ultrasonic treatment on physicochemical, structural and morphological properties of sodium alginate/AgNPs/apple polyphenol films and its preservation effect on strawberry, *Polymers*, 12: 2096.

Laohakunjit, N. and O. Kerdchoechuen (2007). Aroma enrichment and the change during storage of non-aromatic milled rice coated with extracted natural flavor, *Food Chem.*, 101: 339-344.

Le Tien, C., M. Letendre, P. Ispas-Szabo, M.A. Mateescu, G. Delmas-Patterson, H.L. Yu and M. Lacroix (2000). Development of biodegradable films from whey proteins by cross-linking and entrapment in cellulose, *J. Agric. Food Chem.*, 48: 5566-5575.

Lee, M.K., Y.M. Kim, N.Y. Kim, G.N. Kim, S.H. Kim, K.S. Bang and I. Park (2002). Prevention of browning in potato with a heat-treated onion extract, *Biosci. Biotechnol. Biochem.*, 66: 856-858.

Lee, S.L., M.S. Lee and K.B. Song (2005). Effect of gamma-irradiation on the physicochemical properties of gluten films, *Food Chem.*, 92: 621-625.

Lee, H. and S.C. Min (2013). Antimicrobial edible defatted soybean meal-based films incorporating the lactoperoxidase system, *LWT-Food Sci. Technol.*, 54: 42-50.

Lee, E.S., H.G. Song, I. Choi, J.S. Lee and J. Han (2020). Effects of mung bean starch/guar gum-based edible emulsion coatings on the staling and safety of rice cakes, *Carbohydr. Polym.*, 247: 116696.

Leite, L.S.F., C. Pham, S. Bilatto, H.M. Azeredo, E.D. Cranston, F.K. Moreira *et al.* (2021). Effect of tannic acid and cellulose nanocrystals on antioxidant and antimicrobial properties of gelatin films, *ACS Sustain. Chem. Eng.*, 9: 8539-8549.

Ley, J.P. (2008). Masking bitter taste by molecules, *Chemosens. Percept.*, 1: 58-77.

Li, X., S. Li, X. Liang, D.J. McClements, X. Liu and F. Liu (2020). Applications of oxidases in modification of food molecules and colloidal systems: Laccase, peroxidase and tyrosinase, *Trends in Food Sci. Technol.*, 103: 78-93.

Li, M., H. Yu, Y. Xie, Y. Guo, Y. Cheng, H. Qian and W. Yao (2021). Fabrication of eugenol loaded gelatin nanofibers by electrospinning technique as active packaging material, *LWT-Food Sci. Technol.*, 139: 110800.

Licon, C.C., A. Moro, C.M. Librán, A.M. Molina, A. Zalacain, M.I. Berruga and M. Carmona (2020). Volatile transference and antimicrobial activity of cheeses made with Ewes' milk fortified with essential oils, *Foods*, 9: 35.

Lim, J., T. Fujimaru and T.D. Linscott (2014). The role of congruency in taste–odor interactions, *Food Qual. Prefer.*, 34: 5-13.

Linde, G.A., A.L. Junior, E.V. de Faria, N.B. Colauto, F.F. de Moraes and G.M. Zanin (2009). Taste modification of amino acids and protein hydrolysate by α-cyclodextrin, *Food Res. Int.*, 42: 814-818.

Liolios, C.C., O. Gortzi, S. Lalas, J. Tsaknis and I. Chinou (2009). Liposomal incorporation of carvacrol and thymol isolated from the essential oil of *Origanum dictamnus* L. and *in vitro* antimicrobial activity, *Food Chem.*, 112: 77-83.

Littoz, F. and D.J. McClements (2008). Bio-mimetic approach to improving emulsion stability: Cross-linking adsorbed beet pectin layers using laccase, *Food Hydrocoll.*, 22: 1203-1211.

Liu, D., J. Shi, A.C. Ibarra, Y. Kakuda and S.J. Xue (2008). The scavenging capacity and synergistic effects of lycopene, vitamin E, vitamin C, and β-carotene mixtures on the DPPH free radical, *LWT-Food Sci. Technol.*, 41: 1344-1349.

Liu, H., H. Pei, Z. Han, G. Feng and D. Li (2015). The antimicrobial effects and synergistic antibacterial mechanism of the combination of ε-polylysine and nisin against *Bacillus subtilis*, *Food Control*, 47: 444-450.

Liu, F., R.J. Avena-Bustillos, B.S. Chiou, Y. Li, Y. Ma, T.G. Williams *et al.* (2017). Controlled-release of tea polyphenol from gelatin films incorporated with different ratios of free/nanoencapsulated tea polyphenols into fatty food simulants, *Food Hydrocoll.*, 62: 212-221.

Liu, X., C. Le Bourvellec and C.M. Renard (2020a). Interactions between cell wall polysaccharides and polyphenols: Effect of molecular internal structure, *Comp. Rev. Food Sci. Food Saf.*, 19: 3574-3617.

Liu, X., P. Wang, Y.X. Zou, Z.G. Luo and T.M. Tamer (2020b). Co-encapsulation of vitamin C and β-carotene in liposomes: Storage stability, antioxidant activity and *in vitro* gastrointestinal digestion, *Food Res. Int.*, 136: 109587.

Liu, Q., H. Cui, B. Muhoza, E. Duhoranimana, K. Hayat, X. Zhang and C.T. Ho (2021). Mild enzyme-induced gelation method for nanoparticle stabilization: Effect of transglutaminase and laccase cross-linking, *J. Agric. Food Chem.*, 69: 1348-1358.

Lopez, C., C. Mériadec, E. David-Briand, A. Dupont, T. Bizien, F. Artzner *et al.* (2020). Loading of lutein in egg-sphingomyelin vesicles as lipid carriers: Thermotropic phase behaviour, structure of sphingosome membranes and lutein crystals, *Food Res. Int.*, 138: 109770.

Lopez-Rubio, A., R. Gavara and J.M. Lagaron (2006). Bioactive packaging: Turning foods into healthier foods through biomaterials, *Trends Food Sci. Technol.*, 17: 567-575.

Lopresti, F., L. Botta, V. La Carrubba, L. Di Pasquale, L. Settanni and R. Gaglio (2021). Combining carvacrol and nisin in biodegradable films for antibacterial packaging applications, *Int. J. Biol. Macromol.*, 193: 117-126.

Love, J.D. and A.M. Pearson (1971). Lipid oxidation in meat and meat products – A review, *J. Am. Oil Chem. Soc.*, 48: 547-549.

Luo, S., A. Saadi, K. Fu, M. Taxipalati and L. Deng (2021). Fabrication and characterization of dextran/zein hybrid electrospun fibers with tailored properties for controlled release of curcumin, *J. Sci. Food Agric.*, 101: 6455-6357.

Ma, H., P. Forssell, R. Partanen, J. Buchert and H. Boer (2011). Improving laccase catalyzed cross-linking of whey protein isolate and their application as emulsifiers, *J. Agric. Food Chem.*, 59: 1406-1414.

Madene, A., M. Jacquot, J. Scher and S. Desobry (2006). Flavor encapsulation and controlled release – A review, *Int. J. Food Sci. Technol.*, 41: 1-21.

Mahcene, Z., A. Khelil, S. Hasni, F. Bozkurt, M.B. Goudjil and F. Tornuk (2021). Home-made cheese preservation using sodium alginate based on edible film incorporating essential oils, *J. Food Sci. Technol.*, 58: 2406-2419.

Maqsood, S., S. Benjakul and A. Kamal-Eldin (2012). Haemoglobin-mediated lipid oxidation in the fish muscle: A review, *Trends in Food Sci. Tech.*, 28: 33-43.

Marcuzzo, E., A. Sensidoni, F. Debeaufort and A. Voilley (2010). Encapsulation of aroma compounds in hydrocolloidic emulsion based edible films to control flavour release, *Carbohydr. Polym.*, 80: 984-988.

Marín, A., M. Cháfer, L. Atarés, A. Chiralt, R. Torres, J. Usall and N. Teixidó (2016). Effect of different coating-forming agents on the efficacy of the biocontrol agent *Candida sake* CPA-1 for control of *Botrytis cinerea* on grapes, *Biol. Control*, 96: 108-119.

Mariutti, L.R. and N. Bragagnolo (2017). Influence of salt on lipid oxidation in meat and seafood products: A review, *Food Res. Int.*, 94: 90-100.

Mastromatteo, M., G. Barbuzzi, A. Conte and M.A. Del Nobile (2009). Controlled release of thymol from zein-based film, *Innov. Food Sci. Emerg. Technol.*, 10: 222-227.

Mastromatteo, M., M. Mastromatteo, A. Conte and M.A. Del Nobile (2010). Advances in controlled release devices for food packaging applications, *Trends in Food Sci. Technol.*, 21: 591-598.

Matan, N. (2012). Antimicrobial activity of edible film incorporated with essential oils to preserve dried fish (*Decapterus maruadsi*), *Int. Food Res. J.*, 19: 1733-1738.

Mattinen, M.L., K. Kruus, J. Buchert, J.H. Nielsen, H.J. Andersen and C.L. Steffensen (2005). Laccase-catalyzed polymerization of tyrosine-containing peptides, *FEBS J.*, 272: 3640-3650.

Mattinen, M.L., M. Hellman, P. Permi, K. Autio, N. Kalkkinen and J. Buchert (2006). Effect of protein structure on laccase-catalyzed protein oligomerization, *J. Agric. Food Chem.*, 54: 8883-8890.

McClements, D.J., F. Li and H. Xiao (2015). The nutraceutical bioavailability classification scheme: Classifying nutraceuticals according to factors limiting their oral bioavailability, *Annu. Rev. Food Sci. Technol.*, 6: 299-327.

McClements, D.J. and B. Öztürk (2021). Utilization of nanotechnology to improve the handling, storage and biocompatibility of bioactive lipids in food applications, *Foods*, 10: 365.

McCormick, M.L., J.P. Gaut, T.S. Lin, B.E. Britigan, G.R. Buettner and J.W. Heinecke (1998). Electron paramagnetic resonance detection of free tyrosyl radical generated by myeloperoxidase, lactoperoxidase, and horseradish peroxidase, *J. Biol. Chem.*, 273: 32030-32037.

Mehdizadeh, T. and A.M. Langroodi (2019). Chitosan coatings incorporated with propolis extract and Zataria multiflora Boiss oil for active packaging of chicken breast meat, *Int. J. Biol. Macromol.*, 141: 401-409.

Mehdizadeh, T., H. Tajik, A.M. Langroodi, R. Molaei and A. Mahmoudian (2020). Chitosan-starch film containing pomegranate peel extract and *Thymus kotschyanus* essential oil can prolong the shelf-life of beef, *Meat Sci.*, 163: 108073.

Mehdizadeh, A., S.A. Shahidi, N. Shariatifar, M. Shiran and A. Ghorbani-HasanSaraei (2021). Evaluation of chitosan-zein coating containing free and nano-encapsulated

Pulicaria gnaphalodes (Vent.) Boiss extract on quality attributes of rainbow trout, *J. Aquat. Food Prod. Technol.*, 30: 62-75.

Mei, L., A.D. Crum and E.A. Decker (1994). Development of lipid oxidation and inactivation of antioxidant enzymes in cooked pork and beef, *J. Food Lipids*, 1: 273-283.

Mei, J., Q. Guo, Y. Wu and Y. Li (2015). Evaluation of chitosan-starch-based edible coating to improve the shelf-life of bod ljong cheese, *J. Food Protect.*, 78: 1327-1334.

Mild, R.M., L.A. Joens, M. Friedman, C.W. Olsen, T.H. McHugh, B. Law and S. Ravishankar (2011). Antimicrobial edible apple films inactivate antibiotic resistant and susceptible campylobacter jejuni strains on chicken breast, *J. Food Sci.*, 76: M163-M168.

Mileriene, J., L. Serniene, M. Henriques, D. Gomes, C. Pereira, K. Kondrotiene *et al.* (2021). Effect of liquid whey protein concentrate-based edible coating enriched with cinnamon carbon dioxide extract on the quality and shelf-life of Eastern European curd cheese, *J. Dairy Sci.*, 104: 1504-1517.

Millette, M.C.L.T., C. Le Tien, W. Smoragiewicz and M. Lacroix (2007). Inhibition of *Staphylococcus aureus* on beef by nisin-containing modified alginate films and beads, *Food Control.*, 18: 878-884.

Min, B., J.C. Cordray and D.U. Ahn (2010). Effect of NaCl, myoglobin, Fe (II), and Fe (III) on lipid oxidation of raw and cooked chicken breast and beef loin, *J. Agric. Food Chem.*, 58: 600-605.

Min, S., L.J. Harris and J.M. Krochta (2005a). *Listeria monocytogenes* inhibition by whey protein films and coatings incorporating the lactoperoxidase system, *J. Food Sci.*, 70: M317-M324.

Min, S., L.J. Harris and J.M. Krochta (2005b). Antimicrobial effects of lactoferrin, lysozyme, and the lactoperoxidase system and edible whey protein films incorporating the lactoperoxidase system against *Salmonella enterica* and *Escherichia coli* O157: H7, *J. Food Sci.*, 70: M332-M338.

Min, S. and J.M. Krochta (2005c). Inhibition of *Penicillium commune* by edible whey protein films incorporating lactoferrin, lacto-ferrin hydrolysate, and lactoperoxidase systems, *J. Food Sci.*, 70: M87-M94.

Mirshekari, A., B. Madani and J.B. Golding (2019). Aloe vera gel treatment delays postharvest browning of white button mushroom (*Agaricus bisporus*), *J. Food Meas. Charact.*, 13: 1250-1256.

Mishra, B.B., S. Gautam and A. Sharma (2012). Purification and characterization of polyphenol oxidase (PPO) from eggplant (*Solanum melongena*), *Food Chem.*, 134(4): 1855-1861.

Misra, S., P. Pandey and H.N. Mishra (2021). Novel approaches for co-encapsulation of probiotic bacteria with bioactive compounds, their health benefits and functional food product development: A review, *Trends in Food Sci. Technol.*, 109: 340-351.

Miya, S., H. Takahashi, T. Ishikawa, T. Fujii and B. Kimura (2010). Risk of *Listeria monocytogenes* contamination of raw ready-to-eat seafood products available at retail outlets in Japan, *Appl. Environ. Microbiol.*, 76: 3383-3386.

Mokhtar, M., G. Ginestra, F. Youcefi, A. Filocamo, C. Bisignano and A. Riazi (2017). Antimicrobial activity of selected polyphenols and capsaicinoids identified in pepper (*Capsicum annuum* L.) and their possible mode of interaction, *Curr. Microbiol.*, 74: 1253-1260.

Mollica, F., I. Gelabert and R. Amorati (2022). Synergic antioxidant effects of the essential oil component γ-terpinene on high-temperature oil oxidation, *ACS Food Sci. Technol.*, 2: 180-186.

Molina-Hernández, J.B., A. Echeverri-Castro, H.A. Martínez-Correa and M.M. Andrade-Mahecha (2020). Edible coating based on achira starch containing garlic/oregano oils to extend the shelf-life of double cream cheese, *Rev. Fac. Nac. Agron. Medellín.*, 73: 9099-9108.

Monzón-Ortega, K., M. Salvador-Figueroa, D. Gálvez-López, R. Rosas-Quijano, I. Ovando-Medina and A. Vázquez-Ovando (2018). Characterization of Aloe vera-chitosan composite films and their use for reducing the disease caused by fungi in papaya Maradol, *J. Food Sci. Technol.*, 55: 4747-4757.

Moradi, M., H. Tajik, S.M.R. Rohani and A. Mahmoudian (2016). Antioxidant and antimicrobial effects of zein edible film impregnated with *Zataria multiflora* Boiss. essential oil and monolaurin, *LWT-Food Sci. Technol.*, 72: 37-43.

Mostaghimi, M., M. Majdinasab and S.M.H. Hosseini (2021). Characterization of alginate hydrogel beads loaded with thyme and clove essential oils nanoemulsions, *J. Polym. Environ.* https://doi.org/10.1007/s10924-021-02298-w

Mukai, J., E. Tokuyama, T. Ishizaka, S. Okada and T. Uchida (2007). Inhibitory effect of aroma on the bitterness of branched-chain amino acid solutions, *Chem. Pharm. Bull.*, 55: 1581-1584.

Murakami, M., T. Yamaguchi, H. Takamura and T. Matoba (2003). Effects of ascorbic acid and α-tocopherol on antioxidant activity of polyphenolic compounds, *J. Food Sci.*, 68: 1622-1625.

Murdock, C.A., J. Cleveland, K.R. Matthews and M.L. Chikindas (2007). The synergistic effect of nisin and lactoferrin on the inhibition of *Listeria monocytogenes* and *Escherichia coli* O157: H7, *Lett. Appl. Microbiol.*, 44: 255-261.

Mushtaq, M., A. Gani, A. Gani, H.A. Punoo and F.A. Masoodi (2018). Use of pomegranate peel extract incorporated zein film with improved properties for prolonged shelf-life of fresh Himalayan cheese (Kalari/kradi), *Innov. Food Sci. Emerg. Technol.*, 48: 25-32.

Nadarajah, D., J.H. Han and R.A. Holley (2005). Inactivation of *Escherichia coli* O157: H7 in packaged ground beef by allyl isothiocyanate, *Int. J. Food Microbiol.*, 99: 269-279.

Niki, E., N. Noguchi, H. Tsuchihashi and N. Gotoh (1995). Interaction among vitamin C, vitamin E, and beta-carotene, *Am. J. Clin. Nutr.*, 62: 1322S-1326S.

Nilsuwan, K., P. Guerrero, K. de la Caba, S. Benjakul and T. Prodpran (2021). Fish gelatin films laminated with emulsified gelatin film or poly (lactic) acid film: Properties and their use as bags for storage of fried salmon skin, *Food Hydrocoll.*, 111: 106199.

Noori, S., F. Zeynali and H. Almasi (2018). Antimicrobial and antioxidant efficiency of nanoemulsion-based edible coating containing ginger (*Zingiber officinale*) essential oil and its effect on safety and quality attributes of chicken breast fillets, *Food Control*, 84: 312-320.

Ochoa-Velasco, C.E., J.C. Pérez-Pérez, J.M. Varillas-Torres, A.R. Navarro-Cruz, P. Hernández-Carranza, R. Munguía-Pérez *et al.* (2021). Starch edible films/coatings added with carvacrol and thymol: *In vitro* and *in vivo* evaluation against *Colletotrichum gloeosporioides*, *Foods*, 10: 175.

O'Connor, T.P. and N.M. O'Brien (2006). Lipid oxidation, pp. 309-347. *In:* P.F. Fox and P.L.H. McSweeney (Eds.). *Advanced Dairy Chemistry*, vol. 2. Lipids, Springer, Boston. USA.

Okuro, P. K., M. Thomazini, J.C. Balieiro, R.D. Liberal and C.S. Fávaro-Trindade (2013). Co-encapsulation of *Lactobacillus acidophilus* with inulin or polydextrose in solid lipid microparticles provides protection and improves stability, *Food Res. Int.*, 53: 96-103.

Olaimat, A.N., Y. Fang and R.A. Holley (2014). Inhibition of *Campylobacter jejuni* on fresh chicken breasts by κ-carrageenan/chitosan-based coatings containing allyl isothiocyanate or deodorized oriental mustard extract, *Int. J. Food Microbiol.*, 187: 77-82.

Oswell, N.J., H. Thippareddi and R.B. Pegg (2018). Practical use of natural antioxidants in meat products in the US: A review, *Meat Sci.*, 145: 469-479.

Otoni, C.G., R.J. Avena-Bustillos, B.S. Chiou, C. Bilbao-Sainz, P.J. Bechtel and T.H. McHugh (2012). Ultraviolet-B radiation induced cross-linking improves physical properties of cold- and warm-water fish gelatin gels and films, *J. Food Sci.*, 77: E215-E223.

Otoni, C.G., M.R. de Moura, F.A. Aouada, G.P. Camilloto, R.S. Cruz, M.V. Lorevice *et al.* (2014). Antimicrobial and physical-mechanical properties of pectin/papaya puree/cinnamaldehyde nanoemulsion edible composite films, *Food Hydrocoll.*, 41: 188-194.

Oudgenoeg, G., R. Hilhorst, S.R. Piersma, C.G. Boeriu, H. Gruppen, M. Hessing *et al.* (2001). Peroxidase-mediated cross-linking of a tyrosine-containing peptide with ferulic acid, *J. Agric. Food Chem.*, 49: 2503-2510.

Ozdemir, M. and J.D. Floros (2003). Film composition effects on diffusion of potassium sorbate through whey protein films, *J. Food Sci.*, 68: 511-516.

Pabast, M., N. Shariatifar, S. Beikzadeh and G. Jahed (2018). Effects of chitosan coatings incorporating with free or nano-encapsulated Satureja plant essential oil on quality characteristics of lamb meat, *Food Control*, 91: 185-192.

Palafox-Carlos, H., J.F. Ayala-Zavala and G.A. González-Aguilar (2011). The role of dietary fiber in the bioaccessibility and bioavailability of fruit and vegetable antioxidants, *J. Food Sci.*, 76: 6-15.

Paramera, E.I., S.J. Konteles and V.T. Karathanos (2011). Stability and release properties of curcumin encapsulated in *Saccharomyces cerevisiae*, β-cyclodextrin and modified starch, *Food Chem.*, 125: 913-922.

Pattanayaiying, R., H. Aran and C.N. Cutter (2015). Incorporation of nisin Z and lauric arginate into pullulan films to inhibit foodborne pathogens associated with fresh and ready-to-eat muscle foods, *Int. J. Food Microbiol.*, 207: 77-82.

Pavli, F., A.A. Argyri, P. Skandamis, G.J. Nychas, C. Tassou and N. Chorianopoulos (2019). Antimicrobial activity of oregano essential oil incorporated in sodium alginate edible films: Control of *Listeria monocytogenes* and spoilage in ham slices treated with high pressure processing, *Materials*, 12: 3726.

Peng, H., H. Xiong, J. Li, M. Xie, Y. Liu, C. Bai and L. Chen (2010). Vanillin cross-linked chitosan microspheres for controlled release of resveratrol, *Food Chem.*, 121: 23-28.

Peralta-Ruiz, Y., C.G. Tovar, A. Sinning-Mangonez, D. Bermont, A.P. Cordero, A. Paparella and C. Chaves-López (2020). *Colletotrichum gloesporioides* inhibition using chitosan-*Ruta graveolens* L. essential oil coatings: Studies *in vitro* and *in situ* on Carica papaya fruit, *Int. J. Food Microbiol.*, 326: 108649.

Permana, A.W., I. Sampers and P. Van der Meeren (2021). Influence of virgin coconut oil on the inhibitory effect of emulsion-based edible coatings containing cinnamaldehyde against the growth of *Colletotrichum gloeosporioides* (*Glomerella cingulata*), *Food Control*, 121: 107622.

Peters, C.M., R.J. Green, E.M. Janle and M.G. Ferruzzi (2010). Formulation with ascorbic acid and sucrose modulates catechin bioavailability from green tea, *Food Res. Int.*, 43: 95-102.

Petrasch, S., S.J. Knapp, J.A. Van Kan and B. Blanco-Ulate (2019). Grey mould of strawberry, a devastating disease caused by the ubiquitous necrotrophic fungal pathogen *Botrytis cinerea*, *Mol. Plant Pathol.*, 20: 877-892.

Pieretti, G.G., M.P. Pinheiro, M.R. da Silva Scapim, J.M.G. Mikcha and G.S. Madrona (2019). Effect of an edible alginate coating with essential oil to improve the quality of a fresh cheese, *Acta Sci. Technol.*, 41: e36402.

Pirozzi, A., V. Del Grosso, G. Ferrari and F. Donsì (2020). Edible coatings containing oregano essential oil nanoemulsion for improving postharvest quality and shelf-life of tomatoes, *Foods*, 9: 1605.

Prakash, A., R. Baskaran and V. Vadivel (2020). Citral nanoemulsion incorporated edible coating to extend the shelf-life of fresh cut pineapples, *LWT-Food Sci. Technol.*, 118: 108851.

Pranoto, Y., V.M. Salokhe and S.K. Rakshit (2005). Physical and antibacterial properties of alginate-based edible film incorporated with garlic oil, *Food Res. Int.*, 38: 267-272.

Qin, X.S., S.Z. Luo, J. Cai, X.Y. Zhong, S.T. Jiang, Y.Y. Zhao and Z. Zheng (2016). Transglutaminase-induced gelation properties of soy protein isolate and wheat gluten mixtures with high intensity ultrasonic pretreatment, *Ultrason. Sonochem.*, 31: 590-597.

Quan, W., C. Zhang, M. Zheng, Z. Lu and F. Lu (2018). Whey protein isolate with improved film properties through cross-linking catalyzed by small laccase from *Streptomyces coelicolor*, *J. Sci. Food Agric.*, 98: 3843-3850.

Quirós-Sauceda, A.E., H. Palafox-Carlos, S.G. Sáyago-Ayerdi, J.F. Ayala-Zavala, L.A. Bello-Perez, E. Alvarez-Parrilla *et al.* (2014). Dietary fiber and phenolic compounds as functional ingredients: Interaction and possible effect after ingestion, *Food Funct.*, 5: 1063-1072.

Radford, D., B. Guild, P. Strange, R. Ahmed, L.T. Lim and S. Balamurugan (2017). Characterization of antimicrobial properties of *Salmonella* phage Felix O1 and *Listeria* phage A511 embedded in xanthan coatings on Poly (lactic acid) films, *Food Microbiol.*, 66: 117-128.

Randazzo, W., A. Jiménez-Belenguer, L. Settanni, A. Perdones, M. Moschetti, E. Palazzolo *et al.* (2016). Antilisterial effect of citrus essential oils and their performance in edible film formulations, *Food Control*, 59: 750-758.

Raybaudi-Massilia, R.M., J. Mosqueda-Melgar and O. Martín-Belloso (2008). Edible alginate-based coating as carrier of antimicrobials to improve shelf-life and safety of fresh-cut melon, *Int. J. Food Microbiol.*, 121: 313-327.

Recharla, N., M. Riaz, S. Ko and S. Park (2017). Novel technologies to enhance solubility of food-derived bioactive compounds: A review, *J. Funct. Foods*, 39: 63-73.

Revilla, I., M.I. González-Martín, A.M. Vivar-Quintana, M.A. Blanco-López, I.A. Lobos-Ortega and J.M. Hernández-Hierro (2016). Antioxidant capacity of different cheeses: Affecting factors and prediction by near infrared spectroscopy, *J. Dairy Sci.*, 99: 5074-5082.

Rezaei, A., M. Fathi and S.M. Jafari (2019). Nanoencapsulation of hydrophobic and low-soluble food bioactive compounds within different nanocarriers, *Food Hydrocoll.*, 88: 146-162.

Rhee, K.S. and Y.A. Ziprin (1987). Lipid oxidation in retail beef, pork and chicken muscles as affected by concentrations of heme pigments and nonheme iron and microsomal enzymic lipid peroxidation activity, *J. Food Biochem.*, 11: 1-15.

Ribeiro, A., C. Caleja, L. Barros, C. Santos-Buelga, M.F. Barreiro and I.C. Ferreira (2016). Rosemary extracts in functional foods: Extraction, chemical characterization and incorporation of free and microencapsulated forms in cottage cheese, *Food Funct.*, 7: 2185-2196.

Rizzo, V., S. Lombardo, G. Pandino, R.N. Barbagallo, A. Mazzaglia, C. Restuccia *et al.* (2019). Shelf-life study of ready-to-cook slices of globe artichoke 'Spinoso sardo': Effects of anti-browning solutions and edible coating enriched with *Foeniculum vulgare* essential oil, *J. Sci. Food Agric.*, 99: 5219-5228.

Robledo, N., P. Vera, L. López, M. Yazdani-Pedram, C. Tapia and L. Abugoch (2018). Thymol nanoemulsions incorporated in quinoa protein/chitosan edible films; antifungal effect in cherry tomatoes, *Food Chem.*, 246: 211-219.

Rohn, S. (2014). Possibilities and limitations in the analysis of covalent interactions between phenolic compounds and proteins, *Food Res. Int.*, 65: 13-19.

Rojas-Graü, M.A., R.J. Avena-Bustillos, C. Olsen, M. Friedman, P.R. Henika, O. Martín-Belloso et al. (2007). Effects of plant essential oils and oil compounds on mechanical, barrier and antimicrobial properties of alginate-apple puree edible films, *J. Food Eng.*, 81: 634-641.

Rojas-Graü, M.A., R. Soliva-Fortuny and O. Martín-Belloso (2009). Edible coatings to incorporate active ingredients to fresh-cut fruits: A review, *Trends in Food Sci. Technol.*, 20: 438-447.

Roschel, G.G., T.F.F.D. Silveira, L.M. Cajaiba and I.A. Castro (2019). Combination of hydrophilic or lipophilic natural compounds to improve the oxidative stability of flaxseed oil, *Eur. J. Lipid Sci. Technol.*, 121: 1800459.

Rossi-Márquez, G., J.H. Han, B. García-Almendárez, E. Castaño-Tostado and C. Regalado-González (2009). Effect of temperature, pH and film thickness on nisin release from antimicrobial whey protein isolate edible films, *J. Sci. Food Agric.*, 89: 2492-2497.

Rotondo, A., G.L. La Torre, G. Bartolomeo, R. Rando, R. Vadalà, V. Zimbaro and A. Salvo (2021). Profile of carotenoids and tocopherols for the characterization of lipophilic antioxidants in 'Ragusano' cheese, *Appl. Sci.*, 11: 7711.

Sadiq, S., M. Imran, H. Habib, S. Shabbir, A. Ihsan, Y. Zafar and F.Y. Hafeez (2016). Potential of monolaurin based food-grade nano-micelles loaded with nisin Z for synergistic antimicrobial action against *Staphylococcus aureus*, *LWT-Food Sci. Technol.*, 71: 227-233.

Saifullah, M., M.R.I. Shishir, R. Ferdowsi, M.R.T. Rahman and Q. Van Vuong (2019). Micro and nano encapsulation, retention and controlled release of flavor and aroma compounds: A critical review, *Trends in Food Sci. Technol.*, 86: 230-251.

Salminen, W.F. and G. Russotti (2017). Synergistic interaction of ascorbic acid and green tea extract in preventing the browning of fresh cut apple slices, *J. Food Process. Preserv.*, 41: e13192.

Sánchez-González, L., M. Cháfer, A. Chiralt and C. González-Martínez (2010). Physical properties of edible chitosan films containing bergamot essential oil and their inhibitory action on *Penicillium italicum*, *Carbohydr. Polym.*, 82: 277-283.

Sánchez-González, L., M. Cháfer, M. Hernández, A. Chiralt and C. González-Martínez (2011). Antimicrobial activity of polysaccharide films containing essential oils, *Food Control*, 22: 1302-1310.

Sanchís, E., C. Ghidelli, C.C. Sheth, M. Mateos, L. Palou and M.B. Pérez-Gago (2017). Integration of antimicrobial pectin-based edible coating and active modified atmosphere packaging to preserve the quality and microbial safety of fresh-cut persimmon (*Diospyros kaki* Thunb. cv. Rojo Brillante), *J. Sci. Food Agric.*, 97: 252-260.

Saravani, M., A. Ehsani, J. Aliakbarlu and Z. Ghasempour (2019). Gouda cheese spoilage prevention: Biodegradable coating induced by Bunium persicum essential oil and lactoperoxidase system, *Food Sci. Nutr.*, 7: 959-968.

Sartz, L., B. De Jong, M. Hjertqvist, L. Plym-Forshell, R. Alsterlund, S. Löfdahl et al. (2008). An outbreak of *Escherichia coli* O157: H7 infection in southern Sweden associated with consumption of fermented sausage; aspects of sausage production that increase the risk of contamination, *Epidemiol. Infect.*, 136: 370-380.

Savary, G., N. Hucher, O. Petibon and M. Grisel (2014). Study of interactions between aroma compounds and acacia gum using headspace measurements, *Food Hydrocoll.*, 37: 1-6.

Sayadi, M., A.M. Langroodi and K. Pourmohammadi (2021). Combined effects of chitosan coating incorporated with *Berberis vulgaris* extract and *Mentha pulegium* essential oil and MAP in the shelf-life of turkey meat, *J. Food Meas. Charact.*, 15: 5159-5169.

Sebaaly, C., A. Jraij, H. Fessi, C. Charcosset and H. Greige-Gerges (2015). Preparation and characterization of clove essential oil-loaded liposomes, *Food Chem.*, 178: 52-62.

Selinheimo, E., K. Autio, K. Kruus and J. Buchert (2007). Elucidating the mechanism of laccase and tyrosinase in wheat bread making, *J. Agric. Food Chem.*, 55: 6357-6365.

Selinheimo, E., P. Lampila, M.L. Mattinen and J. Buchert (2008). Formation of protein – oligosaccharide conjugates by laccase and tyrosinase, *J. Agric. Food Chem.*, 56: 3118-3128.

Serfert, Y., S. Drusch and K. Schwarz (2010). Sensory odour profiling and lipid oxidation status of fish oil and microencapsulated fish oil, *Food Chem.*, 123: 968-975.

Settier-Ramírez, L., G. López-Carballo, P. Hernández-Muñoz, A. Fontana-Tachon, C. Strub and S. Schorr-Galindo (2022). Apple-based coatings incorporated with wild apple isolated yeast to reduce *Penicillium expansum* postharvest decay of apples, *Postharvest Biol. Technol.*, 185: 111805.

Severino, R., K.D. Vu, F. Donsì, S. Salmieri, G. Ferrari and M. Lacroix (2014). Antimicrobial effects of different combined non-thermal treatments against *Listeria monocytogenes* in broccoli florets, *J. Food Eng.*, 124: 1-10.

Severino, R., G. Ferrari, K.D. Vu, F. Donsì, S. Salmieri and M. Lacroix (2015). Antimicrobial effects of modified chitosan based coating containing nanoemulsion of essential oils, modified atmosphere packaging and gamma irradiation against *Escherichia coli* O157: H7 and *Salmonella typhimurium* on green beans, *Food Control*, 50: 215-222.

Seydim, A.C. and G. Sarikus (2006). Antimicrobial activity of whey protein based edible films incorporated with oregano, rosemary and garlic essential oils, *Food Res. Int.*, 39: 639-644.

Shah, M.A., S.J.D. Bosco and S.A. Mir (2014). Plant extracts as natural antioxidants in meat and meat products, *Meat Sci.*, 98: 21-33.

Shellhammer, T.H. and J.M. Krochta (1997). Whey protein emulsion film performance as affected by lipid type and amount, *J. Food Sci.*, 62: 390-394.

Shen, Y., Z.J. Ni, K. Thakur, J.G. Zhang, F. Hu and Z.J. Wei (2021). Preparation and characterization of clove essential oil loaded nanoemulsion and pickering emulsion activated pullulan-gelatin-based edible film, *Int. J. Biol. Macromol.*, 181: 528-539.

Shi, J., Q. Qu, Y. Kakuda, S.J. Xue, Y. Jiang, S. Koide and Y.Y. Shim (2007). Investigation of the antioxidant and synergistic activity of lycopene and other natural antioxidants using LAME and AMVN model systems, *J. Food Compost. Anal.*, 20: 603-608.

Shishir, M.R.I., L. Xie, C. Sun, X. Zheng and W. Chen (2018). Advances in micro and nano-encapsulation of bioactive compounds using hydrocolloid and lipid-based transporters, *Trends in Food Sci. Technol.*, 78: 34-60.

Shixian, Q., Y. Dai, Y. Kakuda, J. Shi, G. Mittal, D. Yeung and Y. Jiang (2005). Synergistic anti-oxidative effects of lycopene with other bioactive compounds, *Food Rev. Int.*, 21: 295-311.

Shojaee-Aliabadi, S., H. Hosseini, M.A. Mohammadifar, A. Mohammadi, M. Ghasemlou, S. Hosseini *et al.* (2014). Characterization of κ-carrageenan films incorporated plant essential oils with improved antimicrobial activity, *Carbohydr. Polym.*, 101: 582-591.

Shori, A.B. (2017). Microencapsulation improved probiotics survival during gastric transit, *Hayatı J. Biosci.*, 24: 1-5.

Shrestha, S., B.R. Wagle, A. Upadhyay, K. Arsi, I. Upadhyaya, D.J. Donoghue and A.M. Donoghue (2019). Edible coatings fortified with carvacrol reduce *Campylobacter jejuni* on chicken wingettes and modulate expression of select virulence genes, *Front. Microbiol.*, 10: 583.

Silva, Â., A. Duarte, S. Sousa, A. Ramos and F.C. Domingues (2016). Characterization and antimicrobial activity of cellulose derivatives films incorporated with a resveratrol inclusion complex, *LWT-Food Sci. Technol.*, 73: 481-489.

Silva-Weiss, A., M. Quilaqueo, O. Venegas, M. Ahumada, W. Silva, F. Osorio and B. Giménez (2018). Design of dipalmitoyl lecithin liposomes loaded with quercetin and rutin and their release kinetics from carboxymethyl cellulose edible films, *J. Food Eng.*, 224: 165-173.

Sivarooban, T., N.S. Hettiarachchy and M.G. Johnson (2008). Physical and antimicrobial properties of grape seed extract, nisin, and EDTA incorporated soy protein edible films, *Food Res. Int.*, 41: 781-785.

Song, Z., F. Li, H. Guan, Y. Xu, Q. Fu and D. Li (2017). Combination of nisin and ε-polylysine with chitosan coating inhibits the white blush of fresh-cut carrots, *Food Control*, 74: 34-44.

Sørheim, O., T. Aune and T. Nesbakken (1997). Technological, hygienic and toxicological aspects of carbon monoxide used in modified-atmosphere packaging of meat, *Trends in Food Sci. Technol.*, 8: 307-312.

Souza, A.C., G.E.O. Goto, J.A. Mainardi, A.C.V. Coelho and C.C. Tadini (2013). Cassava starch composite films incorporated with cinnamon essential oil: Antimicrobial activity, microstructure, mechanical and barrier properties, *LWT-Food Sci. Technol.*, 54: 346-352.

Soysal, Ç. (2009). Effects of green tea extract on 'golden delicious' apple polyphenoloxidase and its browning, *J. Food Biochem.*, 33: 134-148.

Sozbilen, G.S. and A. Yemenicioğlu (2020). Decontamination of seeds destined for edible sprout production from Listeria by using chitosan coating with synergetic lysozyme-nisin mixture, *Carbohydr. Polym.*, 235: 115968.

Stefańczyk, E., S. Sobkowiak, M. Brylińska and J. Śliwka (2016). Diversity of *Fusarium* spp. associated with dry rot of potato tubers in Poland, *European J. Plant Pathol.*, 145: 871-884.

Sudagidan, M. and A. Yemenicioğlu (2012). Effects of nisin and lysozyme on growth inhibition and biofilm formation capacity of *Staphylococcus aureus* strains isolated from raw milk and cheese samples, *J. Food Protect.*, 75: 1627-1633.

Suhr, K.I. and P.V. Nielsen (2004). Effect of weak acid preservatives on growth of bakery product spoilage fungi at different water activities and pH values, *Int. J. Food Microbiol.*, 95: 67-78.

Sun, X., R.G. Cameron, A. Plotto, T. Zhong, C.M. Ference and J. Bai (2021). The effect of controlled-release carvacrol on safety and quality of blueberries stored in perforated packaging, *Foods*, 10: 1487.

Surendhiran, D., C. Li, H. Cui and L. Lin (2020). Fabrication of high stability active nanofibers encapsulated with pomegranate peel extract using chitosan/PEO for meat preservation, *Food Packag. Shelf-Life*, 23: 100439.

Tabassum, N. and M.A. Khan (2020). Modified atmosphere packaging of fresh-cut papaya using alginate based edible coating: Quality evaluation and shelf-life study, *Sci. Hortic.*, 259: 108853.

Tahir, H.E., Z. Xiaobo, S. Jiyong, G.K. Mahunu, X. Zhai and A.A. Mariod (2018). Quality and postharvest-shelf-life of cold-stored strawberry fruit as affected by gum arabic (*Acacia senegal*) edible coating, *J. Food Biochem.*, 42: e12527.

Tang, S., J.P. Kerry, D. Sheehan, D.J. Buckley and P.A. Morrissey (2001). Antioxidative effect of added tea catechins on susceptibility of cooked red meat, poultry and fish patties to lipid oxidation, *Food Res. Int.*, 34: 651-657.

Tang, H.R., A.D. Covington and R.A. Hancock (2003). Structure-activity relationships in the hydrophobic interactions of polyphenols with cellulose and collagen, *Hydrocolloids: Original Research on Biomolecules*, 70: 403-413.

Tavano, L., R. Muzzalupo, N. Picci and B. de Cindio (2014). Co-encapsulation of antioxidants into niosomal carriers: Gastrointestinal release studies for nutraceutical applications, *Colloid Surface B*, 114: 82-88.

Tayel, A.A., S.H. Moussa, M.F. Salem, K.E. Mazrou and W.F. El-Tras (2016). Control of citrus molds using bioactive coatings incorporated with fungal chitosan/plant extracts composite, *J. Sci. Food and Agric.*, 96: 1306-1312.

Tessaro, L., C.G. Luciano, A.M.Q.B. Bittante, R.V. Lourenço, M. Martelli-Tosi and P.J. do Amaral Sobral (2021). Gelatin and/or chitosan-based films activated with "Pitanga" (*Eugenia uniflora* L.) leaf hydroethanolic extract encapsulated in double emulsion, *Food Hydrocoll.*, 113: 106523.

Theivendran, S., N.S. Hettiarachchy and M.G. Johnson (2006). Inhibition of *Listeria monocytogenes* by nisin combined with grape seed extract or green tea extract in soy protein film coated on turkey frankfurters, *J. Food Sci.*, 71: M39-M44.

Tormos, C.J., C. Abraham and S.V. Madihally (2015). Improving the stability of chitosan–gelatin-based hydrogels for cell delivery using transglutaminase and controlled release of doxycycline, *Drug Deliv. Transl. Res.*, 5: 575-584.

Torpol, K., S. Sriwattana, J. Sangsuwan, P. Wiriyacharee and W. Prinyawiwatkul (2019). Optimising chitosan–pectin hydrogel beads containing combined garlic and holy basil essential oils and their application as antimicrobial inhibitor, *Int. J. Food Sci. Technol.*, 54: 2064-2074.

Torrijos, R., T.M. Nazareth, J. Calpe, J.M. Quiles, J. Mañes and G. Meca (2021). Antifungal activity of natamycin and development of an edible film based on hydroxyethylcellulose to avoid Penicillium spp. growth on low-moisture Mozzarella cheese, *LWT-Food Sci. Technol.*, 154: 112795.

Tran, V.T., P. Kingwascharapong, F. Tanaka and F. Tanaka (2021). Effect of edible coatings developed from chitosan incorporated with tea seed oil on Japanese pear, *Sci. Hortic.*, 288: 110314.

Ucak, I., A.K. Abuibaid, T.M. Aldawoud, C.M. Galanakis and D. Montesano (2021). Antioxidant and antimicrobial effects of gelatin films incorporated with citrus seed extract on the shelf-life of sea bass (*Dicentrarchus labrax*) fillets, *J. Food Process. Preserv.*, 45: e15304.

Ünalan, İ.U., F. Korel and A. Yemenicioğlu (2011). Active packaging of ground beef patties by edible zein films incorporated with partially purified lysozyme and Na_2EDTA, *Int. J. Food Sci. Technol.*, 46: 1289-1295.

Ünalan, İ.U., I. Arcan, F. Korel and A. Yemenicioğlu (2013). Application of active zein based films with controlled release properties to control *Listeria monocytogenes* growth and lipid oxidation in fresh Kashar cheese, *Innov. Food Sci. Emerg. Technol.*, 20: 208-214.

Valero, E., R. Varon and F. Garcia-Carmona (1992). Kinetic study of the effect of metabisulfite on polyphenol oxidase, *J. Agric. Food Chem.*, 40: 904-908.

Vámos-Vigyázó, L. and N.F. Haard (1981). Polyphenol oxidases and peroxidases in fruits and vegetables, *Crit. Rev. Food Sci. Nutr.*, 15: 49-127.

Velderrain-Rodríguez, G.R., L. Salvia-Trujillo and O. Martín-Belloso (2021). Lipid digestibility and polyphenols bioaccessibility of oil-in-water emulsions containing avocado peel and seed extracts as affected by the presence of low methoxyl pectin, *Foods*, 10: 2193.

Vilanova, L., I. Viñas, R. Torres, J. Usall, G. Buron-Moles and N. Teixidó (2014). Acidification of apple and orange hosts by *Penicillium digitatum* and *Penicillium expansum*, *Int. J. Food Microbiol.*, 178: 39-49.

Villeneuve, P., C. Bourlieu-Lacanal, E. Durand, J. Lecomte, D.J. McClements and E.A. Decker (2021). Lipid oxidation in emulsions and bulk oils: A review of the importance of micelles, *Crit. Rev. Food Sci. Nutr.* Doi: 10.1080/10408398.2021.2006138

Vital, A.C.P., A. Guerrero, J.D.O. Monteschio, M.V. Valero, C.B. Carvalho, B.A. de Abreu Filho *et al.* (2016). Effect of edible and active coating (with rosemary and oregano essential oils) on beef characteristics and consumer acceptability, *PloS One*, 11: e0160535.

Vojdani, F. and J.A. Torres (1990). Potassium sorbate permeability of methylcellulose and hydroxypropyl methylcellulose coatings: Effect of fatty acids, *J. Food Sci.*, 55: 841-846.

Wagle, B.R., S. Shrestha, K. Arsi, I. Upadhyaya, A.M. Donoghue and D.J. Donoghue (2019). Pectin or chitosan coating fortified with eugenol reduces *Campylobacter jejuni* on chicken wingettes and modulates expression of critical survival genes, *Poult. Sci.*, 98: 1461-1471.

Wang, D., Y. Dong, X. Chen, Y. Liu, J. Wang, X. Wang *et al.* (2020a). Incorporation of apricot (*Prunus armeniaca*) kernel essential oil into chitosan films displaying antimicrobial effect against *Listeria monocytogenes* and improving quality indices of spiced beef, *Int. J. Biol. Macromol.*, 62: 838-844.

Wang, Q., W. Liu, B. Tian, D. Li, C. Liu, B. Jiang and Z. Feng (2020b). Preparation and characterization of coating based on protein nanofibers and polyphenol and application for salted duck egg yolks, *Foods*, 9: 449.

Wang, Y., Y. Xue, Q. Bi, D. Qin, Q. Du and P. Jin (2021). Enhanced antibacterial activity of eugenol-entrapped casein nanoparticles amended with lysozyme against gram-positive pathogens, *Food Chem.*, 360: 130036.

Weller, C.L., A. Gennadios and R.A. Saraiva (1998). Edible bilayer films from zein and grain sorghum wax or carnauba wax, *LWT-Food Sci. Technol.*, 31: 279-285.

Wen, P., Y. Wen, M.H. Zong, R.J. Linhardt and H. Wu (2017). Encapsulation of bioactive compound in electrospun fibers and its potential application, *J. Agric. Food Chem.*, 65: 9161-9179.

Wessels, B., N. Schulze-Kaysers, S. Damm and B. Kunz (2014). Effect of selected plant extracts on the inhibition of enzymatic browning in fresh-cut apple, *J. Appl. Bot. Food Qual.*, 87: 16-23.

Whiley, H. and K. Ross (2015). Salmonella and eggs: From production to plate, *Int. J. Environ. Res. Public Health*, 12: 2543-2556.

Wu, Y., Z. Chen, X. Li and M. Li (2009). Effect of tea polyphenols on the retrogradation of rice starch, *Food Res. Int.*, 42: 221-225.

Wu, J., H. Liu, S. Ge, S. Wang, Z. Qin, L. Chen *et al.* (2015). The preparation, characterization, antimicrobial stability and *in vitro* release evaluation of fish gelatin films incorporated with cinnamon essential oil nanoliposomes, *Food Hydrocoll.*, 43: 427-435.

Xiao, D., P.M. Davidson and Q. Zhong (2011). Spray-dried zein capsules with coencapsulated nisin and thymol as antimicrobial delivery system for enhanced antilisterial properties, *J. Agric. Food Chem.*, 59: 7393-7404.

Xing, Y., X. Li, Q. Xu, J. Yun and Y. Lu (2010a). Antifungal activities of cinnamon oil against *Rhizopus nigricans, Aspergillus flavus* and *Penicillium expansum in vitro* and *in vivo* fruit test, *Int. J. Food Sci. Technol.*, 45: 1837-1842.

Xing, Y., X. Li, Q. Xu, Y. Jiang, J. Yun and W. Li (2010b). Effects of chitosan-based coating and modified atmosphere packaging (MAP) on browning and shelf-life of fresh-cut lotus root (*Nelumbo nucifera* Gaerth), *Innov. Food Sci. Emerg. Technol.*, 11: 684-689.

Xiong, Y., S. Li, R.D. Warner and Z. Fang (2020). Effect of oregano essential oil and resveratrol nanoemulsion loaded pectin edible coating on the preservation of pork loin in modified atmosphere packaging, *Food Control*, 114: 107226.

Xu, W.T., K.L. Huang, F. Guo, W. Qu, J.J. Yang, Z.H. Liang and Y.B. Luo (2007). Postharvest grapefruit seed extract and chitosan treatments of table grapes to control *Botrytis cinerea, Postharvest Biol. Technol.*, 46: 86-94.

Xu, W., D. Zhu, Z. Li, D. Luo, L. Hang, J. Jing and B.R. Shah (2019). Controlled release of lysozyme based core/shells structured alginate beads with $CaCO_3$ microparticles using pickering emulsion template and in situ gelation, *Colloid. Surface. B*, 183: 110410.

Xu, J., Z. He, M. Zeng, B. Li, F. Qin, L. Wang *et al.* (2017). Effect of xanthan gum on the release of strawberry flavor in formulated soy beverage, *Food Chem.*, 228: 595-601.

Yang, Z., X. Zou, Z. Li, X. Huang, X. Zhai, W. Zhang *et al.* (2019). Improved postharvest quality of cold stored blueberry by edible coating based on composite gum Arabic/roselle extract, *Food Bioproc. Tech.*, 12: 1537-1547.

Yaghoubi, M., A. Ayaseh, K. Alirezalu, Z. Nemati, M. Pateiro and J.M. Lorenzo (2021). Effect of chitosan coating incorporated with *Artemisia fragrans* essential oil on fresh chicken meat during refrigerated storage, *Polymers*, 13: 716.

Ye, L., Z. Li, R. Niu, Z. Zhou, Y. Shen and L. Jiang (2019). All-aqueous direct deposition of fragrance-loaded nanoparticles onto fabric surfaces by electrospraying, *ACS Appl. Polym. Mater.*, 1: 2590-2596.

Yegin, Y., K.L. Perez-Lewis, M. Zhang, M. Akbulut and T.M. Taylor (2016). Development and characterization of geraniol-loaded polymeric nanoparticles with antimicrobial activity against food-borne bacterial pathogens, *J. Food Eng.*, 170: 64-71.

Yemenicioglu, A. (2016). Strategies for controlling major enzymatic reactions in fresh and processed vegetables, pp. 377-388. *In:* Y.H. Hui and Ö. Evranuz (Eds.). *Handbook of Vegetable Preservation and Processing*, 2nd ed. CRC Press, New York, USA.

Yemenicioğlu, A., S. Farris, M. Turkyilmaz and S. Gulec (2020). A review of current and future food applications of natural hydrocolloids, *Int. J. Food Sci. Tech.*, 55: 1389-1406.

Yin, W., R. Su, W. Qi and Z. He (2012). A casein-polysaccharide hybrid hydrogel cross-linked by transglutaminase for drug delivery, *J. Mater. Sci.*, 47: 2045-2055.

Yingyuad, S., S. Ruamsin, D. Reekprkhon, S. Douglas, S. Pongamphai and U. Siripatrawan (2006). Effect of chitosan coating and vacuum packaging on the quality of refrigerated grilled pork, *Packag. Technol. Sci.*, 19: 149-157.

Yordshahi, A.S., M. Moradi, H. Tajik and R. Molaei (2020). Design and preparation of antimicrobial meat wrapping nanopaper with bacterial cellulose and postbiotics of lactic acid bacteria, *Int. J. Food Microbiol.*, 321: 108561.

You, J., J. Zhang, M. Wu, L. Yang, W. Chen and G. Li (2016). Multiple criteria-based screening of *Trichoderma* isolates for biological control of *Botrytis cinerea* on tomato, *Biol. Control*, 101: 31-38.

Yuan, G., H. Lv, W. Tang, X. Zhang and H. Sun (2016). Effect of chitosan coating combined with pomegranate peel extract on the quality of Pacific white shrimp during iced storage, *Food Control*, 59: 818-823.

Zafeiropoulou, T., V. Evageliou, C. Gardeli, S. Yanniotis and M. Komaitis (2012). Retention of selected aroma compounds by gelatin matrices, *Food Hydrocoll.*, 28: 105-109.

Zhang, W., C. Shu, Q. Chen, J. Cao and W. Jiang (2019). The multi-layer film system improved the release and retention properties of cinnamon essential oil and its application as coating in inhibition to *Penicillium expansum* of apple fruit, *Food Chem.*, 299: 125109.

Zhang, L., D.J. McClements, Z. Wei, G. Wang, X. Liu and F. Liu (2020). Delivery of synergistic polyphenol combinations using hydrocolloid-based systems: Advances in physicochemical properties, stability and bioavailability, *Crit. Rev. Food Sci. Nutr.*, 60: 2083-2097.

Zhao, X., L. Chen, J.E. Wu, Y. He and H. Yang (2020). Elucidating antimicrobial mechanism of nisin and grape seed extract against *Listeria monocytogenes* in broth and on shrimp through NMR-based metabolomics approach, *Int. J. Food Microbiol.*, 319: 108494.

Zoffoli, J.P., B.A. Latorre, E.J. Rodriguez and P. Aldunce (1999). Modified atmosphere packaging using chlorine gas generators to prevent *Botrytis cinerea* on table grapes, *Postharvest Biol. Technol.*, 15: 135-142.

Methods of Testing Antimicrobial and Antioxidant Properties of Edible Packaging

6.1 Introduction

As a food preservation method, active packaging aims at one or several of the following effects in the packaged food: (1) increasing microbial safety, (2) delaying microbial spoilage, and (3) inhibiting lipid oxidation or enzymatic browning (Yemenicioğlu, 2017). Thus, a properly designed active edible packaging (film, coating, casing, nanofiber mat, pad, sticker, etc.) should maintain its antimicrobial and/or antioxidant activity during manufacturing processes (e.g. homogenization, solution-casting, extrusion, electrospinning, compression molding, drying, etc.) and storage period until it is used in food packaging application. During food application, the antimicrobials and antioxidants are released mainly from packaging on to the food surface that is the most susceptible part for microbial and oxidative changes. The effectiveness of active agents during storage of food depends on their continuous release from packaging material on to food surface in sufficient amounts, diffusion rate trough depths of food (causing dilution of active agent concentration at the critical food surface) and interactions with the food components (some cause destabilization or neutralization of active agents). Therefore, the design of an active edible packaging needs testing performance and stability of its active properties in different model media or food simulants, and conducting packaging applications by monitoring its effect on microbial and oxidative quality of food during storage. In some cases, to determine the diffusion rate of active agent within food, different layers (obtained by slicing food starting from surface to geometric center) of packaged food are tested for the concentration of released active agent during cold storage. After antimicrobial and antioxidant tests conducted in model media and on target food final adjustments are done in film formulation (concentrations of active agent(s), hydrocolloid(s), plasticizer, emulsifier, cross-linking agent, etc.) to reach the desired release profiles of active agents and to obtain sufficient antimicrobial and/or antioxidant effects in the food. This chapter aims at discussing mainly different testing methods for determination of antimicrobial and antioxidant properties of edible packaging incorporated with natural active compounds. Moreover, methods of determining release profiles of edible films are also briefly discussed at the beginning of this chapter since this is a critical part of most activity determination methods and gives critical information about sustained release properties of packaging and compatibility (solubility and stability) of its active components with packaging.

6.2 Methods of testing release profiles of active agents from edible packaging

The release tests are conducted mainly to determine the amounts and activities of soluble antimicrobials or antioxidants in packaging and to obtain data about their release profiles. In most studies, the release tests are conducted in distilled water, but suitable buffers at a specific pH, food simulants, model gels or the target food itself might also be used to determine the release profiles of an active agent.

6.2.1 Release tests in distilled water or buffer

The release tests conducted in distilled water or suitable buffer (mostly at target food pH) at different temperatures aim to determine soluble antioxidant and/or antimicrobial agent content (or activity) in the packaging, and to compare the release rates of these agents at different film compositions. In general, the tests are conducted by placing packaging in a glass apparatus (e.g. Petri dishes or Erlenmeyer flasks) containing a sufficient volume of distilled water or buffer at the common refrigeration temperature of 4°C, but different temperatures between 0 and 25°C might be used, depending on specific commercial storage and marketing temperatures of target food. Moreover, it is important to shake the release medium horizontally at a slow speed to enable continuous diffusion of active agent from packaging material to the release medium. A fast shaking or mixing (with a stirrer) of the release medium is not suggested since this could disturb the integrity of packaging material. The release test is conducted until reaching equilibrium for the released active agent(s). However, some tests reach equilibrium in a very long time (several weeks) due to the sustained release profiles of active agents. In such cases, the instability of monitored active agent at the release test medium might cause mistakes in determination of its release profiles. Therefore, it is important to ensure the stability of the active agent in release medium. Monitoring of the concentration of released active agent is conducted by collecting samples at different time intervals, and measuring concentration of agent in the release medium. However, if the active agent is an antimicrobial enzyme, such as lysozyme, lactoperoxidase, glucose oxidase, etc., the monitoring should be conducted both for concentration and activity of enzyme to ensure its stability at the film-making, and storage conditions, and during release tests. Similarly, the monitoring of concentration during release test of an antioxidant agent should be combined with antioxidant activity determination using a suitable method. In the release tests, the amount of active agent collected during sampling should be considered during calculation of total released agent content (or activity or antioxidant activity) at each time point. Moreover, the most accurate results are obtained by taking minimum amount of liquid during sampling to minimize the change of volume in the release test medium.

6.2.1.1 Release curve

To form a release curve, the results for each time point are first expressed as amount (or activity) released per cm² (μg/cm²) of films, or as amount (or activity) released per g (μg/g) of all kinds of packaging material. Then, the amounts (or activity) of active agent released per cm² (or per g) *vs.* release times (hr, min or day) are plotted.

6.2.1.2 Calculation of initial release rate

The initial release rate of active agent is determined from the slope of the initial linear portion of release curve as (1) μg/cm²/hr or μg/g/hr (for concentration monitoring); (2) unit/cm²/hr or Unit/g/hr (for enzyme activity monitoring); or (3) μmol Trolox or Vitamin C/cm²/hr or μmol Trolox or Vitamin C/g/hr (for antioxidant activity monitoring).

6.2.1.3 Recovery of active agent

The recovery of active agent from the packaging material is determined from the formula given in Eq. 1

$$\text{Recovery (\%)} = 100 \times [(\text{TC or TA})/(\text{TCI or TAI})] \qquad (1)$$

Where TC and TA are total soluble content and activity (for enzyme or antioxidant activity) released at the equilibrium, respectively while TCI and TAI are total content or activity incorporated into packaging at the beginning. The determination of recovery for activity is particularly important when the active agent is an antimicrobial enzyme that might show dramatic variations in activity during film making. Moreover, recovery value also gives information about the amount of active agent bound by the packaging matrix. An example release curve given in Fig. 6.1 shows the release of lysozyme from zein films. This curve suggests the activation of enzyme in films during film making since total activities released from the films is higher than total activities incorporated into films (recoveries up to 300%) (Mecitoğlu *et al.*, 2006).

6.2.1.4 Release tests of films having a floating problem

The density of the packaging material is a very critical factor affecting the reliability of release tests. For example, during release tests conducted in distilled water or buffer films having a low density might float, thus, upper side of the film remains in the air. This might affect the time necessary to reach the equilibrium for the released active agent since release progresses only from one side of the film. Such film-floating problems occur frequently when composite films are formed by mixing protein or carbohydrate with a lipid. A custom made apparatus might be employed to eliminate the floating problem and to keep the film immersed into the release medium during release test (Fig. 6.2a). This apparatus could be formed by placing films between two squire-shaped glass frame equipped with proper gaskets, fixation of two sides of the frames by inert rubber band (or properly cut parafilm bands), and placing fixed frame into sufficient volume of buffer or water (Arcan and Yemenicioğlu, 2014). Some glass pieces at proper dimensions (e.g.

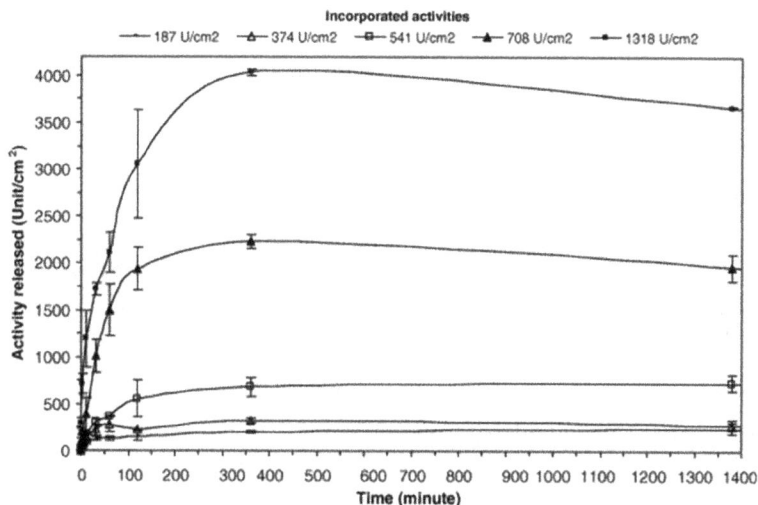

Fig. 6.1: Release profiles of lysozyme from zein films (release test conditions, in distilled water, at +4°C by monitoring of enzyme activity determination); Reprinted with permission from Mecitoğlu *et al.* (2006); copyright © 2006 Elsevier Ltd.

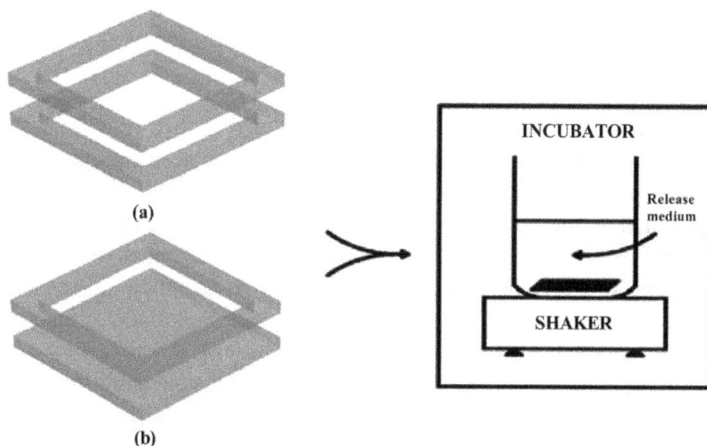

Fig. 6.2: Glass apparatuses used to test release profiles of different films. Apparatus used for films having floating problem (a) and asymmetric surface morphology (b)

H×W×D: 1×1×1 cm) should be placed on the bottom corners of the apparatus to let circulation of water or buffer during shaking. The same apparatus could also be used for release tests of high density films that sink in release medium and adhere on to the bottom surface of glass container used in release test.

6.2.1.5 Release tests of films having an asymmetric surface morphology

A different type of apparatus is needed when tested films are having an asymmetric surface morphology and release kinetics of different sides of this film (dense or porous side) must be studied separately (Gemili *et al.*, 2009). In this case, the bottom frame is replaced by a same sized glass plate. This helps to cut the contact of one side of the film with water or buffer, while letting the other side to contact with the release medium (Fig. 6.2b). To isolate the contact area of the glass frame with glass plate, the film should be squeezed between two waterproof gaskets.

6.2.2 Release tests in different liquid food simulants

Although the distilled water and buffers are extensively used for release tests of edible films, some model liquid media used in migration tests of polymeric packaging are also employed in release tests of edible films (Ribeiro-Santos *et al.*, 2017; Liu *et al.*, 2017; Li *et al.*, 2020). According to the U.S. Food and Drug Administration (FDA), the food simulants are: (1) ethanol at 10% for aqueous and acidic foods; (2) ethanol between 10 and 50% for low- and high-alcoholic foods, respectively, and (3) food oil (e.g. corn oil), HB307 or Miglyol 812, for fatty foods (FDA, 2007). According to European Union regulations (No 10/2011) six different food simulants defined for the migration tests of plastics are: (1) ethanol at 10% (v/v) for aqueous food (simulant A); (2) acetic acid at 3% (w/v) for acidic food (simulant B); (3) ethanol at 20% (v/v) for alcoholic product with ethanol content up to 20% (simulant C); (4) ethanol at 50% (v/v) for fatty food (simulant D1); (5) vegetable oil (simulant D2); and (6) poly(2,6-diphenyl-p-phenylene oxide) with particle size of 60-80 mesh, pore size of 200 nm for dry food (simulant E) (EU Commission Regulation, 2011).

6.2.3 Release tests in gel media

The release tests in model solid gel medium are conducted to simulate the swelling and release properties of films in a potential food application. Such tests are very beneficial when extraction and measurement of active agent concentration (or enzyme activity or antioxidant activity) in food is challenging. A model solid media designed properly could be extracted more easily to quantify the released agent. For example, agar, a basic gelling agent used in microbial media, might be a good alternative to conduct release tests (Sebti *et al.*, 2003; Alkan *et al.*, 2011). The hardness of the agar could be modified by changing concentration of agar and other components (e.g. sugars, acids, salts) in the gel. Tests are conducted simply by placing discs of film (size could vary depending of detection limit of active agent) on to the surface of gel previously cast into a Petri dish. The Petri dishes are closed with their lids to minimize evaporation (sealed additionally with plastic films if needed), and then they are incubated at a suitable temperature and relative humidity to simulate food storage conditions. The gels in the Petri dishes are then extracted with a suitable buffer or distilled water and analyzed for concentration (or activity) of active agent periodically.

6.2.4 Release tests in air

The release tests in air aim mostly at determining release kinetics of natural volatile antimicrobial agents (e.g. essential oils) from packaging materials designed to deliver these agents to the headspace of packaged food. The tests are conducted by monitoring the concentration of the retained antimicrobial within the packaging material or the concentration of released vapors of antimicrobial from the packaging material during storage. For example, Ben Arfa *et al.* (2007) determined the retained carvacrol and cinnamaldehyde in soy protein isolate or modified starch coated papers during storage of these materials in an incubator working at 30°C and 60% RH. These authors assayed the essential oil concentration of antimicrobial papers periodically by a gas chromatographic method following solvent extraction (50:50 (v/v) mixture of water and n-pentane) of retained essential oils from the papers. This test serves mainly to choose best coating formulation for gradual release of volatile active agent from the coating. The second and more challenging method used in such release tests involves periodic monitoring of released antimicrobial's vapors in collected air samples by gas chromatographic methods. This test is useful in determining the release rates of different volatile agents from the packaging material kept in a suitable container at the desired temperature, but such systems cannot reach equilibrium unless they are sealed hermetically (López *et al.*, 2007).

6.2.5 Release and diffusivity tests in food

According to Bhunia *et al.* (2013), the release process of active agents from packaging material to food occurs in four steps: (1) diffusion of active agent through the packaging material; (2) desorption of the diffused active agent molecules from the packaging material's surface; (3) sorption of the active agent molecules at the packaging material–food interface; and (4) desorption of the active agent in the food. The rapid diffusion of active agent into depths of food is undesired since this leaves susceptible food surface unprotected and increases destabilization or neutralization of the active agent due to its interactions forming insoluble or inactive complexes with the food components. For example, phenolic compounds released from packaging materials might form complex with polysaccharides and proteins in the food (Han and Baik, 2008; Palafox-Carlos *et al.*, 2011). The nisin is neutralized in fresh meat since it forms inactive complexes with glutathione by non-enzymatic or enzymatic reactions catalyzed by glutathione S-transferase (Rose *et al.*, 1999; Rose *et al.*, 2002). The action of inherent proteases in raw or smoked fish might also be a factor that affects the stability of NIS (Aasen *et al.*, 2003). Lysozyme forms strong complexes with milk caseins (Antonov *et al.*, 2017). Moreover, stability of lysozyme in different types of meat might also be different. For example, it was reported that the lysozyme released from active packaging on to beef was much more stable than that in lamb meat (Boyacı and Yemenicioğlu, 2018). Thus, monitoring of both concentration and antimicrobial/antioxidant activity of released active agents in food is very challenging. However, the release tests in food are still important since they give information about compatibility of active agent with the target food. The determination of rate of diffusion for active agent within the food is also beneficial to understand the potential effectiveness of active packaging during long storage periods. The rate of diffusion for the active agent in food could be determined by measuring its concentration at different layers (from surface to geometric center) of food periodically during storage. For this purpose, food samples packed with the antimicrobial packaging are stored for different time periods. The samples are than unpacked, sliced equally from surface to geometric center, and concentration (or activity) of active agent released from packaging is analyzed at each slice. However, this method is difficult to apply if detection of low concentrations of antimicrobial is a problem. Thus, a more easily applied method to determine the diffusivity of active agent in target food is that the food sample is immersed into a concentrated solution of active agent, and concentration of active agent after different time periods is determined in different slices of food (Han and Floros, 1998).

6.3 Methods of testing antimicrobial properties of packaging

6.3.1 Antimicrobial properties of packaging in laboratory media

The use of suitable testing methods in the laboratory media is one of the important steps in development of active packaging materials, but these tests cannot simulate the real antimicrobial performance in the food even when they are selected and applied properly. However, it was reported that the films performed well in the properly designed laboratory tests, showing some antimicrobial activity in the target food system with minimal adjustments (Joerger, 2007). There are several critical factors that might mislead workers in the field during antimicrobial tests of their active agent or packaging. For example, it was reported that the films showing potent antimicrobial activity in the laboratory media contained mostly maximum amounts of soluble, but minimum amounts of bound antimicrobial compounds (Boyaci and Yemenicioglu, 2016; Unalan *et al.*, 2013). Therefore, conducting release tests of developed films in suitable media is essential to adjust soluble and bound active agent fractions. Similarly, a film having fast release rates for an antimicrobial might show higher antimicrobial activity in laboratory media than another one having sustained release rates for the same antimicrobial. In contrast, packaging with a sustained release rate mostly gives better antimicrobial performance in foods especially when they are stored for some time. Moreover, low molecular weight antimicrobial agents with high diffusivity might show better antimicrobial performance than high molecular weight antimicrobials in tests conducted on solid media (e.g. zone of inhibition test). Thus, these could cause selection of wrong antimicrobial compound during screening tests designed to choose the most effective active agent to be incorporated into packaging. This section discusses some basic microbiological testing methods of edible packaging by putting some emphasis on critical points that might cause misinterpretation of results.

6.3.1.1 Zone of inhibition test

The zone of inhibition test (also called disc or agar diffusion test) adapted from the classical Kirby-Bauer test is the most frequently used method for testing antimicrobial activity of active edible packaging materials, such as films, mats, and stickers. In this method, discs (rarely squires) prepared from edible film by using a sterile cork-borer (borer diameters between 4 and 23.8 mm are commercially available) are placed on suitable agar medium previously inoculated with a certain number of target microorganism. The microbial load inoculated on to the agar surface could be optimized to improve the appearance of clear zones, but the inoculation level of the target microorganism is mostly set significantly higher than that of possible food contamination levels. It is essential to obtain a dense microbial growth at agar surface since insufficient number of colony formation could cause hardly identifiable zone boarders. In contrast, too dense a microbial growth might interfere with clear zone formation and might cause regrowth of microorganism at the clear zone area before the end of incubation period. In particular, the use of high levels of inoculum in antifungal tests might cause no zone formation even for films incorporated at the highest concentration of antimicrobial agent. Therefore, the proper level of fungal inoculum should be determined by a preliminary test.

The main disadvantages of zone of inhibition test are the incubation times (mostly one to five days) and temperatures (mostly between 25 and 37°C) that are necessary for microbial growth. The active edible packaging is mostly applied at refrigeration temperatures for several weeks to several months. Therefore, the high test temperature probably causes much more release of antimicrobials within a shorter time period than that occurred at normal food storage conditions. Moreover, the high incubation temperatures might cause overestimation of antimicrobial performance for antimicrobial enzyme, like lysozyme that lytic activity increase at elevated incubation temperatures (Sözbilen and Yemenicioğlu, 2021). Another disadvantage of this method is observed during comparison of antimicrobial performances of different film types or formulations. Different types of edible films (e.g. films of zein, starch, pectin, whey, etc.) might show different release rates for the same active compound. Moreover, the same film might show different release profiles for active compounds that vary in molecular weight, isoelectric point, solubility, hydrophobicity, etc. It is important to note that high molecular weight antimicrobials might show low rates of diffusion, both during release from film to agar and within the agar. Thus, keeping the Petri dishes containing film discs for several hours (or longer if it does not interfere with microbial growth) at +4°C before transferring them to microbial incubation temperatures might reduce such disadvantages originating from different release profiles of films. This might help better on seeing the performances of films having sustained release rates that suffer from rapid microbial growth before they release sufficient amounts of antimicrobial. It is also sometimes beneficial to extend the test period (e.g. from 24 to 72 hrs for mesophilic bacteria at 32-37°C) and to conduct periodic multiple measurement of zone diameters (e.g. at 24, 48 and 72 hrs). In general, the microbial incubation conditions are three to five days at 25-30°C for fungi, one to two days at 32-37°C for most mesophilic bacteria, and seven to 10 days between 4 and 10°C for psychrotrophic bacteria (Jay, 2002; Yousef and Carlstrom, 2003). With extension of incubation period, it is easier to identify films having sustained release profiles. It is expected that films having rapid antimicrobial release rates will form fully formed clear zones at the early stages of incubation period (Fig. 6.3). However,

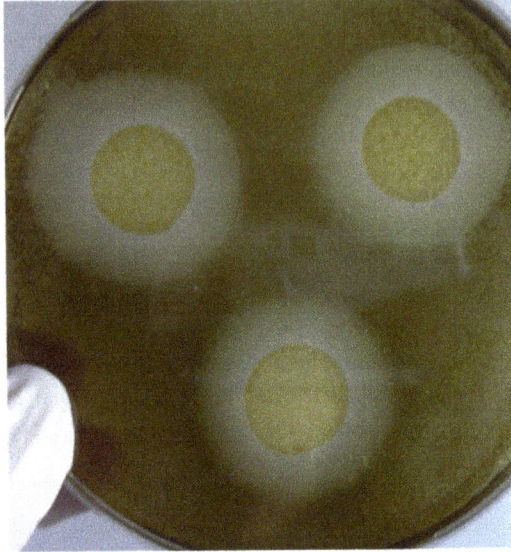

Fig. 6.3: Clear fully formed zones of gallic acid loaded zein film discs against *Campylobacter jejuni* (unpublished data from Yemenicioğlu, 2011)

these clear zones might turn turbid (or tiny individual colonies appear in the clear zone area) or show reduction in clear zone diameter due to microbial regrowth at the later stages of incubation. In contrast, films showing sustained release rates are expected to form smaller clear zones that enlarge or at least maintain their size and clarity by extended incubation. Following measurement of clear zone diameters by a caliper, the results were expressed as average diameter of zone area (mm²). The test gives more accurate results as number of tested discs increases and discs are obtained equally from different parts of the films. The unclear (hazy or turbid) zones are generally not measured for their zone areas, but this might be noted as a limited antimicrobial activity after checking controls and ensuring that they are formed because of reduced microbial growth at the zone area. The test of some active edible films by this method is challenging when they are partially soluble and release some film components that cause haze or turbidity in the zone area. Moreover, the measurement of clear zone area could be difficult when its shape is distorted or clear zone is formed only at one side of the film disc. Such one-sided or distorted zone shape problems occur mostly due to nonhomogeneous distribution of antimicrobials having a solubility problem or poor film manufacturing conditions that cause differences in film thicknesses. The differences in thickness of agar media originating from inappropriate preparation (e.g. low pouring temperature of agar or uneven pouring surface) and test conditions (e.g. uneven incubator shelves that cause leakage of culture and film disc melted at the incubation temperature) might also cause distortions in zone borders. Another significant problem related to zone of inhibition test is local or complete folding of the film disc when it is placed at the agar surface. This problem occurs when developed films are very thin and hydrophobic (e.g. chitosan-based films or composite films incorporated with lipids). This challenging problem could be overcome by placing same sized sterile discs made preferentially from stainless steel (or other suitable inert plastic material) at the top of the film discs placed on to the agar surface.

The testing of films that contain immobilized antimicrobial agent are also not appropriate for zone of inhibition method, but it might be possible to check their limited antimicrobial activity qualitatively by carefully removing films from the agar surface and observing the clarity of film contact surface. A similar evaluation of limited antimicrobial activity at the film disc contact surfaces could also be conducted with antimicrobial films that show sustained release of antimicrobials. However, a clear spot at the same size of the removed film disc might be considered antimicrobial activity only if microbial development is observed in similarly treated control film location (Fig. 6.4). The zone of inhibition method could provide useful data to optimize the film release properties and set the inhibitory concentration of antimicrobial against a target microorganism. However, the evaluation of zone of inhibition test together with other antimicrobial test methods as well as release tests provide a more realistic approach.

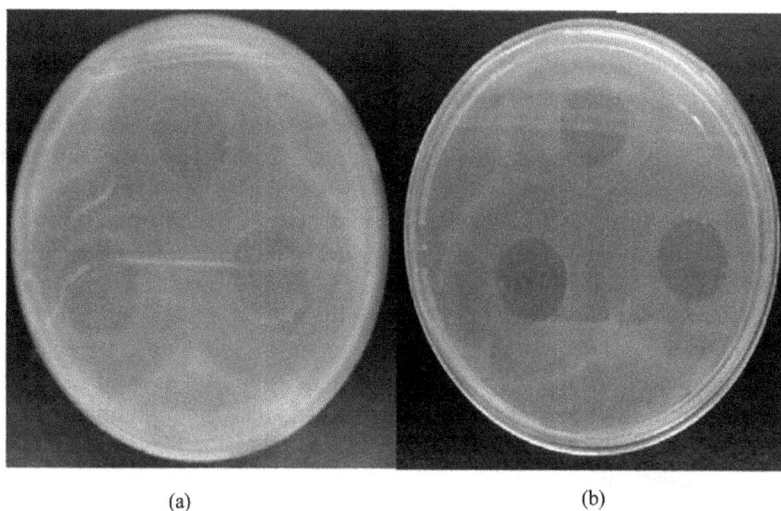

(a)   (b)

Fig. 6.4: Detection of limited antimicrobial activity formed under (+)-catechin loaded zein film discs against *Escherichia coli*. Photos of Petri dishes after removal of control film discs (a) and antimicrobial loaded film discs (b) from the agar surface (Yemenicioğlu and Arcan, 2009)

6.3.1.2 Film inoculation tests

This test is conducted by inoculation of antimicrobial films with a certain number of target microorganism, and counting of remaining microorganism after some time. The microbial load inoculated on to the agar is mostly set significantly higher than that of possible food contamination levels. This is important to achieve countable numbers of colonies in the agar media following dilutions that are necessary to eliminate interference of antimicrobials released from films with the colony counting. The inoculated films are then stored for some time (several days to weeks at room temperature or under refrigeration, depending on storage temperature of target food product), and their remaining microbial load is counted at different time intervals to determine inactivation kinetics of the test microorganism during storage. This test enables comparison of films with different antimicrobials or a film with different concentrations of the same antimicrobial. For isolation of the inoculated bacteria, films could be shaken (or stirred) sufficiently with peptone water (initial inoculation level of control film and its counts might be compared to see effectiveness of isolation). A suitable buffer could also be used in isolation if films contain some acidic component that causes significant drop in pH of isolation medium. The isolate is then diluted properly to obtain countable numbers for the target microorganism on suitable agar media. Counts from higher dilutions are preferred to minimize the possible negative effects of antimicrobial(s) released from films during isolation of microorganism. The microbiological counts are expressed as colony forming unit per gram (CFU/g) of each film. The results are evaluated by determining log reductions compared to control film at different time intervals.

6.3.1.3   Classical shake-flask method

The antimicrobial packaging is mostly applied to inhibit microbial growth at the solid or semi-solid food's surface. Therefore, shake-flask method conducted in broth medium, cannot simulate most food applications except some innovative studies conducted with beverages (Rocha *et al.*, 2017). However, this method could be applied to compare antimicrobial performances of different edible film formulations, particularly those having fast and sustained release properties for the same antimicrobial agent. The films having sustained release rates for incorporated antimicrobials sometimes fail to show any antimicrobial properties in the classical zone of inhibition test that last mostly 24 or 48 h between 25 and 37°C, depending on the growth rate of target microorganism. In contrast, the shake-flask method conducted in the broth medium might be employed even at refrigeration temperatures and extended for several weeks to simulate the performance of films at storage conditions of target food. However, it should be noted that the method is suitable particularly for insoluble films or partially soluble films that show minimal erosion during antimicrobial test.

6.3.1.4   Test of packaging with volatile antimicrobials in the Petri dish

This test is suitable for antimicrobial films and stickers designed particularly to release volatile antimicrobial compounds to the headspace of packaged foods. For this purpose, the film incorporated with volatile antimicrobial is fixed (with adhesive tape from two edges) at the internal side of the lid of a Petri dish. Sometimes the antimicrobial film is cast in the lid of Petri dish (Balaguer *et al.*, 2013). The bottom part of the Petri dish contains agar medium inoculated previously with the target microorganism. Different Petri dishes are similarly prepared for different concentrations of the antimicrobial (e.g. two fold increments of the antimicrobial might be used for different films). After that, the closed Petri dishes (wrapped with plastic packaging films or placed into hermetic jars) are placed in an incubator and incubated for a suitable time for microbial growth. This method gives the minimum inhibitory concentration that prevents the growth of target microorganism on the agar surface. Moreover, the percent inhibition of molds can also be determined by measuring the reduction in diameters of their colonies. This method is very suitable to develop antifungal films, stickers, and sachets for packaging of different foods, such as bakery products, cheeses, and fresh fruits (Winther *et al.*, 2006; Balaguer *et al.*, 2013; Munhuweyi *et al.*, 2017; Lopes *et al.*, 2018).

6.3.2 Antimicrobial properties of packaging in food applications

6.3.2.1 The criterion of success for food application

The antimicrobial tests conducted by inoculation of packed food by pathogenic or spoilage microorganisms, or by monitoring natural microbial flora of non-inoculated packed food, are essential not only to ensure tests conducted in the laboratory media, but also to estimate shelf-life and safety of packaged food. Moreover, food applications provide invaluable information to perform the final adjustments in film's antimicrobial concentration or release rates if this is controlled by a mechanism (e.g. degree of cross-linking, hydrophobicity, morphology, tortuosity, etc.). However, it is important to note that the criterion of success for the antimicrobial edible packaging is not complete inactivation of all viable pathogens and all viable spoilage microorganism that can grow at normal storage temperatures as in the classical preservation methods, like thermal processing. The antimicrobial packaging is mostly employed in combination with refrigeration. Moreover, sometimes it is employed as an important leg of hurdle concept in combination with refrigeration, vacuum packaging, modified atmosphere packaging, or other non-thermal processing methods (e.g. high pressure processing, irradiation, pulsed light, etc.). Therefore, the main objective of antimicrobial packaging might be sufficient inactivation of one or several of the most critical food pathogens that cause frequent or serious health concerns, recalls, and great economic losses. The inhibition of critical spoilage microorganisms (e.g. fungal plant pathogens for fresh fruits, spoilage fungi for some dairy and bakery products, lactic acid spoilage bacteria in different foods) is also frequently targeted in antimicrobial packaging. Thus, this rapidly developing active packaging method provides safer foods within normal shelf-life or improves shelf-life to enable safe transportation of susceptible food to longer distances.

6.3.2.2 Monitoring of specific pathogenic or spoilage microorganisms in inoculated packaged food

The inoculation of food with specific food pathogens or spoilage microorganisms is one of the most frequently used methods to assess effectiveness of antimicrobial packaging. In this method, the target food inoculated with frequently observed pathogenic (e.g. *Listeria monocytogenes*, *S. enterica* serovar Typhimurium, *Escherichia coli* O157:H7, etc.) or spoilage (e.g. *Bacillus subtilis*, *Bacillus mesentericus*, *Botrytis cinerea*, *Penicillium commune*, etc.) microorganisms is packed with the developed antimicrobial packaging. After that, the food is periodically monitored for counts of inoculated microorganism during storage. In such inoculation tests, it is preferred to employ the most pathogenic or antimicrobial-resistant strain of the target pathogen, or cocktail of several strains of the pathogen isolated from food, patients, and food environments. It is highly beneficial to use strains isolated during previous outbreaks of target pathogenic bacteria in antimicrobial tests or food inoculation tests (Tirneta *et al.*, 2010; Alkan *et al.*, 2011). However, the use of nonpathogenic strains or species that show similar antimicrobial resistances with the target pathogenic ones (such as use of nonpathogenic *L. innocua* in place of *L. monocytogenes,* or nonpathogenic *E. coli* in place of pathogenic ones) is also sometimes preferred to reduce risks of laboratory workers. The inoculation method is very suitable to determine effectiveness of developed packaging on a specific food pathogen since it provides the number of log (or decimal) reductions achieved by antimicrobial packaging. In this method, it is very important to inoculate and spread known numbers of a suitable strain of the target microorganism homogenously on the food surface, and to conduct isolation and enumeration of bacteria using proper methods. Moreover, disinfection of food surface prior to inoculation by a suitable method (e.g. irradiation, UV or chemical disinfection) is important for some food applications to prevent suppression of inoculated strains by the native microbial flora. The results of periodically conducted enumerations determined during storage of packaged food are generally reported as logarithm of colony forming units per g of food (log CFU/g).

6.3.2.3 Monitoring of total microbial counts for non-inoculated packaged food

In this method, the target food at the very beginning of its shelf-life (freshly prepared) is packed with the developed antimicrobial packaging without inoculating any microorganism, and then it was monitored for its potential microbial contaminants and natural microflora during storage. It is unlikely to determine the effectiveness of antimicrobial packaging (log reductions by packaging) by this method on pathogenic bacteria due to the low probability of catching high counts of accidental contamination with the target pathogen in a limited number of samples. However, this method provides invaluable data to determine possible inhibition levels of total counts in aerobic mesophilic bacteria (TAMB) (or aerobic plate count, APC; total viable count, TVC), psychrophilic bacteria (TPB), coliform bacteria (TCB), yeast and mold (TYM), *Enterobacteriaceae* (TE), *Pseudomonas* spp. (TP), etc. during storage. The results of periodically conducted enumeration studies are generally reported as logarithm of colony forming units per g of food (log CFU/g). The APC is a particularly important parameter since it can be used as a simple, but effective parameter to determine the microbial quality and shelf-life of foods. Thus, the combination of food inoculation tests with microbial quality monitoring of packaged non-inoculated food provides a more realistic approach to determine shelf-life and safety of food before a commercial food packaging application. However, the national regulations of different countries and international organizations might show some variations for their APC limits of different food categories. For example, in European Union, the APC exceeding the limit of 10^5 CFU/cm^2 for carcasses of cattle, ship, goat, horse and pig, and the limit of 5×10^6 CFU/g for minced meat and mechanically separated meat are considered unsatisfactory microbial quality (EU Commission Regulations, 2005). According to Food and Agriculture Organization of the United Nations (FAO), the upper limit for APC of raw refrigerated meat (carcass or boneless), raw chicken, and fresh fish is 10^7 CFU/g (FAO, 1992). Therefore, the monitoring of microbial food quality should be conducted by following proper regulations.

6.4 Methods of testing antioxidant properties of packaging

6.4.1 Antioxidant properties of packaging in reaction mixtures

The test of antioxidant activity for the developed active edible packaging in reaction mixtures is highly beneficial when tests are designed considering major oxidative changes (e.g. lipid oxidation or enzymatic browning) and factors catalyzing these reactions (e.g. free radicals, peroxides, iron, singlet oxygen, oxidative enzymes, etc.) in the target food system. Different performance tests could be adapted to develop packaging materials capable of neutralizing free radicals or to chelate metal atoms, but the proper determination of total antioxidant capacity needs separate analysis of soluble and bound antioxidant activities. The soluble antioxidant activity originates from release of antioxidant agents incorporated into packaging materials, while bound antioxidant activity is originated from antioxidant agents bound on to film matrix and antioxidant groups of monomeric building blocks of film forming hydrocolloid itself.

The soluble antioxidant activity is important to prevent oxidative changes both on the surface and depths of food, but this fraction might diffuse rapidly from the surface into depths of food, and it might be neutralized due to interactions with the food components. In contrast, the bound antioxidants remain on the food surface, the most susceptible part of food for oxidative changes. The phenolic antioxidants incorporated into packaging contribute significantly to bound antioxidant capacity since they form extensive bonds and interactions with proteins and/or polysaccharides forming the film matrix. The polar groups of carbohydrates could bind incorporated phenolic compounds from their hydroxyl groups with non-covalent interactions, such as hydrogen bonds and van der Waals forces (Palafox-Carlos *et al.*, 2011; Barros *et al.*, 2012; Wu *et al.*, 2009), while carbonyl groups of proteins bind incorporated polyphenols mainly from their hydroxyl groups by hydrogen bonding (Arcan and Yemenicioglu, 2014). The hydrocolloids also contain some bound antioxidants coming from the source of extraction (bound polyphenols of proteins, and bound polyphenols, and proteins of polysaccharides), but most proteins and polysaccharides contain also an inherent antioxidant activity originating reactive group of their monomers. For example, the proteins owe their inherent antioxidant activity to their aromatic, sulfur-containing, and basic amino acids capable of donating protons to free radicals, and basic and acidic amino acids that could chelate metal ions (Je *et al.*, 2005; Rajapakse *et al.*, 2005; Hu *et al.*, 2003). On the other hand, polysaccharides show inherent free radical scavenging and iron chelating capacity due to their reactive groups (e.g. hydroxyl and carboxyl groups of pectin, and hydroxyl and amino groups of chitosan) on polysaccharide monomers (Wang *et al.*, 2016; Gharibzahedi *et al.*, 2019).

6.4.1.1   Determination of soluble antioxidant capacity

The determination of the soluble antioxidant capacity of a packaging material needs organization of a release test conducted in distilled water, food simulant, or buffer at a pH close to that of target food. The release tests can be conducted at +4°C to simulate classical refrigeration temperature used for most food (use of any other specific food storage temperature is also possible) and to prevent loss of antioxidant activity during tests. During the release test, the antioxidant activity of released agent in medium is monitored periodically. It is very critical to conduct the test in sufficient volume of release test medium and to continue monitoring until determination of the final antioxidant activity at the equilibrium. The antioxidant activities of all collected samples are also considered and included in calculation of total soluble antioxidant capacity calculated. The monitoring of antioxidant activity with a suitable method and expression of results with a meaningful unit are essential. The antioxidant activity could be determined by choosing one of the methods that represent possible oxidative reactions in the target food system (Roginsky and Lissi, 2005). However, the ABTS test conducted by radical of chemically oxidized 2,2-azinobis-(3-ethylbenzthiazoline-6-sulfonate) (ABTS) (Re *et al.*, 1999) and DPPH test conducted by radical of 2,2-diphenyl-1-picrylhydrazyl (DPPH) (Molyneux, 2004) are the most popular spectrophotometric methods to determine free radical scavenging-based antioxidant activity of H-atom donors, such as reactive groups in phenolics, proteins, and polysaccharides (Je *et al.*, 2005; Rajapakse *et al.*, 2005; Hu *et al.*, 2003; Roginsky and Lissi, 2005).

The iron chelating capacities (ICC) of released active agents might also be determined by the classical spectrophotometric method that employs ferrozine reagent for detection of iron (Stookey, 1970). However, this method is used when the active agents incorporated into packaging show significant iron-binding capacity (e.g. lactoferrin, protein hydrolysates, phenolic compounds). The iron solution is prepared mostly using $FeCl_2$ that provides Fe^{+2}, an important iron form that generates highly reactive hydroxyl radical ($HO^{\bullet}$) when it reacts with hydrogen peroxide (H_2O_2) by Fenton reaction (Eq. 2).

$$Fe^{+2} + H_2O_2 \rightarrow Fe^{+3} + HO^- + HO^{\bullet} \qquad (2)$$

During antioxidant activity measurements, the reaching of the curvature in % inhibition of ABTS or DPPH free radical vs time plot shows proper measurement period of collected sample's antioxidant activity. The proper dilution of collected sample is important to prevent very rapid and too much inhibition of the free radical solution that causes underestimation of total antioxidant capacity. The incubation period of sample with the iron solution in ICC test should also be optimized to prevent underestimation of sample's capacity. The results of soluble antioxidant capacity (SAC) can be calculated and reported as μmol equivalents of a known antioxidant (Vitamin C or Trolox for ABTS and DPPH tests) or a standard chelating agent (disodium EDTA for ICC test) released per cm^2 or g of an edible packaging material (μmol Trolox, Vitamin C or $EDTA/cm^2$, μmol Trolox, Vitamin C or $EDTA/g$). The amount of soluble antioxidant agent released is also determined at the equilibrium by a proper method, and is expressed as $\mu g/cm^2$ (or $\mu g/g$). This value is used to calculate the percentage of soluble active agent recovered from packaging as given in Eq. 1.

6.4.1.2 Determination of bound antioxidant capacity

The bound antioxidant capacity of packaging can also be determined by using methods, such as ABTS and DPPH (Güçbilmez *et al.*, 2007). However, this method is applicable only to insoluble packaging that shows minimum erosion during the test. Moreover, it is also important to use a material free from soluble antioxidants. Therefore, a piece of packaging obtained from release test that reached an equilibrium for the soluble antioxidant is used in measurements after a final effective washing (with distilled water or buffer) to remove remaining soluble residues. During the test, a piece of packaging material is mixed with a suitable volume of free radical solution (e.g. ABTS, DPPH, etc.). The size of the packaging, the volume of the free radical solution, and the test period should be compatible to neutralize all antioxidant groups in the packaging within a reasonable test period. The reaching of the curvature in % inhibition of free radical vs time plot shows proper measurement period of the material's bound antioxidant capacity. The amount of packaging material used in the test is reduced (or volume of free radical solution increased) when very rapid inhibition of the free radical solution occurs. The bound antioxidant capacity (BAC) can be reported as μmol equivalents of a known antioxidant per cm^2 or g of an edible packaging material (μmol Trolox or Vitamin C/cm^2, μmol Trolox or Vitamin C/g). Following calculation of BAC, the total antioxidant capacity (TAC) of the packaging material can also be calculated by using Equation 3.

$$TAC = SAC + BAC \tag{3}$$

Where TAC is the total antioxidant capacity of the packaging (μmol Trolox or Vitamin C/cm^2; μmol Trolox or Vitamin C/g), SAC is the soluble antioxidant capacity of the packaging, and BAC is the bound antioxidant capacity of the packaging.

6.4.2 Antioxidant properties of packaging in food applications

The release tests, and activity, and concentration measurements, conducted in different laboratory media, are beneficial to characterize the amount of soluble/ bound antioxidant content, release profiles, and antioxidant potential of films. However, optimization of the antioxidant concentration of packaging is not possible without packaging and storage tests with the target food. The food applications are conducted mostly by coating food with the developed antioxidant packaging, or application of self-standing antioxidant edible films, mats or pads in different ways (wrapping of food with edible film, placing food into pouches made of edible film, placing edible film, mat or pad at the bottom or surface of food, or between/among its layers, etc.). After that, the food is stored under refrigeration or at room temperature, and monitored for its lipid oxidation products and sensory properties (mainly color, taste, aroma, and flavor). Different methods exist for the measurement of primary and secondary lipid oxidation products in food (Belitz *et al.*, 2004; McClements and Decker, 2008), but the spectrophotometric measurement of thiobarbituric acid (TBA) reactive substances (TBARS) is the most popular method to monitor lipid oxidation products. The TBARS method measures malondialdehyde (MDA), a dialdehyde produced by a two-step oxidative degradation of fatty acids with three or more double bonds, and aldehydic lipid oxidation products (McClements and Decker, 2008). In general, the results of TBARS correlate well with sensory changes in food associated with lipid oxidation (Rhee and Myers, 2004; Nissen *et al.*, 2004), but monitoring of specific carbonyl compounds (such as hexanal, 2, 4-decadienal, etc.) by gas chromatographic methods might also be employed when food contains some interfering agents that limit the application of TBARS method (Belitz *et al.*, 2004).

Although the lipid oxidation is the primary oxidative change in many processed food products, the minimally processed fruits and vegetables suffer mainly form enzymatic browning catalyzed by polyphenoloxidase (PPO). Thus, the antioxidant edible coatings loaded with active compounds showing antibrowning and/or enzyme inhibitory effects (e.g. ascorbic acid and its derivatives, citric, malic and acetic acids, L-cysteine, etc.) (Castaner *et al.*, 1996; Zhou *et al.*, 2020) are tested on minimally processed fruits and vegetables. The effectiveness of antioxidant coatings on minimally processed fruits and vegetables is evaluated in two ways: (1) determination of the effect of released active agent on inhibition of PPO in the product, and (2) monitoring of browning at the product surface using different instruments, such as tristimulus colorimeters, standard cameras or more sophisticated computer vision system (Quevedo *et al.*, 2009). Moreover, it is also essential to evaluate the effect of antioxidant coating on respiration rates of fruits and vegetables. An applicable edible coating reduces the respiration rate and delays the senescence of minimally processed fruits and vegetables without causing anaerobic respiration of the product (Yemenicioğlu, 2016).

References

Aasen, I.M., S. Markussen, T. Møretrø, T. Katla, L. Axelsson and K. Naterstad (2003). Interactions of the bacteriocins sakacin P and nisin with food constituents, *Int. J. Food Microbiol.*, 87: 35-43.

Alkan, D., L.Y. Aydemir, I. Arcan, H. Yavuzdurmaz, H.I. Atabay, C. Ceylan and A. Yemenicioğlu (2011). Development of flexible antimicrobial packaging materials against *Campylobacter jejuni* by incorporation of gallic acid into zein-based films, *J. Agric. Food Chem.*, 59: 11003-11010.

Antonov, Y.A., P. Moldenaers and R. Cardinaels (2017). Complexation of lysozyme with sodium caseinate and micellar casein in aqueous buffered solutions, *Food Hydrocoll.*, 62: 102-118.

Arcan, I. and A. Yemenicioglu (2014). Controlled release properties of zein–fatty acid blend films for multiple bioactive compounds, *J. Agric. Food Chem.*, 62: 8238-8246.

Balaguer, M.P., G. Lopez-Carballo, R. Catala, R. Gavara and P. Hernandez-Munoz (2013). Antifungal properties of gliadin films incorporating cinnamaldehyde and application in active food packaging of bread and cheese spread foodstuffs, *Int. J. Food Microbiol.*, 166: 369-377.

Barros, F., J.M. Awika and L.W. Rooney (2012). Interaction of tannins and other sorghum phenolic compounds with starch and effects on in vitro starch digestibility, *J. Agric. Food Chem.*, 60: 11609-11617.

Belitz, H.D., W. Grosch and P. Schieberle (2004). *Food Chemistry*, 3rd ed. Springer-Verlag, Berlin Heidelberg, Germany.

Ben Arfa, A., L. Preziosi-Belloy, P. Chalier and N. Gontard (2007). Antimicrobial paper based on a soy protein isolate or modified starch coating including carvacrol and cinnamaldehyde, *J. Agric. Food Chem.*, 55: 2155-2162.

Bhunia, K., S.S. Sablani, J. Tang and B. Rasco (2013). Migration of chemical compounds from packaging polymers during microwave, conventional heat treatment, and storage, *Compr. Rev. Food Sci. Food Saf.*, 12: 523-545.

Boyacı, D. and A. Yemenicioğlu (2018). Expanding horizons of active packaging: Design of consumer controlled release systems helps risk management of susceptible individuals, *Food Hydrocoll.*, 79: 291-300.

Boyacı, D., Korel, F. and Yemenicioglu, A. (2016). Development of activate-at-home-type edible antimicrobial films: An example pH-triggering mechanism formed for smoked salmon slices using lysozyme in whey protein films, *Food Hydrocoll.*, 60: 170-178.

Castaner M., M.I. Gil, F. Artes and F.A. Tomas-Barberan (1996). Inhibition of browning of harvested head lettuce, *J. Food Sci.*, 61: 314-316.

European Union (EU) Commission Regulation (2005). No. 2073/2005 of 15 November 2005 on microbiological criteria for foodstuffs, *Official Journal of the European Union*, 22.12.2005.

European Union (EU) Commission Regulation (2011). No. 10/2011 of 14 January 2011 on plastic materials and articles intended to come into contact with food, *Official Journal of the European Union*, 15.1.2011.

Food and Agricultural Organization (FAO) of the United Nations (1992). *Manual of Food Quality Control: Microbiological Analysis*, 14/4, rev.1, Rome, Italy.

Gemili, S., A. Yemenicioğlu and S.A. Altınkaya (2009). Development of cellulose acetate based antimicrobial food packaging materials for controlled release of lysozyme, *J. Food Eng.*, 90: 453-462.

Gharibzahedi, S.M.T., B. Smith and Y. Guo (2019). Ultrasound-microwave assisted extraction of pectin from fig (*Ficus carica* L.) skin: Optimization, characterization and bioactivity, *Carbohydr. Polym.*, 222: 114992.

Güçbilmez, Ç.M., A. Yemenicioğlu and A. Arslanoğlu (2007). Antimicrobial and antioxidant activity of edible zein films incorporated with lysozyme, albumin proteins and disodium EDTA, *Food Res. Int.*, 40: 80-91.

Han, H. and B.K. Baik (2008). Antioxidant activity and phenolic content of lentils (*Lens culinaris*), chickpeas (*Cicer arietinum* L.), peas (*Pisum sativum* L.) and soybeans (*Glycine max*), and their quantitative changes during processing, *Int. J. Food Sci. Technol.*, 43: 1971-1978.

Han, J.H. and J.D. Floros (1998). Potassium sorbate diffusivity in American processed and Mozzarella cheeses, *J. Food Sci.*, 63: 435-437.

Hu, M., D.J. McClements and E.A. Decker (2003). Lipid oxidation in corn oil-in-water emulsions stabilized by casein, whey protein isolate, and soy protein isolate, *J. Agric. Food Chem.*, 51: 1696-1700.

Jay, J.M. (2002). A review of aerobic and psychrotrophic plate count procedures for fresh meat and poultry products, *J. Food Prot.*, 65: 1200-1206.

Je, J., P. Park and S. Kim (2005). Antioxidant activity of a peptide isolated from Alaska pollack (*Theragra chalcogramma*) frame protein hydrolysate, *Food Res. Int.*, 38: 45-50.

Joerger, R.D. (2007). Antimicrobial films for food applications: A quantitative analysis of their effectiveness, *Packag. Technol. Sci.*, 20: 231-273.

Li, C., J. Pei, X. Xiong and F. Xue (2020). Encapsulation of grapefruit essential oil in emulsion-based edible film prepared by plum (*Pruni domesticae semen*) seed protein isolate and gum acacia conjugates, *Coatings*, 10: 784.

Liu, F., R.J. Avena-Bustillos, B.S. Chiou, B. Li, Y. Ma, T.G. Williams *et al.* (2017). Controlled-release of tea polyphenol from gelatin films incorporated with different ratios of free/nanoencapsulated tea polyphenols into fatty food simulants, *Food Hydrocoll.*, 62: 212-221.

Lopes, L.F., G. Meca, K.C. Bocate, T.M. Nazareth, K. Bordin and F.B. Luciano (2018). Development of food packaging system containing allyl isothiocyanate against Penicillium nordicum in chilled pizza: Preliminary study, *J. Food Process. Preserv.*, 42: e13436.

López, P., C. Sánchez, R. Batlle and C. Nerín (2007). Development of flexible antimicrobial films using essential oils as active agents, *J. Agric. Food Chem.*, 55: 8814-8824.

McClements, D.J. and E.A. Decker (2008). Lipids, pp. 155-216. *In:* S. Damodaran, K.L. Parkin, and O.R. Fennema (Eds.). *Fennema's Food Chemistry*, 4th ed. CRC Press, Taylor and Francis Group, New York, USA.

Mecitoglu, Ç., A. Yemenicioglu, A. Arslanoglu, Z.S. Elmacı, F. Korel and A.E. Çetin (2006). Incorporation of partially purified hen egg white lysozyme into zein films for antimicrobial food packaging, *Food Res. Int.*, 39: 12-21.

Molyneux, P. (2004). The use of the stable free radical diphenylpicrylhydrazyl (DPPH) for estimating antioxidant activity, *Songklanakarin J. Sci. Technol.*, 26: 211-219.

Munhuweyi, K., O.J. Caleb, C.L. Lennox, A.J. van Reenen and U.L. Opara (2017). *In vitro* and *in vivo* antifungal activity of chitosan-essential oils against pomegranate fruit pathogens, *Postharvest Biol. Tech.*, 129: 9-22.

Nissen, L.R., D.V. Byrne, G. Bertelsen and L.H. Skibsted (2004). The antioxidative activity of plant extracts in cooked pork patties as evaluated by descriptive sensory profiling and chemical analysis, *Meat Sci.*, 68: 485-495.

Palafox-Car los, H., J.F. Ayala-Zavala and G.A. González-Aguilar (2011). The role of dietary fiber in the bioaccessibility and bioavailability of fruit and vegetable antioxidants, *J. Food Sci.*, 76: 6-15.

Quevedo, R., O. Díaz, A. Caqueo, B. Ronceros and J.M. Aguilera (2009). Quantification of enzymatic browning kinetics in pear slices using non-homogenous L* color information from digital images, *LWT-Food Sci. Technol.*, 42: 1367-1373.

Rajapakse, N., E. Mendis, W. Jung, J. Je and S. Kim (2005). Purification of a radical scavenging peptide from fermented mussel sauce and its antioxidant properties, *Food Res. Int.*, 38: 175-182.

Re, R., N. Pellegrini, A. Proteggente, A. Pannala, M. Yang and C. Rice-Evans (1999). Antioxidant activity applying an improved ABTS radical cation decolorization assay, *Free Radic. Biol. Med.*, 26: 1231-1237.

Rhee, K.S. and C.E. Myers (2004). Sensory properties and lipid oxidation in aerobically refrigerated cooked ground goat meat, *Meat Sci.*, 66: 189-194.

Ribeiro-Santos, R., N.R. de Melo, M. Andrade and A. Sanches-Silva (2017). Potential of migration of active compounds from protein-based films with essential oils to a food and a food simulant, *Packag. Technol. Sci.*, 30: 791-798.

Rocha, M.A.M., M.A. Coimbra and C. Nunes (2017). Applications of chitosan and their derivatives in beverages: A critical review, *Curr. Opin. Food Sci.*, 15: 61-69.

Roginsky, V. and E. Lissi (2005). Review of methods to determine chain-breaking antioxidant activity in food, *Food Chem.*, 92: 235-254.

Rose, N.L., M.M. Palcic, P. Sporns and L.M. McMullen (2002). Nisin: A novel substrate for glutathione S-transferase isolated from fresh beef, *J. Food Sci.*, 67: 2288-2293.

Rose, N.L., P. Sporns, M.E. Stiles and L.M., McMullen (1999). Inactivation of nisin by glutathione in fresh meat, *J. Food Sci.*, 64: 759-762.

Sebti, I., A.R. Carnet, D. Blanc, R. Saurel and V. Coma (2003). Controlled diffusion of an antimicrobial peptide from a hydrocolloid film, *Chem. Eng. Res. Des.*, 81: 1099-1104.

Sozbilen, G.S. and A. Yemenicioğlu (2021). Antilisterial effects of lysozyme-nisin combination at temperature and pH ranges optimal for lysozyme activity: Test of key findings to inactivate Listeria in raw milk, *LWT-Food Sci. Technol.*, 137: 110447.

Stookey, L.L. (1970). Ferrozine – A new spectrophotometric reagent for iron, *Anal. Chem.*, 42: 779-781.

Trinetta, V., J.D. Floros and C.N. Cutter (2010). Sakacin a-containing pullulan film: An active packaging system to control epidemic clones of *Listeria monocytogenes* in ready-to-eat foods, *J. Food Saf.*, 30: 366-381.

Ünalan, İ.U., I. Arcan, F. Korel and A. Yemenicioğlu (2013). Application of active zein-based films with controlled release properties to control *Listeria monocytogenes* growth and lipid oxidation in fresh Kashar cheese, *Innov. Food Sci. Emerg. Technol.*, 20: 208-214.

US Food and Drug Administration (FDA) (2007). Guidance for industry: Preparation of premarket submissions for food contact substances, *Chemistry Recommendations*, Guidance Compliance Regulatory Information. Docket Number: FDA-2020-D-1925.

Wang, J., S. Hu, S. Nie, Q. Yu and M. Xie (2016). Reviews on mechanisms of in vitro antioxidant activity of polysaccharides, *Oxid. Med. Cell. Longev.*, 2016: 5692852

Winther, M. and P.V. Nielsen (2006). Active packaging of cheese with allyl isothiocyanate, an alternative to modified atmosphere packaging, *J. Food Protect.*, 69: 2430-2435.

Wu, Y., Z. Chen, X. Li and M. Li (2009). Effect of tea polyphenols on the retrogradation of rice starch, *Food Res. Int.*, 42: 221-225.

Yemenicioğlu, A. (2011). *Development of Composite or Blend Active Edible Food Packaging Materials for Controlled Release of Bioactive Substances*. The Scientific and Technical Research Council of Turkey. Project # MAG 108 M 353.

Yemenicioğlu, A. (2016). Strategies for controlling major enzymatic reactions in fresh and processed vegetables, pp. 377-388. *In:* Y.H. Hui and Ö. Evranuz (Eds.). *Handbook of Vegetable Preservation and Processing*, 2nd ed. CRC Press, New York, USA.

Yemenicioğlu, A. (2017). Basic strategies and testing methods to develop effective edible antimicrobial and antioxidant coating, pp. 63-88. *In:* A. Tiwari (Ed.). *Handbook of Antimicrobial Coatings*, 1st ed. Elsevier, Amsterdam, The Netherlands.

Yemenicioğlu, A. and I. Arcan (2009). Controlled release of catechin from edible zein films intended for meat bioactive packaging, *55th International Congress of Meat Science and Technology*, Cophenhagen, Denmark.

Yousef, A.E. and C. Carlstrom (2003). *Food Microbiology: A Laboratory Mannual*. John Wiley & Sons, Hoboken, New Jersey. USA.

Zhou, L., T. Liao, W. Liu, L. Zou, C. Liu and N.S. Terefe (2020). Inhibitory effects of organic acids on polyphenol oxidase: From model systems to food systems, *Crit. Rev. Food Sci. Nutr.*, 60: 3594-3621.

Application of Active Edible Packaging for Different Food Categories

7.1 Introduction

Active edible packaging can be applied to different food categories, such as dairy products, raw and processed beef, pork, fish and poultry, whole or minimally-processed fresh fruits and vegetables, seeds and mushrooms, bakery products, and dough food. In most of the food applications, active edible packaging is combined with refrigeration. Moreover, sometimes active edible packaging is combined with other preservation methods, such as vacuum packaging (VP), modified atmosphere packaging (MAP), high pressure processing (HPP), pulsed light processing (PL), irradiation, etc. In this chapter, food applications of active edible packaging in combination with refrigeration are selected to reflect heavily the current trend of using phenolic compounds in active edible packaging of food with a particular emphasis on their antimicrobial and antioxidant potential, and positive or negative impacts on sensory properties of foods. Some examples are also included to show the beneficial effects of employing nanotechnology (e.g. nanoemulsions, nanoencapsulation) in preparation of phenolic compounds incorporated into films and coatings with respect to product safety and quality, shelf-life, and sensory properties. The recent trends in using antimicrobial enzymes, bacteriocins, and peptides, and protective or probiotic cultures, and prebiotics in active packaging are reflected sufficiently with different examples. In general, most of the active edible packaging studies discuss measuring effectiveness of this application non-specifically by monitoring microbial quality of packaged food during cold storage, based on counting total aerobic mesophilic bacteria (TAMB), total psychrophilic bacteria (TPB), total coliform bacteria (TCB), total yeast and mold (TYM), total *Enterobacteriaceae* (TE), and total *Pseudomonas* spp. (TP). However, some examples have also been included to show the effectiveness of active packaging in inhibiting/controlling more specific critical pathogens, such as *Listeria monocytogenes*, *S. enterica* serovar Typhimurium, *Escherichia coli* O157:H7, *Camphylobacter jejuni*, etc. Moreover, different examples are also included to show the effectiveness of active packaging in preventing food spoilage originating from different fungi. This chapter objectively reflects the details of most applicable and effective recent packaging applications for a great variety of food to attract the interest of researchers and experts working in different food sectors.

7.2 Active edible packaging of cheese

Active edible packaging of cheeses focuses mainly on increasing microbial safety of fresh cheeses and prolonging their shelf-life by retarding fungal growth. The packaging is applied frequently by different coating methods, such as dipping into film-forming solutions or brushing film-forming solutions on to the cheese surface. In different studies, examples of hydrocolloids used in active coating of cheese include Na-alginate coating for fresh home-made cheese (Mahcene *et al.*, 2021); fresh fior di latte cheese (a kind of fresh mozzarella from cow's milk) (Angiolillo *et al.*, 2014); and fresh *paneer* cheese (Raju and Sasikala, 2016); whey protein coating for fresh flamengo cheese (Guimarães *et al.*, 2020); chitosan coating for fresh goat cheese (Cano Embuena *et al.*, 2017); aged mozzarella cheese (Duan *et al.*, 2007); and fresh halloumi cheese (Mehyar *et al.*, 2018); gellan gum coating for fresh, processed cheese (Ordoñez *et al.*, 2021); cassava starch coating for manaba fresh white cheese (Santacruz and Castro, 2018); and galactomannan coating for fresh coalho cheese (Lima *et al.*, 2020). Precast self-standing edible films are also employed in active packaging of cheese by placing films on both sides of cheese slices or by wrapping portioned cheese pieces with films. Some edible films used in active packaging of cheese include whey protein film for fresh kashar cheese (Seydim *et al.*, 2020), and flamengo cheese (Guimarães *et al.*, 2020); Na-alginate film for fresh kashar cheese (Küçük *et al.*, 2020); zein film for fresh kashar cheese (Ünalan *et al.*, 2013; Küçük *et al.*, 2020), fresh Himalayan cheese (Mushtaq *et al.*, 2018) and fresh gouda cheese (zein film or electrospun zein nanofiber mat) (Göksen *et al.*, 2020); chitosan-gelatin composite films for aged prato cheese (Bonilla and Sobral, 2019); chitosan film for aged mozzarella cheese (Duan *et al.*, 2007); and amaranth protein-pullulan nanofiber mat for fresh panala cheese (Soto *et al.*, 2019). Thus, it is evident that Na-alginate and chitosan coatings, and zein-based self-standing films are preferred most frequently in active packaging of different cheeses.

Different natural phenolic compounds incorporated into active packaging include essential oils (EO) from basil, oregano, garlic, Daphne, rosemary, lemon grass, cinnamon, and phenolic-rich extracts, such as Aloe vera, boldo, and pomegranate peel extracts (Table 7.1). From these phenolic compounds, basil EO, used in Na-alginate coating for home-made fresh cheese (Mahcene *et al.*, 2021), lemon grass EO in galactomannan coating for fresh coalho cheese (Lima *et al.*, 2020), oregano EO in chitosan coating for fresh goat cheese (Cano Embuena *et al.*, 2017), cinnamon EO in Na-alginate coating for *paneer* cheese (Raju and Sasikala, 2016), and pomegranate peel extract in zein films for fresh Himalayan cheese (Mushtaq *et al.*, 2018), are preferred or have received acceptable scores by the penellists during sensory analysis while no sensory analysis was performed in the other studies for the specified phenolic compounds. It is important to note that pomegranate peel extract used in zein films is of the rare compounds that showed not only outstanding antibacterial and antifungal activity, but also inhibited lipid

Table 7.1: Examples of active edible packaging for cheese

Film type	Active agent(s)	Product/Application/Storage	Changes in safety shelf-life quality[g]/Sensory properties	References
Na-alginate coating	Basil EO at 1% (v/v)[a]	Fresh home-made cheese/Dipping-draining-drying/10 days at 5-7°C.	Fecal coliform count and TAMB at the end of storage period reduced by 1.8 and ~0.2 log, respectively. Lipid oxidation reduced significantly/Sensory properties (based on color, texture, flavor, and taste) of product with active coating was preferred by panelists.	Mahcene *et al.* (2021)
Na-alginate coating	Cinnamon EO at 2.5% (w/v)[b]	Fresh paneer cheese/Brushing/15 days at 4°C.	Uncoated control cheese and coated cheese had TAMB >6 log CFU/g and 5.39 log CFU/g at 7th and 10th days, respectively/No significant differences were detected between sensory properties (based on taste, color, aroma, texture, appearance) of control and coated cheese.	Raju and Sasikala, (2016)
Na-alginate coating	Pure culture of *Lactobacillus reuteri* at 2% (w/v)[b]	Fior di latte cheese/Dipping-CaCl₂ cross-linking-draining/Seven days at 9°C.	Active coating with plasticizer glycerol caused a significant delay in growth of PC and TE. Incubation of coating solution for 48h before application increased its effectiveness/Cheese with active coating gave the highest sensory scores (based on odor, color, texture, and overall quality).	Angiolillo *et al.* (2014)
Zein or Na-alginate film	Natamycin at 200 to 5000 ppm in film forming solution	Fresh kashar cheese/Films were placed at both surfaces of cheese slices/45 days at 4°C.	Zein films with 200-500 ppm natamycin inhibited the growth of inoculated *P. camemberti* and *A. niger* for 30 days. Na-alginate films with 200-500 natamycin inhibited the growth of inoculated *Aspergillus niger* for 45 days, but they suppressed *P. camemberti* growth only for 15 days/No sensory analysis.	Küçük *et al.* (2020)

Film	Active agent	Food/Conditions	Results	Reference
Zein film or electrospun zein nanofiber	Daphne and rosemary EO at 5 or 10% (w/w)[d]	Fresh gouda cheese (semi-hard)/Nanofibers or films were placed at the cheese surface/28 days at 4°C.	All EO loaded films caused significant reductions in *L. monocytogenes, S. aureus* and TAMB. In general, the zein nanofiber coating loaded with EO at 5% showed comparable antimicrobial activity with zein films loaded with EO at 10%/No sensory analysis.	Göksen *et al.* (2020)
Zein film	Pomegranate peel extract (PPE) at 2.5-7.5% (w/w)[c]	Fresh Himalayan cheese/Cheese samples were wrapped by films/30 days at 4°C.	Active films reduced the TAMB and TYM, and lipid and protein oxidation in cheese in a concentration-dependent manner. TAMB and TYM of cheese wrapped by films with PPE at 7.5% inhibited totally at 30 and 21 days of storage, respectively. However, active films did not show a considerable effect on TLAB of cheese/Active film with 7.5% PPE caused the best sensory properties (based on flavor, aroma, appearance, bitterness, and overall acceptability).	Mushtaq *et al.* (2018)
Zein-carnauba wax composite film	Lysozyme (LYS) at 11.7% (w/w)[c], and catechin (CAT) and gallic acid (GA) at 5% (w/w)[c]	Fresh kashar cheese/Films were placed at both surfaces of cheese slices/56 days at 4°C.	Unpacked and control zein film packed cheese showed 2.6 and 2.2 log increase in *L. monocytogenes* load while zein films with LYS, and LYS-CAT-GA prevented *L. monocytogenes* growth during storage. LYS-CAT-GA loaded film showed sustained release properties and caused significant reduction (~0.4 log) in *L. monocytogenes* load at the end of storage. Polyphenols in film prevented lipid oxidation in cheese/No sensory analysis.	Ünalan *et al.* (2013)
Whey protein isolate film	Oregano EO, Garlic EO or NIS at 2% (w/v)[b]	Fresh kashar cheese (semi-hard)/Films were placed at both surfaces of cheese slices/15 days at 4°C.	*E. coli* O157: H7 counts at the end of storage reduced by 3, 2.8 and 2 log for samples packed with oregano and garlic EO, and NIS loaded films, respectively. All active films caused almost similar reductions (~2 log) in *S. enterica* serovar Enteritidis, *L. monocytogenes* and *S. aureus* at the end of storage/No sensory analysis.	Seydim *et al.* (2020)

(Contd.)

Table 7.1: (*Contd.*)

Film type	Active agent(s)	Product/Application/ Storage	Changes in safety shelf-life quality[g]/Sensory properties	References
Whey protein concentrate coating or film	*Lactobacillus buchneri* UTAD104 (10[11] CFU/mL) at 30% (w/w)[c]	Fresh flamengo cheese (semi-hard)/Dipping-draining-drying or placing film at cheese slice surface/30 days at 25°C.	Lactic acid bacteria loaded coating or film inhibited *P. nordicum* and its ochratoxin A production/No sensory analysis.	Guimarães *et al.* (2020)
Cassava starch edible coating	Encapsulated (Na-alginate-CaCl₂) or free *L. acidophilus* (12 log CFU/g film)	Fresh manaba white cheese/ Dipping-draining-drying/ 30 days at 4°C.	TAMB of cheese coated with free and encapsulated *L. acidophilus* reduced by 3.2 and 7.6 log, respectively/No sensory analysis.	Santacruz and Castro (2018)
Galactomannan coating	Lemon grass EO at 0.2% (v/v)[a]	Fresh coalho cheese/Dipping-draining-drying/ 30 days at 4°C.	TAMB, TCB and thermotolerant coliform count of cheeses with active coating reduced slightly, but remained within spoilage limits during storage/ Coated samples showed acceptable sensory properties (based on flavor, aroma, color, texture, and overall impression).	Lima *et al.* (2020)
Chitosan coating	Oregano or rosemary EO at %0.75 (w/w)[d]	Fresh goat cheese/Dipping-draining-drying (2-3 repeated cycles)/15 days at 10°C.	Oregano EO loaded coatings (three-repeated cycles more effective than two-repeated cycles) reduced incidence of fungal growth (*Mucor* sp. and *Penicillium* sp.). Rosemary EO loaded coatings cannot prevent *Penicillium* sp. growth. Active coating reduced cheese ripening rate/Sensory properties (based on appearance, aroma, flavor, texture, and overall preference) of samples coated two-repeated cycles with oregano EO loaded coating were preferred by the panelists.	Cano Embuena *et al.* (2017)

Chitosan coating	Lysozyme at 60% (w/w)[d]	Fresh halloumi cheese/Dipping-draining-drying/35 days at 3°C.	Lysozyme loaded coatings reduced anaerobic count, TPB, and TYM of cheese stored in 10 and 15% brine. Active coating enabled reduction of NaCl in brine from 15 to 10%/Coatings with or without lysozyme improved the taste and overall acceptability in 10% brine (sensory analysis was based on color, aroma, taste and overall acceptability).	Mehyar et al. (2018)
Chitosan-cassava starch composite film (1:1)	Lysozyme at 0.25% (w/v)[b]	Mongolian cheese/Cheese cubes were wrapped by films/30 days at 4°C.	Lysozyme loaded films reduced (~0.9 log) TAMB and TYM of cheese during cold storage. Edible film reduced the water loss of cheese during storage by 50%/No sensory analysis.	Ma et al. (2021)
Chitosan-gelatin (1:1) composite films	Boldo extract at 1% (v/v)[a]	Aged prato cheese/Precast films were placed between both surfaces of cheese slices/10 days at 4°C.	Composite films with boldo extract reduced TPB and TAMB significantly. TCC reduced significantly by composite films with or without boldo extract. Films with boldo extract prevented lipid oxidation in cheese/No sensory analysis.	Bonilla and Sobral (2019)
Chitosan-lysozyme (60% of chitosan, w/w) composite film or coating	Lysozyme at 60% (w/w)[b]	Aged mozzarella cheese/Films were placed at surface of cheese slices or cheese slices were coated by dipping-drying/14 (bacterial tests) or 30 (antifungal test) days at 4°C.	All active packaging reduced inoculated *P. fluorescens* (~2-2.5 log), *E. coli* (~1.5-2 log) and *L. monocytogenes* (~1-1.5). Active films showed significant reduction in inoculated *Cladosporium* spp., but did not show a considerable antifungal activity against *Candida inconspicua*/No sensory analysis.	Duan et al. (2007)
Gellan gum coating	Aloe vera gel at Brix of 1 (mixed 1:1 v/v with gellan gum solution at 0.9% w/w[c])	Fresh pressed cheese/Dipping-draining/Seven days at 4°C.	Gum gellan coating with Aloe vera gel completely inhibited the radial and mycelial growth of inoculated *P. roqueforti*, but coatings were not very effective on water loss/No sensory analysis.	Ordoñez et al. (2021)

(Contd.)

Table 7.1: (*Contd.*)

Film type	Active agent(s)	Product/Application/ Storage	Changes in safety shelf-life quality[g]/Sensory properties	References
Amaranth protein-pullulan nanofiber mat	Nisin at 20 mg/ml of film forming solution	Fresh panela cheese/Mats were placed at the cheese surface/10 days at 4°C.	Nanofiber mats loaded with nisin caused almost 4 log reduction of *L. mesenteroides*, *L. monocytogenes* and *S. enterica* serovar Typhimurium/No sensory analysis.	Soto *et al.* (2019)

[a] v/v film forming solution; [b] w/v film forming solution; [c] w/w film forming solution; [d] w/w hydrocolloid; [e] v/v film forming solution; [f] v/w hydrocolloid; [g] TAMB: total aerobic mesophilic bacteria; TPB: total psychrophilic bacteria; TCC: total coliform count; TYM: total yeast and mold, TE: Total *Enterobacteriaceae*; PC: *Pseudomonas* spp. count; EO: essential oil

and protein oxidation in actively packed fresh cheeses (Mushtaq *et al.*, 2018). The antifungal activity of oregano EO used in chitosan coating for fresh cheese against *Mucor* sp. and *Penicillium* sp. was also noteworthy (Cano Embuena *et al.*, 2017). Thus, further studies are needed to show widespread applicability of pomegranate peel extract and oregano EO in active packaging of different cheeses.

Lysozyme (LYS) is also one of the natural antimicrobials that has been increasingly used in cheeses to improve their microbial quality without causing changes in their sensory properties. Mehyar *et al.* (2018) controlled microbial load of halloumi cheeses by LYS-loaded chitosan coatings and achieved reduction of salt concentration in brines (from 15 to 10%), used for cheese maturation and storage. Ma *et al.* (2021) also controlled microbial load in Mongolian cheese wrapped by LYS-loaded chitosan-cassava starch composite films. One of the most important benefits of employing LYS in active packaging of cheeses is that it prevents growth of critical pathogen *L. monocytogenes* as demonstrated by Ünalan *et al.* (2013), who applied controlled release strategies with zein films on cold stored kashar cheese. Duan *et al.* (2007) also proved the beneficial effects of LYS in inactivation of *L. monocytogenes* in aged mozzarella packed both with LYS-loaded chitosan film and coating. These authors also showed reduced *P. fluorescens* and *E. coli* load in mozzarella mainly due to the inherent antimicrobial activity of chitosan coatings and films. Thus, it appears that the combination of chitosan with LYS might be a very beneficial strategy to increase the antimicrobial spectrum of packaging used for cheese.

Some innovative studies in active packaging of cheese came by incorporating lactic acid bacteria (LAB) in films and coatings as protective cultures to suppress the cheese spoilage flora. Guimarães *et al.* (2020) employed *Lactobacillus buchneri* in whey protein films and coating as an antifungal to inhibit *Penicillium nordicum* and its ochratoxin A production in flamengo cheese. Moreover, free or encapsulated (by Na-alginate-CaCl$_2$) *L. acidophilus* loaded cassava starch coating was used by Santacruz and Castro (2018) to reduce the bacterial load of manaba fresh white cheese by 3.2 and 7.6 log, respectively. This work showed the beneficial effect of using encapsulation to increase effectiveness (survivability) of LAB in edible coating and cheese.

7.3 Active edible packaging of meat and meat products

Recent studies, related to active edible packaging of meat and meat products, have focused mainly on controlling of microbial load and oxidative changes (lipid and myoglobin oxidation) in fresh beef (ground, slices or pieces) and cooked sausages. Some studies also exist, related to edible active packaging of fresh and cooked pork, fresh or thawed lamb, and fresh mutton meats. Since meat and meat products are highly susceptible to microbial spoilage, the application of inherently antimicrobial chitosan or chitosan-based composite films or coatings loaded with antimicrobial compounds is one of the most frequently used strategies in their active packaging (Pabast *et al.*, 2018; Esmaeli *et al.*, 2020; Mehdizadeh *et al.*,

2020; Wang *et al.*, 2020; Liu *et al.*, 2021; Amor *et al.*, 2021; Alirezalu *et al.*, 2021). However, different active films from whey proteins (Esmaeili *et al.*, 2020), bacterial cellulose (Yordshahi *et at.*, 2020), cassava starch (Caetano *et al.*, 2017), tamarind seed starch-xanthan gum blends (Mohan *et al.*, 2017), pectin-gelatin composites (Bermúdez-Oria *et al.*, 2019), and coatings from Na-alginate (Guerrero *et al.*, 2020) and Balangu seed gum (Behbahani *et al.*, 2020) have also been recently employed in meat preservation.

The use of pure olive polyphenols (hydroxytyrosol and 3,4-dihydroxyphenylglycol); phenolic-rich pomegranate peel extract, pumpkin residue extract, and spice extracts (clove, fennel, thyme, cinnamon, prickly ash and geranium); reduced sized spice powders of clove and cinnamon; and essential oils from oregano, basil, cumin, garlic, thyme, apricot kernel, and *Satureja khuzestanica* have been tested for edible active packaging of meat and meat products (Table 7.2). Sensory analyses of fresh meat are conducted mostly insufficiently by detecting discoloration caused by oxidation of myoglobin to metmyoglobin, and off-odor formation, but it is rare to determine taste, flavor, and texture of packed fresh meat after cooking. Therefore, it is difficult to evaluate exact sensory properties of many phenolic compounds tested in active edible packaging of meats. Pabast *et al.* (2018) reported that significant improvements occurred in microbial and sensory properties of cold-stored lamb steaks treated with *Satureja khuzestanica* essential oil (SAEO) loaded chitosan coatings. According to these authors, the beneficial effects of SAEO on antimicrobial and sensory properties of lamb steaks increased when nanoliposome-encapsulated SAEO was loaded into chitosan coating, instead of free SAEO. Esmaeili (2020) also reported improved sensory properties (mainly in taste and odor) of cold-stored cooked sausages packed with chitosan or whey protein coatings loaded with garlic EO. These authors did not find any considerable difference between lipid oxidation profiles and sensory properties of sausages packed with free or nanoencapsulated garlic EO loaded films, but chitosan films with nanoencapsulated garlic EO provided the highest antimicrobial effect on packed sausages. The positive effects of pomegranate peel extract loaded chitosan-starch composite films and cumin EO loaded Balangu seed gum coatings in sensory properties of raw beef slices were also reported by Mehdizadeh *et al.* (2020) and Behbahani *et al.* (2020), respectively, but no data were given about effects of these active phenolic agents on taste and flavor of beef after cooking. Wang *et al.* (2020) also reported that use of apricot kernel EO loaded chitosan films improved the sensory properties of cold-stored packed minced spiced beef. The active agents and packaging materials that showed noteworthy antimicrobial effect (on TAMB) in meats include basil EO loaded chitosan films for cooked ham (Amor *et al.*, 2021), clove, fennel or geranium extract loaded chitosan coatings for fresh pork meat (Liu *et al.*, 2021), pomegranate peel extract and thyme EO mixture loaded chitosan-starch composite film for beef slices (Mehdizadeh *et al.*, 2020), nanoencapsulated garlic EO loaded chitosan films for cooked sausages (Esmaeili *et al.*, 2020), nanoencapsulated *Satureja khuzestanica* EO loaded chitosan coating for lamb steaks (Pabast *et al.*, 2018), and reduced size clove and cinnamon powder mixture loaded tamarind seed starch-xanthan gum blend films for mutton meat (Mohan *et al.* (2017). The effectiveness of apricot kernel EO loaded chitosan films to reduce

Table 7.2: Examples of active edible packaging for meat and meat products

Film type	Active agent(s)	Product/Application/ Storage	Changes in safety shelf-life quality#/Sensory properties	References
Chitosan coating	Spice extracts (clove, fennel, thyme, cinnamon, pricklyash and geranium)	Fresh pork leg meat/ Dipping-draining/ seven days at 4°C.	Samples coated with clove, fennel and geranium extracts gave the lowest TAMB at the end of storage. Active coatings prevented the discoloration of pork meat/No sensory analysis	Liu *et al.* (2021)
Chitosan film	Encapsulated Basil EO at 1-3% (w/v)[b] (Encapsulant: Na-alginate cross-linking with CaCl₂)	Cooked ham slices/ Wrapping by active films/10 days at 4°C.	Active coating with 1-2% and 3% encapsulated EO kept TAMB of ham below 6 log CFU/g for six to eight days and 10 days, respectively, but controls exceeded 6 log CFU/g within four days/No sensory analysis	Amor *et al.* (2021)
Chitosan film	ε-polylysine at 0.3-0.9% (w/v)[b]	Beef fillet/Wrapping by active films/10 days at 4°C.	Active films loaded with ε-polylysine at 0.9% gave the lowest TAMB, TPB, and TYM. Active films reduced TVB-N and lipid oxidation of fillets/Films with ε-polylysine at 0.9% gave the highest sensory scores (based on flavor, odor, color, texture, and overall acceptability of microwave cooked beef).	Alirezalu *et al.* (2021)
Bacterial cellulose film	Postbiotics of *Lactobacillus plantarum* were impregnated within films	Ground beef/Films were placed at both surfaces of shaped sample/Nine days at 4°C.	Films loaded with postbiotics caused 5 log inhibition of *L. monocytogenes* in inoculated samples within nine days. TAMB and TPB of controls and actively packed samples reached 7 log CFU/g within almost three and six days, and five and eight days, respectively/Control and actively packed samples showed similar sensory properties (based on color, odor, and taste of grilled beef).	Yordshahi *et al.* (2020)

(Contd.)

Table 7.2: (*Contd.*)

Film type	Active agent(s)	Product/Application/Storage	Changes in safety shelf-life quality[g]/Sensory properties	References
Cassava starch film	Oregano EO and pumpkin residue extract (PRE) at 2 and 3% (w/v)[b] respectively	Ground beef/Sample was packed by the films/six days at 4°C.	Films with oregano EO and PRE caused limited reductions in TAMB and TCC of samples at 3rd day of storage, but unpacked controls, samples with control films or active films showed similar TAMB and TCC at 6th day. The films could not prevent the growth of *S. enterica* serovar Enteritidis in inoculated samples in the first day of storage. The active and control films did not reduce the lipid oxidation in samples/No sensory analysis.	Caetano *et al.* (2017)
Chitosan film	Apricot kernel EO at 0.25 to 1% (v/v)[a]	Minced spiced beef/ Samples were placed into plastic cups and films were paste at the internal surface of their lids/24 days at 4°C.	Active films with 0.25, 0.5 and 1% EO caused almost ~2.6, 3.3 and 4.1 log reduction of *L. monocytogenes* counts of inoculated samples within 15 days, respectively. EO reduced the lipid oxidation in a concentration-dependent manner/ Only samples packed with 1% EO loaded films had acceptable sensory properties (based on taste, color, texture, and overall acceptance).	Wang *et al.* (2020)
Balangu (*Lallemantia royleana*) seed gum coating	Cumin EO at 1 to 2% (w/w)[c]	Beef slices/Dipping-draining/Nine days at 4°C.	Active coating with 1, 1.5 and 2% EO showed comparable TAMB and TPB, but coatings with EO at 1.5 and 2% showed the highest inhibitory effect (~2 log) on inoculated *S. aureus* and *E. coli*. Active coatings reduced the lipid oxidation of samples/All samples with active coating showed better sensory properties than controls, but coating with 1% EO gave the best sensory properties (based on odor, color, and overall acceptance).	Behbahani *et al.* (2020)

Chitosan-starch composite film	Pomegranate peel extract (PPE) and thyme EO at 0.5-2% (w/w)[c]	Beef slices/Films were placed at both surfaces of sliced samples/21 days at 4°C.	Active coatings with PPE at 1% + thyme EO at 2%, PPE at 0.5% + thyme EO at 2%, PPE at 1% + thyme EO at 1%, PPE at 0.5% + thyme EO at 1%, and thyme EO at 2% effectively suppressed the TAMB and TPB for 16 to 21 days (controls reached $\geq$ 6 log CFU/g in four days). Active coatings with PPE and thyme EO inhibited lipid oxidation/Active coatings with PPE at 1% + thyme EO at 1%, and PPE at 1% + thyme EO at 2% gave the best sensory properties (based on odor, color, texture, and overall acceptability).	Mehdizadeh *et al.* (2020)
Pectin-gelatin (1:1) film with/without beeswax	Olive polyphenols (Hydroxytyrosol (HTY) or 3,4-dihydroxyphenyl-glycol (DHPG)) at 0.1-0.5% w/w[d]	Fresh beef pieces/Wrapping with films/Seven days at 4°C.	HTY loaded films suppressed the lipid oxidation more effectively than those loaded with DHPG. Antioxidants loaded into films with beeswax more effectively prevented lipid oxidation than those loaded with films lacking beeswax due to reduced oxygen permeability of films/No sensory analysis.	Bermúdez-Oria *et al.* (2019)
Chitosan or whey protein isolate film	Free or nanoencapsulated (NE) garlic EO at 2% (v/v)[a] (Encapsulant: nanoliposomes formed with cholesterol and soybean phosphatidylcholine)	Cooked beef sausages/ Films were placed at both surfaces of sliced samples/50 days at 4°C.	Active films with NE garlic EO more effectively suppressed TAMB, TPB and TLAB (3-4 log CFU/g after 50 days) than films with free garlic EO. All films with EO suppressed lipid oxidation of samples/No considerable differences exist among initial sensory properties of sausages, but actively packed sausages showed superior sensory properties (based on odor, taste, texture, and color) than controls at the end of storage.	Esmaeili *et al.* (2020)

(Contd.)

Table 7.2: *(Contd.)*

Film type	Active agent(s)	Product/Application/ Storage	Changes in safety shelf-life quality[g]/Sensory properties	References
Chitosan coating	Inherent antimicrobial activity of chitosan at 2 or 3% (w/v)[b]	Cooked and smoked Harbin red sausage/ Dipping-drying/Air or vacuum packaging/12 days at 23°C.	TAMB of chitosan coated samples was almost 5.5 log CFU/g after nine days, while uncoated air and vacuum packed controls reached and exceeded 6 log CFU/g threshold within six and nine days, respectively. Vacuum packed samples showed significantly lower lipid oxidation than chitosan coated samples, but the highest lipid oxidation was observed for uncoated air packed controls/No sensory analysis.	Dong *et al.* (2020b)
Natural ovine casing	Casings were impregnated with nisin (NIS) (solutions of NIS at 5 mg/l with or without acidification with 0.1% phosphoric acid)	Cooked sausages/ Casings were used during sausage production/56 days at 4 or 10°C.	TAMB and TLAB of samples with acidified NIS impregnated casing was ~5.5 log CFU/g after 42 days at 4°C while TAMB and TLAB of samples with control casing and non-acidified NIS impregnated casing after 42 days exceeded 6 log CFU/g. The acidic casing alone did not show considerable antimicrobial activity, but it improved antimicrobial activity of NIS. In contrast, active casings did not contribute considerably in controlling microbial load of samples at 10°C/No sensory analysis.	Barros *et al.* (2010)
Na-Alginate coating	Thyme and garlic EOs at 0.05% (w/w)[c]	Thawed lamb steaks/ Dipping-draining-cross-linking with CaCl$_2$-draining/Seven days between 2 and 4°C.	Thyme EO reduced the lipid oxidation of thawed samples considerably while garlic EO did not show considerable effect on lipid oxidation of samples. Coatings with or without the EOs caused a significant reduction in exudate loss during storage. Active coatings with thyme EO gave the best color retention at the end of storage/No sensory analysis.	Guerrero *et al.* (2020)

Chitosan coating	Free or nanoencapsulated (NE) *Satureja khuzestanica* essential oil (SKEO) at 1% (v/v)[a]	Lamb steaks/Dipping-draining-drying/20 days at 4°C.	TAMBs of uncoated control, and samples coated with chitosan, and SKEO loaded chitosan exceeded 7 log CFU/g after nine, 12, and 20 days, respectively while TAMB of samples coated with NE SKEO loaded chitosan after 20 days was 4.8 log CFU/g. Samples coated with NE SKEO loaded chitosan also showed significantly lower TPC and TLAB, and lipid oxidation than all other samples/ Samples coated with NE SKEO loaded chitosan maintained their red color and showed the lowest discoloration and off-odor (sensory properties based on color, discoloration and odor).	Pabast *et al.* (2018)
Tamarind seed starch-xanthan gum (1:0.04) blend film	Reduced sized spice powders of clove and cinnamon at 10 and 20 mg/ml of film forming solution	Mutton leg muscle slices/Wrapping of samples with edible films/34 days at 4 or 10°C.	TAMB of uncoated and spice loaded film wrapped samples were below 7 log CFU/g after eight and 12 days at 10°C, and after 20 and 28 days at 4°C, respectively. Films loaded with spices caused significant reductions in PC, TLAB, TE, TYM, *Brochothrix thermosphacta*, and lipid oxidation and metmyoglobin formation of meat at both temperatures/No sensory analysis.	Mohan *et al.* (2017)

[a] v/v film forming solution; [b] w/v film forming solution; [c] w/v film forming solution; [d] w/w hydrocolloid; [e] v/v hydrocolloid; [f] v/w hydrocolloid; [g] TAMB: total aerobic mesophilic bacteria; TPB: total psychrophilic bacteria; TCC: total coliform count; TYM: total yeast and mold, TE: Total *Enterobacteriaceae*; TLAB: total lactic acid bacteria; PC: *Pseudomonas* spp. count; TVB-N: total volatile basic nitrogen; EO: essential oil

L. monocytogenes by 4.1 log in minced spiced meat is also an important finding that might be adopted to different meat products. Other important findings related to use of polyphenols in meat preservation are outstanding antioxidant effects against lipid oxidation by hydroxytyrosol loaded pectin-gelatin films on fresh beef (Bermúdez-Oria *et al.*, 2019), pomegranate peel extract and thyme EO mixture loaded chitosan-starch composite film on beef slices (Mehdizadeh *et al.*, 2020), free or nanoencapsulated garlic EO loaded whey and chitosan films on cooked sausages (Esmaeili *et al.*, 2020), thyme EO loaded Na-alginate (Guerrero *et al.*, 2020) and *Satureja khuzestanica* EO loaded chitosan coatings (Pabast *et al.*, 2018) on lamb steaks, and reduced size clove and cinnamon powder mixture loaded tamarind seed starch-xanthan gum blend films on mutton meat (Mohan *et al.* (2017).

In addition to phenolic-based antimicrobial compounds, ε-polylysine is also a promising antimicrobial compound that has been employed in chitosan films to increase the shelf-life of beef fillet without affecting its sensory properties (Alirezalu *et al.*, 2021). Moreover, although the use of classical natural antimicrobial nisin suffers from stability problems in fresh meat (Stergiou *et al.*, 2006), it could be used in active packaging of cooked meat (*see* Chapter 4). For example, natural ovine casings impregnated with acidified nisin solutions were used to control microbial load of cooked sausages (Barros *et al.*, 2010). A recent study by Yordshahi *et al.* (2020) is also noteworthy since it showed the possibility of using postbiotics of *Lactobacillus plantarum* in bacterial cellulose film that achieved 5 log inhibition of *L. monocytogenes* in packed ground beef without affecting its sensory properties after grilling.

7.4 Active edible packaging of poultry and poultry products

Active edible packaging of poultry and poultry products is very challenging due to the high susceptibility of this food category to microbial and oxidative changes. Therefore, the use of inherently antimicrobial chitosan or chitosan-based composite films (Mahdavi *et al.*, 2018; Souza *et al.*, 2018), coatings (Huang *et al.*, 2020; Mehdizadeh and Langroodi, 2019; Yaghoubi *et al.*, 2021), or nanofibers (Cui *et al.*, 2018) incorporated with powerful natural antimicrobial and antioxidant compounds is a frequently applied strategy for edible packaging of these products. In addition, the use of two or more antimicrobial compounds within films and coatings to exploit their potential synergies (Bharti *et al.*, 2020; Raeisi *et al.*, 2016; Mehdizadeh and Langroodi, 2019) as well as application of nanoencapsulation strategies (Abdou *et al.*, 2018; Huang *et al.*, 2020) to stabilize and control the release of active agents are applied to maximize the antimicrobial performance of active packaging used for poultry. Some examples of hydrocolloids other than chitosan employed recently for active packaging of poultry are films of gelatin (Fathi-Achachlouei *et al.*, 2021), κ-carrageenan (Farhan and Hani, 2020), starch-carrageenan composite (Bharti *et al.*, 2020), and gelatin-carrageenan blend (Khan *et al.*, 2020); and coatings of pectin (Abdou *et al.*, 2018), alginate (Raeisi *et al.*, 2016), and Na-caseinate (Noori *et al.*, 2018).

Different pure polyphenols, such as curcumin, gallic acid, quercetin; pure essential oils, such as carvacrol and cinnamaldehyde; essential oils obtained from plants, such as *Artemisia fragrans*, *Zataria multiflora*, anise, caraway, nutmeg, garlic, rosemary, ginger, cinnamon, tea tree; and extracts, such as fenugreek and propolis, have been employed for active packaging of poultry and poultry products (Table 7.3). Some of these phenolic compounds contributed positively or at least did not cause significant adverse effects on sensory properties of packaged poultry. For example, positive contributions were reported for *Artemisia fragrans* EO loaded chitosan coatings (Yaghoubi *et al.*, 2021) and anise EO loaded gelatin films (Fathi-Achachlouei *et al.*, 2021) on microbial quality and sensory properties of cold stored chicken breast fillets. Anise EO loaded chitosan films applied on chicken burgers caused some reduction in their odor scores, but slight adjustments in anise EO concentration eliminated the negative effects in the taste of burgers (Mahdavi *et al.*, 2018). It is also important to note that active films with anise caused inactivation (1.3 to ~3 log) of inoculated *P. aeruginosa*, *S. aureus* and *E. coli* in burgers within three days. Noori *et al.* (2018) reported that application of Na-caseinate films with nanoemulsion of ginger EO caused superior antimicrobial activity and more acceptable sensory properties for packed chicken breast than application of films with coarse emulsion of ginger EO. The findings of Abdou *et al.* (2018) were also promising since they reported high antimicrobial activity and quite positive sensory properties for nanoemulsions of curcumin-cinnamon EO mixture loaded pectin emulsions applied on chicken fillets. The chitosan coatings loaded with combination of *Zataria multiflora* essential oil (ZEO) and propolis extract applied on chicken breast during cold storage also gave high antimicrobial activity and satisfactory sensory properties. The outstanding antimicrobial effects of curcumin nanoemulsion loaded gelatin-carrageenan blend films applied on chicken meat by Khan *et al.* (2020) and nanoemulsion of rosemary EO-ε-polylysine mixture loaded chitosan-gelatin blend coating applied on ready-to-eat chicken legs by Huang *et al.* (2020) are also important. Moreover, effective inactivation of inoculated *Listeria monocytogenes* and *Campylobacter jejuni* (3 strains) in chicken breast by nisin-cinnamon EO or nisin-rosemary EO loaded alginate coatings, and cinnamaldehyde loaded pectin-apple puree films are also promising findings by Raeisi *et al.* (2016) and Mild *et al.* (2011), respectively. However, further studies are needed to analyze the effect of these applications on sensory properties of chicken samples before evaluating their potential applicability.

7.5 Active edible packaging of fish and fish products

Active edible packaging of fish and fish products encounters some difficulties due to the rapid microbial growth and proteolysis-mediated decomposition of fish muscle, and related increases in toxic ammonia and toxic and odorous biogenic amines (Sriket, 2014; Prabhakar *et al.*, 2020). The fishy odor associated with fish spoilage is caused by bacterial transformation of trimethylamine oxide (TMAO) into biogenic amine trimethylamine (TMA) (Prabhakar *et al.*, 2020). Therefore,

Table 7.3: Examples of active edible packaging for poultry and poultry products

Film type	Active agent(s)	Product/Application/ Storage	Changes in safety shelf-life quality[g]/Sensory properties	References
Chitosan coating	*Artemisia fragrans* EO at 500, 1000 or 1500 ppm in coating solution	Chicken breast fillet/ Immersion of fillets into coating solution (1 h at 4°C)–draining/12 days at 4°C.	TAMB of samples with 1500 ppm EO loaded chitosan coatings was below 6 log CFU/g while TAMBs of all other samples were ≥6.9 log CFU/g at the end of storage. All samples with active coating showed significantly lower (3 to 4 log) TPB and TYM than controls and they caused significant reductions in lipid oxidation and TVB-N/Active coating with EO at 1500 ppm gave the best sensory properties (based on odor, color, texture, and freshness).	Yaghoubi *et al.* (2021)
Gelatin Film	Anise EO at 0.6 and 0.9% (w/w)[d]	Chicken breast fillet/ Wrapping of samples with films/12 days at 4°C.	Films with 0.6 or 0.9% EO showed similar antimicrobial effectiveness and caused significantly lower TAMB, TPB, TYM, TE, and PC than controls. Active films with 0.6 or 0.9% EO also caused significant reductions in lipid oxidation and TVB-N/Active coating caused superior sensory properties (taste and odor of oven cooked sample) than controls. Samples with 0.6% EO loaded coatings showed better taste and odor than those with 0.9% EO loaded coatings.	Fathi-Achachlouei *et al.* (2021)
Chitosan or chitosan-cellulose nanocrystalcomposite (CNC) film	Inherent antimicrobial activity of chitosan	Chicken breast meat/ Films were placed onto bottom of trays used in samples' packaging/12 days at 4°C.	Chitosan and chitosan-CNC films inhibited TE growth for 14 and seven days, respectively while controls showed significant TE growth at day 2. TPB of samples packed with chitosan and chitosan-CNC films were 4	Costa *et al.* (2021)

Film/coating	Active compound	Food/Application	Results	Reference
Chitosan coating	*Zataria multiflora* essential oil (ZEO) at 1%, or ZEO at 1% (or 0.5%) + propolis extract (PE) at 1% (w/w)[c]	Chicken breast fillet/ Dipping (1-2 min.)-draining/16 days at 4°C.	and 2 log CFU/g lower than that of controls at day two, but counts of all samples got closer at day five. Both types of film were unable to prevent TVB-N increase in samples, but chitosan-CNC films caused lower TVB-N in samples than chitosan films/No sensory analysis.	Mehdizadeh and Langroodi (2019)
			Active coatings with ZEO at 1%, ZEO at 1% (or 0.5%) + PE at 1% gave TAMB less than 6 log CFU/g after 16 days while controls exceeded this limit in four days. Active coatings with ZEO-PE caused the most significant inhibitions (~2.5-3 log) on PC and TPB, and on lipid oxidation/Samples with active coating gave the highest sensory scores (taste, odor, color, and texture of oven cooked samples).	
κ-carra-geenan films	Fenugreek seed extract at 10-20% (w/w)[d]	Chicken breast fillet/ Samples were placed into edible pouches/ Seven days at 4°C.	Active films caused 2-2.4 log reduction in TAMB of samples/Film color darkened as extract concentration increased/No sensory analysis.	Farhan and Hani (2020)
Manihot esculenta starch-carrageenan (3.5:1.5) blend film	Anise EO at 0.5%, or caraway or nutmeg EOs at 1% (v/v)[a]	Chicken nuggets/ Wrapping of samples with films/15 days at 4°C.	Films with or without EOs reduced the TAMB and TPB significantly, but presence of EOs did not affect antibacterial activity considerably. Films with caraway and nutmeg EOs suppressed TYM effectively. Films with EOs showed significantly higher inhibition on lipid oxidation than control film at 15 days/ No sensory analysis.	Bharti *et al.* (2020)

(Contd.)

Table 7.3: *(Contd.)*

Film type	Active agent(s)	Product/Application/ Storage	Changes in safety shelf-life quality[g]/Sensory properties	References
Gelatin-carrageenan (15:1) blend film	Curcumin, gallic acid or quercetin nanomulsions (NE) at 20% (0.1-0.3 mg/ml of NE)	Chicken meat/Samples were placed into pouches made from edible film/15 days at 4°C.	TAMB of samples packed with control film, and films with gallic acid, quercetin or curcumin NE reached ~6 log CFU/g limit at almost 11, 14, 17 and 18 days, respectively. Samples packed with quercetin and curcumin NE loaded films showed minimum pH change during storage/No sensory analysis.	Khan et al. (2020)
Chitosan or chitosan-montmorillonite nanocomposite (CHI-MN) film	Garlic EO at 0.5-2% (v/v)[a]	Minced chicken meat/ Wrapping of samples with films/15 days at 4°C.	TAMB of samples packed with EO loaded chitosan and CHI-MN films were significantly lower than that of unpacked controls, but only samples with 1 or 2% EO loaded chitosan films showed lower TAMB and TPB than samples packed with control chitosan films at the end of 10-days. The CHI-MN prevented EO to show antimicrobial effect. Only chitosan films with EO prevented lipid oxidation at the end of 15 days/No sensory analysis.	Souza et al. (2018)
Chitosan film	Anise EO at 1.5 and 2.0% (v/v)[a]	Chicken burger/Films were placed at both surfaces of burgers/15 days at 4°C.	TAMB of unpacked controls, packed controls, and samples packed with 1.5 or 2% EO loaded films reached 6 log CFU/g within 4, 6, 11 and >12 days, respectively. Films with 1.5 and 2% EO caused ~3 log lower TPB than packed controls. Active films caused inactivation (1.3 to ~3 log) of P. aeruginosa, S. aureus and E. coli in three days, and	Mahdavi et al. (2018)

Coating	Active agent	Food/Application/Storage	Result/Analysis	Reference
			reduced lipid oxidation/Films with EO reduced odor scores of burgers. EO at 1.5% did not affect the taste while EO at 2% reduced taste scores (sensory analysis was based on odor and taste of oven cooked burgers).	Huang et al. (2020)
Chitosan-gelatin (1:5) blend coating	Coarse (CE-REO) or nanoemulsions (NE-REO) of rosemary essential oil at 2% (w/v) + ε-polylysine (ε-PL) at 0.04% (w/v)	Ready-to-eat chicken legs/Dipping-draining/16 days at 4°C.	NE-REO + ε-PL loaded coating caused significantly lower TAMC and TYM than control coating and CE-REO + ε-PL loaded coating. Active coatings with NE-REO + ε-PL or CE-REO + ε-PL caused 2-days delay in growth of TE. All coatings including controls prevented TVB-N increase during storage, but only EO loaded coatings reduced lipid oxidation/No sensory analysis.	Huang et al. (2020)
Na-caseinate coating	Coarse (CE-GEO) or Nanoemulsion (NE-GEO) of Ginger essential oil (GEO) at 3 or 6% (w/w)c	Chicken breast fillet/ Dipping-drying/12 days at 4°C.	The TAMB of uncoated controls, and chickens treated with CE-GEO or NE-GEO loaded coatings after 12 days were >6, 4.5-6 and 3-4 log CFU/g, respectively. NE-GEO loaded coating caused significantly lower TPB and TYM than control coating and CE-GEO loaded coating. Active coatings did not cause considerable reduction in lipid oxidation/Active coating with 6% NE-GEO gave the highest total acceptability in sensory analysis (based on appearance, color, odor, and elasticity).	Noori et al. (2018)

(Contd.)

Table 7.3: *(Contd.)*

Film type	Active agent(s)	Product/Application/ Storage	Changes in safety shelf-life quality*/Sensory properties	References
Alginate coating	Nisin (NIS) (at 10^6 IU/ml of film forming solution), with cinnamon (CEO) or rosemary (REO) essential oils at 0.5% (w/v)[b]	Chicken breast/Dipping-draining-CaCl$_2$ cross-linking/ 15 days at 4°C.	The CEO-REO in coatings caused more significant reductions in TAMB, TPB and PC than CEO-NIS, and REO-NIS. The CEO-REO, and REO-NIS were significantly more effective on TE than CEO-NIS. The combination of two antimicrobials caused significant inhibition of inoculated *L. monocytogenes* in chickens (~2 log). In general, coatings with REO alone showed higher antimicrobial activity than those with CEO or NIS alone/No sensory analysis.	Raeisi *et al.* (2016)
Pectin-apple puree film	Carvacrol (CAR) or cinnamaldehyde (CIN) at 0.5, 1.5, or 3.0% (w/w)[c]	Chicken breast/ Wrapping of samples with films/three days at 4 or 23°C.	*Campylobacter jejuni* (three strains) showed effective inactivation at both 4°C (1.8 to 6.0 log reduction) and 23°C (4.9 to 5.5 log reduction) by 3% CIN loaded film. CAR at 3% was effective on *C. jejuni* at 23°C (2.4 to 5.1 log reduction), but it show limited antimicrobial activity at 4°C (0.5 to 0.9 log reduction)/No sensory analysis.	Mild *et al.* (2011)
Pectin emulsion coatings	Nanoemulsions of curcumin-cinnamon EO (NE-CC), curcumin-garlic EO (NE-CG), or curcumin-sunflower oil (NE-CS).	Chicken fillet/Dipping (10 min.)-draining/12 days at 4°C.	Samples with NE-CC and NE-CG loaded coating caused more considerable reductions in TAMB, TPB, TYM, TVN-B, and lipid oxidation than controls and samples with NE-CS loaded coating. NE-CC was the most potent antimicrobial. Active coatings increased water holding capacity and gave	Abdou *et al.* (2018)

			higher tissue firmness than controls/Samples coated with NE-CC showed the best sensory properties (based on appearance, taste, odor, texture and palatability of fried samples).	
Chitosan-tea tree essential oil (TTO) nanofibers	Liposome encapsulated tea tree oil (TTO) (liposomes contained 6 mg TTO/ml)	Chicken slices/ Wrapping with alu-minimized paper coated with nanofibers/Four days at 4 and 12°C.	Active packaging with chitosan-TTO nanofibers caused almost 1.5 log CFU/g reduction in initial microbial load of samples inoculated with *S. enterica* serovar *Enteritidis* and *S. enterica* serovar *Typhimurium at 4 and 12°C within four days*/Active packaging did not affect overall sensory properties (based on color, odor, juiciness of raw samples).	*Cui et al.* (2018)

[a] v/v film forming solution; [b] w/v film forming solution; [c] w/w film forming solution; [d] w/w hydrocolloid; [e] v/w film forming solution; [f] v/w hydrocolloid; [g] TAMB: total aerobic mesophilic bacteria; TPB: total psychrophilic bacteria; TCC: total coliform count; TYM: total yeast and mold; TE: Total *Enterobacteriaceae*; PC: *Pseudomonas* spp. count; TVB-N: total volatile basic nitrogen; EO: essential oil

the monitoring of total volatile basic nitrogen (TVB-N) is important to understand the quality of packed fish. Moreover, spoilage bacteria synthesize enzymes, such as histidine decarboxylase and lysine decarboxylase that transform certain amino acids released by proteolysis of muscle into toxic biogenic amines, such as histamine, putrescine, and cadaverine (Prabhakar *et al.*, 2020). In addition, the rich polyunsaturated fatty acid content of fish lipids causes fast enzymatic/ nonenzymatic lipid oxidation that leads to formation of off-flavors and off-odors as well as undesired changes in fish color. Hemoglobin is an important catalyst in lipid oxidation and peroxidation of fishes since their blood cannot be removed prior to processing (Maqsood *et al.*, 2012). Therefore, use of active compounds having strong antimicrobial and antioxidant activity is beneficial in edible packaging of fish and fish products. Moreover, it is also important to note that the initial microbial quality of fish is a key factor in the success of active edible packaging (Carrión-Granda *et al.*, 2018).

The edible films of chitosan (Rico *et al.*, 2020; Albertos *et al.*, 2019; Ehsani *et al.*, 2020), gelatin (Ucak *et al.*, 2021; Ehsani *et al.*, 2020), alginate (Ehsani *et al.*, 2020), pectin-gum gellan blend (Pérez-Arauz *et al.*, 2021), and gelatin/ gelatin-palm oil emulsion laminates (Nilsuwan *et al.*, 2021) have been used in active packaging of fish and fish products. Moreover, edible coatings of chitosan (Zarandona *et al.*, 2021; Çoban, 2021; Yuan *et al.*, 2016), chitosan nanoparticles (Zarandona *et al.*, 2021), chitosan-gelatin blends (Xiong *et al.*, 2021), chitosan-zein blends (Mehdizadeh *et al.*, 2021), whey protein (Shokri *et al.*, 2015) and tragacanth gum (Khaledian *et al.*, 2021) have also been recently employed in active edible packaging of fish and fish products.

The examples of natural phenolic compounds recently used in active packaging of fish and fish products include phenolic extracts from propolis, orange seed, lemon seed, lime peel, pomegranate peel, sea fennel, seaweed, red alga and *Pulicaria gnaphaloides*; pure polyphenols, such as gallic acid and epigallocatechin gallate; and essential oils from clove and sage (Table 7.4). From these active compounds, lemon seed extract incorporated into gelatin films contributed quite positively in the increase of antimicrobial and oxidative stability as well as in improving the sensory properties of seabass fillets (Ucak *et al.*, 2021). The application of propolis extract loaded chitosan coatings also improved the microbial, oxidative, and sensory properties of boiled crayfish (Çoban, 2021). The epigallocatechin gallate (EGCG) loaded bilayer laminated pouches (gelatin film inside-gelatin-palm oil emulsion film outside) showed no detectable effects in sensory properties of fried salmon skins stored for 30 days at 28°C, but EGCG effectively reduced the lipid oxidation and formation of volatile secondary lipid oxidation products in this product (Nilsuwan *et al.*, 2021). Fresh Rainbow trout coated with nanoliposome encapsulated *Pulicaria gnaphalodes (Vent.) Boiss* extract containing chitosan-zein solution also showed superior antimicrobial, antioxidant, and sensory properties than those with free extract containing coating solution (Mehdizadeh *et al.*, 2021). Moreover, tragacanth gum coatings with lime peel extract reduced microbial growth, oxidative changes, and melanosis in Pacific white shrimp and gave comparable sensory properties with sulfite-treated samples (Khaledian *et al.*, 2021). However, sulfite loaded tragacanth coatings showed better antimicrobial performance than lime peel extract loaded

Table 7.4: Examples of active edible packaging for fish and fish products

Film type	Active agent(s)	Product/Application/ Storage	Changes in safety shelf-life quality[g]/ Sensory properties	References
Citrus pectin-gum gellan (2:0.4) blend film	Bacteriocin like substances (BLIS) from *Streptococcus infantarius* (at 50 AU/ml of film solution with 1mM EDTA)	Tilapia and trout fillets/ Wrapping of samples with films/Seven days at 4°C.	Tilapia and trout samples packed with BLIS loaded active film caused almost 3 and 1-1.5 log lower TAMB than uncoated and coated control samples, respectively/No sensory analysis.	Pérez-Arauz *et al.* (2021)
High molecular weight chitosan coating	Inherent antimicrobial activity of 800 kDa chitosan	Channel catfish fillets/ Dipping-drying/12 days at 4°C.	Chitosan coating reduced TAMB and TPB, and lipid oxidation of cold stored samples significantly/Solubilization of chitosan into aspartic acid instead of acetic acid prevented pungent acidic odor in coating. Samples treated with chitosan coating prepared by aspartic acid showed acceptable sensory properties (based on appearance, odor, and texture of raw samples).	Karsli *et al.* (2021)
Bilayer laminated film (gelatin film-gelatin-palm oil emulsion film)	Gelatin film part of laminate was incorporated with epigallocatechin gallate (EGCG) at 12% (w/w)[d]	Fried salmon skin/Placing into pouches made from edible gelatin film (active gelatin film was inside)/30 days at 28°C.	Samples packed with EGCG loaded active pouches reduced both lipid oxidation and secondary lipid oxidation products (volatile compounds) significantly/The EGCG showed no significant contribution in fried skin sensory properties (based on color, texture, rancidity, and appearance).	Nilsuwan *et al.* (2021)
Gelatin film	Orange (OSE) or lemon seed extract (LSE) at 2%	Sea bass fillets/Films were placed at both sides of samples/15 days at 4°C.	Active packaging with LSE or OSE loaded films reduced TAMB, TPC, TE, and TYM and lipid oxidation of fillets	Ucak *et al.* (2021)

(Contd.)

Table 7.4: *(Contd.)*

Film type	Active agent(s)	Product/Application/ Storage	Changes in safety shelf-life quality[g]/ Sensory properties	References
			significantly. LSE showed the highest antimicrobial and antioxidant activity/ Samples with active coatings gave the highest sensory properties (based on color, texture, appearance, and odor of raw samples). Coatings with LSE gave higher overall acceptability than those with OSE.	
Chitosan-Gelatin (1:1) coating	Gallic acid (GA) at 0.2% (w/v)[b] and/or clove EO at 0.5% (v/v)[a]	Atlantic salmon fillets/ Dipping-drying/15 days at 4°C.	TAMBs of fillets with clove EO, and GA-clove EO loaded coatings were below 6 log CFU/g at 10 days while control fillets and fillets with GA loaded coating reached/ exceeded this limit in 10 days. Coating with or without phenolic compounds reduced lipid oxidation of fillets similarly/No sensory analysis.	Xiong *et al.* (2021)
Chitosan coating	Propolis extract (PrE) at 0.3 or 0.6% (v/v)[a]	Boiled cryfish meat/ Dipping-drying/16 days at 4°C.	Active coating with 0.3 or 0.6% PrE caused similar significant reductions in TMAB, TPB and PC. However, PrE at 0.6% is more effective on TYM and H₂S producing bacteria than PrE at 0.3%. Active coatings reduced lipid oxidation considerably, but PrE at 0.6% was the most effective antioxidant/ Active coating with PrE at 0.6% and 0.3% gave the first and second highest sensory properties, respectively (based on odor, taste and firmness).	Çoban, 2021

Coating	Active agent	Food/method/duration	Results	Reference
Chitosan or chitosan nanoparticles (CN) coating	Gallic acid at 10% (w/w)[d]	Defrost Atlantic horse mackerel fillets/Dipping-draining/13 days at 4°C.	Chitosan and CN coating improved the microbial quality of fillets, but GA did not contribute to microbial quality (TAMB, PC, TE, TLAB) considerably. Chitosan coating, and chitosan and CN coatings loaded with GA prevented formation of TVB-N similarly, but only GA loaded chitosan coatings prevented lipid oxidation effectively/No sensory analysis.	Zarandona et al. (2021)
Chitosan-zein (1:1) composite coating	Free (PG) or nanoliposome encapsulated (NE-PG) *Pulicaria gnaphalodes (Vent.) Boiss.* extract	Fresh rainbow trout fillets/Dipping/14 days at 4°C.	Active coatings caused significant improvements in microbial and oxidative quality of fillets, but coatings with NE-PG caused more significant reductions in TAMB, PC, TLAB and lipid oxidation than those with PG/Active coatings with NE-PG showed the best sensory properties (based on texture, odor, color).	Mehdizadeh et al. (2021)
Tragacanth gum (TG) coating	Lime peel extract (LPE) at 1 or 2% (v/v)[a]	Pacific white shrimp/Dipping/10 days at 4°C.	TAMB of controls and samples with LPE loaded coating reached almost 6 log CFU/g within eight and 10 days, respectively. Samples with LPE loaded coating showed significantly lower PC, TPB, TLAB, H_2S producing bacteria, and lipid oxidation and melanosis than controls. However, control coatings with sulfite showed the highest antimicrobial activity/Active coatings gave acceptable sensory properties (based on appearances, odor, texture, melanosis, orange head, loosening of head) comparable to sulfite treated ones.	Khaledian et al. (2021)

(Contd.)

Table 7.4: (*Contd.*)

Film type	Active agent(s)	Product/Application/ Storage	Changes in safety shelf-life quality[g]/ Sensory properties	References
Chitosan coating	Pomegranate peel extract (PPE) at 1% (w/w)[c]	Pacific white shrimp/ Dipping/10 days at 4°C.	TAMB of uncoated and coated controls and sulfite treated samples exceeded 6 log CFU/g within 10 days while that of samples with PPE loaded coating reached ~5.5 log CFU/g at the end of 10 days. Samples with PPE coating showed significantly lower TVB-N than controls and sulfite treated samples/Active coating with PPE and sulfite treatment reduced melanosis and loss of sensory properties (based on appearance, odor, color, and texture) similarly.	Yuan *et al.* (2016)
Chitosan film	Aqueous sea fennel extract (ASFE) was used in film preparation	Atlantic horse mackerel burger/Wrapping of samples with films/10 days at 4°C.	Coatings with ASFE did not affect the TAMB, TPB, and TLAB of samples considerably, but they effectively reduced the lipid oxidation of burgers/Active films reduced the fishy off-odors, but they caused negative effects on burger color (sensory analysis was based on fishy odor, aromatic odor, and color).	Rico *et al.* (2020)
Chitosan film	Aqueous seaweed extracts of *Himanthalia elongate* (HE) or *Palmaria palmate* (PP) were used in film preparation	Aquaculture rainbow trout burger/Samples were wrapped with films/Seven days at 4°C.	Coatings with HE and PP extracts both inhibited the TAMB of samples for 5 days while samples coated with control coating showed significant increase in their TAMB. Only coatings with HE inhibited the growth of TPB in samples for 5 days while other samples showed significant increase in their	Albertos *et al.* (2019)

Chitosan, alginate or gelatin film	Lactoperoxidase (LPS) system at 10% (v/v)[a] (solution of incubated reaction mixture with thiocyanate and H₂O₂) or sage EO at 0.5% (v/v)[a]	Common carp burger/ Films were placed at both sides of samples/20 days at 4°C.	TPBs. Coatings with HE inhibited the lipid oxidation of burgers more effectively than the others. Coatings did not cause significant color change in samples/No sensory analysis. Samples with sage EO or LPS loaded chitosan films and LPS loaded alginate films showed significantly lower TAMB, TPB and PC than packed and unpacked controls and other active films. However, chitosan films with LPS showed the highest antimicrobial activity. The active films caused only limited reductions in lipid oxidation of burgers/Only LPS loaded chitosan films gave acceptable odor at the end of storage, but no differences exist among color of samples except controls (sensory analysis was based on odor and color).	Ehsani *et al.* (2020)
Whey protein coating	Lactoperoxidase (LPS) system at 7.5% (v/v)[a] (solution of incubated reaction mixture with thiocyanate and H₂O₂)	Rainbow trout fillet/ Dipping-draining/16 days at 4°C.	Active films with LPS caused significant reductions in TAMB, TPB, PC, *Pseudomonas fluorescens* and *Shewanella putrefaciens* loads of samples during storage. The TAMB loads of coated controls exceeded 7 log CFU/g threshold within 12 days while TAMB of samples with active coating were below this limit at 16 days. Active coating reduced the TVB-N, but had no effect on lipid oxidation/ Active coating gave superior sensory properties (based on color, odor, and texture)	Shokri *et al.* (2015)

[a] v/v film forming solution; [b] w/v film forming solution; [c] w/w film forming solution; [d] w/w hydrocolloid; [e] v/v hydrocolloid; [f] v/w hydrocolloid; [g] TAMB: total aerobic mesophilic bacteria; TPB: total psychrophilic bacteria; TCC: total coliform count; TYM: total yeast and mold; TE: Total Enterobacteriaceae; TLAB: total lactic acid bacteria; PC: *Pseudomonas* spp. count; TVB-N: total volatile basic nitrogen; EO: essential oil

coatings. One more promising study related to Pacific white shrimp is that of Yuan *et al.* (2016), who employed pomegranate peel extract loaded chitosan coating in active packaging. The findings of this study showed that the pomegranate peel extract loaded chitosan coatings had better antimicrobial performance and control over TVB-N than sulfite treatment, and caused comparable positive improvements with sulfite treatment in sensory properties of shrimp. These findings are important since they suggest that active coatings with natural phenolic extracts might be used as sulfite alternative in shrimp preservation. All these studies employed phenolic compounds effectively without interfering the sensory quality of different fish and fish products, but it is important to note that the most outstanding antimicrobial performances were observed for propolis extract, lemon and pomegranate peel extracts, and nanoliposome encapsulated *Pulicaria gnaphalodes (Vent.) Boiss* extract. Other studies that employed gallic acid and/or clove EO loaded chitosan-gelatin blend coatings on Atlantic salmon fillets (Xiong *et al.*, 2021), gallic acid loaded chitosan or chitosan nanoparticle coatings on defrost Atlantic horse mackerel fillets (Zarandona *et al.*, 2021), and seaweed (*Himanthalia elongate* or *Palmaria palmate*) extract loaded chitosan films on Aquaculture Rainbow trout burger (Albertos *et al.*, 2019) did not report sensory analysis of packaged products. However, antimicrobial and antioxidant activity of *Himanthalia elongate* seaweed extract loaded chitosan loaded films was noteworthy. In contrast, gallic acid did not work as an effective antimicrobial and antioxidant agent in coatings obtained from chitosan-gelatin blends (Xiong *et al.*, 2021) and chitosan nanoparticles (Zarandona *et al.*, 2021). The GA did not also contribute to the antimicrobial activity of chitosan coatings, but improved the antioxidant activity of films, and inhibited lipid oxidation of coated Atlantic horse mackerel fillets (Zarandona *et al.*, 2021). Moreover, it is also important to note that the use of chitosan nanoparticles in development of gallic acid loaded coating instead of common chitosan did not induce any considerable advantages. Instead, use of chitosan nanoparticles masked the antioxidant activity of gallic acid due to increased interactions between gallic acid and chitosan (Zarandona *et al.*, 2021).

Some other active components employed in edible packaging of fish and fish products also exist, such as lactoperoxidase (LPS) loaded chitosan, gelatin, alginate films (Ehsani *et al.*, 2020), and whey protein coatings (Shokri *et al.*, 2015). The LPS incorporated chitosan films showed the highest antimicrobial activity in controlling microbial load of cold stored common carp burgers, while lower antimicrobial activities were observed in applications with LPS loaded alginate and gelatin films (Ehsani *et al.*, 2020). The LPS loaded whey protein coatings were also highly effective in controlling microbial load and inhibiting pathogenic bacteria, such as *P. fluorescens* and *Shewanella putrefaciens* on cold stored Rainbow trout fillets (Shokri *et al.*, 2015). In general, the LPS loaded films and coatings contribute positively in sensory properties (especially odor) of cold stored fish and fish products since they reduce microbiological spoilage effectively, but these films are not very effective in lipid oxidation (Ehsani *et al.*, 2020; Shokri *et al.*, 2015). Thus, the combination of LPS incorporated edible packaging with antioxidants and MAP might be beneficial to control oxidative changes in fish and fish products. A recent study has also brought a different approach in fish preservation by incorporating bacteriocin-like substances (BLIS) from *Streptococcus infantarius* into pectin-gum gellan blend films. The developed films with BLIS were effective in reducing the microbial load of tilapia and trout fillets significantly during cold storage, but further studies are needed to characterize the constituents of BLIS and to determine their effects on sensory properties of fish and fish products.

7.6 Active edible packaging of fresh fruits and vegetables, seeds, and mushrooms

Active edible packaging of whole or minimally processed fruits and vegetables, seeds, and mushrooms needs control of multiple factors, such as respiration rate, microbial load, enzymatic browning, loss of firmness, flavor changes, etc. Coating is the primary method applied in active packaging of respiring products, but some rare applications are also conducted by wrapping edible films on to the surface of suitable products. It is also important to note that active edible coating is mostly applied on fruits, while applications on vegetables focus mainly on minimally-processed fresh-cut root vegetables. Other applications discussed in this section are related to mushrooms and some emerging applications, such as antimicrobial coating of seeds destined for sprouting.

The primary objectives of active edible coating of whole fresh fruits and vegetables are reduction of their respiration rate to delay senescence symptoms (e.g. changes in sugars and acids, aroma, and flavor, browning or pigment degradation, softening or toughening, etc.) and control of undesired microbial changes (e.g. mainly to prevent growth of contaminated fungal/bacterial plant pathogens, and bacterial/viral human pathogens). Active edible coating of minimally processed fruits and vegetables is much more challenging than that of intact ones since peeling, slicing, chopping, etc. applied during processing increase the respiration rate of plant tissues and cause enzymatic browning due to decompartmentation of polyphenoloxidases (PPOs) in cytoplasm and their phenolic substrates located in vacuoles (Yemenicioğlu, 2016). Therefore, edible coatings employed for minimally processed fruits and vegetables are mostly incorporated by PPO inhibitors, such as ascorbic acid and derivatives and organic acids. It was generally accepted that hydrocolloids, such as cellulose, casein, zein, soy protein, and chitosan are suitable for classical edible coating (lacking active agents) of whole fruits and vegetables as their films show desired gas barrier/permeation properties, and they are odorless, tasteless, and transparent (Park, 1999). On the other hand, hydrocolloids, such as alginate, gellan, whey protein, pectin, and chitosan are mostly incorporated with antimicrobials and/or PPO inhibitors, before being used in active edible coating of minimally processed fruits and vegetables (Rojas-Graü *et al.*, 2009). However, these are not rigid rules in selection of hydrocolloids since recent applications also use blends or composites of different materials, or hydrocolloids or gums from alternative sources. Some of the hydrocolloids recently used in active edible coating of whole fruits and vegetables and other respiring products, such as seeds and mushrooms, include chitosan (Nair *et al.*, 2018; Raigond *et al.*, 2019; Peralta-Ruiz *et al.*, 2020; Sozbilen and Yemenicioğlu, 2020; Dong *et al.*, 2020a; Riaz *et al.*, 2021; Saidi *et al.*, 2021), chitosan-beeswax composite (Sultan *et al.*, 2021), chitosan-pullulan blend (Kumar *et al.*, 2021), Na-alginate (Gundewadi *et al.*, 2018; Nair *et al.*, 2018; Zhu *et al.*, 2019; Bambace *et al.*, 2019; Dong *et al.*, 2020a: Pirozzi *et al.*, 2020; Louis *et al.*, 2021), gelatin (Temiz and Özdemir, 2021), cassava starch-CMC composite (Li *et al.*, 2020), high amylose corn starch (Ochoa-Velasco *et al.*, 2021), starch coated paper (Shao *et al.*, 2021), banana starch-chitosan-Aloe vera blend (Pinzon *et al.*, 2020), carrageenan (Dong *et al.*, 2020a), and CMC-stearic acid composite (Saidi *et al.*, 2021), and tragacanth gum (Nasiri *et al.*, 2018). Some of the active edible coating materials used for minimally processed fruits and vegetables are chitosan (Basaglia *et al.*, 2021; Kurek *et al.*, 2020; Prakash *et al.*, 2020), Na-alginate (Li *et al.*, 2017; Alvarez *et al.*, 2021; Shigematsu *et al.*, 2018), Kognac glucomannan gum (Hashemi and Jafarpour, 2021), pectin, gum Arabic and CMC (Kurek *et al.*, 2020), and locust bean gum (Rizzo *et al.*, 2019). Thus, it is clear that there is a good demand for use of chitosan and its composite or blend coatings, and alginate coatings in active edible coating of respiring products.

Some of the natural phenolic compounds used in active edible packaging of fresh fruits and vegetables, and mushrooms, include phenolic extracts from apple and pomegranate peels, olive leaf; Aloe vera gel extracts; essential oil components, such as citral, carvacrol, cinnamaldehyde and thymol; essential oils of cinnamon, oregano, thyme, basil, fennel, *Satureja khuzistanica*, rue herb, *Cinnamomum cassia*; and phenolic-rich bee products, such as pollen grain (Table 7.5). Amino acids, such as L-cysteine and phenylalanine, are also used to prevent enzymatic browning and reduce decay incidences in fruits, respectively. Kumar *et al.* (2021) reported that bell peppers with pomegranate peel extract (PPE) loaded chitosan-pullulan (1:1) coating showed minimum firmness reduction and weight loss, and acceptable sensory properties. Basaglia *et al.* (2021) showed that cinnamon EO loaded chitosan coatings reduced the fungal growth and contributed positively to preventing weight loss and in improving sensory properties of pineapple pieces. The pineapple slices were also coated by citral nanoemulsion loaded Na-alginate coatings to reduce their microbial load and browning, and to improve their sensory properties during cold storage (Prakash *et al.*, 2020). It was also reported that alginate, chitosan, and carrageenan films incorporated with *Cinnamomum cassia* EO were effective in reducing microbial load of whole strawberries without showing a negative effect in their sensory properties (Dong *et al.*, 2020a). However, none of these active coatings affected the maturity index of strawberries. In contrast, the rue herb (*Ruta graveolens*) EO loaded chitosan coatings were highly effective in reducing the maturity index, and in controlling the decay rate and severity of *Colletotrichum gloeosporioides* infection in whole papayas without causing undesirable changes in their sensory properties (Peralta-Ruiz *et al.*, 2020). It is also important to note the remarkable antifungal performances of cinnamon EO, rue herb EO, *Cinnamomum cassia* EO, and phenylalanine loaded chitosan coatings without causing sensory problems in fruits. Saidi *et al.* (2021) reported that CMC-stearic acid coatings loaded with phenylalanine (an elicitor of fruit defense response) caused significantly higher reductions in decay incidence and stem-end rot of avocadoes than chitosan coatings loaded with phenylalanine at the end of 12-days at 22°C. In contrast, chitosan coating with phenylalanine was more effective than CMC-stearic acid coatings with

Table 7.5: Examples of active edible packaging for fresh fruits and vegetables, and seeds and mushrooms

Film type	Active agent(s)	Product/Application/ Storage	Changes in safety shelf-life quality/ Sensory properties	References
Chitosan coating	Apple peel phenolic extract (APPE) at 0.25-1% (w/v)[b]	Whole Strawberries/ Dipping-draining-drying/six days at 20°C.	Application of active coating reduced the decay rate at the end of storage depending on APPE concentration (decay rate: 93, ~65, ~27, and 19% for uncoated, chitosan coated, and 0.25 or 1% APPE loaded chitosan coated fruits, respectively). Active coating reduced weight loss and anthocyanin loss, but it increased total soluble solids content and firmness/No sensory analysis.	Riaz *et al.* (2021)
Na-alginate, chitosan or carrageenan coating	*Cinnamomum cassia* EO at minimum fungicidal concentration (MFC) or half MFC (122 or 61 ppm in coating solution, respectively)	Whole strawberries/Dipping into coating solutions (additional KCl and CaCl$_2$ dipping were applied for carrageenan and alginate coatings, respectively)/19 days at 4°C.	All edible coatings with or without EO reduced total soluble solids content of fruits, but did not affect maturity index and titratable acidity at the end of storage. All edible films with EO at 122 ppm caused significant reductions in TAMB and TYM of samples. Alginate coatings with 61 ppm of EO reduced TAMB of fruits significantly while chitosan coatings with 61 ppm of EO reduced both TAMB and TYM of fruits significantly/Films with 66 ppm of EO did not affect the sensory properties (based on tasting samples) of strawberries.	Dong *et al.* (2020a)
Gelatin coating	*Lactobacillus rhamnosus* at 11 log CFU/ml of coating with or without inulin at 2.5% (w/w)[c]	Probiotic whole strawberries/Dip-coating-drying/16 days at 4°C.	*L. rhamnosus* in coating with or without prebiotics are stable during cold storage (7.4 and 7 log CFU/g at the end of 16 days, respectively). Presence of inulin increased stability of *L. rhamnosus* in coating. Inulin did not affect TYM of strawberries, but caused significant reductions in their TAMB. Samples treated with active coating showed higher phenolic contents/No sensory analysis.	Temiz and Özdemir (2021)

(Contd.)

Table 7.5: (*Contd.*)

Film type	Active agent(s)	Product/Application/ Storage	Changes in safety shelf-life quality[a]/ Sensory properties	References
Banana starch-chitosan-Aloe vera gel blend coating	Aloe vera gel (AVG) at 10 or 20% (w/v)[b] (Inherent anti-microbial activity of Aloe vera polyphenols)	Whole strawberries/ Dipping-drying/19 days at 4°C.	Decay rates of uncoated controls, coated controls, or samples with 10 or 20% AVG loaded coatings were 100 (in nine days), 55, 30, and 18% (in 14 days), respectively. TAMB and TYM of samples with control coating and with 10 and 20% AVG loaded coatings were almost 1.3, 1.6, 1.7 log and 2.3, 3.4, 2.6 log lower than those of uncoated controls, respectively. AVG loaded films showed the lowest weight loss and color change/No sensory analysis.	Pinzon *et al.* (2020)
High amylose corn starch coating	Carvacrol-thymol mixture (F1:750 + 750 or F2:1125 + 375 mg/l of coating solution)	Whole mango and papaya/ Dipping-draining-drying/18 days at 20°C.	Active coating with formulation F1 and F2 reduced softening, maturation, number of lesions, and color change of mango and papaya fruits. Formulation F2 showed the highest antifungal activity on *Colletotrichum gloeosporioides*/No sensory analysis.	Ochoa-Velasco *et al.* (2021)
Chitosan coating	Rue herb (*Ruta graveolens*) essential oil (REO) at 0.5-1.5%	Whole papaya/Dipping-drying/12 days at 20°C.	Disease incidence of *Colletotrichum gloeosporioides* in uncoated, chitosan coated, 0.5, 1 or 1.5% EO loaded chitosan coated fruits were 100, 100, 60, 0, and 0%, respectively. The severity of disease in actively coated fruits was significantly lower than those of uncoated and chitosan coated fruits. The EO loaded coatings caused greater reductions in fruit maturation index and softening than chitosan coated and uncoated fruits/Edible	Peralta-Ruiz *et al.* (2020)

Coating	Active ingredient	Food/Method/Storage	Results	Reference
Chitosan-pullulan (1:1) blend film	Pomegranate peel extract (PPE) at 5% (w/w)[c]	Bell pepper/Dipping-drying (two-cycles)/18 days at 4 or 23°C.	films with REO did not cause an undesirable change in sensory properties (based on aroma, flavor, brightness, color, and texture). The samples treated with PPE loaded coating better maintained their color, firmness, phenolic content, and antioxidant activity at both temperatures, but coating treatment caused some reduction in their total soluble solids content and titratable acidity. All coated samples showed less decay than uncoated ones at the end of storage/The samples treated with PPE loaded coating gave the best sensory properties (based on freshness, color, texture, taste).	Kumar et al. (2021)
Chitosan or Na-alginate coating	Pomegranate peel extract (PPE) at 1% (w/v)[b]	Bell pepper/Dipping-drying/25 days at 10°C.	Samples coated with chitosan coating with or without PPE showed significantly lower softening, water loss, degradation rates for vitamin C and chlorophyll, and total color change than those with control and PPE loaded alginate coatings and uncoated controls. The PPE loaded chitosan coatings caused significantly higher reductions in TAMB and TYM of samples than those with PPE loaded alginate coatings and uncoated controls/ Samples with PPE loaded chitosan coatings and control chitosan coatings gave the highest and second highest overall acceptability (based on color, aroma, taste, and texture).	Nair et al. (2018)
Chitosan-beeswax (1:0.5) composite or beeswax coating	Pollen grains (PG) at 0.5% (w/v)[b]	Whole pear/Dipping-drying/105 days at 0°C.	Coatings with PG reduced weight loss and fruit decay rate during cold storage, but beeswax coating with PG was the most effective on these parameters	Sultan et al. (2021)

(Contd.)

Table 7.5: *(Contd.)*

Film type	Active agent(s)	Product/Application/ Storage	Changes in safety shelf-life quality/ Sensory properties	References
			(decay rate of two season trial at the end of storage were almost 10-15% for beeswax, 22-35% for chitosan-beeswax and 40-50% for uncoated controls). Coated fruits had a significantly firmer texture and higher total soluble solids content than controls after ≥90 days/No sensory analysis.	
Na-alginate coating	*L. rhamnosus* (at 9 Log CFU/ml of coating solution) with or without prebiotics (inulin + oligofructose each at 8%, w/w^c)	Probiotic and probiotic-prebiotic fresh blueberries/ Dipping-draining-CaCl$_2$ cross-linking/21 days at 5°C.	Blueberries coated with *L. rhamnosus* loaded alginate or *L. rhamnosus* and prebiotic loaded alginate coatings maintained their *L. rhamnosus* counts at ≥6 log CFU/g for less than two and almost three weeks, respectively. *L. rhamnosus* with or without prebiotic in coating increased the TAMB, TPB, and TYM of fruits, but microbial loads of all samples were below 7 log CFU/g after 21 days. The *L. rhamnosus* loaded coatings increased fungal fruit decay rate after two weeks (decay rate of ~60% within three weeks). *L. rhamnosus* and prebiotics loaded coating reduced inoculated *L. innocua* significantly while coatings with *L. rhamnosus* showed growth inhibitory effect on *Listeria*. However, none of the coatings caused inhibition of inoculated *E. coli* O157: H7/Probiotic and prebiotics contributed positively in sensory properties (based on odor, flavor, and appearance) of coated blueberries.	Bambace *et al.* (2019)

Cassava starch-CMC (1:0.05) composite film	L. plantarum or P. pentosaceus at 0.5-2% (w/w)[c]	Whole banana/Wrapping of samples with films/Seven days at 30°C.	Addition of lactic acid bacteria with high exopolysaccharide production capacity increased the free radical scavenging activity of films considerably. Films with 2% of L. plantarum or P. pentosaceus reduced the brown spot formation at banana surface significantly/No sensory analysis.	Li et al. (2020b)
Na-alginate coating	Oregano EO nanoemulsion (NE) at 0.17% (w/w)	Cherry tomatoes/Dipping-coating-drying/15 days at 24°C.	Coating with NE of oregano EO caused significantly higher reductions in TAMB of samples than uncoated and coated control sample. Coatings with or without NE of oregano EO caused similar significant reductions in TYM of samples/No sensory analysis.	Pirozzi et al. (2020)
Chitosan or CMC-Stearic acid (1:0.6) composite coating	Phenylalanine (Phe) at 4 mM in coating solution	Whole avocado/Brushing-drying/12 days at 22°C (12-day storage procedure), or 18 days at 2 or 5°C + seven days at 22°C (25-day storage procedure).	CMC-stearic acid coatings with Phe caused most significant reductions in decay incidence and stem end rot of avocadoes during 12-day storage procedure. Chitosan coating with Phe caused the lowest surface pitting and stem end rot during 25-day storage procedure/For fruits stored by 12-day storage procedure, highest flavor scores were obtained for all Phe loaded coatings. For fruits stored by 25-day storage procedure, the best flavor was obtained for Phe loaded chitosan coatings.	Saidi et al. (2021)
Chitosan coating	Cinnamon EO at 0.5 or 1% (v/v)[a]	Pineapple pieces/Dipping-draining/15 days at 5°C.	Chitosan coating with or without cinnamon EO caused a significant reduction in weight loss and TYM of stored fruits. All coatings reduced fruit firmness. EO made a limited contribution on antifungal activity/Samples with EO loaded coatings showed better sensory properties (based on texture, color, and aroma) than controls during storage.	Basaglia et al. (2021)

(Contd.)

Table 7.5: *(Contd.)*

Film type	Active agent(s)	Product/Application/ Storage	Changes in safety shelf-life quality[g]/ Sensory properties	References
Alginate coating	Citral nanoemulsion (NE) at 0.1-1% (v/v)[a] was combined with passive MAP	Pineapple slices/ Dipping-draining-drying/ packaging/12 days at 4°C.	Alginate coating with or without NE of citral improved fruit firmness and vitamin C retention (min. four-fold) during cold storage, but increased citral concentration (at 1%) caused loss of gained firmness by coating. Coating alone had no beneficial effect on inhibition of fruit browning, but active coating reduced browning at a citral concentration dependent manner. All coatings caused significant reductions in TAMB and TYM of samples, but coatings with 1% citral were the most effective antimicrobial films/Samples treated with 0.5% coating caused the best sensory properties (based on color, appearance, texture, odor, and taste)	Prakash *et al.* (2020)
Konjac glucomannan gum coating	*Lactobacillus plantarum* (3 strains) at 9.4 log CFU/ml	Probiotic kiwi slices/ Dipping/five days at 4°C.	TLAB of fruits were greater than 6 log CFU/g during storage period. Active packaging with *L. plantarum* loaded coatings reduced fruit color change, degradation of polyphenol, chlorophyll and vitamin C, decay rate, and TYM/Samples with probiotic loaded coatings showed the best sensory properties (based on overall acceptability).	Hashemi and Jafarpour (2021)
Na-alginate coating	*L. rhamnosus* or *B. animalis* subsp. *lactis* (each at 5×10^{11} CFU/ml of	Probiotic fresh cut apple cubes/Dipping-draining-CaCl₂ cross-linking in solution with ascorbic acid	The probiotic bacteria in coatings were stable during cold storage (9.1-9.5 log CFU/g at the end of eight days). Probiotic loaded coatings with or without prebiotics did not considerably affect the TPB and TYM of samples. Probiotic and prebiotic	Alvarez *et al.* (2021)

Coating type	Composition	Food/Method/Duration/Temperature	Results	Reference
	coating solution) with or without prebiotics (inulin + oligofructose each at 8% (w/w)c)	at 1% (w/v)/eight days at 4°C.	loaded coating caused significant inhibition of inoculated *L. innocua* and *E. coli* O157: H7/ Cold stored samples with *L. rhamnosus* and prebiotics loaded coatings showed inacceptable sensory properties while coatings with *B. animalis* subsp. *lactis* and prebiotics gave acceptable sensory properties (based on appearance, odor, and flavor).	
Na-alginate coating	Polylysine (PL) at 0.05-0.15% (w/v)b	Kiwi slices/Dip-coating-draining-drying/14 days at 4°C.	Coatings with PL at 0.05% minimized reductions in soluble solids, titratable acidity, chlorophyll degradation, electrolyte leakage, and malonaldehyde content. The coatings with 0.05% PL reduced TAMB and TYM of samples significantly while PL at higher concentrations had no effect on microbial load and quality of fruits/No sensory analysis.	Li *et al.* (2017)
Na-alginate coating	*Lactobacillus acidophilus* (7.36 log CFU/g of coating)	Carrot slices/Dip-coating-$CaCl_2$ cross-linking/19 days at 8°C.	Samples with *L. acidophilus* loaded coating maintained LAB viability for 19 days (7.1 log CFU/g). Coating with *L. acidophilus* caused significantly higher TAMB (4.5 log CFU/g) than uncoated controls (3.0 log CFU/g), but TYM is absent in both samples. Control samples showed a higher orange tonality than coated ones/No sensory analysis.	Shigematsu *et al.* (2018)
Chitosan coating	Inherent antimicrobial activity of chitosan at 0.1, 0.25 or 0.5% (w/w)c	Whole potatoes (two cultivars)/Dip-coating-draining-drying/170 days at 12°C.	Coated samples had slightly (at 0.5% chitosan) to moderately (0.1 and 0.25% chitosan) lower final reducing sugar content than uncoated controls. Coating did not affect major potato flavor compounds and vitamin C content. Coated and uncoated potatoes gave acceptable chips color after	Raigond *et al.* (2019)

(Contd.)

Table 7.5: *(Contd.)*

Film type	Active agent(s)	Product/Application/ Storage	Changes in safety shelf-life quality[g]/ Sensory properties	References
			frying. Chitosan coating reduced the weight loss and initial *Fusarium* population (except in one cultivar at 0.25% chitosan) at the peel surface of potatoes/No sensory analysis.	
Chitosan, pectin, gum Arabic, or CMC coatings	Olive leaf extract (OLE) or sodium ascorbate (SA) at 1% (w/v)[b]	Fresh-cut potatoes/Dipping-draining-drying/Vacuum packaging for seven days at 10°C/Frying.	Chitosan coating was found inappropriate since it caused browning at the product surface during storage. Pectin coating with OLE also gave potatoes with inferior surface color. CMC and gum Arabic coatings with OLE or SA gave acceptable color and decreased the fat content of fried samples up to 45% in comparison to uncoated samples/No sensory analysis.	Kurek *et al.* (2020)
Locust bean gum coating	Fennel EO at 0.75% (v/v)[a]	Sliced artichoke heads/ Dipping into ascorbic acid-citric acid or L-cysteine)/ Dipping into coating-draining/Hermetic plastic packaging (trays + semi-permeable plastic films/11 days at 4°C.	Samples with fennel EO loaded coating caused limited reductions in TAMB, TPC, TYM, TE, PC, and inoculated *E. coli* (0.5 to 2 log CFU/g). Coating with fennel EO applied after L-cysteine dipping caused the highest vitamin C and polyphenol retention among stored samples. L-cysteine is a more effective PPO inhibitor than ascorbic acid-citric acid combination/Samples with fennel EO loaded coating showed the best overall sensory properties (based on brightness, browning, off-odor, firmness).	Rizzo *et al.* (2019)

Alginate coating	Nanoemulsion (NE) of basil EO at 0.5% (v/v)[a] was prepared by using Tween-20 or saponin (natural surfactant) rich *Sapindus* sp. extract	Okra/Dipping-CaCl$_2$ crosslinking-drying (20 min.)/11 days at 5 or 24°C.	The NE of basil EO obtained with natural surfactant is comparable with NE of basil EO obtained with Tween-20. Alginate coating with NE of basil EO increased resistance to fungal decay, and reduced water loss, browning index and toughening of okra pods during storage/Active coatings gave the highest sensory properties (based on overall acceptability or intent to purchase).	Gundewadi *et al.* (2018)
Chitosan coating	Lysozyme and nisin at 3.5 and 0.5 mg/cm^2, respectively	Lentil, mung bean and wheat destined for sprouting/Dipping-dying/No storage.	Chitosan coating with or without LYS-NIS caused 3.3 and 2.5, 3.4 and 2.8, and > 4.1 and 3.6 log reduction in initial *L. innocua* loads of seeds for mung beans, lentils, and wheat, respectively. Edible coating had no significant effect on seed germination rate/No sensory analysis.	Sozbilen and Yemenicioğlu (2020)
Na-alginate coating	Combination of thyme EO at 1% (v/v)[a], L-cysteine at 0.3 g/L and nisin at 0.4 g/L	*Pholiota nameko* mushroom/Dip-coating-CaCl$_2$ cross-linking-Drying (30 min.)/Nine days at 20°C.	Coating with or without active agents reduced the cap opening and vitamin C loss during storage, but only coating with active agents reduced browning of mushrooms. Coating with or without active agents reduced the accumulation of phenolic compounds during storage. Active coating caused significantly lower TAMB than coated and uncoated controls. The overall results showed that samples treated by active coating and stored at 20°C had comparable quality with refrigerated uncoated control mushrooms/No sensory analysis.	Zhu *et al.* (2019)
Tragacanth gum coating	*Satureja khuzistanica* essential oil (SKEO) at 100, 500 or 1000 ppm in coating solution	Button (*Agaricus bisporus*) mushroom/Dipping (5 min.)-draining-drying (30 min.)/16 days at 4°C.	All coatings caused significant reductions in TAMB, TPB, PC, and TYM, and weight loss, loss of firmness, browning, degradation of polyphenols and vitamin C. Coatings with 500 and 1000 ppm SKEO were the most effective on PC and TYM. Coatings with 500 ppm SKEO were the most effective in	Nasiri *et al.* (2018)

(Contd.)

Table 7.5: *(Contd.)*

Film type	Active agent(s)	Product/Application/ Storage	Changes in safety shelf-life quality[g]/ Sensory properties	References
			reducing weight loss and browning/Coatings with 500 ppm SKEO gave the best sensory properties (based on gill color, dark zone formation, off-odor, gill and cap uniformity).	
Na-alginate coating	Cinnamaldehyde (CIN) nanoemulsion (NE) at 0.025, 0.05, 0.1% (v/v)[a]	Button (*Agaricus bisporus*) mushroom/Spraying/ Draining-drying/16 days at 4°C.	Active coating with NE of CIN reduced weight loss, loss of firmness, degradation of polyphenols and polyphenoloxidase activity in mushrooms more effectively than uncoated and coated controls. Coatings with or without NE of CIN reduced the browning of mushrooms. Coatings with 0.1% NE of CIN reduced the PC of mushrooms considerably (~2 log) while lower concentrations of NE caused limited antimicrobial activity/No sensory analysis.	Louis *et al.* (2021)
Starch coated paper	Microencapsulated cinnamon EO (β-cyclodextrin was used as encapsulant)-starch mixtures (1:1, 3:1, 5:1) were coated onto paper	Button (*Agaricus bisporus*) mushroom/Mushrooms were placed into a box and a sheet of active paper covered on the top of the mushroom box was sealed with an adhesive tape.	Active paper reduced the weight loss, loss of firmness and limited the increase in membrane permeability, but it showed limited antimicrobial effect on TAMB, TYM, TPB and PC of mushrooms. The highest antimicrobial effect was observed for the paper with the highest cinnamon EO content (at 5:1 ratio). Application of active paper increased the antioxidant enzymes (glutathione reductase and ascorbate peroxidase) in mushrooms. Antioxidant enzyme activity increased as EO concentration was increased/No sensory analysis.	Shao *et al.* (2021)

[a] v/v film forming solution; [b] w/v film forming solution; [c] w/w film forming solution; [d] w/w hydrocolloid; [e] v/v hydrocolloid; [f] v/w hydrocolloid; [g] TAMB: total aerobic mesophilic bacteria; TPB: total psychrophilic bacteria; TCC: total coliform count; TYM: total yeast and mold; TE: Total *Enterobacteriaceae*; TLAB: total lactic acid bacteria; PC: *Pseudomonas* spp. count; EO: essential oil

phenylalanine in reducing surface pitting and stem-end rot in avocadoes during 18-days of cold storage (at 2 or 5°C), followed by seven-days' storage at 22°C (Saidi *et al.*, 2021). It was also reported that carvacrol-thymol mixture loaded high amylose corn starch coatings applied on whole mango and papaya (Riaz *et al.*, 2021), and apple peel phenolic extract loaded chitosan coatings applied on whole strawberries (Ochoa-Velasco *et al.*, 2021) were also effective in reducing fungal decay in these fruits, but these applications were not supported with sensory analysis.

The active coating is also applied on root vegetables, such as whole or fresh-cut potatoes. The cold stored whole potatoes suffer mainly from fungal attacks and water loss that cause economic losses. Moreover, increase in reduced sugar content during cold storage is a major problem in potatoes since this is the primary reason for undesired browning and formation of acrylamide by Maillard reaction during frying of french-fries (Biedermann-Brem *et al.*, 2003). It was reported that chitosan coating of whole potatoes during cold storage did not affect their major flavor compounds and vitamin C content, but it was beneficial to control their reduced sugar content and weight loss, and to reduce *Fusarium* population at their peels (Raigond *et al.*, 2019). Kurek *et al.* (2020) applied olive leaf extract or sodium ascorbate loaded edible coatings of chitosan, pectin, gum Arabic, or CMC to increase the quality of fresh-cut potatoes. These workers found that chitosan is an inappropriate coating for fresh-cut potatoes since it causes browning on the product surface during cold storage. Pectin coatings also failed to maintain the desired surface color. However, olive leaf extract or sodium ascorbate loaded gum Arabic and CMC coatings were found successful in controlling browning of fresh-cut potatoes for one week and in reducing their oil uptake (up to 45%) during frying (Kurek *et al.*, 2020). Another vegetable whose quality could be improved by active edible coating is artichoke. Rizzo *et al.* (2019) successfully applied antioxidant dipping (into ascorbic acid-citric acid combination or L-cysteine solutions) and fennel EO loaded locust bean gum coating to control browning and microbial load, and to maintain sensory properties of sliced artichoke heads during cold storage. Gundewadi *et al.* (2018) also effectively used basil EO nanoemulsion loaded alginate coatings to reduce water loss, browning index, and toughening, and to improve fungal decay resistance and sensory properties of stored (at cold or ambient temperatures) okra.

Active edible coatings have also been applied to mushrooms to improve their microbial quality, to prevent their enzymatic browning, and weight loss, and to maintain their firmness. Nasiri *et al.* (2018) achieved significant reductions in microbial load, enzymatic browning, loss of firmness and weight loss of cold stored button mushrooms by applying *Satureja khuzistanica* essential oil (SKEO) loaded tragacanth gum coating. These authors also reported that the SKEO loaded coatings gave the best sensory properties based on gill color, dark zone, and off-flavor formation, and gill and cap uniformity. Zhu *et al.* (2019) incorporated a combination of thyme EO, nisin, and L-cysteine into Na-alginate coatings and reduced microbial load, browning, cap opening, and vitamin C loss of *Pholiota nameko* mushrooms stored for nine days at 20°C. Louis *et al.* (2021) also improved the microbial quality and color, and reduced loss of firmness in button mushrooms by using cinnamaldehyde nanoemulsion loaded Na-alginate coatings.

Moreover, recently Shao *et al.* (2021) employed active paper coated with a starch layer incorporated with microencapsulated cinnamon EO in packaging of button mushrooms. The active papers covered at the surface of boxed mushrooms reduced weight loss and loss of firmness, controlled increase of mushroom membrane permeability, and activated the antioxidant enzymes in mushrooms, but showed a limited antimicrobial effect.

Some examples of coating applications conducted by using non-phenolic active agents also exist, such as polylysine loaded Na-alginate coatings applied on kiwi slices (Li *et al.*, 2017). It was reported that the use of suitable concentrations of polylysine is beneficial in reducing microbial load of kiwi slices, and in minimizing reduction of their soluble solid contents, titratable acidity, electrolyte leakage, and chlorophyll degradation. The active edible coatings were also recently used for decontamination of *Listeria* from seeds destined for edible sprout production. Sozbilen and Yemenicioğlu (2019) showed that application of lysozyme-nisin combination loaded chitosan coating on seeds, such as lentils, mung beans, and wheat caused 3.3 to 3.6 log reduction in inoculated *L. innocua* in these seeds. It was also reported that the active chitosan coating of these seeds did not affect their germination rates.

The recent trend of incorporating probiotic/protective LAB and prebiotics into edible packaging also found some application in fresh fruits and vegetables. Hashemi and Jafarpour (2021) employed *Lactobacillus plantarum* (three strains) loaded konjac glucomannan gum coatings in suppressing the fungal growth and in improving the quality of strawberries. These workers reported that the strawberries maintained their LAB count at 6 log CFU/g, and showed improvement in their sensory properties during five-day cold storage. Bambace *et al.* (2019) applied *L. rhamnosus* and/or prebiotic (inulin-oligofructose mixture) loaded alginate coatings on fresh blueberries without causing a negative effect on fruit sensory properties (appearance and odor for two weeks, and flavor for minimum one week). It was reported that coatings with *L. rhamnosus* and combination of *L. rhamnosus* with prebiotics increased the TAMB, TPB, and TYM of blueberry fruits, but caused growth-inhibiting effect and significant reduction in inoculated *E. coli* O157:H7 and *Listeria innocua* in fruits, respectively (Bambace *et al.*, 2019). Moreover, *L. rhamnosus* counts of blueberries with *L. rhamnosus* and prebiotic or *L. rhamnosus* loaded coatings remained $\geq 10^6$ CFU/g for three and one weeks, respectively (Bambace *et al.*, 2019). Alvarez *et al.* (2021) also employed *L. rhamnosus* or *B. animalis* subsp. *Lactis,* and prebiotics (*inulin*-oligofructose mixture) loaded alginate coatings and obtained probiotic-prebiotic apple cubes with stable LAB counts above 10^6 CFU/g for eight days at 4°C. The *L. rhamnosus* and prebiotic loaded alginate coatings did not give apple cubes any acceptable sensory properties. However, *B. animalis* subsp. *lactis* and prebiotic loaded alginate coatings gave apples acceptable sensory properties, and caused almost 1.7 and 2.5 log reduction in *E. coli* O157: H7 and *L. innocua* inoculated on the surface of fruits at the end of eight-day cold storage, respectively (Alvarez *et al.*, 2021). Some other LAB and prebiotic coating studies also exist, such as *Lactobacillus rhamnosus* and inulin loaded gelatin coatings applied on whole strawberries (Temiz and Özdemir, 2021), *L. plantarum* and *P. pentosaceus* loaded cassava starch-CMC films applied on whole banana (Li *et al.*, 2020b), and *L. acidophilus* loaded alginate coatings applied on carrot slices (Shigematsu *et al.*, 2018). These studies also provided promising preservation effects of probiotic cultures, but they were not supported by sensory analysis.

7.7 Active edible packaging of bread and other dough food

The challenges associated with active edible coating of baked products and dough food involve undesired textural changes (mainly firming) caused by staling and moisture loss (Cauvain, 1998), and microbiological spoilage caused mostly by fungi, such as *Aspergillus* and *Penicillium* spp. (Garcia *et al.*, 2019). The major problem of staling is a result of starch gelatinization during the baking process. The gelatinization transforms ordered (crystalline) structure of starch originating from its amylopectin fraction into a disordered state (Cauvain, 1998). The staling observed, as in the firming of bread crumb, occurs gradually during storage when disordered starch starts to reorder or to retrograde (Cauvain, 1998). The poor water-vapor barrier properties of most edible films are a great disadvantage that limit the use of bags made from edible films in packaging of bakery product (Kõrge *et al.*, 2020; Oliveira *et al.*, 2020), but active edible coatings applied on product surface or on the surface of other plastic packaging materials could find different applications in inhibiting fungal growth, or in developing functional (e.g. prebiotic or probiotic) bakery products. Some examples of hydrocolloids and mixtures (blend or composites of hydrocolloids) employed for active edible coating of bread and other dough food include methylcellulose (Otoni *et al.*, 2014), starch (Ju *et al.*, 2020), mung bean starch-guar gum-sunflower oil (Lee *et al.*, 2020), mung bean starch-soy protein isolate (Li *et al.*, 2020a), Na-alginate-whey protein concentrate (Soukoulis *et al.*, 2014), Na-alginate-whey, high amylose starch-gelatin (Gregirchak *et al.*, 2020), and gelatin-gum Arabic (Gonçalves *et al.*, 2017). Moreover, films from gliadin (Balaguer *et al.*, 2013), gelatin-cashew gum (Oliveira *et al.*, 2020), and chitosan (Kõrge *et al.*, 2020) are also used in active edible coating of bread and dough food.

Since it is important to suppress the fungal growth in bread and other dough food, phenolic compounds preferred in active edible packaging of this food category involve mainly volatile essential oils obtained from lemongrass, clove, oregano and thyme, and volatile essential oil components, such as cinnamaldehyde, eugenol, citral (Table 7.6). However, although it is really highly critical to balance antimicrobial benefits and sensory effects of EO in bakery products, only a limited number of studies are supported by sensory analysis. Ju *et al.* (2020) placed small sachets containing eugenol and citral loaded starch microparticles into plastic bags employed for active packaging of bread. These workers delayed mold development in packed bread for almost two weeks at room temperature, but the overall sensory attributes of bread received 'moderate liking' on day 12. The plastic bags coated with cinnamaldehyde loaded gliadin (Balaguer *et al.*, 2013), antimicrobial edible bags from lemongrass EO loaded gelatin-cashew gum films (Oliveira *et al.*, 2020), and metalized plastic bags coated with clove or oregano EO nanoemulsion

Table 7.6: Examples of active edible packaging for bread and other dough food

Film type	Active agent(s)	Product/Application/ Storage	Changes in safety shelf-life quality[g]/Sensory properties	References
Gliadin film	Cinnamaldehyde at 5% (w/w)[d]	Bread/A piece of film was placed into plastic bag used for packaging of bread/30 days at 23°C.	Inoculated *P. expansum* did not grow at surface of bread slices with EO loaded film within 30 days while it started to grow in controls within seven days. Naturally present fungi growth in actively packed and control bread slices occurred within 30 and seven days, respectively/No sensory analysis.	Balaguer *et al.* (2013)
Gelatin-cashew gum (1:1) film (cross-linked with oxidized ferulic acid)	Lemongrass EO at 1% (w/w)[c]	Bread/Samples were placed into bags from gelatin-cashew gum film/Eight days at room temperature.	Mold growth in breads packed with plastic bags (control) and active edible bags was observed in four and seven days, respectively. Bread in active edible bag showed significantly greater hardening (as a result of excessive moisture loss) than that in plastic film/No sensory analysis.	Oliveira *et al.* (2020)
Methylcellulose coating	Clove or oregano EO at 4% (w/v)[b] (coarse emulsions or nanoemulsions were formed by mixing or ultrasonication, respectively	Bread/Sliced bread was placed into metalized plastic bags coated inside with active edible film/ 15 days at 25°C.	Bread packed in bag with active coating showed significantly lower TYM than controls during storage. Most effective suppression of fungi was observed for EO nanoemulsion loaded coatings/ No sensory analysis.	Otoni *et al.* (2014)
Starch microparticles in sachets	Eugenol and citral were encapsulated into starch microparticles (SMP)	Bread/SMP with EOs were placed into sachets/ Sachets and bread were placed into plastic bags/12 days at 25°C.	Breads bagged with or without active sachets showed mold development after 15 and six days, respectively/Bread packed together with active sachets showed moderate liking (based on overall sensory attributes).	Ju *et al.* (2020)

Konjac glucomannan coating	L. casei (9 log CFU/ml of film forming solution) or L. casei (10 log CFU/ml of film forming solution)-inulin (1% w/v)ᵇ	Bread buns/Coating was brushed onto bun surface after baking/15 days at 25°C.	Coating with L. casei prevented mold growth in bread for seven days, while breads with control coating and coating with L. casei in initiated mold growth in five days. L. casei in coatings survived by 74-78% at the end of seven days/No sensory analysis.	Pruksarojanakul et al. (2020).
Na-alginate-whey (AW) or high amylose corn starch-gelatin (SG) coating	Commercial lactic acid bacteria (LAB) starter culture at ~0.9% (w/v)ᵇ (*The starter culture contained S. salivarius subsp. thermophilus, L. delbrueckii subsp. bulgaricus and L. acidophilus*)	Bread/Spreading at the surface/86 h at 23°C.	LAB in AW coating showed significantly higher stability than that in SG coating. Coatings with LAB caused significant reduction (min. 1 log) in TAMB, and suppressed the growth of inoculated A. niger and P. chrysogenum in samples/Coated and uncoated samples gave similar sensory properties (based on color, aroma, and taste).	Gregirchak et al. 2020
Na-alginate or Na-alginate-whey protein concentrate (WPC) (1:4) blend coating	Lactobacillus rhamnosus	Bread/Brushing of coating at bread surface after baking/Drying/Seven days at 25°C.	L. rhamnosus is more stable in Na-alginate-WPC than in Na-alginate coating. Probiotic coating had no effect on bread staling, texture, and flavor compounds. Coated bread slice (30–40 g) delivered 6.5–6.9 log CFU/portion after in-vitro digestion/No sensory analysis.	Soukoulis et al. (2014)
Mung bean starch-guar gum-sunflower oil coating (2:0.75:1.5)	Grapefruit seed extract at 0.8% (w/v)ᵇ	Traditional Asian rice cake/Dipping-draining-drying/48 hours at 25°C.	Active coating reduced inoculated B. cereus and P. citrinum by ~0.7 and ~1.2 log within 48h, respectively. The coating delayed the hardening and staling of starch cakes/No sensory analysis.	Lee et al. (2020)

(Contd.)

Table 7.6: *(Contd.)*

Film type	Active agent(s)	Product/Application/ Storage	Changes in safety shelf-life quality[g]/Sensory properties	References
Chitosan film	Chestnut extract at 1% (w/v)[b]	Fresh pasta/Samples were placed into bags from chitosan films/60 days at 8°C.	TAMB, TYM, and TE of coated samples were under detection limits during storage period. The active packaging did not prevent moisture loss of pasta. No significant phenolic release occurred from films to pasta/No sensory analysis.	Kõrge *et al.* (2020)
Mung bean starch-soy protein isolate coating	Nanoemulsion (NE) of clove EO at 0.6 and 0.9% (w/w)[c]	Steamed buns/ Brushing/10 days at 10°C.	Active coatings kept the TAMB of buns below 5 log CFU/g for 10 days while TAMB of controls exceeded 6 log CFU/g in seven days. Active coating suppressed growth of yeast and mold in buns, and reduced their hardening/Buns with active coating showed acceptable sensory properties (based on flavor, appearance, texture).	Li *et al.* (2020a)
Gelatin-gum Arabic macroparticles coating	Thyme EO was encapsulated by the macroparticles (0.125 or 0.6 mg EO/ml of coating solution in ethanol)	Baked cake/Spraying on hot sample surface after baking/30 days at 25°C.	Active coating at 0.125 and 0.6 mg EO/ml coating caused almost 3 and 4 log reductions in mold counts of cakes stored for 30 days, respectively/ No sensory analysis.	Gonçalves *et al.* (2017)

[a] v/v film forming solution; [b] w/v film forming solution; [c] w/w film forming solution; [d] w/w hydrocolloid; [e] v/v film forming solution; [f] v/w hydrocolloid; [g]TMAC: total mesophilic aerobic count; TPC: total psychrophilic count; TCC: total coliform count; TYMC: total yeast and mold count; PC: *Pseudomonas* spp. count

loaded methylcellulose films (Otoni *et al.*, 2014) have also been used to inhibit mold growth in bread. However, further studies are needed to evaluate the sensory properties of these applications on packaged bread. Li *et al.* (2020a) reported that clove EO loaded mung bean starch-soy protein isolate coatings were effective in reducing hardening and microbial load (TAMB) of cold stored steamed buns without causing any undesired sensory problems. The application of non-volatile phenolic compounds to bread and dough food is more limited than volatile ones. However, it is important to note that application of grapefruit seed extract loaded mung bean starch-guar gum blend coating on Asian rice cake was beneficial in reducing *B. cereus* and *P. citrinum* load, and in delaying hardening and staling of this traditional product without causing any undesired effect in its sensory attributes (Lee *et al.*, 2020). Chitosan bags incorporated with chestnut extract also successfully kept the bacterial and fungal counts of cold stored fresh pasta at uncountable levels for 60 days, but the active packaging failed to prevent the moisture loss in pasta (Kõrge *et al.*, 2020).

The coatings loaded with probiotics or protective LAB cultures and prebiotics have also been used for preservation of bakery products. Pruksarojanakul *et al.* (2020) reported that konjac glucomannan coating with *L. casei* delayed mold growth in bread (two more days than control), while same coating with *L. casei* and inulin failed to delay mold growth. Gregirchak *et al.* (2020) successfully used a starter culture (a mixture of 3 LAB) loaded Na-alginate-whey blend coating to reduce the microbial load (TAMB) and to suppress growth of inoculated *A. niger* and *P. chrysogenum* in bread samples. Soukoulis *et al.* (2014) obtained probiotic pan bread by using *Lactobacillus rhamnosus* loaded Na-alginate-whey protein concentrate coatings. According to these authors, an individual consuming 30-40 g bread slice can deliver approximately 6.5-6.9 log CFU/portion after *in-vitro* digestion. However, the application of probiotic coating did not change the staling mechanism in bread (Soukoulis *et al.*, 2014).

7.8   Examples of active edible packaging of food in combination with other preservation methods

Active edible food packaging is not a self-standing method and needs mostly a combination with refrigeration. However, when the product is very susceptible to microbiological spoilage or a marginal extension in normal product shelf-life is the main target, active packaging and refrigeration are supported by a third preservation method, such as modified atmosphere (MAP) or vacuum packaging (VP), high pressure processing (HPP), γ-irradiation, or pulsed light treatment. It is important to note that active chitosan-based coatings are the most frequently combined packaging materials with other methods since they help in maximizing the obtained antimicrobial activity. Some examples of active edible packaging combined with MAP are coatings from whey protein (Carrión-Granda *et al.*, 2018), pectin (Xiong *et al.*, 2020), chitosan (Fang *et al.*, 2018; Langroodi *et al.*, 2018; Cao *et al.*, 2019), Na-alginate (Tabassum and Khan, 2020), and Na-caseinate (Caillet *et al.*, 2006) while VP is frequently combined with active coatings from chitosan (Yingyuad *et al.*, 2006; Günlü and Koyun, 2013; Duran and Kahve, 2020). Caillet *et al.* (2006) also combined active Na-caseinate coatings with both of MAP and γ-irradiation. The γ-irradiation alone was also applied in combination with N-palmitoyl chitosan (Severino *et al.*, 2014) and Na-alginate coating (Ben-Fadhel *et al.*, 2017), and chitosan film (Dini *et al.*, 2020). HPP was also combined with active edible films from gelatin-chitosan blend (Gómez-Estaca *et al.*, 2018) and chitosan (Albertos *et al.*, 2015), and active coatings from chitosan (Martillanes *et al.*, 2021; Pavli *et al.*, 2019), Na-alginate (Pavli *et al.*, 2019) and N-palmitoyl chitosan (Donsì *et al.*, 2015). Other less common combinational application involves use of pulsed light treatment with N-palmitoyl chitosan (Donsì *et al.*, 2015) or chitosan coating (Koh *et al.*, 2017).

Similar to all other active edible packaging applications, phenolic compounds are also extensively used in active packaging combined with other methods. Examples of phenolic compounds employed in the combinational applications include mainly essential oils from *Zataria multiflora Boiss*, cumin, clove, oregano, thyme, lemongrass, and mandarin; essential oil component cinnamaldehyde; pure polyphenols, such as gallic acid; and phenolic extracts from sumac and rice bran. Other natural antimicrobials, such as nisin, γ-polylysine, and natamycin are also used alone or in combination with polyphenols to obtain a synergy, or to broaden the antimicrobial spectrum of films, or to reduce concentrations of polyphenols that affect the sensory properties of food (Table 7.7).

HPP is combined with active packaging especially in microbiologically susceptible food, such as fish and fish products and ham. Some forms of application include a combination of clove EO loaded fish gelatin-chitosan film with HPP (15 min. at 250 MPa) for salmon carpaccio (Gómez-Estaca *et al.*, 2018), clove EO loaded chitosan film with HPP (10 min. at 300 MPa) for rainbow trout fillet (Albertos *et al.*, 2015), and oregano EO loaded Na-alginate film with HPP (2 min. at 500 MPa) for ham slices (Pavli *et al.*, 2019). Moreover, Donsì *et al.* (2015) also applied mandarin EO nanoemulsion loaded N-palmytoyl chitosan coating in combination with HPP (5 min. at 300 MPa) for green beans. In all these applications, HPP increased the antimicrobial performance of active packaging considerably. Martillanes *et al.* (2021) also combined nisin or rice bran extract (RBE) loaded chitosan films with HPP (8 min. at 600 MPa) for preservation of ham slices. However, these authors reported that chitosan films loaded with nisin-RBE combination showed synergetic effect and HPP contributed only slightly to overall antimicrobial performance. This finding clearly shows that benefits of employing synergetic mixture of natural antimicrobials should be evaluated in packaging alone before considering a combination with more expensive methods, like HPP.

Pulsed light treatment (PL) is another decontamination method that can be combined with active packaging. Donsì *et al.* (2015) combined mandarin EO nanoemulsion loaded N-palmytoyl chitosan coating with a single dose of PL (12 J/cm^2 at each side of sample) applied at the beginning of storage, but they did not observe any antimicrobial benefits in this treatment. In contrast, Koh *et al.* (2017) combined chitosan coating of cantaloupe slices with repeated PL (13 times in every 48 h to achieve cumulative 11.7 J/cm^2) and achieved effective reduction

Table 7.7: Examples of combining active edible food packaging/coating with other preservation methods

Film type	Active agent(s) and combination method(s)	Product/Application/ Storage	Changes in safety shelf-life quality*/Sensory properties	References
Fish gelatin-chitosan film	Clove EO (at 0.75 ml/g gelatin or gelatin-chitosan) was combined with HPP (15 min at 250 MPa and 7°C)	Salmon carpaccio/ Covering of sample with film/HPP/ 11 days at 5°C.	AP reduced the TAMB, H$_2$S producing bacteria, PC, TE, TLAB of samples significantly. AP-HPP caused more significant reductions in microbial counts than AP alone. AP alone increased TVB-N of sample, but AP-HPP gave TVB-N comparable to that of control. AP and AP-HPP inhibited lipid oxidation and reduced free fatty acid formation in samples/No sensory analysis.	Gómez-Estaca *et al.* (2018)
Chitosan film	Clove EO at 2% (w/w)^c was combined with HPP (10 min. at 300 MPa and 12°C).	Rainbow trout fillet/ Wrapping of samples with films/HPP/22 days at 4°C.	AP-HPP combination showed better antimicrobial activity than AP alone. AP-HPP kept TAMB < 2 and TCC < 3 log CFU/g and it inhibited TLAB at 22nd day while TAMB and TCC of uncoated controls exceeded 8 log CFU/g at eighth day/No sensory analysis.	Albertos *et al.* (2015)
Na-alginate film (CaCl$_2$ cross-linked)	Oregano EO at 1% (v/v)^a was combined with HPP (2 min at 500 MPa and 20°C)	Ham slices/Films were placed on both sides of samples/HPP/40, 47, and 66 days at 4, 8 or 12°C, respectively.	AP with EO loaded films caused almost 1.5 log reduction in *L. monocytogenes* (cocktail of four strains) at 8 or 12°C, and almost 2.5 log reduction at 4°C by storage. The AP-HPP combination caused almost 1 log additional *Listeria* reduction at all treatments. Moreover, AP-HPP increased inactivation of *Listeria* and reduced growth rates of TAMB and TLAB at all temperatures/AP-HPP gave superior sensory properties (based on aroma, taste, and appearance) than AP or HPP alone.	Pavli *et al.* (2019)

(Contd.)

Table 7.7: (*Contd.*)

Film type	Active agent(s) and combination method(s)	Product/Application/ Storage	Changes in safety shelf-life quality#/Sensory properties	References
Chitosan film	NIS at 0.15% (w/w)[c] and/ or rice bran extract (RBE) at 0.25% (w/w)[c] was combined with HPP (8 min at 600 MPa)	Dry-cured sliced ham/ Films were placed on both sides of samples/ HPP/36 h at 4°C.	AP with NIS or RBE loaded films caused 1-1.5 log reduction in *L. monocytogenes* counts of samples, but AP-HPP combination increased *L. monocytogenes* reduction of these films by 4-4.5 log. AP with NIS-RBE loaded chitosan coating caused almost 3.5 log reduction in *L. monocytogenes*, but combination of HPP did not considerably improve the antilisterial effect of this coating (~1 log increase in bacterial reduction)/No sensory analysis.	Martillanes *et al.* (2021)
N-almitoyl chitosan coating	Nanoemulsion of mandarin EO at 0.05% (w/v)[b] combined with HPP (5 min. at 300 MPa and 25°C) or pulsed light (PL) (12 J/cm² for each side)	Green beans/Spray-coating at each side of the samples-drying/HPP or PL/14 days at 4°C.	AP and AP-HPP combination reduced *L. innocua* of samples 2.9 and 4.9 log at the end of storage, respectively. PL exhibited an antagonistic effect with antimicrobial coating. AP did not affect the firmness of samples while HPP and PL caused significant increases in green bean firmness during cold storage. The HPP caused some undesired changes in color of samples while PL did not show a negative impact on sample color/No sensory analysis, but authors reported negative impact of EO on taste and flavor.	Donsi *et al.* (2015)
Chitosan coating	Inherent antimicrobial activity of chitosan was combined with repetitive PL (13 repetitive PL treatment	Cantaloupe slices/ Dipping-alkali treatment was applied for gelation of coating-drying/	AP-PL combination inhibited growth of TAMB and TYM of samples for 28 days. TMAB and TYM of uncoated controls, and samples treated with AP, PL, or AP-PL at the end of storage were >8, ~6, ~5, ~1	Koh *et al.* (2017)

	in every 48h. Total cumulative PL of 11.7 J/cm²)	Repetitive PL at each 48 h/28 days at 4°C	log CFU/g, respectively. AP-PL combination did not affect phenolic content of fruits, but it improved their firmness. AP caused significantly higher weight loss and vitamin C loss than uncoated fruits/ No sensory analysis.	Duran and Kahve (2020)
Chitosan coating	Inherent antimicrobial activity of chitosan was combined with VP	Beef loin stripes/ Wrapping of samples with films/VP/45 days at 4°C.	Chitosan coating in combination with VP caused almost 0.56 and 2.35 log lower TAMB and TLAB than VP alone, respectively. VP applied samples maintained their initial *S. aureus* load (2-2.3 log CFU/g) during storage, while samples treated with coating and VP had uncountable number of *S. aureus* during storage. All samples showed similar lipid oxidation profiles, but sampled treated with active coating and VP showed lower TVB-N than other samples/No sensory analysis.	
Chitosan coating	Inherent antimicrobial activity of chitosan was combined with VP	Fresh seabass fillet/ Wrapping of samples with films/VP/30 days at 4°C.	The TAMB and TPC results showed that the shelf-life of control samples and VP groups ended within five days (both counts > 6 log CFU/g) whereas samples treated with AP-VP combination maintained their microbial quality (TAMB and TPC counts < 6 log CFU/g) almost 25 days/ Antimicrobial coating reduced TMA-N and TVB-N formation significantly/No sensory analysis.	Günlü and Koyun (2013)
Chitosan coating	Inherent antimicrobial activity of chitosan was combined with VP	Grilled pork/Marination with Thai souse/Dipping-drying/VP/35 days at 2°C.	AP-VP caused almost 2.5 log lower TMAB than VP alone, and gave a better pork color/AP-VP combination improved the sensory properties (based on color, odor, and overall acceptability) of pork.	Yingyuad et al. (2006)

(Contd.)

Table 7.7: *(Contd.)*

Film type	Active agent(s) and combination method(s)	Product/Application/Storage	Changes in safety shelf-life quality[g]/Sensory properties	References
Whey protein coating	Oregano or thyme EOs at 1 or 3% (w/w)[c] were combined with active MAP (at 50% CO_2 + 45% N_2 + 5% O_2)	Hake fish fillets/Dipping-draining-drying (repeated two cycles)-MAP/16 days at 4°C.	AP-MAP combination gave significantly better microbial quality than AP alone. Coatings with 3% thyme or oregano EO in combination with MAP significantly reduced TAMB, TLAB, TE, TPB of samples/No sensory analysis.	Carrión-Granda et al. (2018)
Pectin coating	Emulsions or nanoemulsions (NE) of γ-polylysine (γ-PL) at 2% (w/v)[b] and Oregano EO at 0.5% (w/v)[b] was combined with active MAP (at 20% CO_2 and 80% O_2)	Fresh pork loin/Dipping-drying/MAP/20 days at 4°C.	The TAMB of uncoated and pectin coated samples exceeded 7 log CFU/g threshold after 15 and 20 days, respectively, while TAMB of γ-PL and oregano EO emulsion loaded pectin coated samples, and γ-PL-oregano EO NE loaded pectin coated samples were below this threshold after 20 days. The coatings with NE of antimicrobials showed higher antimicrobial activity than emulsion of antimicrobials. The active coatings reduced the lipid oxidation and protein oxidation of pork significantly. Pork treated with active coatings showed the highest tenderness after 20 days/No sensory analysis.	Xiong et al. (2020)
Chitosan coating	Gallic acid (GA) at 0.2 or 0.4% (w/w)[c] was combined with MAP (at 20% CO_2 and 80% O_2)	Fresh pork loin/Dipping-drying/MAP/20 days at 4°C.	Coatings with GA (0.2 or 0.4%) and control chitosan coatings showed ~1.8 and ~1.2 log lower TAMB than uncoated controls under MAP, respectively. Coatings with GA effectively prevented lipid oxidation of pork samples. Chitosan coating with GA at 0.4% gave firmer pork samples than those with GA at 0.2% due to pro-oxidant activity of GA on meat proteins/No sensory analysis.	Fang et al. (2018)

Chitosan coating	Gallic acid (GA) and nisin (NIS) at 0.2% (w/w)[c] were combined with MAP (at 20% CO_2 and 80% O_2)	Fresh pork loin/Dipping-drying/MAP/20 days at 2°C.	Control coatings and coatings with GA, NIS, or GA-NIS combined with MAP showed 0.42, 1, 0.9, and 2.2 log reduction in TAMB of pork samples at the end of storage, respectively. Chitosan coating alone reduced the lipid oxidation significantly, but addition of GA caused further reductions in lipid oxidation. Coatings with or without active compounds reduced the firmness of samples significantly/No sensory analysis.	Cao *et al.* (2019)
Chitosan coating	Sumac extract (SE) at 2 or 4% (w/w)[c] and/or *Zataria multiflora* Boiss EO (ZEO) at 1% (w/w)[c] were combined with MAP (at 20% CO_2 and 80% O_2)	Fresh beef/Dipping-draining-drying/MAP/20 days at 4°C.	MAP alone did not considerably affect the microbial load of samples. Control coating alone, or active coating with SE or SE-ZEO combined with MAP caused significant reductions in TAMB, TLAB, EC, and PC of samples. Coatings with SE at 2% and ZEO at 1%, and SE at 4%-ZEO at 1% in combination with MAP suppressed TAMB and TLAB for 16 days, EC for 20 days, PC for 12 days, and TYM for eight days. Coatings with SE at 4%-ZEO at 1% were most effective in preventing lipid oxidation/Coatings with SE at 2%-ZEO at 1%, and SE at 4%-ZEO at 1% gave the best sensory properties (based on texture, color, odor, and overall accessibility).	Langroodi *et al.* (2018)
Na-alginate coating	Oregano or thyme EOs at 0.5, 1 or 2% (v/v)[c] were combined with passive MAP	Papaya cubes/Dipping-draining/packaging/12 days at 4°C.	Fruits with oregano and thyme EO loaded coatings (except coating with thyme EO at 0.5%) showed minimum 7.8 and 4.8 log, and 5.1 and 3.2 log lower TAMB and TYM than controls, respectively. Active coating reduced fruit respiration rate and weight loss/Active coatings reduced taste and	Tabassum and Khan (2020)

(Contd.)

Table 7.7: (*Contd.*)

Film type	Active agent(s) and combination method(s)	Product/Application/ Storage	Changes in safety shelf-life quality[g]/Sensory properties	References
			aroma scores of fruits, but showed the highest overall acceptability (sensory analysis was based on color, texture, taste, aroma, juiciness, and overall acceptability).	
Na-caseinate coating	Cinnamaldehyde at 0.025% (w/v)[b] was combined with active MAP (60% O_2+30% CO_2+10% N_2) with or without γ-irradiation (0.25 or 0.50 kGy)	Peeled mini carrots/Spray coating (each side twice)/ MAP/ γ-irradiation/21 days at 4°C.	Coating alone has no antimicrobial effect on *L. innocua* while EO loaded coating alone or MAP alone caused almost 1.3 and 1 log reduction of *L. innocua* at the end of storage, respectively. *Listeria* was completely inhibited (>3 log inactivation) in carrots treated with AP-MAP and AP-MAP-γ-irradiation (0.25 kGy) within 14 and seven days, respectively/No sensory analysis.	Caillet *et al.* (2006)
Chitosan film	Nanoemulsion of cumin EO at 1% (v/v)[a] combined with γ-irradiation at 0.25 kGy	Beef loins/Wrapping of samples with films/γ-irradiation/23 days at 3°C.	Control coating-γ-irradiation, and AP-γ-irradiation caused ~6–7 log reduction of *E. coli* O157:H7, *S. enterica* serovar Typhimurium and *L. monocytogenes* in 12 and four days, respectively. Control coating-γ-irradiation and AP-γ-irradiation caused similar reductions in TAMB, TPB, TLAB, TE, and TVB-N/AP-γ-irradiation gave the highest overall acceptability (sensory analysis was based on color, odor, texture, and overall acceptability).	Dini *et al.* (2020)
N-palmitoyl chitosan coating	Nanoemulsion of mandarin EO at 0.05% (w/v)[b] combined with γ-irradiation at 0.25 kGy	Broccoli florets/ Spraying-drying/ γ-irradiation/13 days at 4°C.	Samples treated with AP and AP-γ-irradiation combination showed 1.1 and 2.6 log CFU/g lower *L. monocytogenes* counts than uncoated controls, respectively/No sensory analysis.	Severino *et al.* (2014)

Na-alginate coating	Mixture of lemongrass EO at 300 ppm, Na-diacetate at 5000 ppm and natamycin at 80 ppm with or without γ-irradiation at 0.4 or 0.8 kGy	Broccoli florets/Dipping-CaCl$_2$ cross-linking-drying (24 h at 4°C)/γ-irradiation/14 days at 4°C.	AP-γ-irradiation combination (at 0.4 kGy) effectively inhibited inoculated *E. coli, L. monocytogenes, and S. enterica* serovar Typhimurium during cold storage and gave samples with minimum 3.5-4 log lower bacterial loads. However, inactivation of *A. niger* (min. 2 log) needed AP-γ-irradiation (0.8 kGy)/No sensory analysis.	Ben-Fadhel et al. (2017)

[a] v/v film forming solution; [b] w/v film forming solution; [c] w/w film forming solution; [d] w/w hydrocolloid; [e] v/w film forming solution; [f] v/w hydrocolloid; FFG: Film forming solution; gTAMB: total aerobic mesophilic bacteria; TPB: total psychrophilic bacteria; TCC: total coliform count; TYM: total yeast and mold; TE: Total *Enterobacteriaceae*; TLAB: total lactic acid bacteria; PC: *Pseudomonas* spp. count; TVB-N: Total volatile basic nitrogen; TMA-N: Trimethylamine nitrogen; MAP: Modified atmosphere packaging; PL: Pulsed light; HPP: High pressure processing, VP: Vacuum packaging; EO: Essential oil; AP: Active packaging

(almost 5 log greater than coating alone) of microbial load (TAMB and TYM) for the fruit samples.

The active edible coating has also been successfully combined with vacuum packaging (VP) and modified atmosphere packaging (MAP). The inherently antimicrobial chitosan coatings have been combined effectively with VP to improve microbial quality of beef loin stripes (Duran and Kahve, 2020), marinated grilled pork (Yingyuad *et al.*, 2006), and fresh seabass fillets (Günlü and Koyun, 2013). The results of Yingyuad *et al.* (2006) also confirmed the positive contribution of chitosan coating-VP combination in the sensory properties of grilled pork. The combination of active edible packaging with MAP has also been studied for different foods, such as fresh meat and pork, fish, and fruits and vegetables. For example, γ-polylysine and oregano EO loaded pectin coating (Xiong *et al.*, 2020), gallic acid loaded chitosan coatings (Fang *et al.*, 2018), and gallic acid and nisin loaded chitosan coatings (Cao *et al.*, 2019) have been employed for preservation of fresh pork loin in combination with MAP at the gas composition of 20% CO_2 and 80% O_2. All these applications were effective in reducing the lipid oxidation of pork loins, but considerable antimicrobial effect was obtained only for those that employed pectin films with polylysine and oregano EO, and chitosan coatings with gallic acid and nisin. The MAP gas composition of 20% CO_2 and 80% N_2 was also employed for application of sumac extract and *Zataria multiflora Boiss* essential oil (ZEO) loaded chitosan coating on fresh beef (Langroodi *et al.*, 2018). This study is important as in that, it suppressed the microbial load and prevented lipid oxidation of cold-stored fresh beef without causing undesired changes in its sensory properties. Another example of active packaging-MAP combination is related to oregano or thyme EO loaded whey protein coating applied to Hake fish fillets with a gas composition of 50% CO_2 + 45% N_2 + 5% O_2 (Carrión-Granda *et al.*, 2018). These workers used increased CO_2 concentrations in MAP and clearly showed that the key factor in success of combinational treatment was initial microbial load of fish fillets. Oregano or thyme EO was also incorporated into Na-alginate coating that was combined with passive MAP for minimally processed papaya cubes (Tabassum and Khan, 2020). This work showed that the active coating combined with MAP was effective in suppressing bacterial and fungal growth as well as in reducing respiration rates of fruits. However, the sensory tests conducted in this work also proved the extreme care needed to balance antimicrobial and physiological benefits of EOs and their undesired effects on taste and aroma of fruits (Tabassum and Khan, 2020). Caillet *et al.* (2006) reported that cinnamaldehyde loaded Na-caseinate coating or MAP (30% CO_2 + 10% N_2 + 60% O_2) showed only a limited inactivation (1.3 and 1 log, respectively) on *L. innocua* inoculated on to peeled mini carrots cold stored for 21 days. In contrast, these workers showed that combination of cinnamaldehyde loaded Na-caseinate coating with MAP (30% CO_2 + 10% N_2 + 60% O_2) caused a 3 log *L. innocua* inactivation in peeled mini carrots within 14 days of cold storage. Moreover, the time to observe the 3 log *Listeria* inactivation in peeled mini carrots was further reduced to seven days by supporting active edible coating-MAP combination with γ-irradiation (at 0.25 kGy) (Caillet *et al.*, 2006). This work clearly showed the possibility of employing active edible packaging as part of a hurdle concept that targets inactivation of critical pathogens in food. However, combinational approaches employing three different methods need careful evaluation of their economic feasibility for the food market. The effectiveness of γ-irradiation to boost antimicrobial performance of active packaging has also been demonstrated by Dini *et al.* (2020). These authors reported that chitosan coating-γ-irradiation, and cumin nanoemulsion loaded chitosan coating-γ-irradiation combinations caused ~6-7 log reduction of inoculated *E. coli* O157:H7, *S. enterica* serovar Typhimurium, and *L. monocytogenes* on beef loins on 12th and 4th days of storage, respectively. The sensory analysis conducted by these authors also proved that the active coating-γ-irradiation combination gave the highest overall acceptability for beef samples (Dini *et al.*, 2020). Severino *et al.* (2014) also showed that mandarin EO loaded N-palmitoyl chitosan coating-γ-irradiation (0.25 kGy) combination was capable of inhibiting *L. monocytogenes* inoculated on broccoli florets by almost 2.6 log. Moreover, Ben-Fadhel *et al.* (2017) achieved almost 3.5-4 log reduction in *E. coli, L. monocytogenes, and S. enterica* serovar Typhimurium inoculated on broccoli florets by using lemongrass EO-Na-diacetate-natamycin mixture loaded Na-alginate coating-γ-irradiation (0.4 kGy) combination. Therefore, it is clear that active edible coating-γ-irradiation combination provides a very effective hurdle to control pathogenic bacteria in different food products.

This chapter clearly shows that active edible packaging can be used as a very effective method to control undesired microbial and oxidative changes in different foodstuff, such as cheeses, raw and processed meat, pork, poultry, fish, bread and dough food, whole or minimally processed fresh fruits and vegetables, and mushrooms and seeds. The use of natural phenolic compounds, essential oils and their active constituents, phenolic extracts and pure phenolic compounds as active components of films and coatings is the predominant trend in food applications. However, to develop industrially relevant applications, it is essential to consider the thin balance between benefits of phenolic compounds and their undesired effects on aroma and flavor of food. Other natural compounds, such as lysozyme, nisin, natamycin, and ε-polylysine, etc. have also been employed in active edible films and coatings. However, these compounds are frequently combined with polyphenols to increase their antimicrobial spectrum and to attain an antioxidant effect for the developed active packaging. Finally, this chapter also clearly reflects the emerging trend of employing probiotic and protective LAB cultures as well as prebiotics in active packaging. The combinational application of active edible packaging with different strategies (e.g. nanoencapsulation, nanoemulsification) and methods, such as HPP, MAP and VP, repeated pulsed light treatment, and γ-irradiation provides alternative opportunities to extend the shelf-life and to increase the microbial safety of food, and to minimize the concentration of natural active agents used in food applications.

References

Abdou, E.S., G.F. Galhoum and E.N. Mohamed (2018). Curcumin loaded nanoemulsions/ pectin coatings for refrigerated chicken fillets, *Food Hydrocoll.*, 83: 445-453.

Albertos, I., A.B. Martin-Diana, M. Burón and D. Rico (2019). Development of functional bio-based seaweed (Himanthalia elongata and Palmaria palmata) edible films for extending the shelflife of fresh fish burgers, *Food Packag. Shelf-Life*, 22: 100382.

Albertos, I., D. Rico, A.M. Diez, L. González-Arnáiz, M.J. García-Casas and I. Jaime (2015). Effect of edible chitosan/clove oil films and high-pressure processing on the microbiological shelf-life of trout fillets, *J. Sci. Food Agric.*, 95: 2858-2865.

Alirezalu, K., S. Pirouzi, M. Yaghoubi, M. Karimi-Dehkordi, S. Jafarzadeh and A.M. Khaneghah (2021). Packaging of beef fillet with active chitosan film incorporated with ε-polylysine: An assessment of quality indices and shelf-life assessment, *Meat Sci.*, 176: 108475.

Alvarez, M.V., M.F. Bambace, G. Quintana, A. Gomez-Zavaglia and M. del Rosario Moreira (2021). Prebiotic-alginate edible coating on fresh-cut apple as a new carrier for probiotic lactobacilli and bifidobacteria, *LWT-Food Sci. Technol.*, 137: 110483.

Amor, G., M. Sabbah, L. Caputo, M. Idbella, V. De Feo, R. Porta *et al.* (2021). Basil essential oil: Composition, antimicrobial properties, and microencapsulation to produce active chitosan films for food packaging, *Foods*, 10: 121.

Angiolillo, L., A. Conte, A.V. Zambrini and M.A. Del Nobile (2014). Biopreservation of Fior di Latte cheese, *J. Dairy Sci.*, 97: 5345-5355.

Balaguer, M.P., G. Lopez-Carballo, R. Catala, R. Gavara and P. Hernandez-Munoz (2013). Antifungal properties of gliadin films incorporating cinnamaldehyde and application in active food packaging of bread and cheese spread foodstuffs, *Int. J. Food Microbiol.*, 166: 369-377.

Bambace, M.F., M.V. Alvarez and M. del Rosario Moreira (2019). Novel functional blueberries: Fructo-oligosaccharides and probiotic lactobacilli incorporated into alginate edible coatings, *Food Res. Int.*, 122: 653-660.

Barros, J.R.D., L. Kunigk and C.H. Jurkiewicz (2010). Incorporation of nisin in natural casing for the control of spoilage microorganisms in vacuum packaged sausage, *Braz. J. Microbiol.*, 41: 1001-1008.

Basaglia, R.R., S. Pizato, N.G. Santiago, M.M.M. de Almeida, R.A. Pinedo and W.R. Cortez-Vega (2021). Effect of edible chitosan and cinnamon essential oil coatings on the shelf-life of minimally processed pineapple (Smooth cayenne), *Food Biosci.*, 41: 100966.

Behbahani, B.A., M. Noshad and H. Jooyandeh (2020). Improving oxidative and microbial stability of beef using Shahri Balangu seed mucilage loaded with cumin essential oil as a bioactive edible coating, *Biocatal. Agric. Biotechnol.*, 24: 101563.

Ben-Fadhel, Y., S. Saltaji, M.A. Khlifi, S. Salmieri, K.D. Vu and M. Lacroix (2017). Active edible coating and γ-irradiation as cold combined treatments to assure the safety of broccoli florets (*Brassica oleracea* L.), *Int. J. Food Microbiol.*, 241: 30-38.

Bermúdez-Oria, A., G. Rodríguez-Gutiérrez, F. Rubio-Senent, Á. Fernández-Prior and J. Fernández-Bolaños (2019). Effect of edible pectin-fish gelatin films containing the olive antioxidants hydroxytyrosol and 3, 4-dihydroxyphenylglycol on beef meat during refrigerated storage, *Meat Sci.*, 148: 213-218.

Bharti, S.K., V. Pathak, T. Alam, A. Arya, V.K. Singh, A.K. Verma and V. Rajkumar (2020). Materialization of novel composite bio-based active edible film functionalized with essential oils on antimicrobial and antioxidative aspect of chicken nuggets during extended storage, *J. Food Sci.*, 85: 2857-2865.

Biedermann-Brem, S., A. Noti, K. Grob, D. Imhof, D. Bazzocco and A. Pfefferle (2003). How much reducing sugar may potatoes contain to avoid excessive acrylamide formation during roasting and baking? *Eur. Food Res. Technol.*, 217: 369-373.

Bonilla, J. and P.J. Sobral (2019). Gelatin-chitosan edible film activated with Boldo extract for improving microbiological and antioxidant stability of sliced Prato cheese, *Int. J. Food Sci. Technol.*, 54: 1617-1624.

Caetano, K.D.S., C.T. Hessel, E.C. Tondo, S.H. Flôres and F. Cladera-Olivera (2017). Application of active cassava starch films incorporated with oregano essential oil and pumpkin residue extract on ground beef, *J. Food Saf.*, 37: e12355.

Caillet, S., M. Millette, S. Salmieri and M. Lacroix (2006). Combined effects of antimicrobial coating, modified atmosphere packaging, and gamma irradiation on Listeria innocua present in ready-to-use carrots (*Daucus carota*), *J. Food Prot.*, 69: 80-85.

Cano Embuena, A.I., M. Cháfer Nácher, A. Chiralt Boix, M.P. Molina Pons, M. Borrás Llopis, M.C. Beltran Martínez and C. González Martínez (2017). Quality of goat's milk cheese as affected by coating with edible chitosan-essential oil films, *Int. J. Dairy Technol.*, 70: 68-76.

Cao, Y., R.D. Warner and Z. Fang (2019). Effect of chitosan/nisin/gallic acid coating on preservation of pork loin in high oxygen modified atmosphere packaging, *Food Control*, 101: 9-16.

Carrión-Granda, X., I. Fernández-Pan, J. Rovira and J.I. Maté (2018). Effect of antimicrobial edible coatings and modified atmosphere packaging on the microbiological quality of cold stored hake (Merluccius merluccius) fillets, *J. Food Qual.*, 2018: 6194906.

Cauvain, S.P. (1998). Improving the control of staling in frozen bakery products, *Trends in Food Sci. Technol.*, 9: 56-61.

Çoban, M.Z. (2021). Effectiveness of chitosan/propolis extract emulsion coating on refrigerated storage quality of crayfish meat (*Astacus leptodactylus*). CyTA-J, *Food*, 19: 212-219.

Costa, S.M., D.P. Ferreira, P. Teixeira, L.F. Ballesteros, J.A. Teixeira and R. Fangueiro (2021). Active natural-based films for food packaging applications: The combined effect of chitosan and nanocellulose, *Int. J. Biol. Macromol.*, 177: 241-251.

Cui, H., M. Bai, C. Li, R. Liu and L. Lin (2018). Fabrication of chitosan nanofibers containing tea tree oil liposomes against Salmonella spp. in chicken, *LWT-Food Sci. Technol.*, 96: 671-678.

Dini, H., A.A. Fallah, M. Bonyadian, M. Abbasvali and M. Soleimani (2020). Effect of edible composite film based on chitosan and cumin essential oil-loaded nanoemulsion combined with low-dose gamma irradiation on microbiological safety and quality of beef loins during refrigerated storage, *Int. J. Biol. Macromol.*, 164: 1501-1509.

Dong, L.M., N.T.T. Quyen and D.T.K. Thuy (2020a). Effect of edible coating and antifungal emulsion system on *Colletotrichum acutatum* and shelf-life of strawberries, *Vietnam J. Chem.*, 58: 237-244.

Dong, C., B. Wang, F. Li, Z.Q. Hong, X. Xia and B. Kong (2020b). Effects of edible chitosan coating on Harbin red sausage storage stability at room temperature, *Meat Sci.*, 159: 107919.

Donsì, F., E. Marchese, P. Maresca, G. Pataro, K.D. Vu, S. Salmieri *et al.* (2015). Green beans preservation by combination of a modified chitosan based-coating containing nanoemulsion of mandarin essential oil with high pressure or pulsed light processing, *Postharvest Biol. Technol.*, 106: 21-32.

Duan, J., S.I. Park, M.A. Daeschel and Y. Zhao (2007). Antimicrobial chitosan-lysozyme (CL) films and coatings for enhancing microbial safety of mozzarella cheese, *J. Food Sci.*, 72: M355-M362.

Duran, A. and H.I. Kahve (2020). The effect of chitosan coating and vacuum packaging on the microbiological and chemical properties of beef, *Meat Sci.*, 162: 107961.

Ehsani, A., M. Hashemi, A. Afshari, M. Aminzare, M. Raeisi and T. Zeinali (2020). Effect of different types of active biodegradable films containing lactoperoxidase system or sage essential oil on the shelf-life of fish burger during refrigerated storage, *LWT-Food Sci. Technol.*, 117: 108633.

Esmaeili, H., N. Cheraghi, A. Khanjari, M. Rezaeigolestani, A.A. Basti, A. Kamkar and E.M. Aghaee (2020). Incorporation of nanoencapsulated garlic essential oil into edible films: A novel approach for extending shelf-life of vacuum-packed sausages, *Meat Sci.*, 166: 108135.

Fang, Z., D. Lin, R.D. Warner and M. Ha (2018). Effect of gallic acid/chitosan coating on fresh pork quality in modified atmosphere packaging, *Food Chem.*, 260: 90-96.

Farhan, A. and N.M. Hani (2020). Active edible films based on semi-refined κ-carrageenan: Antioxidant and color properties and application in chicken breast packaging, *Food Packag. Shelf-Life*, 24: 100476.

Fathi-Achachlouei, B., N. Babolanimogadam and Y. Zahedi (2021). Influence of anise (*Pimpinella anisum* L.) essential oil on the microbial, chemical, and sensory properties of chicken fillets wrapped with gelatin film, *Food Sci. Technol. Int.*, 27: 123-134.

Garcia, M.V., A.O. Bernardi and M.V. Copetti (2019). The fungal problem in bread production: Insights of causes, consequences, and control methods, *Curr. Opin. Food Sci.*, 29: 1-6.

Göksen, G., M.J. Fabra, H.I. Ekiz and A. López-Rubio (2020). Phytochemical-loaded electrospun nanofibers as novel active edible films: Characterization and antibacterial efficiency in cheese slices, *Food Control*, 112: 107133.

Gómez-Estaca, J., M.E. López-Caballero, M.Á. Martínez-Bartolomé, A.M.L. de Lacey, M.C. Gómez-Guillen and M.P. Montero (2018). The effect of the combined use of high pressure treatment and antimicrobial edible film on the quality of salmon carpaccio, *Int. J. Food Microbiol.*, 283: 28-36.

Gonçalves, N.D., F. de Lima Pena, A. Sartoratto, C. Derlamelina, M.C.T. Duarte, A.E.C. Antunes and A.S. Prata (2017). Encapsulated thyme (*Thymus vulgaris*) essential oil used as a natural preservative in bakery product, *Food Res. Int.*, 96: 154-160.

Gregirchak, N., O. Stabnikova and V. Stabnikov (2020). Application of lactic acid bacteria for coating of wheat bread to protect it from microbial spoilage, *Plant Foods Hum. Nutr.*, 75: 223-229.

Guerrero, A., S. Ferrero, M. Barahona, B. Boito, E. Lisbinski, F. Maggi and C. Sañudo (2020). Effects of active edible coating based on thyme and garlic essential oils on lamb meat shelf-life after long-term frozen storage, *J. Sci. Food Agric.*, 100: 656-664.

Guimarães, A.C., Ó. Ramos, M. Cerqueira, A. Venâncio and L. Abrunhosa (2020). Active whey protein edible films and coatings incorporating *Lactobacillus buchneri* for *Penicillium nordicum* control in cheese, *Food Biproc. Tech.*, 13: 1074-1086.

Gundewadi, G., S.G. Rudra, D.J. Sarkar and D. Singh (2018). Nanoemulsion based alginate organic coating for shelf-life extension of okra, *Food Packag. Shelf-Life*, 18: 1-12.

Günlü, A. and E. Koyun (2013). Effects of vacuum packaging and wrapping with chitosan-based edible film on the extension of the shelf-life of sea bass (*Dicentrarchus labrax*) fillets in cold storage (4°C), *Food Bioprocess Tech.*, 6: 1713-1719.

Hashemi, S.M.B. and D. Jafarpour (2021). Bioactive edible film based on Konjac glucomannan and probiotic *Lactobacillus plantarum* strains: Physicochemical properties and shelf-life of fresh-cut kiwis, *J. Food Sci.*, 86: 513-522.

Huang, M., H. Wang, X. Xu, X. Lu, X. Song and G. Zhou (2020). Effects of nanoemulsion-based edible coatings with composite mixture of rosemary extract and ε-poly-L-lysine on the shelf-life of ready-to-eat carbonado chicken, *Food Hydrocoll.*, 102: 105576.

Ju, J., Y. Xie, H. Yu, Y. Guo, Y. Cheng, R. Zhang and W. Yao (2020). Synergistic inhibition effect of citral and eugenol against Aspergillus niger and their application in bread preservation, *Food Chem.*, 310: 125974.

Karsli, B., E. Caglak and W. Prinyawiwatkul (2021). Effect of high molecular weight chitosan coating on quality and shelf-life of refrigerated channel catfish fillets, *LWT-Food Sci. Technol.*, 142: 111034.

Khaledian, S., S. Basiri and S.S. Shekarforoush (2021). Shelf-life extension of Pacific white shrimp using tragacanth gum-based coatings containing Persian lime peel (*Citrus latifolia*) extract, *LWT-Food Sci. Technol.*, 141: 110937.

Khan, M.R., M.B. Sadiq and Z. Mehmood (2020). Development of edible gelatin composite films enriched with polyphenol loaded nanoemulsions as chicken meat packaging material, CyTA-J, *Food*, 18: 137-146.

Koh, P.C., M.A. Noranizan, Z.A.N. Hanani, R. Karim and S.Z. Rosli (2017). Application of edible coatings and repetitive pulsed light for shelf-life extension of fresh-cut cantaloupe (*Cucumis melo* L. reticulatus cv. Glamour), *Postharvest Biol. Technol.*, 129: 64-78.

Kõrge, K., M. Bajić, B. Likozar and U. Novak (2020). Active chitosan–chestnut extract films used for packaging and storage of fresh pasta, *Int. J. Food Sci. Technol.*, 55: 3043-3052.

Küçük, G.S., Ö.F. Çelik, B.G. Mazi and H. Türe (2020). Evaluation of alginate and zein films as a carrier of natamycin to increase the shelf-life of kashar cheese, *Packag. Technol. Sci.*, 33: 39-48.

Kumar, N., A. Ojha, A. Upadhyay, R. Singh and S. Kumar (2021). Effect of active chitosan-pullulan composite edible coating enrich with pomegranate peel extract on the storage quality of green bell pepper, *LWT-Food Sci. Technol.*, 138: 110435.

Kurek, M., M. Repajić, M. Marić, M. Ščetar, P. Trojić, B. Levaj and K. Galić (2020). The influence of edible coatings and natural antioxidants on fresh-cut potato quality, stability and oil uptake after deep fat frying, *J. Food Sci. Technol.*, 58: 3073-3085.

Langroodi, A.M., H. Tajik, T. Mehdizadeh, M. Moradi, E.M. Kia and A. Mahmoudian (2018). Effects of sumac extract dipping and chitosan coating enriched with *Zataria multiflora* Boiss oil on the shelf-life of meat in modified atmosphere packaging, *LWT-Food Sci. Technol.*, 98: 372-380.

Lee, E.S., H.G. Song, I. Choi, J.S. Lee and J. Han (2020). Effects of mung bean starch/guar gum-based edible emulsion coatings on the staling and safety of rice cakes, *Carbohydr. Polym.*, 247: 116696.

Li, S., L. Zhang, M. Liu, X. Wang, G. Zhao and W. Zong (2017). Effect of poly-ε-lysine incorporated into alginate-based edible coatings on microbial and physicochemical properties of fresh-cut kiwifruit, *Postharvest Biol. Technol.*, 134: 114-121.

Li, K., M. Zhang, B. Bhandari, J. Xu and C. Yang (2020a). Improving storage quality of refrigerated steamed buns by mung bean starch composite coating enriched with nano-emulsified essential oils, *J. Food Process Eng.*, 43: e13475.

Li, S., Y. Ma, T. Ji, D.E. Sameen, S. Ahmed, W. Qin et al. (2020b). Cassava starch/carboxymethylcellulose edible films embedded with lactic acid bacteria to extend the shelf-life of banana, *Carbohydr. Polym.*, 248: 116805.

Lima, A.E.F., P.L. Andrade, T.L.G. de Lemos, D.E.D.A. Uchoa, M.C.A. Siqueira, A.S. do Egito et al. (2020). Development and application of galactomannan and essential oil-based edible coatings applied to 'coalho' cheese, *J. Food Process. Preserv.*, 45: e15091.

Liu, T., L. Liu, X. Gong, F. Chi and Z. Ma (2021). Fabrication and comparison of active films from chitosan incorporating different spice extracts for shelf-life extension of refrigerated pork, *LWT- Food Sci. Technol.*, 135: 110181.

Louis, E., R. Villalobos-Carvajal, J. Reyes-Parra, E. Jara-Quijada, C. Ruiz, P. Andrades *et al.* (2021). Preservation of mushrooms (*Agaricus bisporus*) by an alginate-based-coating containing a cinnamaldehyde essential oil nanoemulsion, *Food Packag. Shelf-Life*, 28: 100662.

Ma, S., Y. Zheng, R. Zhou and M. Ma (2021). Characterization of chitosan films incorporated with different substances of konjac glucomannan, cassava starch, maltodextrin and gelatin, and application in mongolian cheese packaging, *Coatings*, 11: 84.

Mahcene, Z., A. Khelil, S. Hasni, F. Bozkurt, M.B. Goudjil and F. Tornuk (2021). Home-made cheese preservation using sodium alginate based on edible film incorporating essential oils, *J. Food Sci. Technol.*, 58: 2406-2419.

Mahdavi, V., S.E. Hosseini and A. Sharifan (2018). Effect of edible chitosan film enriched with anise (*Pimpinella anisum* L.) essential oil on shelf-life and quality of the chicken burger, *Food Sci. Nutr.*, 6: 269-279.

Maqsood, S., S. Benjakul and A. Kamal-Eldin (2012). Haemoglobin-mediated lipid oxidation in the fish muscle: A review, *Trends in Food Sci. Tech.*, 28: 33-43.

Martillanes, S., J. Rocha-Pimienta, J. Llera-Oyola, M.V. Gil, M.C. Ayuso-Yuste, J. García-Parra and J. Delgado-Adámez (2021). Control of *Listeria monocytogenes* in sliced dry-cured Iberian ham by high pressure processing in combination with an eco-friendly packaging based on chitosan, nisin and phytochemicals from rice bran, *Food Control*, 124: 107933.

Mehdizadeh, T. and A.M. Langroodi (2019). Chitosan coatings incorporated with propolis extract and *Zataria multiflora* Boiss oil for active packaging of chicken breast meat, *Int. J. Biol. Macromol.*, 141: 401-409.

Mehdizadeh, T., H. Tajik, A.M. Langroodi, R. Molaei and A. Mahmoudian (2020). Chitosan-starch film containing pomegranate peel extract and Thymus kotschyanus essential oil can prolong the shelf-life of beef, *Meat Sci.*, 163: 108073.

Mehdizadeh, A., S.A. Shahidi, N. Shariatifar, M. Shiran and A. Ghorbani-Hasan Saraei (2021). Evaluation of chitosan-zein coating containing free and nano-encapsulated *Pulicaria gnaphalodes* (Vent.) Boiss. extract on quality attributes of rainbow trout, *J. Aquat. Food Prod. Technol.*, 30: 62-75.

Mehyar, G.F., A.A. Al Nabulsi, M. Saleh, A.N. Olaimat and R.A. Holley (2018). Effects of chitosan coating containing lysozyme or natamycin on shelf-life, microbial quality, and sensory properties of Halloumi cheese brined in normal and reduced salt solutions, *J. Food Process. Preserv.*, 42: e13324.

Mild, R.M., L.A. Joens, M. Friedman, C.W. Olsen, T.H. McHugh, B. Law and S. Ravishankar (2011). Antimicrobial edible apple films inactivate antibiotic resistant and susceptible Campylobacter jejuni strains on chicken breast, *J. Food Sci.*, 76: M163-M168.

Mohan, C.C., S. Babuskin, K. Sudharsan, V. Aafrin, P. Mariyajenita, K. Harini *et al.* (2017). Active compound diffusivity of particle size reduced *S. aromaticum* and *C. cassia* fused starch edible films and the shelf-life of mutton (*Capra aegagrus* hircus) meat, *Meat Sci.*, 128: 47-59.

Mushtaq, M., A. Gani, A. Gani, H.A. Punoo and F.A. Masoodi (2018). Use of pomegranate peel extract incorporated zein film with improved properties for prolonged shelf-life of fresh Himalayan cheese (Kalari/kradi), *Innov. Food Sci. Emerg. Technol.*, 48: 25-32.

Nair, M.S., A. Saxena and C. Kaur (2018). Characterization and antifungal activity of pomegranate peel extract and its use in polysaccharide-based edible coatings to extend the shelf-life of capsicum (*Capsicum annuum* L.), *Food Bioprocess Tech.*, 11: 1317-1327.

Nasiri, M., M. Barzegar, M.A. Sahari and M. Niakousari (2018). Application of Tragacanth gum impregnated with Satureja khuzistanica essential oil as a natural coating for enhancement of postharvest quality and shelf-life of button mushroom (*Agaricus bisporus*), *Int. J. Biol. Macromol.*, 106: 218-226.

Nilsuwan, K., P. Guerrero, K. de la Caba, S. Benjakul and T. Prodpran (2021). Fish gelatin films laminated with emulsified gelatin film or poly (lactic) acid film: Properties and their use as bags for storage of fried salmon skin, *Food Hydrocoll.*, 111: 106199.

Noori, S., F. Zeynali and H. Almasi (2018). Antimicrobial and antioxidant efficiency of nanoemulsion-based edible coating containing ginger (*Zingiber officinale*) essential oil and its effect on safety and quality attributes of chicken breast fillets, *Food Control*, 84: 312-320.

Ochoa-Velasco, C.E., J.C. Pérez-Pérez, J.M. Varillas-Torres, A.R. Navarro-Cruz, P. Hernández-Carranza, R. Munguía-Pérez *et al.* (2021). Starch edible films/coatings added with carvacrol and thymol: *in vitro* and *in vivo* evaluation against *Colletotrichum Gloeosporioides*, *Foods*, 10: 175.

Oliveira, M.A., M.L. Gonzaga, M.S. Bastos, H.C. Magalhães, S.D. Benevides, R.F. Furtado *et al.* (2020). Packaging with cashew gum/gelatin/essential oil for bread: Release potential of the citral, *Food Packag. Shelf-Life*, 23: 100431.

Ordoñez, R., C. Contreras, C. González-Martínez and A. Chiralt (2021). Edible coatings controlling mass loss and *Penicillium roqueforti* growth during cheese ripening, *J. Food Eng.*, 290: 110174.

Otoni, C.G., S.F. Pontes, E.A. Medeiros and N.D.F. Soares (2014). Edible films from methylcellulose and nanoemulsions of clove bud (*Syzygium aromaticum*) and oregano (*Origanum vulgare*) essential oils as shelf-life extenders for sliced bread, *J. Agric. Food Chem.*, 62: 5214-5219.

Pabast, M., N. Shariatifar, S. Beikzadeh and G. Jahed (2018). Effects of chitosan coatings incorporating with free or nano-encapsulated Satureja plant essential oil on quality characteristics of lamb meat, *Food Control*, 91: 185-192.

Park, H.J. (1999). Development of advanced edible coatings for fruits, *Trends in Food Sci. Tech.*, 10: 254-260.

Pavli, F., A.A. Argyri, P. Skandamis, G.J. Nychas, C. Tassou and N. Chorianopoulos (2019). Antimicrobial activity of oregano essential oil incorporated in sodium alginate edible films: Control of *Listeria monocytogenes* and spoilage in ham slices treated with high pressure processing, *Materials*, 12: 3726.

Peralta-Ruiz, Y., C.G. Tovar, A. Sinning-Mangonez, D. Bermont, A.P. Cordero, A. Paparella and C. Chaves-López (2020). Colletotrichum gloesporioides inhibition using chitosan-*Ruta graveolens* L. essential oil coatings: Studies *in vitro* and *in situ* on Carica papaya fruit, *Int. J. Food Microbiol.*, 326: 108649.

Pérez-Arauz, Á.O., A.I. Rodríguez-Hernández, M. del Rocío López-Cuellar, V.M. Martínez-Juárez and N. Chavarría-Hernández (2021). Films based on Pectin, Gellan, EDTA, and bacteriocin-like compounds produced by *Streptococcus infantarius* for the bacterial control in fish packaging, *J. Food Process. Preserv.*, 45: e15006.

Pinzon, M.I., L.T. Sanchez, O.R. Garcia, R. Gutierrez, J.C. Luna and C.C. Villa (2020). Increasing shelf-life of strawberries (*Fragaria* ssp) by using a banana starch-chitosan-Aloe vera gel composite edible coating, *Int. J. Food Sci. Technol.*, 55: 92-98.

Pirozzi, A., V. Del Grosso, G. Ferrari and F. Donsì (2020). Edible coatings containing oregano essential oil nanoemulsion for improving postharvest quality and shelf-life of tomatoes, *Foods*, 9: 1605.

Prabhakar, P.K., S. Vatsa, P.F.P. Srivastav and S.S. Pathak (2020). A comprehensive review on freshness of fish and assessment: Analytical methods and recent innovations, *Food Res. Int.*, 33: 109157.

Prakash, A., R. Baskaran and V. Vadivel (2020). Citral nanoemulsion incorporated edible coating to extend the shelf-life of fresh cut pineapples, *LWT-Food Sci. Technol.*, 118: 108851.

Pruksarojanakul, P., C. Prakitchaiwattana, S. Settachaimongkon and C. Borompichaichartkul (2020). Synbiotic edible film from konjac glucomannan composed of Lactobacillus casei-01® and Orafti® GR, and its application as coating on bread buns, *J. Sci. Food Agric.*, 100: 2610-2617.

Raeisi, M., A. Tabaraei, M. Hashemi and N. Behnampour (2016). Effect of sodium alginate coating incorporated with nisin, *Cinnamomum zeylanicum*, and rosemary essential oils on microbial quality of chicken meat and fate of *Listeria monocytogenes* during refrigeration, *Int. J. Food Microbiol.*, 238: 139-145.

Raigond, P., V. Sagar, T. Mishra, A. Thakur, B. Singh, V. Kumar *et al.* (2019). Chitosan: a safe alternative to synthetic fungicides to manage dry rot in stored potatoes, *Potato Res.*, 62: 393-409.

Raju, A. and M.S. Sasikala (2016). Natural antimicrobial edible film for preservation of paneer, *Biosci. Biotechnol. Res. Asia*, 13: 1083-1088.

Riaz, A., R.M. Aadil, A.M.O. Amoussa, M. Bashari, M. Abid and M.M. Hashim (2021). Application of chitosan-based apple peel polyphenols edible coating on the preservation of strawberry (*Fragaria ananassa* cv Hongyan) fruit, *J. Food Process. Preserv.*, 45: e15018.

Rico, D., I. Albertos, O. Martinez-Alvarez, M.E. Lopez-Caballero and A.B. Martin-Diana (2020). Use of sea fennel as a natural ingredient of edible films for extending the shelf-life of fresh fish burgers, *Molecules*, 25: 5260.

Rizzo, V., S. Lombardo, G. Pandino, R.N. Barbagallo, A. Mazzaglia, C. Restuccia *et al.* (2019). Shelf-life study of ready-to-cook slices of globe artichoke 'Spinoso sardo': Effects of anti-browning solutions and edible coating enriched with Foeniculum vulgare essential oil, *J. Sci. Food Agric.*, 99: 5219-5228.

Rojas-Graü, M.A., R. Soliva-Fortuny and O. Martín-Belloso (2009). Edible coatings to incorporate active ingredients to fresh-cut fruits: A review, *Trends in Food Sci. Tech.*, 20: 438-447.

Saidi, L., D. Duanis-Assaf, O. Galsarker, D. Maurer, N. Alkan and E. Poverenov (2021). Elicitation of fruit defense response by active edible coatings embedded with phenylalanine to improve quality and storability of avocado fruit, *Postharvest Biol. Technol.*, 174: 111442.

Santacruz, S. and M. Castro (2018). Viability of free and encapsulated *Lactobacillus acidophilus* incorporated to cassava starch edible films and its application to Manaba fresh white cheese, *LWT-Food Sci. Technol.*, 93: 570-572.

Seydim, A.C., G. Sarikus-Tutal and E. Sogut (2020). Effect of whey protein edible films containing plant essential oils on microbial inactivation of sliced Kasar cheese, *Food Packag. Shelf-Life*, 26: 100567.

Severino, R., K.D. Vu, F. Donsì, S. Salmieri, G. Ferrari and M. Lacroix (2014). Antimicrobial effects of different combined non-thermal treatments against *Listeria monocytogenes* in broccoli florets, *J. Food Eng.*, 124: 1-10.

Shao, P., J. Yu, H. Chen and H. Gao (2021). Development of microcapsule bioactive paper loaded with cinnamon essential oil to improve the quality of edible fungi, *Food Packag. Shelf-Life*, 27: 100617.

Shigematsu, E., C. Dorta, F.J. Rodrigues, M.F. Cedran, J.A. Giannoni, M. Oshiiwa and M.A. Mauro (2018). Edible coating with probiotic as a quality factor for minimally processed carrots, *J. Food Sci. Technol.*, 55: 3712-3720.

Shokri, S., A. Ehsani and M.S. Jasour (2015). Efficacy of lactoperoxidase system-whey protein coating on shelf-life extension of rainbow trout fillets during cold storage (4°C), *Food Bioproc. Tech.*, 8: 54-62.

Soto, K.M., M. Hernández-Iturriaga, G. Loarca-Piña, G. Luna-Bárcenas and S. Mendoza (2019). Antimicrobial effect of nisin electrospun amaranth: Pullulan nanofibers in apple juice and fresh cheese, *Int. J. Food Microbiol.*, 295: 25-32.

Soukoulis, C., L. Yonekura, H.H. Gan, S. Behboudi-Jobbehdar, C. Parmenter and I. Fisk (2014). Probiotic edible films as a new strategy for developing functional bakery products: The case of pan bread, *Food Hydrocoll.*, 39: 231-242.

Souza, V.G., J.R. Pires, É.T. Vieira, I.M. Coelhoso, M.P. Duarte and A.L. Fernando (2018). Shelf-life assessment of fresh poultry meat packaged in novel bionanocomposite of chitosan/montmorillonite incorporated with ginger essential oil, *Coatings*, 8: 177.

Sozbilen, G.S. and A. Yemenicioğlu (2020). Decontamination of seeds destined for edible sprout production from Listeria by using chitosan coating with synergetic lysozyme-nisin mixture, *Carbohydr. Polym.*, 235: 115968.

Sriket, C. (2014). Proteases in fish and shellfish: Role on muscle softening and prevention, *Int. Food Res. J.*, 21: 433.

Stergiou, V.A., L.V. Thomas and M.R. Adams (2006). Interactions of nisin with glutathione in a model protein system and meat, *J. Food Prot.*, 69: 951-956.

Sultan, M., O.M. Hafez, M.A. Saleh and A.M. Youssef (2021). Smart edible coating films based on chitosan and beeswax–pollen grains for the postharvest preservation of Le Conte pear, *RSC Adv.*, 11: 9572-9585.

Tabassum, N. and M.A. Khan (2020). Modified atmosphere packaging of fresh-cut papaya using alginate based edible coating: Quality evaluation and shelf-life study, *Sci. Hortic.*, 259: 108853.

Temiz, N.N. and K.S. Özdemir (2021). Microbiological and physicochemical quality of strawberries (Fragaria× ananassa) coated with *Lactobacillus rhamnosus* and inulin enriched gelatin films, *Postharvest Biol. Technol.*, 173: 111433.

Ucak, I., A.K. Abuibaid, T.M. Aldawoud, C.M. Galanakis and D. Montesano (2021). Antioxidant and antimicrobial effects of gelatin films incorporated with citrus seed extract on the shelf-life of sea bass (*Dicentrarchus labrax*) fillets, *J. Food Process. Preserv.*, 45: e15304.

Ünalan, İ.U., I. Arcan, F. Korel and A. Yemenicioğlu (2013). Application of active zein-based films with controlled release properties to control *Listeria monocytogenes* growth and lipid oxidation in fresh Kashar cheese, *Innov. Food Sci. Emerg. Technol.*, 20: 208-214.

Wang, D., Y. Dong, X. Chen, Y. Liu, J. Wang, X. Wang *et al.* (2020). Incorporation of apricot (Prunus armeniaca) kernel essential oil into chitosan films displaying antimicrobial effect against *Listeria monocytogenes* and improving quality indices of spiced beef, *Int. J. Biol. Macromol.*, 62: 838-844.

Xiong, Y., S. Li, R.D. Warner and Z. Fang (2020). Effect of oregano essential oil and resveratrol nanoemulsion loaded pectin edible coating on the preservation of pork loin in modified atmosphere packaging, *Food Control*, 114: 107226.

Xiong, Y., M. Kamboj, S. Ajlouni and Z. Fang (2021). Incorporation of salmon bone gelatin with chitosan, gallic acid and clove oil as edible coating for the cold storage of fresh salmon fillet, *Food Control*, 125: 107994.

Yaghoubi, M., A. Ayaseh, K. Alirezalu, Z. Nemati, M. Pateiro and J.M. Lorenzo (2021). Effect of chitosan coating incorporated with Artemisia fragrans essential oil on fresh chicken meat during refrigerated storage, *Polymers*, 13: 716.

Yemenicioğlu, A. (2016). Strategies for controlling major enzymatic reactions in fresh and processed vegetables, pp. 377-391. *In:* Y.H. Hui and E.O. Evranuz (Eds.). *Handbook of Vegetable Preservation and Processing*, 2nd ed. CRC Press, Boca Raton, USA.

Yingyuad, S., S. Ruamsin, D. Reekprkhon, S. Douglas, S. Pongamphai and U. Siripatrawan (2006). Effect of chitosan coating and vacuum packaging on the quality of refrigerated grilled pork, *Packag. Technol. Sci.*, 19: 149-157.

Yordshahi, A.S., M. Moradi, H. Tajik and R. Molaei (2020). Design and preparation of antimicrobial meat wrapping nanopaper with bacterial cellulose and postbiotics of lactic acid bacteria, *Int. J. Food Microbiol.*, 321: 108561.

Yuan, G., H. Lv, W. Tang, X. Zhang and H. Sun (2016). Effect of chitosan coating combined with pomegranate peel extract on the quality of Pacific white shrimp during iced storage, *Food Control*, 59: 818-823.

Zarandona, I., M.E. López-Caballero, M.P. Montero, P. Guerrero, K. de la Caba and M.C. Gómez-Guillén (2021). Horse mackerel (*Trachurus trachurus*) fillets biopreservation by using gallic acid and chitosan coatings, *Food Control*, 120: 107511.

Zhu, D., R. Guo, W. Li, J. Song and F. Cheng (2019). Improved postharvest preservation effects of Pholiota nameko mushroom by sodium alginate-based edible composite coating, *Food Bioproc. Tech.*, 12: 587-598.

Index

For Product Safety Concerns and Information please contact our EU
representative GPSR@taylorandfrancis.com
Taylor & Francis Verlag GmbH, Kaufingerstraße 24, 80331 München, Germany

www.ingramcontent.com/pod-product-compliance
Lightning Source LLC
Chambersburg PA
CBHW060808220326
41598CB00022B/2568

9 781032 371122